T0419929

Nanoethics and Nanotoxicology

P. Houdy M. Lahmani F. Marano
(Eds.)

Nanoethics and Nanotoxicology

With 112 Figures and 31 Tables

Philippe Houdy, PhD
Université d'Évry
Boulevard François Mitterrand, 91025 Évry Cédex, France
E-mail: philippe.houdy@univ-evry.fr

Marcel Lahmani, PhD
Club Nano-Micro-Technologie de Paris
Boulevard François Mitterrand, 91025 Évry Cédex, France
E-mail: marcel.lahmani@univ-evry.fr

Francelyne Marano
Université Paris Diderot-Paris 7, Laboratoire des RMCX
Bâtiment Buffon
rue Marie-Andrée Lagroua Weill-Hall 4, 75205 Paris Cédex, France

Translation from the French language edition of
"Les Nanosciences – 4. Nanotoxicologie et nanoéthique"
© 2010 Editions Belin, France

ISBN 978-3-642-20176-9 e-ISBN 978-3-642-20177-6
DOI 10.1007/978-3-642-20177-6
Springer Heidelberg Dordrecht London New York

Library of Congress Control Number: 2011935632

Cover design: WMXDesign GmbH, Heidelberg

Printed on acid-free paper

Springer is part of Springer Science+Business Media (www.springer.com)

All is poison, nothing is poison,
only the dose matters

Paracelsus (1493–1541)
Swiss alchemist and physician

Foreword to the French Edition

Science moves forward inexorably, and the ever more powerful means of investigation at its disposal provide a continuous supply of new discoveries and novel applications. The same is also true in the nanosciences. History has shown that scientific progress can lead to worldwide benefits, but sometimes also to unprecedented human disasters. This observation is not restricted to the products of the nanotechnologies, but the development here may be taken as an opportunity for a documented raising of awareness regarding their risks, consulting the general public on broad scale, and above all, informing the whole of society as objectively as possible.

Indeed, the very real problems raised by asbestos and the questions posed by the unknown long term impacts of genetically modified organisms (GMO) – more and more common in our everyday food supply – have caused considerable discord between the scientists who develop them, the companies who wish to industrialise them, and the consumer organisations. This kind of controversy is broadly covered by the media, with a view either to warn the public on behalf of competent and responsible organisations, or to achieve more or less dubious political ends.

The real problem here is to carry out a rigorous investigation of the potential dangers and risks raised by the use, development, and commercialisation of current or new products likely to threaten people's health or the environment, with harmful consequences for future generations. There is no doubt that the specific properties of nanoscale objects can radically enhance their chemical reactivity, and transform their electronic or magnetic behaviour, sometimes increasing their capacity to enter deep into living systems.

This book is the last of four volumes providing as complete a picture as possible of the current state of our knowledge in the nanosciences. The vocation of the series is didactic, aimed at graduate students, research scientists, and engineers. From a scientific standpoint, it reviews the state of the art in nanobiotechnology and nanotoxicology. But what is novel in this fourth volume is that it emphasises the efforts made by researchers to cater for the consequences of their work, and the ways engineers set up safety systems when potentially dangerous products go into mass production.

To my knowledge, this is the first book in which scientific knowledge and ethical and social recommendations can be found side by side, along with specific policies developed by national and international authorities to handle the potential problems of nanotechnology.

President of the French National Ethical Committee
Hôpital Armand-Trousseau, Paris
University Pierre and Marie Curie, Paris

Alain Grimfeld
August 2007

Preface to the French Edition

Nanotoxicology studies the toxicity of nanomaterials, nanoparticles, and more generally, any naturally occurring or man-made objects with dimensions in the range 1–100 nanometers. Such small dimensions induce specific properties, making these objects much more reactive, for example, than larger ones, and in particular, they allow them to pass through certain natural biological barriers. The potentially harmful effects that may result thus constitute one of the quite legitimate reasons for the concern they inspire.

A key objective of this book is to set out some up-to-date scientific studies of nanotoxicity, and exemplify the preventive measures taken during fabrication or manipulation of nano-objects. Another is to describe the way the public is informed about these new scientific discoveries, and also the legal arrangements currently under preparation for regulating their use.

Considering the controversy to which the nanosciences have given rise – as witnessed for all forms of scientific innovation – it seems important to expose the ethical considerations taken into account in the context of nanotechnology. Indeed, scientists have been questioned, sometimes forcefully, about the social consequences of their research, and it seems opportune to set up a responsible debate between the so-called hard sciences and the social sciences from the very beginning of any scientific project with wide-ranging industrial and social impacts. Such attempts to raise public awareness are of course relevant both on the national and international level.

The book is divided into five main parts. The first two concern nanotoxicology, and are purely scientific, providing specific examples of the potential or proven impacts of nanoparticles on humans and on the environment. The last three concern nanoethics. After a brief introduction to the basic ethical issues, there is a fairly exhaustive discussion of the implications for national and international authorities regarding the way the public demand for information is being treated, and also regarding the degree of transparency with which current developments in nanotechnology are being presented, as well as the need for rigour, responsibility, and caution in their use.

The reader will thus find one of the first books to combine both scientific and societal aspects of an emerging field, containing many references for each of its disciplines.

Acknowledgements

We would like to thank all members of the French nanoscience community who gave a very favourable welcome to the writing of these four pedagogical introductions to nanotechnology and nanophysics, nanomaterials and nanochemistry, nanobiotechnology and nanobiology, and nanotoxicology and nanoethics, without whom they would have been impossible. Special thanks go, of course, to all those who contributed to these books.

Given the current debate over the nanosciences, it seemed important to mention in the context of a scientific textbook that research scientists do indeed take into account the potential risks involved in their innovations, and give considerable thought to all the necessary precautions before they are industrialised.

From the ethical standpoint, it is more and more important to inform the general public, and the book discusses the various authorities concerned with the citizen's demand for information when new technologies with suggested or proven risks are launched on the market. This part of the book was put together by Françoise Roure, president of the information technology department of the French Ministry for Trade and Industry. Her contribution played a key role in harmonising with the prior scientific chapters.

The latter were coordinated by Francelyne Marano, co-editor of the book, for the first part and by Jean-Yves Bottero for the second. We would like to express our most sincere gratitude for his scientific actions and for organising the working groups at the CEREGE.

We are particularly grateful to the late Professor Hubert Curien, who supported this undertaking from the start, and to Jean-Marie Lehn, Axel Kahn, and Alain Grimfeld, who have also given it their backing.

We warmly acknowledge the material and financial support of the French Ministry of Research, the French national science research organisation (CNRS), and the French atomic energy authority (CEA), and especially Jean Therme, Director of the CEA, Grenoble, who supported this project from the beginning.

The editors of the nanobooks would like to express their gratitude to Stephen N. Lyle for his excellent translation of the four nanovolumes of this Nanoscience series: *Nanotechnolgies and Nanophysics*, *Nanomaterials and Nanochemistry*, *Nanobiotechnology and Nanobiology*, and *Nanotoxicology and Nanoethics*.

Marcel Lahmani
Francelyne Marano
Philippe Houdy

Contents

Part VI Nanoethics and Social Issues

List of Contributors

Wafa Achouak
Laboratoire d'écologie microbienne de la rhizosphere (LEMiR)
UMR 6191 CNRS CEA Université Aix Marseille II
CEA Cadarache
13108 Saint Paul lez Durance
wafa.achouak@cea.fr

Olivier Aguerre-Chariol
Innovation pour la mesure
Direction des risques chroniques
Institut national de l'environnement industriel et des risques (INERIS)
Parc technologique ALATA
60550 Verneuil-en-Halatte
olivier.aguerre-chariol@ineris.fr

Mélanie Auffan
Centre européen de recherche et d'enseignements des géosciences de l'environnement (CEREGE)
UMR 6635 CNRS Aix/Marseille Université
Equipe SE3D
Europole de l'Arbois BP 80
13545 Aix-en-Provence Cedex 4
auffan@cerege.fr

Armelle Baeza-Squiban
Université Paris Diderot-Paris 7
Laboratoire des réponses moléculaires et cellulaires aux xénobiotiques (RMCX)
Unité de biologie fonctionnelle et adaptative (BFA), EAC CNRS 4413
Case 7073
5 rue Thomas Mann
75205 Paris Cedex 13
baeza@univ-paris-diderot.fr

Denis Bard
Département d'épidémiologie
Ecole des hautes études en santé publique
Avenue du Professeur Léon-Bernard
CS 74312
35043 Rennes Cedex
denis.bard@ehesp.fr

Laïla Benameur
Laboratoire de biogénotoxicologie et mutagenèse environnementale
EA 1784, FR 3098 ECCOREV
Faculté de médicine
Université d'Aix-Marseille
27 Bd Jean-Moulin
13385 Marseille Cedex 05
laila.benameur@univmed.fr

Daniel Bernard
ARKEMA
420 Rue D'Estienne D'Orves
92705 Colombes
daniel.bernard@arkema.com

Daniel Bloch
Medical Advisor on Nanomaterials
Commissariat à l'énergie atomique (CEA)
17 rue des Martyrs
38054 Grenoble Cedex 9
daniel.bloch@cea.fr

Frédéric Y. Bois
Technological University
of Compiègne (UTC)
Chair of Mathematical Modelling
for Systems Toxicology
Royallieu Research Center
BP 20529
60205 Compiègne Cedex
frederic.bois@ineris.fr

Jacques Bordé
Comité d'éthique des sciences du CNRS (COMETS)
jacques.borde@polytechnique.org

Alain Botta
CHU Timone
13385 Marseille Cedex 5
Laboratoire de biogénotoxicologie et mutagenèse environnementale
EA 1784, FR 3098 ECCOREV
Faculté de médicine
Université d'Aix-Marseille
27 Bd Jean-Moulin
13385 Marseille Cedex 05
alain.botta@univmed.fr

Jean-Yves Bottero
Centre européen de recherche et d'enseignements des géosciences de l'environnement (CEREGE)
UMR 6635, CNRS/Université d'Aix-Marseille
Europole de l'Arbois BP 80
13545 Aix-en-Provence Cedex 4
bottero@cerege.fr

Daniel Boy
Centre de recherches politiques de Sciences Po
CEVIPOF
98, rue de l'Université
75007 Paris
daniel.boy@sciences-po.fr

Roberta Brayner
Nanomaterials Team
University of Paris 7 Paris-Diderot
Interfaces, traitements, organisation et dynamique des systèmes (ITODYS)
CNRS UMR 7086
Bât. Lavoisier, 8ème étage, salle 820
15 rue Jean de Baïf
75205 Paris Cedex 13
roberta.brayner@univ-paris-diderot.fr

Jean-Marc Brignon
Economics and Decision Tools
Chronic Risks Division
Institut national de l'environnement industriel et des risques (INERIS)
Parc Technologique ALATA
60550 Verneuil-en-Halatte
jean-marc.brignon@ineris.fr

Patrick Brochard
Laboratoire santé, travail, environnement (EA 3672)
Université Victor Segalen Bordeaux 2
146, rue Léo Saignat
Case 11
33076 Bordeaux Cedex
patrick.brochard@chu-bordeaux.fr

José Cambou
Santé-Environnement de France
Nature Environnement
6 rue Dupanloup
45000 Orléans
jose.cambou@fne.asso.fr

Marie Carrière
Equipe toxicologie humaine et environnementale
UMR 3299 CEA/CNRS, SIS2M
91191 Gif-sur-Yvette Cedex
carriere@drecam.cea.fr

Corinne Chanéac
Laboratoire de chimie de la matière condensée
UMR 7574 UPMC/CNRS
Collège de France, Bât. C-D
11 place Marcelin Berthelot
75231 Paris Cedex 05
corinne.chaneac@upmc.fr

Sonia Desmoulin-Canselier
Centre de recherche Droit, Sciences et techniques (CRDST)
Bureau 407
UMR 8103 CNRS/Université Paris 1
9, rue Malher
75004 Paris
Sonia.Desmoulin@univ-paris1.fr

François Ewald
Conservatoire national des arts et métiers (CNAM)
Observatoire du principe de précaution
2, rue Conté, 75003 Paris
francois.ewald@cnam.fr

Fernand Fievet
Laboratoire interfaces, traitements, organisation et dynamique des systèmes (ITODYS)
CNRS UMR 7086, Université Paris 7 Diderot
Bâtiment Lavoisier
15, rue Jean Antoine de Baïf
75205 Paris Cedex 13
fernand.fievet@univ-paris-diderot.fr

Emmanuel Flahaut
Centre Interuniversitaire de Recherche et d'Ingénierie des Matériaux (CIRIMAT)
UMR CNRS 5085
University Paul Sabatier
Bâtiment 2R1
31062 Toulouse Cedex 9
NAUTILE (Nanotubes et écotoxicologie)
ARKEMACNRS/UPS/INPT
flahaut@chimie.ups-tlse.fr

Eric Gaffet
Nanomaterials Research Group (NRG)
CNRS UMR 5060
Site de Sévenans (UTBM)
90010 Belfort Cedex
eric.gaffet@utbm.fr

Philippe Galiay
European Commission
Directorate General for Research
SDME 7/68
B-1049 Brussels, Belgium
philippe.galiay@ec.europa.eu

Jean-Gabriel Ganascia
Laboratoire d'informatique de Paris 6 (LIP6)
Université Paris 6 Pierre and Marie Curie
Boîte courrier 169
Couloir 25-26, Etage 5, Bureau 510
4 place Jussieu
75252 Paris Cedex 05
Jean-Gabriel.Ganascia@LIP6.fr

Jeanne Garric
CEMAGREF
Laboratory of Ecotoxicology
Freshwater systems, ecology and pollution research unit
3 bis, quai Chauveau CP 220
69336 Lyon Cedex 09
jeanne.garric@cemagref.fr

Laury Gauthier
University of Toulouse III
Laboratoire d'écologie fonctionnelle (Ecolab, UMR 5245)
CNRS/UPS/INPT
Avenue de l'Agrobiopole
BP 32607
Auzeville tolosane
31326 Castanet-Tolosan
NAUTILE (Nanotubes et écotoxicologie), ARKEMA-CNRS/UPS/INPT
lgauthie@cict.fr

Christine O. Hendren
Civil and Environmental Engineering Department
Duke University
Hudson Hall Box 90287
Durham, North Carolina 27708, USA
christineohendren@gmail.com

Christian Huard
Association de défense, d'éducation et d'information du consommateur (Adéic)
3, rue de la Rochefoucauld
75009 Paris
christian.huard@adeic.asso.fr

Philippe Hubert
Risques chroniques
Institut national de l'environnement industriel et des risques (INERIS)
Parc technologique ALATA
60550 Verneuil-en-Halatte
philippe.hubert@ineris.fr

Marie-Claude Jaurand
Institut national de la santé et de la recherche médicale (INSERM U674)
27 rue Juliette Dodu
75010 Paris
marie-claude.jaurand@inserm.fr

Jean-Pierre Jolivet
Laboratoire de chimie de la matière condensée
UMR 7574 UPMC/CNRS, Paris
Collège de France, Bât. C-D
11 place Marcelin Berthelot
75231 Paris Cedex 05
jean-pierre.jolivet@upmc.fr

Béatrice L'Azou
EA3672 Université Victor Segalen Bordeaux 2
146, rue Léo Saignat
33076 Bordeaux Cedex
beatrice.lazou@u-bordeaux2.fr

Jérôme Labille
Centre européen de recherche et d'enseignements des géosciences de l'environnement (CEREGE)
UMR 6635 CNRS Aix/Marseille Université
Europole de l'Arbois BP 80
13545 Aix-en-Provence Cedex 4
labille@cerege.fr

Stéphanie Lacour
Normativités et nouvelles technologies
Centre d'etudes sur la coopération juridique internationale (CECOJI)
UMR 6224 CNRS
Université de Poitiers
Faculté de Droit
Hôtel Aubaret
15, rue Sainte-Opportune
86022 Poitiers Cedex
lacour@ivry.cnrs.fr

Ghislaine Lacroix
Experimental Toxicology Unit
Institut national de l'environnement industriel et des risques (INERIS)
Parc technologique ALATA
60550 Verneuil-en-Halatte
ghislaine.lacroix@ineris.fr

Sophie Lanone
Institut National de la Santé et de la Recherche Médicale (INSERM U955)
Groupe Hospitalier Universitaire Albert Chenevier–Henri Mondor
51, avenue du Maréchal-de-Lattre-de-Tassigny
94010 Créteil Cedex
sophie.lanone@inserm.fr

Sophie Larrieu
Cellule de l'InVS en région (CIRE) Océan Indien
2 bis, Av. G. Brassens
97400 Saint Denis La Réunion
sophie.larrieu@ars.sante.fr

Olivier Le Bihan
Direction des risques chroniques, Institut national de l'environnement industriel et des risques (INERIS)
Parc technologique ALATA
60550 Verneuil-en-Halatte
olivier.le-bihan@ineris.fr

Agnès Lefranc
Institut de veille sanitaire (InVS)
12 rue du Val d'Osne
94415 Saint-Maurice Cedex
a.lefranc@invs.sante.fr

Corinne Mandin
Centre Scientifique et Technique du Bâtiment
84 avenue Jean Jaurès
Champs-sur-Marne
77447 Marne-la-Vallée Cedex 2
corinne.mandin@cstb.fr

Francelyne Marano
Université Paris Diderot-Paris 7
Laboratoire des réponses moléculaires et cellulaires aux xénobiotiques (RMCX)
Unité de biologie fonctionnelle et adaptative (BFA)
CNRS EAC 4413 case 7073 Bâtiment Buffon
4 rue Marie-Andrée Lagroua Weill-Hallé
75205 Paris Cedex 13
marano@univ-paris-diderot.fr

Solange Martin
Sciences-Po, Paris
27 rue Saint-Guillaume
75337 Paris Cedex 07
Agence de l'Environnement de la Maîtrise de l'Energie (ADEME)
Paris
solange.martin@sciences-po.org

Armand Masion
Centre européen de recherche et d'enseignements des géosciences de l'environnement (CEREGE)
UMR 6635 CNRS Aix/Marseille Université
Europole de l'Arbois BP 80
13545 Aix-en-Provence Cedex 4
masion@cerege.fr

Mark Morrison
Chief Executive Officer
Institute of Nanotechnology
Strathclyde University Incubator
Graham Hills Building
50 Richmond Street
Glasgow G1 1XP
Scotland
mark.morrison@nano.org.uk

Florence Mouchet
Laboratoire d'écologie fonctionnelle (ECOLAB)
UMR 5245 CNRS/UPS/INPT

Laboratoire commun NAUTILE (Nanotubes et Ecotoxicologie)
CNRS/UPS/ INPT, ARKEMA
florence.mouchet@orange.fr

Shamila Nair-Bedouelle
INSERM, Paris
snairbedouelle@gmail.com

Jean-Claude Pairon
Institut National de la Santé et de la Recherche Médicale (INSERM U955)
Groupe Hospitalier Universitaire Albert Chenevier–Henri Mondor
51, avenue du Maréchal-de-Lattre-de-Tassigny
94010 Créteil Cedex
jc.pairon@chicreteil.fr

Corine Pelluchon
University of Poitiers,
Centre de recherche Hegel et l'idéalisme allemand (CRHIA)
8 rue Descartes
86022 Poitiers Cedex
Centre de recherche sens, éthique et société (CERSES)
University of Paris 5 René Descartes
45 rue des Saints-Pères
75006 Paris
corine.pelluchon@univ-poitiers.fr

Alain Pompidou
Honorary Member of the European Parliament
Former president of the European Patents Office
President of the French National Academy of Technologies
President of COMEST (UNESCO)
apompidou@skynet.be

Dominique Proy
Directoire du réseau Santé-Environnement de France Nature Environnement
6 rue Dupanloup
45000 Orléans
proydomi@yahoo.fr

Myriam Ricaud
Institut national de recherche et de sécurité (INRS)
30 rue Olivier Noyer
75680 Paris Cedex 14
myriam.ricaud@inrs.fr

Jérôme Rose
Centre européen de recherche et d'enseignements des géosciences de l'environnement (CEREGE)
UMR 6635 CNRS Aix/Marseille Université
Europole de l'Arbois BP 80
13545 Aix-en-Provence Cedex 4
rose@cerege.fr

Françoise Roure
Conseil général de l'industrie, de l'énergie et des technologies
Ministère de l'Economie, de l'Industrie et de l'Emploi
francoise.roure@finances.gouv.fr

Antoine Thill
Laboratoire interdisciplinaire sur l'organisation nanométrique et supramoléculaire (LIONS)
(IRAMIS), UMR 3299 CEA/CNRS
CEA Saclay, Bâtiment 125
91191 Gif-sur-Yvette
antoine.thill@cea.fr

Eric Thybaud
Institut national de l'environnement industriel et des risques (INERIS)
Parc technologique ALATA
60550 Verneuil-en-Halatte
eric.thybaud@ineris.fr

Bernard Umbrecht
Association de défense, d'éducation et d'information du consommateur (Adéic)
3, rue de la Rochefoucauld
75009 Paris
bernard.umbrecht@adeic.asso.fr

Mark R. Wiesner
Wiesner Research Group
Department of Civil and Environmental Engineering
Duke University
Box 90287 Hudson Hall
Durham, North Carolina 27708
USA
Center for the Environmental Implications of NanoTechnology (CEINT)
wiesner@pratt.duke.edu

Olivier Witschger
Laboratoire de métrologie des aérosols
Institut national de recherche et de sécurité (INRS)
Rue du Morvan
CS 60027
54519 Vandoeuvre-lès-Nancy Cedex
olivier.witschger@inrs.fr

Introduction. Nanomaterials and Nanoproducts: World Markets and Human and Environmental Impacts

Eric Gaffet

Nanotechnology and nanoscience lie at the meeting point of many disciplines, from physics to chemistry, biology, and mechanics. Today they have become established as one of the main fields of research for the coming years. The first volume of this series, entitled *Nanoscience: Nanotechnologies and Nanophysics*, shows how useful it can be to structure matter on the nanoscale, with implications in fields as disparate as magnetism, data storage, biology, and electronics, with the development of completely new components, e.g., near-field techniques, lithographic processes, fullerenes, and spin electronics.

The second volume, *Nanomaterials and Nanochemistry*, presents a complete overview of nanomaterials, their fundamental properties, and novel applications that may come from fullerenes, carbon nanotubes, and other previously unimagined materials. This book provides a broad panorama of the main methods used to synthesise nanomaterials, and the resulting production processes, not forgetting the self-assembly of complex structures, one of the most promising channels of investigation opened up by nanochemistry.

The third book in the series, *Nanoscience: Nanobiotechnology and Nanobiology* provides an exhaustive and accessible overview of biological nano-objects, building blocks for existing and future constructions. After detailing the methods used for investigation in nanobiotechnology, there is a review of the many current and potential applications here, such as the synthesis of activatable nanoparticles able to accurately target cancer cells.

Coming directly from laboratory work in a highly accelerated way compared with other fields of research, the introduction and implementation of nanomaterials in nanoproducts has already become an industrial and economic reality. As in other industrial sectors, it is important to consider the social consequences (nanoethics) and the impact of these novel products on both human health and the environment (nanotoxicity), in order to avoid possible risks in the future. This is indeed a crucial issue to guarantee a responsible development of nanomaterials and nanotechnology, and it is the subject of the present volume.

1 Nanotechnologies

This field of research and development consists in building structures, devices, and systems using processes for structuring matter on the atomic, molecular, and supramolecular level, with characteristic length scales of 1–100 nanometres (nm). These so-called building blocks form a relatively small fraction of nanomaterials in terms of the quantity produced. In the field of nanomaterials, one must also consider nanoparticles and nanostructured coatings, but also dense bulk materials and nanocomposites (with organic, inorganic, or metallic matrices).

Matter is expected to behave in new ways, owing to the relative importance of the laws of quantum physics that find their full expression on this length scale. Many industrial and medical applications are currently being developed at a tremendous rate, and some have already been fully implemented.

For these reasons, one may consider the advent of the nanosciences (nanotechnology and nanomaterials) as a turning point in the industrial development of the twenty-first century.

2 Nanomaterials

A nanomaterial can be defined as a material made up of nano-objects, for which at least one of the three physical dimensions lies in the range 1–100 nm, and displaying specific nanoscale properties. These nano-objects may be particles, fibres, or tubes (one speaks of fillers and strengtheners), or structural constituents.

These nano-objects are used either as-is, e.g., catalysts for chemical reactions, vectors for carrying medicines to target cells, substances for polishing wafers and hard disks in microelectronics, etc., or for synthesising nanomaterials. The latter fall into three categories:

1. *Nanostrengthed Materials.* The nanobjects are incorporated or produced in a matrix in order to ensure some new functionality or modify its physical properties. A good example is provided by nanocomposites, where these modifications improve resistance to wear.
2. *Surface-Nanostructured Materials.* Here the nano-objects are used to constitute a surface coating. Fabrication procedures for these coatings exploit techniques of physical deposition like physical vapour deposition (PVD), electron beams, laser ablation, and so on, or techniques of chemical deposition, like chemical vapour deposition (CVD), epitaxy, sol–gel, and so on.
3. *Bulk-Nanostructured Materials.* Nano-objects can also be constituents of bulk materials which, through features of their intrinsic nanometric structure, like porosity, microstructure, or nanocrystalline lattice, display specific physical properties.

3 Social and Economic Aspects

3.1 Nanotechnology and Nanomaterials Markets

The European Commission estimates that the world nanotechnology market was slightly above 40 billion euros in 2001. But by 2008, the global market for products resulting from nanotechnology was expected to reach more than 700 billion euros. In 2010–2015, the economic consequences of nanotechnologies should weigh in at around 1 000 billion euros per year, if all sectors are included, according to the US National Science Foundation, with some 340 billion euros of this specifically in the area of nanomaterials (Hitachi Research Institute). As a consequence, nanotechnological enterprises may directly employ 2 to 3 million people in the world.

3.2 Financing in France

The study [1] assessed public investment in the nanotechnologies. It shows that France is making a considerable effort in this direction. According to this study, taking into account the whole range of credits made available and the means allocated in this field by the French national research organisation (CNRS) and the French atomic energy authority (CEA), including staffing costs, the final figure is 551.6 million euros without including tax, or 637 million euros all included, for 2003. However, although it is sometimes possible to identify a specifically nanotechnological activity, such a distinction cannot always be made, and would generally have little meaning.

Quantitatively speaking, there is a very significant level of public funding in the field of nanotechnology and nanomaterials in France, both in absolute value and also in relation to France's main European partners, i.e., Germany and Great Britain. But this financing is well below the level in Japan and the United States.

In France, since 2005, calls for national programmes (ANR, A2I, etc.) have supported coordinated assessment of the effects nanoparticles may have on health. Eighteen months after the recommendations made by reports from the French agency for health and safety at work and in the environment (*Agence française de sécurité sanitaire de l'environnement et du travail* AFSSET) [2] and the French commission for prevention and safeguards (*Comité de la prévention et de la précaution* CPP) [3], which stress the need to coordinate ways of controlling risks on a national, if not European, level, a panel of experts was set up within the French public health authority (*Haut Conseil à la Santé Publique* HCSP), with the title *Groupe de veille sur les impacts sanitaires des nanotechnologies* (GVISN).

This interministerial watchdog is briefed to provide analyses and make recommendations with regard to relevant questions raised either within or outside the group, in order to provide the government (the Ministry of Health and other ministries that may be concerned by this subject, such as environment, agriculture, research, and industry) with the support and advice it will need to define policy and handle new issues raised by nanomaterials and nanotechnology with regard to health and safety. At the beginning of 2009, on the basis of information provided by the GVISN, the HCSP made a pronouncement regarding carbon nanotubes [4].

At the end of 2005, all governments taken together had spent some 18 billion dollars to finance the nanotechnology and nanomaterials sector. With close to 6 billion dollars more in 2006, it was estimated that this world level of funding had equalled the whole of the Apollo programme which took men to the Moon.

3.3 Production and Applications of Nanomaterials

World Market

There are many applications of nanomaterials, as can be seen from Table 1 [5]. The world nanoparticle market for energy applications was estimated as around 54.5 million euros in 2000 and was expected to reach 77 million euros in 2005, i.e., a mean annual growth rate of 7%. This market has been driven by increasing awareness of the need to protect the environment. Nanoparticles are used for catalysis applications in the car industry, ceramic membranes, fuel cells, photocatalysis, propellants and explosives, antiscratch coatings, structural ceramics, and thermal spray coatings.

The world nanoparticle market for biomedical, pharmaceutical, and cosmetic applications [6] was estimated at 85 million euros in 2000 and was expected to reach 126 million euros in 2005, i.e., a mean annual growth rate of 8.3%. This is the market represented by the inorganic particles used to produce antibacterial agents, biological tags for research and diagnostics, biomagnetic separation processes, drug carriers, contrast media for magnetic resonance imaging (MRI), orthopedic devices, and solar protection screens.

The worldwide annual production of nanocomposites currently amounts to just a few thousand tonnes, mainly in cabling and packaging. However, by 2010, this production is expected to leap to 500 000 tonnes per year. Markets have been identified in the transport, engineering, and high technology sectors, due to the potential these materials have for strengthening structures while making them lighter, together with different design possibilities, e.g., reduction of thickness.

By 1995, the production of carbon black had already reach around 6 million tonnes per year worldwide. By 2005, global production was estimated at 10 million tonnes. The production of silica is around 300 000 tonnes per year, while titanium oxide has reached some 3.5 million tonnes for particles with micrometric dimensions, and close to 3 800 tonnes of nanoparticles were produced in 2000. The volume of aluminium nanoparticules is estimated at around 100 tonnes per year worldwide.

Table 1. Applications of different types of nanomaterial

Nanomaterial	Field of application
Nanoceramics	Structural composite materials Anti-UV components Mechanochemical polishing of wafers in microelectronics Photocatalysis applications
Nanometals	Antibacterial and or catalysis sectors Conducting films for screens, sensors, or energy generating materials
Nanoporous materials	Aerogels for thermal insulation in electronics, optics, and catalysis Biomedical applications to drug carriers and implants
Nanotubes	Electrically conducting nanocomposites Structural materials Single-sheet nanotubes for electronics and screens
Bulk nanomaterials	Hard coatings Structural components for the aeronautic industry, cars, ducts in the petroleum and gas industries, the sports sector, and anticorrosion applications
Dendrimers	Medical applications, including administration of medicines, fast detection techniques Cosmetics
Quantum dots	Optoelectronics (screens) Photovoltaic cells Inks and paints for anticounterfeit tagging
Fullerenes	Sports (nanocomposites) and cosmetics sectors
Nanowires	Applications in conducting layers of screens or solar cells and in electronic devices

French Production of Nanomaterials and Their Current Uses

According to a report published in 2007 by the French *Institut national de recherche et de sécurité* (INRS) [7], a first general survey of French nanoparticle production could already be drawn up. This information was consolidated in 2008 by elements from an AFSSET report entitled *Nanomaterials and Safety at Work* [2].

The main themes of the INRS report regarding the different nanoparticles produced in France can be summarised as follows:

- *Titanium Dioxide.* The French production of TiO_2 is around 240 000 tonnes. Different sizes of particle are used, depending on the sector, in the range 150–400 nm as pigment or opacifier in the paint and plastics industries, positioning them at the upper end of the nanoparticle range. The production of nanometric titanium dioxide is all carried out by 270 workers at one site, and reaches some 10 000 tonnes per year for three applications: architectonics, cosmetics, and air purification systems.

- *Silica.* With a production of 200 000 tonnes of SiO_2, France is the second largest producer of natural silica in the world, extracting from one particular rock called diatomite. This production occurs at two extraction sites and involves about a hundred workers. As far as synthetic silica is concerned, i.e., precipitated silica, pyrogenic silica, and fumed silica, the annual production is greater than 100 000 tonnes and involves some 300 people. The main use is rubber reinforcement for tyres (where it is associated with carbon black in a 1:1 ratio), shoe soles, and rubber technical parts for wires and cables. In the food industry, these silicas are used as substrates for vitamins, acidifiers, and anticaking agents. The paint industry uses them as matting agents, while toothpaste manufacturers use them as thickeners and mild abrasives.
- *Nanoclays.* Two countries share the whole market here, Germany and the US. One site is currently under development for nanoclay production in France. A volume of around 100 tonnes is planned for 2007. About 50 people should be employed there.
- *Single-Wall Carbon Nanotubes (SWCNT).* The production capacity for this category of nanotubes is between a few grams and a few tens of kilograms per day. At the present time, the maximal capacity is produced by an American company with 40 kg/day, using chemical vapour deposition (CVD) and a gaseous mixture of $Fe(CO)_5$ and CO (the HiPCO process). French production of SWCNT is currently limited to university research laboratories. Several sites are equipped to produce quantities of around 10 g, either using a similar, low temperature process (CVD or catalytic CVD), or using a high temperature process (arc or plasma).
- *Multiwall Carbon Nanotubes (MWCNT).* These have produced by one French company since 2006. This production unit, with a capacity of 10 tonnes per year, was a pilot project, involving about 10 people. Production there will be increased to several hundred tonnes per year by the end of the decade.
- *Carbon Black.* This is essentially composed of spheres with diameters in the range 10–500 nm, in aggregations of between ten and a few hundred particles. French production was 240 000 tonnes in 2005. It is carried out at four production sites, and involves a workforce of around 350 people. Seventy percent of carbon black is used by the tyre industry. The proportion, which may reach 30% of the weight of a tyre, is tending to fall, being replaced by precipitated silica. The rubber industry also uses it to make protective sheaths for cables and in the composition of conveyor belts, drive belts, and joints.
- *Aluminas.* A single production site in France produces ultrahigh purity aluminas. These are made using an alum process, i.e., aluminium sulfate with multihydrated ammonia. Two horizontal units are being set up at the site to take production to 1 000–1 700 tonnes per year from 2008. Other

producers share the alumina market. This so-called speciality alumina is synthesised by the Bayer process which uses bauxite as raw material. French production of speciality aluminas represented 468 000 tonnes in 2004. This includes a proportion of ultrafine and nanostructured alumina on top of the traditional range.

Future Developments and Markets

According to the investigation by Rocco in 2004 [8], there are four main stages in the development of nanotechnology and nanomaterial production: passive nanostructures, active nanostructures, systems of nanosystems, and molecular nanosystems. Currently commercialised nanoproducts belong mainly to the first category of passive nanostructures.

An active nanostructure is one that can modify its own state, e.g., morphology, shape, and mechanical, electronic, magnetic, optical, or biological properties, and so on, during its use. As an illustration, a mechanical actuator might change size, while the morphology and/or chemical composition of nanoparticles used as drug carriers in medicine might evolve in order to get through biological barriers, for example. These novel states of nanostructures might in turn evolve, in particular, to make them harmless at the end of their life cycle. Such changes will be all the more complex as the structures and systems are required to become more bulky and to implement several functions.

Examples of such active nanostructures are the nanoelectromechanical systems (NEMS), biological nanodevices, transistors, amplifiers, pharmaceutical and chemical carriers, molecular machines, light-activated molecular motors, nanofluidic systems, sensors, and radiofrequency identification devices (RFID).

In the field of tagging and identification, it should be stressed that some systems are already operational. As an illustration, in 2006, Hitachi presented the smallest RFID chip ever made. With dimensions $0.05 \times 0.05\,\mathrm{mm}^2$, it has been referred to as smart dust. It contains a 128 bit read-only memory (ROM) that can stock a 38 digit identity number and it is easily integrated into a sheet of paper, for example. Recall that an RFID chip is used for automatic identification of whatever it tags. RFID chips can contain all kinds of information and are found on a great many different items, from passports to labels on products on sale at the supermarket, not to mention concert tickets. The advent of RFIDs in the form of a dust makes it easy to integrate them into ever more varied items. In parallel with these developments, a UK company has developed a device that can locate such RFID chips at up to a distance of 180 m [9] and to an accuracy of 2 cm in a 3D region. The possibility of such a high degree of miniaturisation has raised questions about tracking and checking up on individuals without them knowing.

In medicine, nanomaterials are already used in commercialised medical equipment, such as bandages, implants, prosthetics, and others. Medical biology uses nanoelements for in vitro diagnosis of infectious diseases, immunological disorders, and cancers. Some devices for day-to-day medical observation of biological parameters, e.g., glycaemia, will usefully benefit from the extreme miniaturisation made possible by nanotechnology. The medical imaging sector is also investigating the possibilities for improving the contrast and resolution of MRI images by placing nanoparticles in target organs. Pharmacological research has long been exploring the possibility of carrying therapeutic drugs as close as possible to lesions, using nanoparticles designed to target sick cells. Therapeutic trials are under way, especially in the field of cancer treatment. Nanotechnologies may one day be able to customise drugs. Some nanoparticle contrast agents and drugs have already been accepted by the relevant regulatory bodies.

One application in particular has seen rapid development and no doubt benefits significantly from progress in miniaturisation as procured by nanotechnology, and that is deep brain stimulation by microelectrodes placed in the brain [10]. Since the 1980s, a team in Grenoble (France) led by Professor A.L. Benabid has discovered that electrical stimulation of a certain part of the brain can reduce or completely remove the shaking symptoms of those suffering from Parkinson's disease. Since then, in collaboration with several international teams, applications to other medical problems have been proposed, including acute dystonia (a neurological movement disorder), epilepsy, and others. The technique was then tested – with the agreement of the French *Comité consultatif national d'éthique pour les sciences de la vie* (CCNE)[1] – in the treatment of obsessive–compulsive disorders (OCD) and depressive syndromes that could not be relieved by conventional medical treatments. At the present time, some 35 000 people are being treated by this 'brain pacemaker', including about 1 000 in France. Other research is investigating the brain–machine interface, with a view to controlling muscular movements either by acting directly on the nerve or muscle fibres, or by going through the central nervous system. These are promising applications for people suffering from paralysis or anomalous movements (tics).

However, these techniques carry the risk of side-effects, in particular when the electrode is implanted, since it may be rejected or cause brain hemorrhage. This is why the idea of using nanoscale electrodes came into being. In France, the CLINATEC project is a biomedical research center devoted to nanomedicine, focussing primarily on implanted devices and the brain–machine interface. It is important to debate the possible abuses of nanotechnology in medical applications, in particular for specific medical applications like deep brain stimulation.

It should also be stressed that there is a very clear dual development of nanotechnology and nanomaterials for specific defence applications. For example, items for personal protection such as bullet-proof vests and helmets incorporating carbon nanotubes, ultrafast and ultrasensitive detection devices,

---

[1] National Consultative Committee for the Life Sciences.

chemical and bacteriological carriers, exoskeletons (powered mobile frameworks worn by the soldier and interfaced on the human brain, which should make it possible to carry 80 kg for 80 km at more than 50 km/h, developed by DARPA, USA), not to mention the development of thermobaric bombs which use pyrophoric nanoparticles and produce equivalent blast waves to a nuclear weapon [11] – the first tests were carried out by the United States and Russia in 2007. This all-pervading dual aspect of nanotechnological applications raises the question of whether we should renew international negotiations about the proliferation of weapons of mass destruction with a view to drawing up new treaties.

3.4 Nanomaterials and Safety

Our current understanding of the effects of micro- and nanoparticles in atmospheric pollution has raised fears regarding the consequences of man-made nanoparticles for human health. While very few reliable data are available in this field, studies published about the interactions of nanoparticles at the cellular level suggest that we should be cautious. Recently, work by Donaldson et al. [4] tends to show that some carbon nanotubes can induce similar effects to asbestos fibres, inducing mesothelioma.

As shown as early as 2005 in a summary note [12], the important scientific questions regarding nanoparticles and health must be concerned with the whole life cycle and must consider the following specific features:

- The physicochemical characterisation and classification of nanoparticles according to their level of surface reactivity, a good indicator of potential biological effects.
- The detection and characterisation of exposure to these particles by everyone from factory employees to users.
- Their potential biological effects on humans.

These issues concern workers in the nanotechnology and nanomaterials sectors, who may be exposed to high concentrations of nanoparticles, but also the population at large, whose exposure to these nanoparticles is less direct and related to the life cycle of the nano-object in question.

Finally, the risks associated with nanoparticle explosions must also be given due attention. At the present time, little has been done, e.g., with regard to staff involved in the production of nanomaterials from such nanoparticles.

Given the importance of these issues, some websites have been set up to monitor publications in this area:

- *The Virtual Journal of Nanotechnology Environment, Health and Safety* [13].
- *Nanotechnology: Health and Environmental Implications – An Inventory of Current Research* [14].
- *Safe Production and Use of Nanomaterials* [15].

4 The Need for Studies in Nanotoxicity and Nanoethics

As discussed by Roure [16], the highly diverse industrial economy of nanotechnology and nanomaterials is well under way. Given the speed with which laboratory research is transformed into nanoproducts, some already commercialised, and given their all-pervading tendencies [17], the time has come to assess our current understanding of nanotoxicity, and also to address the relevant ethical questions. Indeed, research in nanoscience and nanotechnology stands out by the difficulty in distinguishing the fundamental from the technological. Synergies arise through the NBIC convergence (nanotechnology, biotechnology, information technology, and cognitive science) and their effects are difficult to quantify in the mid to long term. This fourth volume of the *Nanoscience* series aims to present the state of the art, both in the field of nanotoxicity and with regard to what we shall define as nanoethics. We hope it will contribute to a responsible and safe use of nanomaterials and nanotechnology.

References

1. *Le financement des nanotechnologies et des nanosciences. L'effort des pouvoirs publics en France: comparaisons internationales*, published in January 2004 on behalf of the French *Ministère de la Jeunesse, de l'Education nationale, et de la Recherche* (MJENR)
2. *Agence française pour la sécurité sanitaire de l'environnement et du travail* (AFSSET): Report on *Nanomatériaux: effets sur la santé de l'homme et sur l'environnement*, www.afsset.fr/?pageid=&newsid=105&MDLCODE=news&search=yes&txtSearch=nano; *Nanomatériaux et sécurité au travail*, www.afsset.fr/upload/bibliotheque/258113599692706655310496991596/afsset-nanomateriaux-2-avis-rapport-annexes-vdef.pdf
3. Conclusions of the *Comité de la prévention et de la précaution* (CPP): *Nanotechnologies, nanoparticules: quels dangers? quels risques?*, www.ecologie.gouv.fr/Avis-du-Comite-de-la-Prevention-et.html
4. Conclusions of the *Haut Conseil de la Santé Publique* (January 2009), www.hcsp.fr/hcspi/docspdf/avisrapports/hcspa20090107_ExpNanoCarbone.pdf
5. Report *Etude prospective sur les nanomatériaux*, on behalf of DIGITIP/MineFi by the French company *Développement & Conseil* (2004)
6. Report produced by the *Direction générale de l'industrie, des technologies de l'information et des postes* (2004)
7. B. Honnert, R. Vincent: Production et utilisation industrielle des particules nanostructurées. Hygiène et Sécurité du Travail, ND 2277 (2007)
8. M.C. Rocco: Nanoscale science and engineering: Unifying and transforming tools. AIChE Journal **50**, 890–897 (2004)
9. www.tfot.info/pod/152/rfid-loc8tor.html
10. NanoForum – Nanomedicine session (5 June 2008)
11. news.bbc.co.uk/2/hi/europe/6990815.stm

12. J. Boczkowski, E. Gaffet, A. Lombard: Note de synthèse pour le PNSE – Les déterminants environnementaux: nanoparticules et santé (2005)
13. The Virtual Journal of Nanotechnology Environment, Health and Safety, www.icon.rice.edu/virtualjournal.cfm
14. Nanotechnology: Health and Environmental Implications – An inventory of current research, www.nanotechproject.org/18/esh
15. Safe Production and Use of Nanomaterials: www.nanosafe.org
16. F. Roure: Economie internationale des nanotechnologies et initiatives politiques. Annales des Mines, 5–12 (2004)
17. Report by the *Office parlementaire des choix scientifiques et technologiques. Audition publique sur les nanotechnologies: risques potentiels, enjeux éthiques.* France (2006), no. Assemblée Nationale 3658, no. Sénat 208

Part I

Nanotoxicity: Experimental Toxicology of Nanoparticles and Their Impact on Humans

The fast developing nanotechnology markets have led to an increasing risk of human exposure to nanoparticles, through the lungs or the skin, but also by ingestion, or by injection in the form of medicines.[2] Indeed the applications of nanotechnology are many and varied. They are already present in our day-to-day lives in such everyday products as cosmetics and body lotions, electronic and household goods, food packaging, and clothes, and the list gets longer all the time. There are particularly interesting medical applications. Many of these products provide no clear indications of such contents and an assessment of the risks for humans and the environment is only required for certain nanoparticles, depending on the level of production and their usage, in particular in the medical area. Discussions are currently under way to define a regulatory framework that would allow us to control the risks of disseminating these products, both for the consumer and for the environment.

Exposure to nanoparticles may be professional, environmental, or medical, and it concerns a variety of nanoparticles used in a broad range of different forms. However, the rapid evolution of this technology, expected to constitute the industrial revolution of the twenty-first century, makes it essential to evaluate the risks and hazards as early on as possible, on the basis of a better understanding of their biological effects.

The following chapters, written by recognised specialists in the field of nanotoxicology, the toxicology and metrology of atmospheric particles and fibres, and the assessment of environmental risks for humans, relate the current state of understanding in this area. They describe the results of experimental work in toxicology. Indeed, toxicologists were the first to warn public health authorities of the increased risk due to particles with a given chemical composition when they come into contact with living systems in a nanometric form. While our understanding regarding human exposure and its consequences remains highly uncertain, experimental results on other animals or in vitro on cell cultures are sufficiently clear in some cases to allow an evaluation of the risks, even in a situation of uncertainty. From the beginning, nanotoxicology has followed a different approach to the one traditionally used in regulatory toxicology. In particular, it incorporates the latest models and techniques of modern molecular and cellular biology, allowing a systemic approach. This evolution, one might even say revolution, in toxicology is especially relevant when the type of exposure is difficult to characterise, doses are very low, and effects are varied. The data presented in the next eleven chapters provide a snapshot of a particularly fast evolving field of research, but they will nevertheless serve as a solid foundation for the reader who wishes to familiarise herself or himself with this complex area, an area where society as a whole is especially interested in a proper evaluation of the risks and hazards.

---

[2] Introduction by Francelyne Marano, President of the *Groupe de veille sur les impacts sanitaires des nanotechnologies* and member of the *Haut Conseil de Santé publique*.

1

Toxicity of Particles: A Brief History

Marie-Claude Jaurand and Jean-Claude Pairon

Over the last few decades, a certain number of pathologies have been directly linked to various kinds of inorganic dust affecting subjects exposed to these substances in the workplace. As a consequence, public health authorities have become increasingly interested in determining the effects particles can have on health. But analytical investigations of pollution, along with other evidence, has shown that exposure to dusts is not limited to workers in specific sectors. In fact, it may also affect the population at large, a finding that has led to the setting up of think tanks, closer assessment of different types of pollution, and research specifically devoted to the toxicology of dusts. Note also that the terminology itself has evolved. In particular, the word 'dust' has gradually been replaced by 'particulate matter', although both terms refer to the solid fraction in aerosols.

The aim of the present chapter is to summarise the context and the facts that have led to the area of investigation we now call particulate toxicology. We begin with the sociological and technical features that have made the study of toxicity what it is today. For chronological reasons, we then define the pathologies caused by exposure to inorganic dusts, before going on to describe the particles responsible for pathogenic effects. To explain the mechanisms by which particles can act, we discuss methods for investigating toxicity and the ways they have evolved. On this basis, hypotheses are formulated about the mechanisms leading to observed biological effects. Before concluding, we shall consider the results of toxicological studies carried out up to now to assess the toxicity of particles occurring in professional and general environments, or likely to be generated in such environments, but also some questions which have not yet been answered, or which have arisen from earlier experiments.

P. Houdy et al. (eds.), *Nanoethics and Nanotoxicology*,
DOI 10.1007/978-3-642-20177-6_1,

1.1 Sociological and Technical Factors Conditioning the Study of Particle Toxicity

Concern over the toxicology of inorganic dusts was first inspired by the discovery of pulmonary pathologies in workers with occupational exposure to such substances, e.g., in mines [1,2]. In an interesting chapter on pneumoconioses in mines in the north of France, Amoudru summarises the steps leading to the recognition of these illnesses as work-related [3]. Note that it took over a hundred years to fully recognise this fact. Indeed, while lung diseases had been observed in several European countries, including France, at the beginning of the nineteenth century, it was not until 1945 that a statute was published recognising silicosis as a work-related disease. There were several reasons for this delay: medical controversy which delayed the undeniable recognition of the risk, factors leading to confusion, e.g., silicosis favours infectious pathologies like tuberculosis, and world political events which postponed application of the decision to recognise silicosis as a work-related disease until after the war. In the case of asbestos, purely economic motives of industrial protectionism hampered progress on the research front [4,5].

Up to now, the toxicology of inorganic dusts has mainly been concerned with respiratory problems, because the principal route by which particles could enter the organism was of course inhalation. Once the particles causing pathological effects had been identified, experiments could be devised to determine the consequences of inhalation in tissues and cells, in order to understand the relevant mechanisms and determine the factors involved in the cell's response. In parallel, other research was investigating the physical, chemical, and physicochemical properties of the particles that led to this biological activity.

The relationship between exposure and pulmonary pathology in humans were identified by the end of World War II, but the experimental work was only published around the end of the 1960s. Returning to the example of dust in the coal mines, experimental data only became available in the 1950s. It is interesting to note that, in his autobiography, Dulbecco mentions that at one point, somewhere around 1946, he wanted to investigate certain illnesses caused by dust inhalation which it seemed concerned mainly miners. But the results of morphological analyses based on lung sections were insufficient to explain the causes of the lesions he observed, and unfortunately this eminent scientist did not pursue his efforts in this field [6]. A particle toxicology journal presents a list of European research programmes launched under the auspices of the *Communauté européenne du charbon et de l'acier* (CECA),[1] founded in 1951, to evaluate the effects of carbon and silica dusts. Note that the first dates back to 1955, and refers to medical issues [7].

The significant technological developments of the nineteenth and twentieth centuries, fuelled by the requirements of war, led to several groups of workers

---

[1] European Carbon and Steel Community.

being exposed to pollutants. Asbestos provides a second example of a harmful agent for certain groups of workers, and which further stimulated research into particle toxicology. The considerable increase in the use of asbestos fibres during World War II and the emergence of the associated afflictions only instigated research after a delay of about 20 years. The first consequences to be recorded were of pulmonary asbestosis. The role played by asbestos in the development of this pulmonary disease was described as early as the 1920s by several authors who had observed asbestos workers [8, 9]. Regarding the relationship with lung cancer, the inaugural study is often taken to be the work by Lynch and Smith [9, 10]. An increase in the incidence of lung cancer in the case of pulmonary asbestosis was subsequently reported by Doll in 1955, and then a few years later, an article was published about the high level of mesothelioma affecting workers in asbestos mines in South Africa, as well as other inhabitants of the region [11, 12]. The impact of this paper was twofold, because it identified the harmful effects of asbestos, but it also recognised this cancer as a primitive pleural tumour. Indeed, the reality of these tumours was debated [13].

With the case of asbestos, toxicology acquired a new feature. Whereas for carbon and silica, it was mainly in the production sector that the pathological consequences of exposure were being felt, in the case of asbestos, similar pathologies were being observed in a second sector, namely, people using asbestos-based materials (several trades, notably the building trade). This new wave of asbestos-related illnesses, sometimes called the second wave, led some to raise the question of a possible third wave, one which might reach people exposed for non-professional reasons in a general environment, e.g., buildings containing asbestos, natural pollution in regions with outcrops of asbestos-containing rocks, etc. [14, 15].

The evolution in the type of population affected by the harmful consequences of dusts, going from production to applications and the general environment, continues today with regard to the populations exposed for non-professional reasons. Indeed, we are now concerned about exposure to dusts of human origins, viz., fine particles (FP) and ultrafine particles (UFP). Examples of sudden increases in mortality associated with pollution peaks have raised concern in this area [16–18]. In parallel with the greater number of different populations affected by respiratory problems, other pathologies related to exposure to particles have had to be taken into consideration, such as cardiovascular disease [19]. With the emergence of nanoparticles (NP) resulting from nanotechnological developments, this diversification in the nature and origin of the particles on the one hand, and in the consequences for the health on the other, also affects sectors that are not directly in contact. In the case of NPs, apart from the production sector, applications, and the general environment, an ecological dimension has also come into being [20]. Concerning pathologies, the respiratory and cardiovascular systems are no longer the only ones to be affected, since the question of toxicity is now raised with regard to other sites, like the nasopharynx, brain, and kidneys [21].

Another important point must be taken into account, related to the technology, when considering the long term consequences of particle toxicology. In the 1950s, the relevant journals were mainly medical. But from the 1970s to 1980s, the possibilities for publication increased significantly, facilitating the exchange of information between different groups involved in this line of research. Furthermore, international exchanges between the various research centers became commonplace with the proliferation of conferences and seminars. Availability of information through numerous publications meant that results of studies and inquiries in the workplace became known to research teams who could then develop toxicological studies on animals, and subsequently, in vitro approaches on cell cultures. The fact that independent groups were now involved was a key factor in the development of this research. By comparing results on the international level, hypotheses could be formulated regarding the mechanisms involved, while warnings could be sent out regarding the potential toxicity of new products, and in some cases, help could be provided to formulate regulatory measures. The improved means of communication was an important factor in the evolution of research in particle toxicology. It should be borne in mind that Internet is a recent source of exchange between research scientists. Although this network was established by the end of the 1980s, email and file transfer protocols were only available to a few at the beginning, and it was only around 1995 that they became widely accessible [22].

1.2 Pathologies Caused by Inorganic Dusts

1.2.1 Pneumoconiosis

Pneumoconiosis is a pathology due to the presence of exogenous particles. These particles accumulate in the lungs, causing a tissue reaction associating cell inflammation and macrophages. Together these can form so-called foreign-body granulomas, containing giant multinuclear cells. Carbon dusts cause various lung diseases, including fibrosis, emphysema, chronic bronchitis, and an alteration of lung function [2, 23, 24]. Exposure may be associated with crystalline silica, depending on how the carbon is extracted.

Silicosis is a common form of pneumoconiosis caused by the deposition of silica particles in the lungs. Apart from activities associated with coal mining, it is encountered across a broad range of occupations related to metal mining and quarrying, the building industry, etc. [25]. On a histological level, it is a nodular fibrosis formed by fibrohyaline tissue [26]. This respiratory disease is still observed today, despite the decline in coal mining [27].

Asbestosis is caused by exposure to asbestos, and occurs in the form of a diffuse interstitial fibrosis [28]. In countries where the use of asbestos has been forbidden, it has become much less common, because it is a pathology that generally develops only after exposure to very high doses. Pleural plaques are benign lesions caused by asbestos. This fibrosis is usually localised in the parietal pleura and evolves by calcifying [28].

Forms of pneumoconiosis induced by exposure to other types of dust have also been identified. These are mainly due to accumulation of dusts, some of which may evolve into fibrosis. They are caused by metal compounds of iron, aluminium, tin, or barium, but also by beryllium, and are manifested in the form of granuloma or fibrosis, e.g., siderosis, berylliosis, aluminosis, etc. [29].

Fibrosis results from increased synthesis of fibrous tissue, collagen, and proteins of the conjunctive tissue. It occurs in the pulmonary parenchyma, around the bronchi and in the pleura, and is associated with a tissue repair process. The increased amount of collagen and fibronectin come from a greater synthesis of proteins by fibroblasts in the extracellular matrix and/or an increase in the number of these cells. This in turn occurs in response to tissue lesions and factors emitted during inflammation by macrophages and neutrophils, both in the lungs and in the pleura [30, 31]. Studies carried out so far to investigate the fibrosing effects of particles have been based on this fibrogenesis mechanism [24]. Recent work discusses the role of inflammation in these tissue repair processes, attributing a role to the pneumocytes of the respiratory epithelium in the initiation and development of this process of pulmonary fibrogenesis [32].

1.2.2 Cancer

Cancer results from proliferation of cells exhibiting genetic alterations acquired over successive divisions. This neoplastic transformation involves several stages. During the process, various modifications occur in the cell's genetic material (mutations, deletions, translocations, etc.) and in the regulation of functions (control over DNA integrity, proliferation, recycling, and apoptosis). These modifications have several consequences in their turn:

- on gene expression, which is deregulated both qualitatively and quantitatively, and in terms of the expression rate;
- on the equilibrium between cell proliferation and mortality;
- on the relationship between the cell and the extracellular environment.

On a molecular level, the neoplastic transformation has been described as a mechanism of oncogene activation together with the silencing of tumour suppressor genes, and this model has been validated by observations and experimental studies [33]. Other studies have led to the proposal of a more general mechanism underlying the neoplastic process, this time involving several genes [34]. Hanahan and Weinberg have suggested that cells should acquire six hallmarks indicating their neoplastic character, viz., growth autonomy, resistance to antiproliferation signals, resistance to apoptosis, the potential for unlimited replication, sustained angiogenesis, invasion, and metastasis [35]. Recently, a seventh indicator of neoplastic transformation has been put forward, viz., inflammatory conditions [36]. The evolution of the cell during tumor growth is sustained by chromosomal instability. This generates an alteration of the genetic material which is transmitted to the daughter cells during successive divisions [37, 38]. The chromosomal aberrations found in tumours are indicators of genetic deregulation and chromosomal instability.

Neoplastic evolution is accompanied by phenotypic modifications, such as loss of contact inhibition between cells, independence from growth factors for proliferation, abnormal karyotype, genetic disequilibrium, and differential gene expression as compared with the normal cell. Toxicological tests are based on these modifications in the field of oncogenesis. They use techniques for revealing one or more phenotypic changes associated with this neoplastic evolution.

1.3 Particles Causing Pathogenic Effects in the Airways and Respiratory System

Particle toxicity is often discussed in cases where exposure is by inhalation in the professional environment, because the associated pathologies are generally discovered in the workplace. But exposure may happen in the environment as a whole, and not just in the vicinity of the sources of contamination, but some distance away, transported by atmospheric currents.

1.3.1 Origin of Particles

The origin of particles that have undergone toxicological studies has already been mentioned. These are particles generated by human industrial activities, industrial applications, or products used in everyday life. Stocks of samples have sometimes been established to be distributed to different research groups for the purposes of toxicological studies. This was the case for silica, by setting up stocks from different mines, particularly in Germany, and also for asbestos, through the action of the *Union internationale contre le cancer* (UICC), which prepared samples of each kind of fibre [39, 40]. More recently, samples of artificial inorganic fibres have been distributed for various studies. These were glass, rock, slag, and refractory ceramic fibres [41]. Regarding fine and ultrafine particles (or nanoparticles), such as PM_{10} and $PM_{2.5}$, carbon black or titanium oxide and carbon nanotubes, this has not been done systematically. Although certain sources have been favoured for these particles, for others, they turn out to be very varied. It should be borne in mind that, in the general case, the samples used for experimentation are not always fully representative of the particles to which subjects have been exposed, owing to the great diversity of possible sources, whether they be natural or synthetic. Moreover, it would be difficult to test this diversity, due to lack of information regarding the nature of the exposure and the wide range of different particles.

1.3.2 Types of Particle

Carbon Dusts

These are complex compounds whose composition depends on the mine. These carbon-containing dusts contain various silicates, carbonates, and sulfates.

Silica

Silica is composed of silicon dioxide (SiO_2). There are several types, with the same chemical composition, but with different crystal structure and cytotoxic activity, viz., tridymite, crystobalite, and quartz. There is also a non-crystalline form, viz., amorphous silica, opal.

Asbestos

There are several types of asbestos. The term covers fibrous silicates in hydrated crystalline form, with various possible cation compositions, e.g., Mg, Ca, Na, Fe [42]. There are several forms of asbestos, with different structures and chemistry, mainly used in industry: band silicates such as the amphiboles, e.g., crocidolite, amosite, and anthophyllite, among others, which contain different cations Mg, Fe et Na, and sheet silicates such as phyllosilicate, and chrysotile, which is a hydrated magnesium silicate [43]. The sheets are rolled up around a central axis, giving the elementary fibril a multilayer, hollow tube structure. Today, asbestos is by far the most widely studied type of particle.

Other Fibres

According to the definition provided by the World Health Organisation (WHO), a fibre is a solid particle, either natural or artificial, with an elongated shape and parallel edges, with length greater than 5 μm and aspect ratio (length to diameter) greater than 3.

Among the other natural inorganic fibres, the main one to attract the attention of toxicological studies has been erionite, a zeolite mineral, which is basically aluminium silicate containing Na, K, and Ca. This is due to epidemiological observations associating mesothelioma with environmental exposure to these fibres [44]. In addition, many kinds of synthetic inorganic fibre such as rock wool, slag wool, and glass wool, or again refractory ceramic fibres, have been studied owing to their broad spectrum of applications, including their use as an asbestos substitute. These are silicates with different aluminium contents, and also different alkali metal and alkaline earth cations. Their chemical composition is very varied, depending on the application [45]. Research in this field has led to a notion of biopersistence, to be defined and discussed later.

For the record, one should also mention that synthetic organic fibres have been studied, e.g., para-aramid and aramid, used as strengthening materials

[25]. Recently, carbon nanotubes have attracted some attention, owing to the similarity of their physical characteristics with those of asbestos, which raises a doubt over their potential toxicity [46].

Fine and Ultrafine Particles. Nanoparticles

FPs and UFPs began to attact attention with the problems arising over pollution peaks. These terms cover a wide range of particles, from chimney smoke to carbon black and titanium oxide, not to mention particles contained in the surrounding air. Their chemical composition and structure are thus highly diverse. More recently, the development of nanotechnology has witnessed an expansion of the world of nanoparticles, which have applications across a very broad range of situations.

Studies of atmospheric pollution require the collection of particles according to their aerodynamic characteristics. Sampled particles are separated by reference to their aerodynamic diameters (AD): PM_{10} and $PM_{2.5}$ (AD less than 10 μm and 2.5 μm, respectively). Ultrafine toxicity studies suggest that effects were probably related to the ultrafine fraction, leading to the hypothesis that these UFPs are potentially more toxic than the FPs [47].

The development of nanotechnology created further sources of UFPs, with the synthesis of nanomaterials or the use of UFPs in a wide range of products. Nanoparticles are particles with at least one dimension of nanometric size, so they can be included with the UFPs. However, at the present time, the term UFP is generally reserved for naturally occurring particles, or man-made particles that have been unintentionally produced, while NP is used for particles produced by or resulting from the field of nanotechnology. Recent AFSSET reports in France survey the current situation with regard to toxicity and health risks raised by nanomaterials [48, 49]. According to one of these reports, nanomaterials are classified into four families depending on the form in which they are used [48]:

- in dispersed form, either random or ordered,
- in the form of nanowires or nanotubes,
- in the form of a thin film,
- in compact form.

The small size of NPs bestows special physicochemical properties upon them, which can make them highly reactive in a biological context. Nanoparticles resulting from nanotechnological activities can thus be produced as such or result from the manipulation of larger samples of material, e.g., by milling, degradation, etc. As far as chemical composition is concerned, nanoparticles may be metals, metal oxides, polymers, composite materials, or even biomolecules. Although of nanometric size, these particles tend to agglomerate and form aggregates, thereby increasing their overall size.

1.4 Evolution in the Methods for Investigating Toxicity

The methods for investigating particle toxicity have changed enormously since the first studies to be found in the literature, which date roughly from the middle of the twentieth century. Although earlier work is mentioned in several papers, it was only from the 1950s that a continuous and coherent literature came into being with the studies on silica, and later asbestos.

Subsequent work aimed not only to determine physiopathological effects, but also to understand the mechanisms of particle action. This was achieved through anatomopathological studies which, associated with other techniques like immunohistochemistry and histochemistry, then the main methods for characterising lesions, were able to identify the proteins involved in these mechanisms. Specific effects were observed in certain species (alveolar lipoproteinosis) [50]. Morphological observations led to in vitro studies on growing or surviving cells, and in particular on alveolar macrophages and fibroblasts. On these cells, the particles having exhibited cytotoxicity, revealed by cell viability assays, associated with the internalisation of the particles by the cells, work was then carried out to determine the nature of the interactions between particle and cell. Since the viability assays were based on examination of alterations in membrane integrity, 'model' cells were used. These were red blood cells with no capacity to internalise the particles, but providing information about the interactions between the cytoplasmic membranes and the particles. As data accumulated, efforts were made to understand the physiopathological mechanisms, reaction to foreign bodies, inflammation, and fibrosis, by determining the responses of specialised cells, macrophages, and fibroblasts. Then, with confirmation of the connection between cancer and exposure to certain types of particle, others investigated the effects on and responses of epithelial cells. It should also be noted that these methods were greatly speeded up by the commercialisation of cell culture equipment such as cell culture flasks, throwaways, and industrially produced culture medium.

1.4.1 Studies on Animals

To begin with, the means available for studying particle toxicity consisted in exposing animals, usually rats, but to some extent also guinea pigs and mice, in inhalation chambers. Intratracheal instillation was also used early on [51,52]. This methodology brought in the possibility of anatomopathological studies which, depending on the exposure time, evaluated the fibrosing or oncogenic potential of the particles. Exposure by inhalation is a method requiring a large amount of costly equipment, e.g., systems for aerosolisation of the particles, dedicated inhalation chambers, large quantities of the particle, etc. It is assumed to reproduce a type of exposure that can be taken as realistic as far as human exposure is concerned. However, exposure doses are sometimes difficult to specify, e.g., dusts from animal fur. Another commonly used method is intratracheal injection of particles in suspension in a physiological solution. This technique is not physiological, but the exposure dose is then clearly determined. On the other hand, there are various disadvantages, because the dose may be poorly distributed in the lungs, and a phenomenon of partial rejection can sometimes occur (see [53] for a review). But this method nevertheless proves to be useful and is often used to investigate the short to mid term effects of particles [54, 55]. It is then associated with analysis of the liquid resulting from broncho-alveolar lavage (BAL) (injecting physiological solution into the lungs by the bronchial route, then recovering the liquid) to identify and quantify the cellular and humoral inflammatory response caused by the particles. The methods most recently developed for inhalation exposure favour better knowledge of the exposure dose, to the detriment of the physiological situation, using the so-called nose-only method, where the animals are immobilised and exposed in individual chambers [56]. This technique, currently widely used, may also have a biological impact due to stress [57].

All these inhalation methods have led to a way of determining the dose of particles deposited in the lungs, the clearance rate, the retention level, and in the case of fibres, the evolution of dimensional characteristics. The technique involves extracting the particles after clearance from the pulmonary tissue [58, 59]. It is observed that fibres may break, and that bundles of fibres may split up to some extent, thereby altering the density of fibres of given dimensions as time goes by. Several groups used radio tagged fibres to determine the migration and bioavailability of particles in various pulmonary locations [60,61]. Many studies have shown preferential clearance of short fibres, thanks to macrophage purification, while longer fibres tend to be retained by the pulmonary parenchyma [62, 63]. Note that the nose-only method sometimes returns contradictory results, a point still in need of explanation [64].

To determine the long term consequences for the serous membranes (pleura and peritoneum) of exposure to particles, methods involve intracavitary injection or implantation of particles. In the context of this historical survey, it is interesting to note that earlier work on the implantation of solid substances led to a 'solid state' theory of carcinogenesis, inspired by polymer implantation experiments [65]. This terminology was used to distinguish tumours produced by a solid agent from those produced by a chemical agent. The use of such experimental systems for particle toxicology can be related to the emergence of questions over asbestos exposure [66, 67]. These methods have been criticised for their non-physiological nature. However, inhalation methods cause very little pleural reaction in rats, even with forms of asbestos that are considered to be highly carcinogenic, e.g., crocidolite, and they are not sensitive enough to assess the fibrosing and carcinogenic effect. (The life expectation of the animal is too short compared with humans, where pleural pathologies materialise only 30–40 years after the beginning of exposure.)

Over the last few years, work has been carried out using genetically modified mice and rats. These are animals in which certain genes have been deactivated (silenced), or new genes introduced by genetic manipulation into the animals' genomes. Studies made on these animals help to get a better understanding of the mechanisms through which particles act, but also give insights into the regulatory channels that are stimulated or altered in response to the particles, identifying the factors and genes involved in the development of pathologies. Work has been done on fibrogenesis and carcinogenesis [68–73].

1.4.2 Isolated Cells

Many cell types have been used in toxicological studies of solid particles. As mentioned above, the first cytotoxicity studies were carried out with red blood cells, using hemolysis as evaluation criterion, attesting as it does to the destruction of the cell membrane by the particles. These studies were done in the context of research on silica and asbestos [74]. It is unrealistic to extrapolate the results to pathologies caused by these particles, but these studies give insights into the factors modulating cytotoxicity. They revealed an electrostatic type of interaction and adsorption of membrane phospholipids. These simple models have had interesting spin-offs as regards reflection on the interactions between a solid surface and the biological medium. They were used to show that physicochemical features, e.g., charge, redox status, surface defects, were particularly important [75–77]. They also showed that the particles were not totally inert with regard to the cells, since they had adsorption properties with respect to biological macromolecules. Current work involving nanoparticles confirms that these interactions must be taken into consideration [78].

Alveolar and peritoneal macrophages were also put to use early on, in two different approaches, either ex vivo after BAL recovery from animals exposed in vivo, or in vitro after culturing the BAL obtained from untreated animals and incubating the cells in the presence of particles. With these systems, one can study the cytotoxicity of the particles and the response of the cells, usually an inflammatory reaction, i.e., production of cytokines, growth factors, chemokines, and so on [79].

Another cell type has been used, namely the fibroblasts, exploiting their function of producing the molecules of the conjunctive tissue. These cells are used to study fibrogenesis. The response of these cells is determined either after direct exposure to particles, or in response to inflammatory factors produced by other cells exposed to particles, e.g., macrophages, epithelial cells.

Legislation to reduce dust levels in the workplace, thereby reducing the number of cases of pulmonary fibrosis, has focused concern on cancer. However, as mentioned earlier, this illness is always present in certain situations. Silica is currently used as a research tool in studies of the inflammatory mechanism and the immune reaction [80, 81]. Such observations will improve our understanding of the role played by inflammation in particle-related pulmonary pathologies. It is also due to this interest in cancer that studies on macrophages on the one hand, and epithelial cells on the other, has been pursued: the first, because they can produce factors able to interact with other cell types, and the second, because these are the cells producing tumours in the lungs and pleura (bronchial and mesothelial cells, respectively).

Owing to difficulties in obtaining broncheal cell differentiation in cultures, these cells have been replaced by cultures of tracheal explants [82, 83]. Work is also done on 'immortalised' cells, obtained by transfer of a gene allowing these cells to divide in culture. For studies of pleural toxicology, the first cells cultured to study effects relating to pleural cancer were mesothelial pleural cells from rats [84, 85]. Human cells are also used by various groups [86, 87]. Many other types of cell, epithelial or otherwise, have been used since then to identify the effects of carcinogenic chemical molecules. Examples are bacteria (Ames test) and Chinese hamster ovary (CHO) cells, Syrian hamster embryo (SHE) cells, and mouse embryo fibroblasts (NIH3T3). For these cells, data was available about their response to carcinogenic chemicals, making it easier to interpret results obtained with particles. Depending on the cell type, different tests have been adapted to particles: mutagenesis (bacteria and mammal cells), genotoxicity (CHO, SHE), and transformation (SHE, BALB/3T3, CH310T1/2) tests. Models specifically devised to investigate the mutagenic effects of ionising radiation (A1 cells) have been applied [88]. Note that, in order to tackle the issue of mutagenic potential, the bacterial systems that proved so useful for chemical substances turned out to be less relevant for particles, since particles only have effects when internalised (phagocytosis), but they are unable to cross the bacterial wall.

Ames Test. This is a biological assay for identifying mutagenic substances. It was developed in the 1960s by B. Ames to determine the mutagenic potential of chemical substances. The idea is to examine mutations in bacteria. To do this, one uses mutant bacterial strains of *Salmonella typhimurium* which cannot grow without the availability of certain nutritive elements (more specifically, histidine), owing to a mutation on a gene regulating the use of these elements. The bacteria are incubated with the substance whose mutagenic potency is to be tested. The effect of mutagenic substances is characterised by a phenotypic reversal (reverse mutation) which allows the bacteria to grow without access to the nutritive elements needed for the growth of untreated bacteria.

Apart from assessing the genotoxic and transforming effects on target cells, work has also tried to determine the effects particles have on cell functions and specific regulatory channels (see Sect. 1.5).

1.4.3 Molecular Epidemiology

Although not a part of experimental toxicology, it is interesting to mention some methods used over the past few years to look for exposure biomarkers in cancers. In this field, studies have compared molecular characteristics of cancers in subjects exposed or not exposed to asbestos in the workplace. They have focused on the status of several genes that are important in carcinogenic mechanisms through their oncogenic role or tumour-suppressing genes: *k-RAS*, *RASSF1*, *TP53*, *EGF*, and *P16/CDKN2A*. However, there is no definitive data on the differences between subjects exposed or not exposed to asbestos [89–94].

Mutation of TP53. Several categories of genes are associated with the mechanism underlying cancerous transformation of cells. The oncogenes have increased activity in cancers as compared with their activity in normal cells. This modification occurs, for example, due to a point mutation, amplification, or translocation. In contrast, tumour-suppressing genes (TSG) are silenced in tumoral cells, e.g., by mutation or deletion. The gene *TP53* (tumor protein p53) is a TSG, coding for the protein p53. *TP53* has several functions in the regulation of cell proliferation, apoptosis, and DNA damage repair. This gene is silenced in many types of cancer. The gene *NF2* (neurofibromin 2) is another TSG, coding for the protein Nf2. This gene is well known because germinal mutations affect patients suffering from type II neurofibromatosis. This pathology is associated with benign tumours of the central nervous system (schwannomas). At the present time, few types of malignant tumour are known to silence this gene. Mesothelioma seems to be an exception since the gene *NF2* is silenced in a high proportion of cases (about 50% of cases).

1.5 Results on Particle Toxicity Mechanisms

Our understanding of the way cells work has progressed enormously due to recent advances in molecular biology and analytical tools. As a result, toxicology has turned more toward fundamental research to determine the mechanisms whereby particles act rather than work that might be more directly applicable to the problem of assessing toxicological risk. In the latter camp, research structures were rather poorly developed. This kind of research, which tries to understand mechanisms by studying cell functions or identified alterations of cells (response to stress, alterations to genetic material, to the regulation of proliferation, to the control of cell division, etc.), is opening up today to large scale global analyses of DNA and gene or protein expression. These methods are likely to develop over the coming years. The various systems used in silica- and asbestos-related research have served as a heuristic model for the study of atmospheric particles and NPs. Other systems must be imagined to improve the level of understanding and adapt to the specificities of particles. In this chapter, we shall consider only particles that have been the subject of many studies owing to the questions they raise with regard to public health.

1.5.1 Proven Major Risk Factors: Silica and Asbestos

Silica

Studies investigating the mechanisms underlying the effects of silica have shown that surface structure and physicochemical properties play a role. In the 1960s, experimental studies revealed that injections of the polymer poly-2-vinylpyridine-*N*-oxide (PVNO) could inhibit fibrogenesis produced by introducing silica into the peritoneum or the lungs [95]. This work was a continuation of other studies in which it had been observed that the toxicity of silica particles was reduced by treating them with aluminium. Assays had even been undertaken in vivo, i.e., on living animals, to try to control the silicotic process by exposure to aluminium, and aluminium dust was considered to have a prophylactic effect against silicosis [23].

Many in vitro studies then showed that the cytotoxicity of silica was actually connected to its surface reactivity. These conclusions were reached by observing the inhibitive effect of pretreating the particles with various agents, including PVNO or proteins [75]. Using the model provided by red blood cells, the toxicity of silica could be attributed to the formation of hydrogen bonds between a donor (silicic acid formed at the particle surface) and the surface molecules of the cells, e.g., phospholipids. Subsequent work focused on surface activity.

The surface of quartz carries silanol groups (SiOH) and siloxane bridges (Si–O–Si) that get broken when water is present [96, 97]. Apart from the formation of hydrogen bonds, the surface of quartz can produce reactive oxygen species (ROS), such as the superoxide anion ($O_2^{\bullet -}$) or the hydroxyl radical ($OH^\bullet$) [98, 99]. These species can have toxic effects, depending on the level of production and specific features of the cells, by causing peroxidation of membrane lipids, DNA damage (both nuclear and mitochondrial), and alteration of proteins and mitochondrial functions. Oxidative stress is the name for the cell's response to this agression (activation of defence channels, reduced synthesis of oxidising agents).

Later on, the surface reactivity of silica was studied in vitro, using acellular systems. It was found that ROS production depended on the surface state of the particles, which could be modified by mechanical milling, thermal treatment (heating), or chemical treatment (with acid or by adsorption) [100]. These observations confirmed the role of silanol groups, the number and availability of such groups being modified by these treatments [98, 100]. According to these results, it is reasonable to suggest that the effect of silica on cells may depend on experimental conditions, since the surface state can modulate the cell response, either directly, or by influencing physiological phenomena, e.g., phagocytosis, a function which is itself ROS-producing. We thus understand why the nature of the dusts alone cannot explain their pathological effects. Their level of activity will depend on the different possible origins of the silica and the varied circumstances of the workers producing or using them.

The Red Blood Cell Model. Red blood cells are cells with no nucleus, produced by medullary erythroblasts. Their function is to fix oxygen by means of their intracellular hemoglobin and to carry that oxygen from the lungs to the body tissues. These cells were chosen to study the initial interactions between particles and the cell membrane. Lesions of the membrane can be evaluated by the release of hemoglobin into the extracellular medium.

However, the particulate mechanism is not only explained by surface reactivity. In parallel, cell cytotoxicity studies, mainly on macrophages, have demonstrated a production of reactive oxygen and reactive nitrogen species (ROS and RNS, respectively), related to phagocytosis of the particles [101, 102].

Phagocytosis begins by sequestering the particles in phagocytic vacuoles (phagosomes), into which the contents of the lysosomes are poured, associated with an acidification of the phagosomes. This may destabilise the phagosome membrane and lead to cell death.

Inflammation-related factors have been sought to understand fibrogenesis. The mechanism put forward involves phagocytosis of the silica particles by macrophages, and then, depending on the toxicity of the particles, cell death. In this case, the cell contents are released and the proteins can enter the extracellular medium, ready for further internalisation by macrophages. This cycle can continue, and it is felt that this mechanism could explain the increased autoimmune reactions observed in subjects exposed to silica [23, 80].

During silica phagocytosis without cell death, there may be macrophage activation and production of inflammatory molecules (ROS, RNS, cytokines, chemokines, growth factors), leading to neutrophil recruitment and activation of signalling channels [101, 103]. Recent studies on macrophages have helped to determine the different stages of this mechanism. The results support the assumptions about the role of phagocytic processes (phagosome destruction), by identifying the molecules involved in the cell response [80]. Note that the observed effects cannot necessarily be generalised to all cell types. In addition, several groups have demonstrated cooperation between macrophages and fibroblasts following exposure to silica particles, both in vivo and in vitro. The macrophages produce factors stimulating the proliferation of fibroblasts and collagen production [104–106].

To sum up, in certain forms, silica can produce ROS either directly or indirectly through the cell response. The latter stimulates activation of signalling channels that can cause apoptosis or the expression of genes favouring fibrosis. More recent work emphasises the role of cell–cell interactions, in particular between alveolar macrophages and epithelial cells (the role of pulmonary surfactant adsorbed at the surface of the silica particles), and the mechanisms whereby particles are recognised by the macrophages. There are receptors called scavenger receptors at the surface of the macrophages, and these bind to a wide range of ligands, including particles. Certain studies have shown, using so-called null mutants, i.e., not expressing these receptors, that apoptosis does indeed depend on the presence of these receptors. The review by Hamilton et al. [80] weighs up the strengths and weaknesses of these hypotheses about the action of silica.

Asbestos

Studies of the interactions between silica particle surfaces and cells had repercussions for investigations into the way asbestos fibres achieve their effects, and the same type of research on these fibres led to similar conclusions, revealing the role of electrical, redox, and adsorption properties of the fibre surfaces [98]. However, differences were found between chrysotile fibres and the various kinds of amphibole. On the one hand, the surface charge and adsorption capacity are different, and on the other, the concentration of metal elements available for redox reactions is also different, including between the various kinds of amphibole. This is true in particular of the iron concentration [75].

Due to their shape, asbestos fibres have special properties. The fibrous shape, particularly of long fibres, measuring a few tens of microns, allows them to deposit themselves in the deep lung. As far as globular fibres are concerned, those with AD greater than 5 μm are retained in the upper airways and cannot reach the alveolar region. But fibres can reach these regions owing to their smaller diameters. These may vary from a few hundred nanometers down to nanometric order for chrysotile, depending on the number of elementary fibrils. As with silica, these interactions can lead to cell death or cell activation. Macrophages are not the only cells able to internalise particles. The epithelial and mesothelial cells can also do this. The result is phagocytosis of the longer fibres and a difficulty for internalisation, associated with extracellular regurgitation of intracellular factors (possible phenomenon of frustrated phagocytosis). There may also be an abnormal chromosome segregation during mitosis, as described in the literature [88, 107].

In the field of carcinogenesis, hypotheses about underlying mechanisms are based on data obtained from animal experiments and from different cell culture systems including the target cells [107, 108]. To sum up what is known about the toxicity of asbestos fibres, two non-exclusive mechanisms have been identified. One is associated with the inflammatory reaction accompanying the deposition of fibres in the airways and lungs, with a rush of inflammatory cells producing ROS, RNS, and cytokine factors. These molecules can have a genotoxic effect and favour cell proliferation. Base oxidation, in particular, of 8-hydroxy-deoxyguanosine (8-OHdG), and single-strand breaks in DNA have been detected in cells exposed to asbestos, and these might be explained by this mechanism [109, 110]. DNA damage is also suggested indirectly by the discovery that DNA repair mechanisms are activated and that the cell cycle is sometimes arrested in cells exposed to asbestos [88]. The proliferation of epithelial cells whose DNA displays damage that is either poorly repaired or not repaired at all will of course lead to an increased risk of neoplastic transformation.

Another mechanism underlying asbestos fibre toxicity, non-exclusive with regard to the last, results from the ability of epithelial and mesothelial cells to internalise the asbestos fibres. It has been shown that the phagocytosis of asbestos fibres is also associated with ROS and RNS generation, and that cell division is considerably altered by exposure to asbestos [88, 107, 109, 111]. The fibres do not seem to enter directly into the cell nuclei. However, they can end up there after mitosis, given that the nuclear membrane is destroyed during cell division and reforms within the daughter cells. Many studies on different cell types, including pleural mesothelial cells, have shown that mitosis is perturbed and chromosomes altered. In fact, various alterations have been observed, e.g., breaks in the chromosomes, abnormalities in chromosome segregation, loss of heterozygosity [88, 112–115]. These different aberrations in the structure and number of chromosomes are not necessarily caused by mechanical effects, but may result from DNA damage or loss of control over mitosis. These effects have serious consequences for the genetic resources of the cells, in terms of both quantity of genetic material and gene expression (deletions, translocations, deregulated expression, etc.), and form part of the general mechanism of oncogenesis.

Studies carried out on animals have reproduced the pathologies observed for humans, viz., fibrosis and cancers, using different means of exposure by inhalation, intraperitoneal (IP) injection, intrapleural injection, intratracheal instillation, and intrathoracic implantation [116, 117]. A specific role played by fibre dimensions has been found in vivo in inhalation studies and intracavitary inoculation studies, as well as in culture cell experiments. In these different investigations, when comparisons are made between samples of different dimensions, it is generally observed that long fibres are more active than short ones [118]. The first work was published by Stanton et al. [119], who used intrathoracic implantation. The authors found that the highest likelihood of pleural tumours was observed for fibres of length greater than 8 μm and diameter less than 0.25 μm.

As for silica particles, the surface properties of the fibres constitute another parameter affecting their reactivity. Concerned here are the redox properties associated with the presence of metals, especially iron, playing the role of catalyst and ROS generator, adding to the ROS generated by the cells. The role played by iron turns out to be complex. Its activity depends on its oxidation state and bioavailability [75]. Adsorption of proteins like vitronectin or serum proteins on the fibre surface can modify their reactivity in cell cultures, affecting phagocytosis and ROS production. DNA can also be adsorbed onto the fibre surface. Note that asbestos fibres are efficient for transfection of genome sequences, attesting to their interaction with DNA [120–122]. The potential consequences of this property in fibre oncogenesis mechanisms have not yet been scrutinised in detail, but future studies of the interactions between DNA and nanoparticles can be expected to provide useful information in this area.

Organic molecules such as polycyclic aromatic hydrocarbons (PAH) have also been detected at the surface of these fibres, where they constitute a carcinogenic cofactor. This may explain the multiplicative effect of tobacco smoking when smokers are exposed to asbestos [123, 124].

The chemical composition of asbestos fibres also enters the equation when accounting for their carcinogenic potential, as attested by the lower tumorigenicity of chrysotile fibres when their magnesium content is reduced by acid leaching [125]. However, this same treatment modifies other fibre parameters, e.g., dimensions, surface charge, and increases the specific surface area, emphasising the importance of particle characteristics in toxicological studies.

Genuine biochemical reactions can take place between fibres and biological medium, such as the formation of asbestos bodies between asbestos fibres and cells. These formations, discovered by Marchand in 1906, comprise an asbestos fibre core surrounded by a ferrous protein sheath [126]. This sheath seems complex, forming mainly around long fibres inside giant cells by deposition of mucopolysaccharides and calcium phosphate (apatite), and associated with a ferritin impregnation that can be converted to hemosiderin by oxidation [126].

To investigate the role of certain enzymes involved in inflammation and fibrogenesis, genetically modified mice have been used, in which a gene for modulating the inflammatory reaction has been turned off (knockout mice). The difference observed between the responses of normal and knockout subjects can be used to determine the gene's involvement in the given biological process [71, 72, 127].

Some studies have considered mutagenesis in vivo using BigBlue transgenic rats expressing the gene *lacI*. This is a way of revealing mutations, but the method has seen little development so far. An increase in the mutation rate of pulmonary DNA has been observed in BigBlue rats exposed to crocidolite by inhalation, and likewise for the DNA in peritoneal cells, after intracavitary inoculation [128, 129]. Mutations have also been detected in mice made susceptible to the development of cancers by germ cell mutation of a tumour-suppressing gene [73, 130]. Moreover, the reproduction in mice of human cancers related to a given carcinogen constitutes an interesting method for studying the mechanisms of neoplastic transformation and identifying the genes involved in oncogenesis. Knowing certain genes that are altered in human tumours, these cancers can be reproduced by a rational strategy. For example, a mutation of *TP53* has been observed in human mesothelioma in a limited number of cases, along with frequent silencing of *NF2* and genes at the locus *INK4* [131]. Exposure by intraperitoneal injection of mice that are hemizygous for a mutation of the gene *NF2* has shown that the mesotheliomas obtained with mice did indeed reproduce the characteristics of mesotheliomas described in humans. These mice were also more sensitive to mesothelioma development than non-mutant mice [130]. The mesothelioma cells obtained in this way are useful for subsequent investigation of other molecular alterations and identification of genes altered during this process.

BigBlue Rats. The genome of these rats has been modified by adding a gene *lacZ* coding for a bacterial enzyme (β-galactosidase). Each cell of these animals thus carries this gene, which serves as a tool for detecting mutations. When inserted in a cloning vector, the gene *lacZ* serves as a reporter gene on which the search for mutations will be operated. The animals are exposed to the agent under investigation, whereupon the DNA is extracted from the relevant tissues, e.g., the lungs in the case of animals that have inhaled fibres. A multistage process is then implemented to isolate the gene *lacZ* and express β-galactosidase in bacteria. The activity of this enzyme is revealed by a coloured reaction, and the β-galactosidase may or may not be functional, depending on whether the gene has mutated or not.

1.5.2 Suspected Risk Factors: Artificial Mineral Fibres

Many carcinogenicity studies on animals have focused on artificial mineral fibres (AMF), such as glass wool, rock wool, slag wool, specialty glass fibres, and refractory ceramic fibres (RCF), with the same exposure methods as for asbestos. Inhalation studies carried out before the end of the 1980s proved negative, but the results were debated for several reasons: either because the control animals exposed to asbestos did not develop pulmonary tumours, or because the fibres used were of too high a diameter, incompatible with deposition in the lungs of the animals. Toward the end of the 1980s and the beginning of the 1990s, studies were carried out on rats and hamsters using the nose-only method. A certain number of samples (RCF) produced a significant increase in the incidence of pulmonary tumours in rats and mesotheliomas in hamsters. Exposure by intracavitary injection produced a significantly higher rate of tumours in animals treated with the fibres as compared with control animals. Recall that one of the first articles to suggest the lower toxicity of short fibres as compared with long ones was published by Stanton et al. [119], and in this study, the authors implanted 70 samples of glass fibres with various granulometric size distributions in rat pleuras.

However, this dimensional parameter could not alone explain the differences in carcinogenic potential of the various samples. Work on AMFs focused on the biodurability of these fibres, a term referring to their tendency to resist dissolving or disintegrating in the biological medium. This notion led to the idea of biopersistence which takes into account both biodurability and clearance, referring to the ability of a fibre to perdure in the lungs while conserving its chemical and physical characteristics. This in turn inspired a classification of fibre toxicity in terms of their biopersistence, the most durable fibres being considered as potentially the most carcinogenic. These studies were used to classify AMFs by the *Centre de recherche sur le cancer* (CIRC) in France and subsequently to set up a European directive (see Sect. 1.6).

One study used intratracheal instillation to investigate genotoxicity in vivo in BigBlue rats. It showed a significant increase in the mutation rate for a rock wool sample and a non-significant one for glass fibres [132].

Various cell systems have been used to study the effects of AMFs. Some samples had genotoxic effects, including DNA damage and induction of chromosomal aberrations, nuclear abnormalities, and mutations, together with a transformation of mammalian cells. In addition, fibres can cause an inflammatory reaction producing ROS, growth factors, and cytokines. ROS production by fibres does not seem to be an important characteristic of these particles.

A discussion of AMF carcinogenicity for all the different types of fibres would go beyond the scope of this review. The INSERM reports contain a discussion of the different results, while the CIRC document provides experimental details [133, 134].

1.5.3 Unknown Risk Factors: Nanoparticles

Studies on the effects of nanoparticles (NP) are flourishing. An overview of the general state of the art has been published recently [135]. There are several reviews of the latest work [46, 89, 136–138]. The experimental setups devised for silica and asbestos studies have been applied to NPs. Widely different particles have been investigated, e.g., titanium oxide, carbon black, polystyrene, metals, metal salts, diesel smoke products, and particles from the surrounding atmosphere. Tests focused on migration and translocation, inflammatory reactions (production of inflammatory cells and factors in animals by BAL analysis), and in vitro on culture cells (inflammation, genotoxicity) [137–139].

Results showed an inflammatory response and oxidative stress in the lungs, but this response varies, and depends on the samples. The reason for these differences has not yet been identified. A lot of studies have demonstrated the genotoxic potential of NPs, but it is not yet possible to draw definitive conclusions about the parameters and factors producing these effects [140].

The penetration of NPs into cells is an important process to be taken into consideration, as for all other particles. NPs can enter cells by endocytosis, but it seems that they can also cross the cytoplasmic membrane. They may be able to enter the nucleus by transfer via the nuclear pores, or as suggested for asbestos, after mitosis [112, 140–142]. Further studies will be needed to find support for these hypotheses.

With these particles, there is some discussion over the best parameter to use for relating observed effects: mass concentration, number, surface area, and/or surface activity. The tendency is to express effects in terms of the surface area of the particles. However, a glance at the literature shows that this idea is difficult to generalise to all NPs [143]. Exposure by cutaneous NP delivery did not reveal notable effects, while systemic administration gave variable results, depending on the type of particles, characterised largely by morphological abnormalities located in the liver, the kidney, and the spleen [139]. As with asbestos and silica, knockout mice have been used, in particular to study the role of certain enzymes involved in inflammation and fibrogenesis [144].

Carbon nanotubes (CNT) have particularly interesting properties in a range of different fields of application. Many studies, including genomic methods, have demonstrated a capacity to cause oxidative stress, pulmonary inflammation, and mesotheliomas in mice. The similarity between the pathogenic properties of multiwall CNTs and asbestos fibres is currently under discussion [46, 145].

1.6 Results and Further Questions

The results of toxicological studies deserve comment in this chapter, and the historical context is relevant here. Regarding the main types of particle discussed above, the work on silica and asbestos confirms the effects on humans observed earlier on, and provides an insight into what is going on, but the exact mechanisms remain to be clarified. In the light of recent findings concerning the interactions between cells and fibres, particles cause a range of pathological consequences depending on their nature, even if they are particles with the same chemical composition as silica. Indeed, differences in tissue and cell response are observed depending on their mineralogical nature, the surface state, and the extent of interactions between the particle surface and the cell membrane. Research in this area has also demonstrated the role played by shape and dimensions. The dependence of the effects on physical and physicochemical properties has been confirmed by studies on asbestos fibres, which justify a generalisation of these hypotheses. The incidence of the particle characteristics on biological effects is also confirmed by comparative studies of the effects of FPs and UFPs of the same chemical nature. The present understanding of the interactions between cells and silica or asbestos has influenced studies of synthetic mineral fibres, leading to the definition of biopersistence as one of the key elements determining pathological effects.

Furthermore, the data that has been built up has drawn attention to particle dynamics, and in particular, their migration toward and translocation within different organs and their chemical and dimensional evolution within the organism, avoiding the idea that they might somehow be inert as was sometimes suggested in reports on earlier observations. Bearing in mind the many applications of NPs, it is safe to predict that these questions will remain pertinent, given the tendency of NPs to aggregate and the importance of their surface properties.

In the case of asbestos, the various studies have supported epidemiological surveys, and experimental demonstrations finally led to its being outlawed in certain countries. Furthermore, studies on these minerals have stimulated research on the toxicity of other fibres, be they synthetic, inorganic, or organic, allowing us to anticipate the harmful effects resulting from these particles.

Studies carried out on silica led to certain forms being classified as carcinogenic (group 1) by the CIRC in 1996 [25]. RCFs and certain specialty glass fibres have been classified in group 2B (possibly carcinogenic for humans), whereas insulating wools have been put into group 3 (unclassifiable with regard to carcinogenicity for humans due to lack of data) [133]. Despite epidemiological studies showing various results, carbon black and titanium oxide have been classified as possibly human carcinogenic for altered clearance under high pulmonary contaminant levels, on the basis of experimental studies and effects compatible with a carcinogenic mechanism [146].

Note that, before 2006, experimental studies on animals were taken into account for this classification of carcinogenic potential, while mechanism studies carried out on isolated cells carried no weight in the final decision, but were considered only as indicators. Today studies to determine the underlying mechanisms have become a central part of CIRC assessments, while epidemiological studies are not conclusive with regard to either an absence of proof or a sufficient proof of carcinogenicity. Since 2006, mechanistic data are taken into consideration in evaluations, and can provide strong evidence for carcinogenic potential [147].

Toxicological studies of AMFs have led to the formulation of European directives for carcinogenicity tests on artificial vitrous silicate fibres. These authorise exemption from classification as 'carcinogenic' on the basis, for example, of the biopersistence (half-life) of fibres in the lungs, in a short term inhalation or intratracheal instillation study [148]. Furthermore, a model has been made, using experimental data obtained with RCFs, to define an estimate of the increase in the risk of cancer associated with exposure to these fibres [149]. And hypotheses have also been formulated regarding the possible mechanisms whereby these fibres act. Using a two-stage clonal expansion model, with the stages being initiation and promotion, it has been suggested that the best fit to RCF data has the fibres as initiators [150, 151].

It may also be considered that data acquired on FPs and UFPs have stimulated interest in environmental pollution, and they have undoubtedly had consequences for investigation of the effects of NP toxicity, a subject of major importance at the present time. In addition, the protocols and methods already devised with particles in the field of inhalation toxicology will speed up investigations in other fields of exposure presently emerging with NPs, even if some adaptation will be needed. Indeed, given that these particles are present not only in aerosols, but also in other products, e.g., foods and cosmetics, other exposure routes must be taken into consideration.

These studies on particle toxicity raise a range of different questions on both the cognitive and methodological levels:

- The validity of the notion of biopersistence as an indicator of the carcinogenic potential of AMFs is still debated, and the generalisation of this notion as a means for assessing the carcinogenic potential of all fibre types has not yet been established. Note that the biopersistence of a carcinogenic agent is not a necessary factor for it to have a carcinogenic effect. Biopersistence modulates the dose rate, and introduces a time factor into the cumulative dose. Questions have been raised about the limitations of short term biopersistence studies, used to exempt AMFs from classification as carcinogens [148].
- Just as surface reactivity cannot alone explain pathological cell response, so inflammation is unable to account fully for carcinogenic effects. A recent analysis of data in the literature suggests that cancer is not necessarily related to inflammatory reaction and oxidative stress [152]. The inflammatory reaction is a natural defence process and the lungs have an antioxidant defence potential. The level of production of these reactive species must therefore be a determining factor for toxicity. Alteration of the genetic material (genetic and chromosomal mutations) is an important indicator of the carcinogenic process.
- The way particles enter into cells, and what happens to them thereafter with regard to interaction with genetic material and cell regulatory channels, need to be explored further in order to define the mechanisms of particle action and identify criteria for evaluating toxicity endpoints. To assess the potential for damage repair, it is also important to explore associated mechanisms: genetic (DNA repair), cellular (apoptosis, repopulation, etc.), and tissue (scar formation, etc.) mechanisms.

- Research carried out up to now has favoured certain mechanisms, introducing a bias toward a general understanding of cell response, focusing on one process or one mechanism. In the future, genome-wide studies should make it easier to identify regulatory channels that are activated or inhibited in cells responding to these particles, and this in a dynamic way that takes into account the microscopic surroundings of the cells.
- We should also be concerned about the best strategy or strategies to adopt to study the most representative particles in terms of risk factors, and try to identify the biological systems that are best suited to assessing hazards and risks.

1.7 Conclusions and Prospects

Particle toxicology has come into being thanks to the experimental data acquired mainly during the second half of the last century. Research brought out several mechanisms and physiological routes to be explored when examining the potential toxicity of solid particles. Points to be analysed concern not only biological aspects, such as inflammation, effects on systems regulating cell homeostasis, cell integrity, cell cooperation, and interactions between the cell and its micro-environment, but also particle aspects, i.e., physical and physicochemical characteristics. The bioavailability of the particles, their penetration into the cells, and their stability in the biological medium are important factors to take into consideration. One should expect new biological aspects and new factors to become relevant, so that new characteristics will have to be taken into account. Upstream, in order to make toxicological studies as relevant as possible, we need to ask about the context of exposure, not only with regard to the kind of particles likely to enter the organism (chemical nature, shape, and dimensions), but also with regard to the population at risk and the environmental and ecological extent of the risk. This understanding is essential for setting up the best expert systems, modelling particle–cell interactions, and determining the probability of physiopathological response. When discussing particle toxicology, the current situation is very different from previous ones, and it is essential to take this into account. Indeed, in the past, experimental studies came after the pathologies had been identified, whereas today, the new materials will precede the pathologies. Let us hope that data already made available will not simply be ignored, delaying the benefits of knowing about them when identifying and characterising new risk factors.

The significant development in the means for analysing cell functioning and the rapid expansion of means of communication make it possible today to analyse a huge volume of data, ensuring fast progress in our understanding of the life of the cell and the way it can respond to exogenous factors, in an integrated system that will take into account both biological and molecular interactions. Up to now, mechanistic studies have observed isolated responses in a general biological and physiological context, and often under conditions that do not justify extrapolation to assess the level of risk. This integrated approach will result from large scale analyses of the various features of cell function, with benefits for toxicology and the possibility of limiting experiments on animals. Their use should be strongly encouraged to promote the rapid development of research into the effects of particles on structure, and the genetic and epigenetic modifications of genome activity. Present and future data might also be used to construct algorithms that would assist in the evaluation of toxicity, taking into account the parameters of the toxicity related to particle characteristics, biological mechanisms of the pathologies, and exposure conditions. In order to ask the relevant questions and offer efficient solutions, information must be supplied on two levels: upstream, regarding the nature of particles likely to produce health risks (role of manufacturers and safety specialists, environmental data), and downstream, regarding potential or proven risks (factory doctors, public authorities, registers, early warning systems). Considerable progress will be made in the coming years and we must acknowledge the role played by earlier investigations which, with the means available to them, laid the foundations for modern particle toxicology. Research in this area will necessarily be multidisciplinary, associating groups specialising in the physicochemical characterisation of particles and all the different aspects of biology (pathology, and cellular and molecular biology). Let us hope that the forces needed to tackle all these aspects of the research so necessary today will be successfully set in motion to achieve positive and efficient management of future health and safety requirements.

References

1. C.M. Fletcher: Pneumoconiosis of coal-miners. Br. Med. J. **1**, 1065–1074 (1948)
2. J.E. Martin: Coal miners' pneumoconiosis. Am. J. Public Health Nations Health **44**, 581–591 (1954)
3. C. Amoudru: Pneumoconioses: l'exemple des Houillères du Nord Pas-de-Calais (1944–1990). In *L'émergence des risques*, J.-M. Mur (ed.), pp. 41–71. EDP Sciences, Les Ulis, France (2008)
4. P.W. Bartrip: History of asbestos related disease. Postgrad. Med. J. **80**, 72–76 (2004)
5. R. Lenglet: *L'affaire de l'amiante*. Editions de la Découverte, Paris (1996)
6. R. Dulbecco: *Aventurier du vivant*. Plon, Paris (1989)
7. P.J.A. Borm: Particle toxicology: From coal mining to nanotechnology. Inhal. Toxicol. **14**, 311–324 (2002)
8. R.R. Sayers, W.C. Dreessen: Asbestosis. Am. J. Public Health Nations Health **29**, 205–214 (1939)
9. I.J. Selikoff: Historical developments and perspectives in inorganic fiber toxicity in man. Environ. Health Perspect. **88**, 269–276 (1990)
10. K.M. Lynch, W.A. Smith: Pulmonary asbestosis. III. Carcinoma of lung in asbestos-silicosis. Am. J. Cancer **24**, 56 (1935)

11. R. Doll: Mortality from lung cancer in asbestos workers. Br. J. Ind. Med. **12**, 81–86 (1955)
12. J.C. Wagner, C.A. Sleggs, P. Marchand: Diffuse pleural mesothelioma and asbestos exposure in the North Western Cape Province. Br. J. Ind. Med. **17**, 260–271 (1960)
13. J.C. Wagner: Historical background and perspectives of mesothelioma. In *The Mesothelial Cell and Mesothelioma*, J.J. Bignon (ed.), pp. 1–17. Marcel Dekker, New York, Basel (1994)
14. INSERM: *Effets sur la santé des principaux types d'exposition à l'amiante.* Editions INSERM, Paris (1997)
15. M.M. Maule, C. Magnani, P. Dalmasso, D. Mirabelli, F. Merletti, A. Biggeri: Modeling mesothelioma risk associated with environmental asbestos exposure. Environ. Health Perspect. **115**, 1066–1071 (2007)
16. I. Annesi-Maesano, W. Dab: Air pollution and the lung: Epidemiological approach. Med. Sci. (Paris) **22**, 589–594 (2006)
17. W.P. Logan: Mortality in the London fog incident, 1952. Lancet **1**, 336–338 (1953)
18. W.P. Logan: Mortality from fog in London, January, 1956. Br. Med. J. **1**, 722–725 (1956)
19. A. Maitre, V. Bonneterre, L. Huillard, P. Sabatier, R. de Gaudemaris: Impact of urban atmospheric pollution on coronary disease. Eur. Heart J. **27**, 2275–2284 (2006)
20. R.D. Handy, F. von der Kammer, J.R. Lead, M. Hassellov, R. Owen, M. Crane: The ecotoxicology and chemistry of manufactured nanoparticles. Ecotoxicology **17**, 287–314 (2008)
21. P.J. Borm, D. Robbins, S. Haubold, T. Kuhlbusch, H. Fissan, K. Donaldson, R. Schins, V. Stone, W. Kreyling, J. Lademann, J. Krutmann, D. Warheit, E. Oberdorster: The potential risks of nanomaterials: A review carried out for ECETOC. Part. Fibre Toxicol. **3**, 11 (2006)
22. C. Huitema: *Et Dieu créa l'Internet* .... Eyrolles, Paris (1995)
23. G.W. Schepers: Lung disease caused by inorganic and organic dust. Dis. Chest. **44**, 133–140 (1963)
24. R.P. Schins, P.J. Borm: Mechanisms and mediators in coal dust induced toxicity: A review. Ann. Occup. Hyg. **43**, 7–33 (1999)
25. IARC: Silica, some silicates, coal dusts and para-aramid fibrils. In *Monographs on the Evaluation of Carcinogenic Risk of Chemicals to Humans*, Vol. 68. IARC Press, Geneva (1997)
26. C. Voisin: Silicose et pneumoconioses à poussières mixtes renfermant de la silice. In *Pneumologie*, M. Aubier, R. Pariente (eds.), pp. 777–788. Flammarion Médecine-Sciences, Paris (1996)
27. R.A. Cohen, A. Patel, F.H. Green: Lung disease caused by exposure to coal mine and silica dust. Semin. Respir. Crit. Care Med. **29**, 651–661 (2008)
28. J. Pairon, P. Brochard, J. Bignon: Pathologies respiratoires de l'amiante. In *Pneumologie*, M. Aubier, R. Pariente (eds.), pp. 789–798. Flammarion Médecine-Sciences, Paris (1996)
29. P. de Vuyst: Pathologies respiratoire liées à l'inhalation de particules métalliques. In *Pneumologie*, M. Aubier, R. Pariente (eds.), pp. 799–802. Flammarion Médecine-Sciences, Paris (1996)

30. I.Y.R. Adamson, J. Bakowska, D.H. Bowden: Mesothelial cell proliferation: A nonspecific response to lung injury associated with fibrosis. Am. J. Respir. Cell. Mol. Biol. **10**, 253–258 (1994)
31. E. Crouch: Pathobiology of pulmonary fibrosis. Am. J. Physiol. **259**, L159–L184 (1990)
32. W.D. Hardie, S.W. Glasser, J.S. Hagood: Emerging concepts in the pathogenesis of lung fibrosis. Am. J. Pathol. **175**, 3–16 (2009)
33. C.M. Croce: Genetic approaches to the study of the molecular basis of human cancer. Cancer Res. **51**, 5015s–5018s (1991)
34. A.C. Schinzel, W.C. Hahn: Oncogenic transformation and experimental models of human cancer. Front. Biosci. **13**, 71–84 (2008)
35. D. Hanahan, R.A. Weinberg: The hallmarks of cancer. Cell **100**, 57–70 (2000)
36. F. Colotta, P. Allavena, A. Sica, C. Garlanda, A. Mantovani: Cancer-related inflammation, the seventh hallmark of cancer: Links to genetic instability. Carcinogenesis **30**, 1073–1081 (2009)
37. D.G. Albertson, C. Collins, F. McCormick, J.W. Gray: Chromosome aberrations in solid tumors. Nat. Genet. **34**, 369–376 (2003)
38. A. Masuda, T. Takahashi: Chromosome instability in human lung cancers: Possible underlying mechanisms and potential consequences in the pathogenesis. Oncogene **21**, 6884–6897 (2002)
39. K. Robock: Standard quartz dq12 greater than 5 micro m for experimental pneumoconiosis research projects in the Federal Republic of Germany. Ann. Occup. Hyg. **16**, 63–66 (1973)
40. V. Timbrell, J.C. Gibson, I. Webster: UICC standard reference samples of asbestos. Int. J. Cancer **3**, 406–408 (1968)
41. T.W. Hesterberg, G.A. Hart: Synthetic vitreous fibers: A review of toxicology research and its impact on hazard classification. Crit. Rev. Toxicol. **31**, 1–53 (2001)
42. The Encyclopedia of Earth: www.eoearth.org/article/Geology_of_asbestos
43. R.E. Rendall: Physical and chemical characteristics of UICC reference samples. IARC Sci. Publ. 87–96 (1980)
44. I. Baris, L. Simonato, M. Artvinli, F. Pooley, R. Saracci, J. Skidmore, C. Wagner: Epidemiological and environmental evidence of the health effects of exposure to erionite fibres: A four-year study in the Cappadocian region of Turkey. Int. J. Cancer. **39**, 10–17 (1987)
45. AFSSET: Les fibres minérales artificielles siliceuses. Fibres céramiques réfractaires – Fibres de verre à usage spécial. In *Avis de l'AFSSET et rapport du groupe d'experts*. Agence française de sécurité sanitaire de l'environnement et du travail (2007); www.afsset.fr/index.php?pageid=718&parentid=424&search=yes&txtSearch=fma
46. M.C. Jaurand, A. Renier, J. Daubriac: Mesothelioma: Do asbestos and carbon nanotubes pose the same health risk? Part. Fibre Toxicol. **6**, 16 (2009)
47. A. Seaton, W. MacNee, K. Donaldson, D. Godden: Particulate air pollution and acute health effects. Lancet **345**, 176–178 (1995)
48. AFSSET: Les nanomatériaux. Effets sur la santé de l'homme et surl'environnement. In *Avis de l'AFSSET et rapport du groupe d'experts*. Agence française de sécurité sanitaire de l'environnement et du travail (2006); www.afsset.fr/index.php?pageid=619&newsid=105&MDLCODE=news

49. AFSSET: Les nanomatériaux. Sécurité au travail. In *Avis de l'AFSSET, rapport d'expertise collective et annexes*. Agence française de sécurité sanitaire de l'environnement et du travail (2008); www.afsset.fr/index.php?pageid=619&newsid=398&MDLCODE=news
50. K.M. Reiser, T.W. Hesterberg, W.M. Haschek, J.A. Last: Experimental silicosis. I. Acute effects of intratracheally instilled quartz on collagen metabolism and morphologic characteristics of rat lungs. Am. J. Pathol. **107**, 176–185 (1982)
51. E.J. King, G.P. Mohanty, C.V. Harrison, G. Nagelschmidt: Effect of modifications of the surface of quartz on its fibrogenic properties in the lungs of rats. 1. Quartz leached with ringer's solution. 2. Quartz etched with hydrofluoric acid. 3. Quartz coated with coal extract. AMA. Arch. Ind. Hyg. Occup. Med. **7**, 455–477 (1953)
52. E.J. King, B.M. Wright, S.C. Ray, C.V. Harrison: Effect of aluminium on the silicosis-producing action of inhaled quartz. Br. J. Ind. Med. **7**, 27–36 (1950)
53. K.E. Driscoll, D.L. Costa, G. Hatch, R. Henderson, G. Oberdorster, H. Salem, R.B. Schlesinger: Intratracheal instillation as an exposure technique for the evaluation of respiratory tract toxicity: Uses and limitations. Toxicol. Sci. **55**, 24–35 (2000)
54. G. Oberdorster, C. Cox, R. Gelein: Intratracheal instillation versus intratracheal inhalation of tracer particles for measuring lung clearance function. Exp. Lung Res. **23**, 17–34 (1997)
55. V. Vu, J.C. Barrett, J. Roycroft, L. Schuman, D. Dankovic, P. Bbaro, T. Martonen, W. Pepelko, D. Lai: Chronic inhalation toxicity and carcinogenicity testing of respirable fibrous particles. Workshop report. Regul. Toxicol. Pharmacol. **24**, 202–212 (1996)
56. R.F. Phalen, R.C. Mannix, R.T. Drew: Inhalation exposure methodology. Environ. Health Perspect. **56**, 23–34 (1984)
57. E.M. Thomson, A. Williams, C.L. Yauk, R. Vincent: Impact of nose-only exposure system on pulmonary gene expression. Inhal. Toxicol. **21**, 74–82 (2009)
58. T.W. Hesterberg, G. Chase, C. Axten, W.C. Miller, R.P. Musselman, O. Kamstrup, J. Hadley, D.M. Morscheidt, D.M. Berstein, P. Thevenaz: Biopersistence of synthetic vitreous fibers and amosite asbestos in the rat lung following inhalation. Toxicol. Applied Pharmacol. **151**, 262–275 (1998)
59. E.E. McConnell: Synthetic vitreous fibers – Inhalation studies. Regul. Toxicol. Pharmacol. **20**, S22–S34 (1994)
60. A. Morgan, J.C. Evans, A. Holmes: Deposition and clearance of inhaled fibrous minerals in the rat. Studies using radioactive tracer techniques. Inhaled Part. **4** Pt 1, 259–274 (1975)
61. P.E. Morrow, F.R. Gibb, H. Beiter, R.W. Kilpper: Pulmonary retention of neutron-activated coal dust. Arch. Environ. Health. **34**, 178–183 (1979)
62. P.G. Coin, V.L. Roggli, A.R. Brody: Persistence of long, thin chrysotile asbestos fibers in the lungs of rats. Environ. Health Perspect. **102** (Suppl. 5), 197–199 (1994)
63. A. Morgan, R.J. Talbot, A. Holmes: Significance of fibre length in the clearance of asbestos fibres from the lung. Br. J. Ind. Med. **35**, 146–153 (1978)
64. AFSSET: Les fibres courtes et les fibres fines d'amiante. Prise en compte du critère dimensionnel pour la caractérisation des risques sanitaires liés à l'inhalation d'amiante. In *Avis de l'AFSSET et rapport d'expertise collective*. Agence française de sécurité sanitaire de l'environnement et du travail (2009), www.afsset.fr/index.php?pageid=717&parentid=424

65. F. Bischoff, G. Bryson: Carcinogenesis through solid state surfaces. Prog. Exp. Tumor. Res. **5**, 85–133 (1964)
66. F. Pott, F. Huth, K.H. Friedrichs: Tumorigenic effect of fibrous dusts in experimental animals. Environ. Health Perspect. **9**, 313–315 (1974)
67. M.F. Stanton, C. Wrench: Mechanisms of mesothelioma induction with asbestos and fibrous glass. J. Natl. Cancer Inst. **48**, 797–821 (1972)
68. M. Dorger, A.M. Allmeling, R. Kiefmann, A. Schropp, F. Krombach: Dual role of inducible nitric oxide synthase in acute asbestos-induced lung injury. Free Radic. Biol. Med. **33**, 491–501 (2002)
69. F. Gao, J.R. Koenitzer, J.M. Tobolewski, D. Jiang, J. Liang, P.W. Noble, T.D. Oury: Extracellular superoxide dismutase inhibits inflammation by preventing oxidative fragmentation of hyaluronan. J. Biol. Chem. **283**, 6058–6066 (2008)
70. C. Lecomte, P. Andujar, A. Renier, L. Kheuang, V. Abramowski, L. Mellottee, J. Fleury-Feith, J. Zucman-Rossi, M. Giovannini, M.C. Jaurand: Similar tumor suppressor gene alteration profiles in asbestos-induced murine and human mesothelioma. Cell Cycle. **4**, 1862–1869 (2005)
71. A. Shukla, K.M. Lounsbury, T.F. Barrett, J. Gell, M. Rincon, K.J. Butnor, D.J. Taatjes, G.S. Davis, P. Vacek, K.I. Nakayama, K. Nakayama, C. Steele, B.T. Mossman: Asbestos-induced peribronchiolar cell proliferation and cytokine production are attenuated in lungs of protein kinase C-delta knockout mice. Am. J. Pathol. **170**, 140–151 (2007)
72. D.E. Sullivan, M. Ferris, D. Pociask, A.R. Brody: The latent form of TGF-beta(1) is induced by TNFalpha through an ERK specific pathway and is activated by asbestos-derived reactive oxygen species in vitro and in vivo. J. Immunotoxicol. **5**, 145–149 (2008)
73. C.A. Vaslet, N.J. Messier, A.B. Kane: Accelerated progression of asbestos-induced mesotheliomas in heterozygous p53 (+/−) mice. Toxicol. Sci. **68**, 331–338 (2002)
74. G. Macnab, J.S. Harington: Haemolytic activity of asbestos and other mineral dusts. Nature **214**, 522–523 (1967)
75. M. Gulumian: An update on the detoxification processes for silica particles and asbestos fibers: Successes and limitations. J. Toxicol. Environ. Health B Crit. Rev. **8**, 453–483 (2005)
76. M.C. Jaurand, L. Magne, J. Bignon: Inhibition by phospholipids of haemolytic action of asbestos. Br. J. Ind. Med. **36**, 113–116 (1979)
77. S.V. Singh, P.N. Viswanathan, Q. Rahman: Interaction between erythrocyte plasma membrane and silicate dusts. Environ. Health Perspect. **51**, 55–60 (1983)
78. G. Oberdorster, A. Maynard, K. Donaldson, V. Castranova, J. Fitzpatrick, K. Ausman, J. Carter, B. Karn, W. Kreyling, D. Lai, S. Olin, N. Monteiro-Riviere, D. Warheit, H. Yang: Principles for characterizing the potential human health effects from exposure to nanomaterials: Elements of a screening strategy. Part. Fibre Toxicol. **2**, 8 (2005)
79. A. Shukla, M. Ramos-Nino, B. Mossman: Cell signaling and transcription factor activation by asbestos in lung injury and disease. Int. J. Biochem. Cell Biol. **35**, 1198–1209 (2003)
80. R.F. Hamilton, S.A. Thakur, A. Holian: Silica binding and toxicity in alveolar macrophages. Free Radic. Biol. Med. **44**, 1246–1258 (2008)

81. V. Hornung, F. Bauernfeind, A. Halle, E.O. Samstad, H. Kono, K.L. Rock, K.A. Fitzgerald, E. Latz: Silica crystals and aluminum salts activate the NALP3 inflammasome through phagosomal destabilization. Nat. Immunol. **9**, 847–856 (2008)
82. J.M. Davis: The effects of chrysotile asbestos dust on lung macrophages maintained in organ culture. An electron-microscope study. Br. J. Exp. Pathol. **48**, 379–385 (1967)
83. B.T. Mossman, J.B. Kessler, B.W. Ley, J.E. Craighead: Interaction of crocidolite asbestos with hamster respiratory mucosa in organ culture. Lab. Invest. **36**, 131–139 (1977)
84. S. Bryks, F.D. Bertalanffy: Cytodynamic reactivity of the mesothelium. Pleural reaction to chrysotile asbestos. Arch. Environ. Health. **23**, 469–472 (1971)
85. M.C. Jaurand, H. Kaplan, J. Thiollet, M.C. Pinchon, J.F. Bernaudin, J. Bignon: Phagocytosis of chrysotile fibers by pleural mesothelial cells in culture. Am. J. Pathol. **94**, 529–538 (1979)
86. B. Burmeister, T. Schwerdtle, I. Poser, E. Hoffmann, A. Hartwig, W.U. Muller, A.W. Rettenmeier, N.H. Seemayer, E. Dopp: Effects of asbestos on initiation of DNA damage, induction of DNA-strand breaks, P53-expression and apoptosis in primary, SV40-transformed and malignant human mesothelial cells. Mutat. Res. **558**, 81–92 (2004)
87. J.F. Lechner, T. Tokiwa, M. LaVeck, W.F. Benedict, S. Banks-Schlegel, H. Yeager, A. Barnerjee, C.C. Harris: Asbestos-associated chromosomal changes in human mesothelial cells. Proc. Natl. Acad. Sci. USA **82**, 3884–3888 (1985)
88. M.C. Jaurand: Mechanisms of fiber-induced genotoxicity. Environ. Health Perspect. **105**, 1073–1084 (1997)
89. P. Andujar, S. Lanone, P. Brochard, J. Boczkowski: Respiratory effects of manufactured nanoparticles. Rev. Mal. Respir. **26**, 625–637 (2009)
90. R. Dammann, M. Strunnikova, U. Schagdarsurengin, M. Rastetter, M. Papritz, U.E. Hattenhorst, H.S. Hofmann, R.E. Silber, S. Burdach, G. Hansen: CpG island methylation and expression of tumour-associated genes in lung carcinoma. Eur. J. Cancer **41**, 1223–1236 (2005)
91. K. Husgafvel-Pursiainen, A. Karjalainen, A. Kannio, S. Anttila, T. Partanen, A. Ojajärvi, H. Vainio: Lung cancer and past occupational exposure to asbestos. Role of p53 and K-ras mutations. Am. J. Respir. Cell. Mol. Biol. **20**, 667–674 (1999)
92. D.H. Kim, H.H. Nelson, J.K. Wiencke, S. Zheng, D.C. Christiani, J.C. Wain, E.J. Mark, K.T. Kelsey: p16(INK4a) and histology-specific methylation of CpG islands by exposure to tobacco smoke in non-small cell lung cancer. Cancer Res. **61**, 3419–3424 (2001)
93. A. Lamy, R. Sesboue, J. Bourguignon, B. Dautreaux, J. Metayer, T. Frebourg, L. Thiberville: Aberrant methylation of the CDKN2a/p16INK4a gene promoter region in preinvasive bronchial lesions: A prospective study in high-risk patients without invasive cancer. Int. J. Cancer **100**, 189–193 (2002)
94. H.H. Nelson, D.C. Christiani, J.K. Wiencke, E.J. Mark, J.C. Wain, K.T. Kelsey: k-ras mutation and occupational asbestos exposure in lung adenocarcinoma: Asbestos-related cancer without asbestosis. Cancer Res. **59**, 4570–4573 (1999)
95. A.G. Heppleston: Pulmonary toxicology of silica, coal and asbestos. Environ. Health Perspect. **55**, 111–127 (1984)

96. B. Fubini: Surface chemistry and quartz hazard. Ann. Occup. Hyg. **42**, 521–530 (1998)
97. T. Nash, A.C. Allison, J.S. Harington: Physico-chemical properties of silica in relation to its toxicity. Nature **210**, 259–261 (1966)
98. B. Fubini, A. Hubbard: Reactive oxygen species (ROS) and reactive nitrogen species (RNS) generation by silica in inflammation and fibrosis. Free Radic. Biol. Med. **34**, 1507–1516 (2003)
99. V. Vallyathan, J.H. Kang, K. Van Dyke, N.S. Dalal, V. Castranova: Response of alveolar macrophages to in vitro exposure to freshly fractured versus aged silica dust: The ability of Prosil 28, an organosilane material, to coat silica and reduce its biological reactivity. J. Toxicol. Environ. Health **33**, 303–315 (1991)
100. M. Gulumian: The role of oxidative stress in diseases caused by mineral dusts and fibres: Current status and future of prophylaxis and treatment. Mol. Cell. Biochem. **196**, 69–77 (1999)
101. V. Castranova: Role of nitric oxide in the progression of pneumoconiosis. Biochemistry **69**, 32–37 (2004)
102. S. Zhu, M. Manuel, S. Tanaka, N. Choe, E. Kagan, S. Matalon: Contribution of reactive oxygen and nitrogen species to particulate-induced lung injury. Environ. Health Perspect. **106** (Suppl. 5), 1157–1163 (1998)
103. B.T. Mossman, K.M. Lounsbury, S.P. Reddy: Oxidants and signaling by mitogen-activated protein kinases in lung epithelium. Am. J. Respir. Cell. Mol. Biol. **34**, 666–669 (2006)
104. I.Y.R. Adamson, H.L. Letourneau, D.H. Bowden: Comparison of alveolar and interstitial macrophages in fibroblast stimulation after silica and long or short asbestos. Lab. Invest. **64**, 339–344 (1991)
105. S.C. Benson, J.C. Belton, L.G. Scheve: Regulation of lung fibroblast proliferation and collagen synthesis by alveolar macrophages in experimental silicosis. I. Effect of macrophage conditioned medium from silica instilled rats. J. Environ. Pathol. Toxicol. Oncol. **7**, 87–97 (1986)
106. J.S. Harington: Fibrogenesis. Environ. Health Perspect. **9**, 271–279 (1974)
107. M.C. Jaurand, F. Levy: Effets cellulaires et moléculaires de l'amiante. Med. Sci. **15**, 1370–1378 (1999)
108. S. Mohr, G. Keith, B. Rihn: Amiante et mésothéliome malin: Aspects moléculaires, cellulaires et physiopathologiques. Bull. Cancer. **92**, 959–976 (2005)
109. A.B. Kane: Mechanisms of mineral fibre carcinogenesis. In *Mechanisms of Fibre Carcinogenesis*, A.B. Kane, P. Boffetta, R. Saracci, J.D. Wilboum (eds.), pp. 11–34. IARC Scientific Publications, no. 140 (1999)
110. D. Upadhyay, D.W. Kamp: Asbestos-induced pulmonary toxicity: Role of DNA damage and apoptosis. Exp. Biol. Med. **228**, 650–659 (2003)
111. J. Wu, W. Liu, K. Koenig, S. Idell, V.C. Broaddus: Vitronectin adsorption to chrysotile asbestos increases fiber phagocytosis and toxicity for mesothelial cells. Am. J. Physiol. Lung Cell. Mol. Physiol. **279**, L916–L923 (2000)
112. B.A. Cortez, G.M. Machadosantelli: Chrysotile effects on human lung cell carcinoma in culture: 3D reconstruction and DNA quantification by image analysis. BMC Cancer **8**, 181 (2008)
113. C.G. Jensen, M. Watson: Inhibition of cytokinesis by asbestos and synthetic fibres. Cell Biol. Int. **23**, 829–840 (1999)
114. C.G. Jensen, L.C.W. Jensen, C.L. Rieder, R.W. Cole, J.G. Ault: Long crocidolite asbestos fibers cause polyploidy by sterically blocking cytokinesis. Carcinogenesis **17**, 2013–2021 (1996)

115. R.P. Schins: Mechanisms of genotoxicity of particles and fibers. Inhal. Toxicol. **14**, 57–78 (2002)
116. M. Kannerstein, J. Churg: Mesothelioma in man and experimental animals. Environ. Health Perspect. **34**, 31–36 (1980)
117. S. Toyokuni: Mechanisms of asbestos-induced carcinogenesis. Nagoya J. Med. Sci. **71**, 1–10 (2009)
118. J.C. Barrett, P. Lamb, R.W. Wiseman: Multiple mechanisms for the carcinogenic effects of asbestos and other mineral fibres. Environ. Health Perspect. **81**, 81–92 (1989)
119. F. Stanton, M. Layard, A. Tegeris, E. Miller, M. May, E. Kent: Tumorigenicity of fibrous glass: Pleural response in the rat in relation to fiber dimension. J. Natl. Cancer Inst. **58**, 587–603 (1977)
120. J.D. Appel, T.M. Fasy, D.S. Kohtz, J.D. Kohtz, E.M. Johnson: Asbestos fibers mediate transformation of monkey cells by exogenous plasmid DNA. Proc. Natl. Acad. Sci. USA **85**, 7670–7674 (1988)
121. G.R. Dubes, L.R. Mack: Asbestos-mediated transfection of mammalian cell cultures. In Vitro Cell Dev. Biol. **24**, 175–182 (1988)
122. N. Yoshida, T. Ikeda, T. Yoshida, T. Sengoku, K. Ogawa: Chrysotile asbestos fibers mediate transformation of *Escherichia coli* by exogenous plasmid DNA. FEMS Microbiol. Lett. **195**, 133–137 (2001)
123. P. Gerde, P. Scholander: A hypothesis concerning asbestos carcinogenicity: The migration of lipophilic carcinogens in adsorbed lipid bilayers. Ann. Occup. Hyg. **31**, 1–6 (1987)
124. I.J. Selikoff, E.C. Hammond: Asbestos and smoking. JAMA **242**, 458–459 (1979)
125. G. Monchaux, J. Bignon, M.C. Jaurand, J. Lafuma, P. Sebastien, R. Masse, A. Hirsch, J. Goni: Mesotheliomas in rats following inoculation with acid-leached chrysotile asbestos and other mineral fibres. Carcinogenesis **2**, 229–236 (1981)
126. J.M. Davis: Further observations on the ultrastructure and chemistry of the formation of asbestos bodies. Exp. Mol. Pathol. **13**, 346–358 (1970)
127. C.L. Fattman, R.J. Tan, J.M. Tobolewski, T.D. Oury: Increased sensitivity to asbestos-induced lung injury in mice lacking extracellular superoxide dismutase. Free Radic. Biol. Med. **40**, 601–607 (2006)
128. B.H. Rihn, S. Mohr, S.A. McDowell, S. Binet, J. Loubinoux, F. Galateau, G. Keith, G.D. Leikauf: Differential gene expression in mesothelioma. FEBS Lett. **480**, 95–100 (2000)
129. K. Unfried, C. Schürkes, J. Abel: Distinct spectrum of mutations induced by crocidolite asbestos: Clue for 8-hydroxydeoxyguanosine-dependent mutagenesis in vivo. Cancer Res. **62**, 99–104 (2002)
130. J. Fleury-Feith, C. Lecomte, A. Renier, M. Matrat, L. Kheuang, V. Abramowski, F. Levy, A. Janin, M. Giovannini, M.C. Jaurand: Hemizygosity of Nf2 is associated with increased susceptibility to asbestos-induced peritoneal tumours. Oncogene **22**, 3799–3805 (2003)
131. A. De Rienzo, J.R. Testa: Recent advances in the molecular analysis of human malignant mesothelioma. Clin. Ther. **151**, 433–438 (2000)
132. J. Topinka, P. Loli, M. Dusinska, M. Hurbankova, Z. Kovacikova, K. Volkovova, A. Kazimirova, M. Barancokova, E. Tatrai, T. Wolff, D. Oesterle, S.A. Kyrtopoulos, P. Georgiadis: Mutagenesis by man-made mineral fibres in the lung of rats. Mutat. Res. **595**, 174–183 (2006)

133. IARC: Man-made mineral fibres. *IARC Monographs on the Evaluation of Carcinogenic Risks to Humans*, Vol. 81 (2002)
134. INSERM: Effets sur la santé des fibres de substitution à l'amiante. Report on behalf of the *Direction générale de la santé* and the *Direction des relations du travail* (1999)
135. B. Hervé-Bazin: *Les Nanoparticules – Un enjeu pour la santé au travail?* EDP Sciences, Les Ulis (2007)
136. F. Prosie, F.X. Lesage, F. Deschamps: Nanoparticles: Structures, utilizations and health impacts. Presse Med. **37**, 1431–1437 (2008)
137. A.A. Shvedova, E.R. Kisin, D. Porter, P. Schulte, V.E. Kagan, B. Fadeel, V. Castranova: Mechanisms of pulmonary toxicity and medical applications of carbon nanotubes: Two faces of Janus? Pharmacol. Ther. **121**, 192–204 (2009)
138. V. Stone, H. Johnston, M.J. Clift: Air pollution, ultrafine and nanoparticle toxicology: Cellular and molecular interactions. IEEE Trans. Nanobioscience **6**, 331–340 (2007)
139. S.T. Stern, S.E. McNeil: Nanotechnology safety concerns revisited. Toxicol. Sci. **101**, 4–21 (2008)
140. N. Singh, B. Manshian, G.J. Jenkins, S.M. Griffiths, P.M. Williams, T.G. Maffeis, C.J. Wright, S.H. Doak: NanoGenotoxicology: The DNA damaging potential of engineered nanomaterials. Biomaterials. **30**, 3891–3914 (2009)
141. N.S. Wang, M.C. Jaurand, L. Magne, L. Kheuang, M.C. Pinchon, J. Bignon: The interactions between asbestos fibers and metaphase chromosomes of rat pleural mesothelial cells in culture. A scanning and transmission electron microscopic study. Am. J. Pathol. **126**, 343–349 (1987)
142. M. Yegles, L. Saint-Etienne, A. Renier, X. Janson, M.C. Jaurand: Induction of metaphase and anaphase/telophase abnormalities by asbestos fibers in rat pleural mesothelial cells in vitro. Am. J. Respir. Cell. Mol. Biol. **9**, 186–191 (1993)
143. M.C. Jaurand: Etude critique du rôle des paramètres physiques dans l'activité biologique. In *Les Nanoparticules – Un enjeu pour la santé au travail?* B. Hervé-Bazin (ed.), pp. 530–560, EDP Sciences, Les Ulis (2007)
144. N.R. Jacobsen, P. Moller, K.A. Jensen, U. Vogel, O. Ladefoged, S. Loft, H. Wallin: Lung inflammation and genotoxicity following pulmonary exposure to nanoparticles in ApoE−/− mice. Part. Fibre Toxicol. **6**, 2 (2009)
145. A. Poma, M.L. Di Giorgio: Toxicogenomics to improve comprehension of the mechanisms underlying responses of in vitro and in vivo systems to nanomaterials: A review. Curr. Genomics **9**, 571–585 (2008)
146. R. Baan, K. Straif, Y. Grosse, B. Secretan, F. El Ghissassi, V. Cogliano: Carcinogenicity of carbon black, titanium dioxide, and talc. Lancet Oncol. **7**, 295–296 (2006)
147. V.J. Cogliano, R.A. Baan, K. Straif, Y. Grosse, B. Secretan, F. El Ghissassi: Use of mechanistic data in IARC evaluations. Environ. Mol. Mutagen. **49**, 100–109 (2008)
148. Directive 97/69/CE of the Commission of 5 December 1997 carrying the twenty-third adaptation to the technical progress of the directive 67/548/CEE of the Council, concerning the reconciliation of legislative arrangements relating to the classification, packaging, and labelling of hazardous substances, eur-lex.europa.eu/LexUriServ/LexUriServ.do?uri=CELEX:31997L0069:FR:HTML (1997)

149. D.L. Maxim, C.P. Yu, G. Oberdorster, M.J. Utell: Quantitative risk analyses for RCF: Survey and synthesis. Regul. Toxicol. Pharmacol. **38**, 400–416 (2003)
150. S.H. Moolgavkar, E.G. Luebeck, J. Turim, R.C. Brown: Lung cancer risk associated with exposure to man-made fibers. Drug Chem. Toxicol. **23**, 223–242 (2000)
151. S.H. Moolgavkar, E.G. Luebeck, J. Turim, L. Hanna: Quantitative assessment of the risk of lung cancer associated with occupational exposure to refractory ceramic fibers. Risk Anal. **19**, 599–611 (1999)
152. K. Donaldson, P.J. Borm, V. Castranova, M. Gulumian: The limits of testing particle-mediated oxidative stress in vitro in predicting diverse pathologies; relevance for testing of nanoparticles. Part. Fibre Toxicol. **6**, 13 (2009)

2

Exposure, Uptake, and Barriers

Armelle Baeza-Squiban and Sophie Lanone

The nanotechnologies market is booming, e.g., in the food industry (powder additives, etc.) and in medical applications (drug delivery, prosthetics, diagnostic imaging, etc.), but also in other industrial sectors, such as sports, construction, cosmetics, and so on. In this context, with an exponential increase in the number of current and future applications, it is particularly important to evaluate the problem of unintentional (i.e., non-medical) exposure to manufactured nanoparticles (so excluding nanoparticles found naturally in the environment). In this chapter, we begin by discussing the various parameters that must be taken into account in any serious assessment of exposure to man-made nanoparticles. We then list the potential routes by which nanoparticles might enter into the organism, and outline the mechanisms whereby they could get past the different biological barriers. Finally, we describe the biodistribution of nanoparticles in the organism and the way they are eliminated.

2.1 Exposure

Many factors enter into the problem of unintentional exposure to artificial nanoparticles. The parameters that need to be taken into consideration to characterise potential exposure are:

- The environmental compartment in which the nanoparticles occur (water, air, soil).
- The shape of nanoparticle (primary, secondary).
- The exposure context (workers, users/consumers).
- The dose of nanoparticles to which we are potentially exposed.

These points will be discussed in the following sections.

P. Houdy et al. (eds.), *Nanoethics and Nanotoxicology*,
DOI 10.1007/978-3-642-20177-6_2, © Springer-Verlag Berlin Heidelberg 2011

2.1.1 Which Environmental Compartment?

Owing to the ever increasing production of manufactured nanoparticles (sometimes of the order of several tonnes), their release into the environment becomes more and more likely, and there is therefore a potential risk of exposure that needs to be understood. Such exposure may occur in different environmental compartments, i.e., air, water, soil, and it may be intentional or otherwise.

Indeed, unintentional emissions of artificial nanoparticles into the atmosphere are one possibility. Moreover, water or soil in the vicinity of nanoparticle production areas may be contaminated by effluents.

Exposure may also result from intentional use of certain manufactured nanoparticles. For example, iron nanoparticles are used to decontaminate the water table [1], which is thus directly exposed to nanoparticles. The use of titanium dioxide nanoparticles (TiO_2) in sunscreen creams to improve the spreading quality of the cream on the skin and protect against the sun's ultraviolet radiation, may lead to the dispersal of these nanoparticles in water when people bathe. The use of TiO_2 or zinc oxide (ZnO) nanoparticles in paints to improve their appearance of whiteness or to make them self-cleaning may cause ground contamination after repeated washing of the treated surfaces of buildings, cars, and so on, by rainfall. Finally, if the ground is contaminated, the nanoparticles may be returned to the atmosphere, with subsequent risk of atmospheric exposure.

2.1.2 What Kind of Particles?

As already mentioned, a particle is described as a nanoparticle if one of its dimensions is of nanometric order. Nanoparticles fall into two main groups for the purposes of toxicology: primary nanoparticles, i.e., deliberately synthesised, and secondary nanoparticles. The latter may be produced through the degradation of a material during some mechanical or thermal process, but also through the transformation of primary nanoparticles by interaction with other compounds. Among the secondary nanoparticles are diesel combustion products, tobacco smoke, welding fumes, and others.

Whether they are primary or secondary, nanoparticles very quickly form aggregates or agglomerates. We speak of an aggregate when the elementary particles are bound by strong forces, and an agglomerate when the assembly of elementary particles is held together by relatively weak and easily broken bonds, such as van der Waals or electrostatic forces [2]. So aggregates and agglomerates of elementary nanoparticles may be nanostructured (made up of nanoparticles) and have nanometric dimensions or reach sizes of a few microns. In every case, their nanoscale structure confers novel surface properties on them, one of the motivations for their industrial applications. It is thus clear that the range of nanoparticles to which we are potentially exposed is very broad and cannot be predefined.

2.1.3 Exposure Context

Exposure to nanoparticles can occur in an occupational or private context. Indeed, workers in the nanotechnology sector may be exposed during manufacture, during transport, or during storage of the nanoparticles. These workers involved in the synthesis and use of nanoparticles currently represent some 20 000 people around the world, and this number is on the increase. Estimates by the National Science Foundation suggest that around 2 million workers will be employed in the nanotechnologies sector within the next fifteen years or so [3].

As mentioned earlier, manufactured nanoparticles may be released into the environment at some point during their life cycle, and thereby reach the general population. Exposure of the general public can also arise through the use of products containing nanoparticles. The latter are already available on the open market. The example of sunscreen creams containing TiO_2 has already been cited, but one should also mention food additives used to improve the dispersion of powders such as salt, chocolate powder, and so on, and there are other applications in clothing, sports equipment, and so on. Since 2005, the Woodrow Wilson International Center for Scholars has published an inventory of commercially available nanotechnological derivatives [4]. In 2010, more than 1000 products were identified, and the list gets longer every day.

2.1.4 Dose

At the present time there is no factual data concerning the concentrations of manufactured nanoparticles present in the environment, since the many different sources are not properly controlled and coordinated. Regarding exposure levels in the workplace, these vary depending on the amount of nanoparticles produced and the post occupied by the worker, e.g., production worker on the shop floor, maintenance technician, storage or transport agent.

A quantitative evaluation of the potential exposure to nanoparticles will require nanoparticle metrology, to be discussed in Chaps. 7 and 8.

2.2 Uptake

Nanoparticles are conventionally considered to be able to enter into direct contact with the organism via three main routes: respiratory, digestive, and cutaneous. Indeed, these three systems are permanently exposed to the environment and hence likely to come into direct contact with nanoparticles.

2.2.1 Respiratory Route

The respiratory system is particularly exposed to man-made nanoparticles, not only because it is the entry route for inhaled particles, but also because the respiratory system receives the whole of the cardiac output. For this reason, there is an exposure risk for the respiratory system whenever nanoparticles are first taken up systemically, e.g., as a consequence of cutaneous exposure, ingestion, or systemic administration through nanomedical use.

Structure of the Respiratory System

The respiratory system can be considered as a system of ducts, the airways, whereby air enters via the nose and mouth, then passes into the lungs and pulmonary alveoli during breathing, as shown in see Fig. 2.1A.

We distinguish the upper airways, which are extrathoracic, comprising the nose, mouth, pharynx, and larynx, from the lower, intrathoracic airways. The latter includes an air conduction zone, made up schematically by the trachea and bronchi (stem bronchi which subdivide into lobar bronchi and then bronchioli), and a respiratory zone which handles gaseous exchanges between the air and the blood, and which consists mainly of pulmonary alveoli, rather like little bags at the end of each respiratory duct.

The pulmonary alveoli, about 300 million in number in an adult human, provide a huge area for exchanges to take place, in fact about $140\,\mathrm{m}^2$, roughly the area of a tennis court! The alveoli have a very thin wall, less than $0.5\,\mu\mathrm{m}$ thick (alveolar epithelium), and are covered with very fine vessels called capillaries. It is at the location of the alveolo-capillary barrier (see Fig. 2.1C) that gas exchange takes place between air and blood, whereby the alveoli fulfill their double role of transferring oxygen from the air to the blood and extracting carbon dioxide from the blood into the air.

The alveolar epithelium stands upon a continuous basal membrane. It is made up of two types of epithelial cell, the type I and type II pneumocytes, which meet at tight junctions. The type I pneumocyte is a highly flattened cell about $0.2\,\mu\mathrm{m}$ thick, spread out against the basal membrane. There are about 100 type I pneumocytes per alveolus, which represents 40% of the total number of epithelial cells, but 90% of the total epithelial surface area. Indeed, the type II pneumocytes, numbering about 150 per alveolus, cover only 10% of the total alveolar surface area. These are massive cells, encased between the cytoplasmic veils of the type I pneumocytes. They have an apical pole with short microvilli, and their cytoplasm contains cytoplasmic vesicles that are extremely rich in phospholipids, with a layered structure (lamellar bodies). It is the type II pneumocytes that synthesise the main components of the pulmonary surfactant.

Deposition

Deposition of nanoparticles in the respiratory system, i.e., the interaction of these particles with the various structures of the pulmonary surface, can occur at different points of the breathing apparatus depending on factors intrinsic or extrinsic to the nanoparticles.

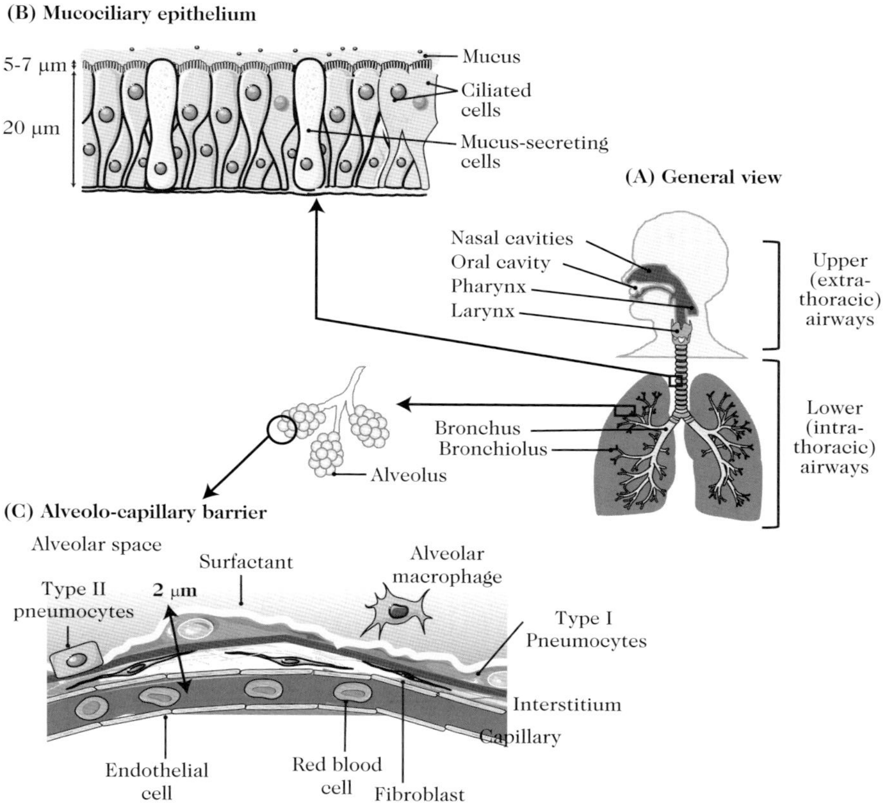

Fig. 2.1. Respiratory system (see colour plate). Illustration produced using Servier Medical Art, www.servier.fr. (**A**) General view showing the anatomy of the respiratory system. (**B**) Details of the mucociliary epithelium lining the upper and lower airways. The mucus produced by secreting cells traps particles and is moved up to the pharynx by ciliary beating to be expectorated or ingested. (**C**) Detail of the alveolar epithelium in the alveolo-capillary barrier. The alveolar epithelium is made up of type II pneumocytes involved in surfactant synthesis and type I pneumocytes, which are extremely fine cells covering 90% of the alveolar surface. The air–blood distance is about 2 µm. Macrophages in the alveolar lumen ensure particle phagocytosis

When they are in suspension in the air, the particles constitute an aerosol. The behaviour of this aerosol will depend to a large extent on the size of the particles, and this then determines the mode of deposition of the particles. Generally speaking, particles may be subjected to various forces, namely, inertia, gravity, or diffusion. As far as nanoparticles are concerned, diffusion forces tend to dominate. Indeed, as the particle size approaches the molecular level, which is the case for nanoparticles, their dynamical behaviour

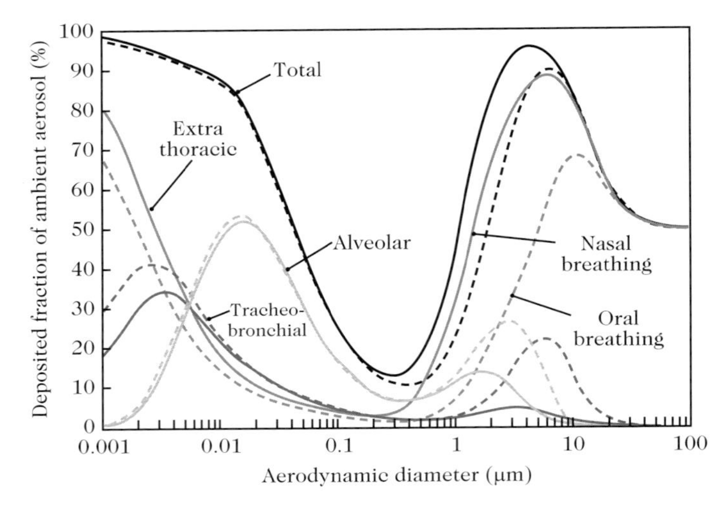

Fig. 2.2. Predicted total and regional deposition of particles in the human respiratory tract as a function of particle size. The deposited fraction includes the probability of inhalation. The subject is assumed to breathe mainly through the mouth (*dotted curve*) or the nose (*continuous curve*), while carrying out a standard physical effort (see colour plate). From [5]

tends to be superposed on the dynamical behaviour of a gas, and hence to obey the gas diffusion laws. Nanoparticles thus enter into collisions with the gas molecules of the surrounding air (Brownian motion) and are themselves carried along by random diffusive motion. The speeds of these motions are inversely proportional to the diameters of the particles, i.e., the smaller the particles, the faster the motions. The tendency of nanoparticles to aggregate and agglomerate must also be considered, as explained earlier, noting that this phenomenon will be favoured by the possibility of collisions. The formation of aggregates and agglomerates will largely determine the subsequent deposition of the nanoparticles in the respiratory tract.

What exactly happens to nanoparticles in the respiratory tree is not perfectly understood. This is why predictive models are employed. In order to model the deposition of nanoparticles in the respiratory tract, the latter is divided up schematically into three regions, viz., the nasopharyngeal (upper airways), tracheobronchial, and alveolar regions. Predictive models have been set up using data provided by the International Commission on Radiological Protection (ICRP), describing the probability of total deposition of particles measuring up to 100 μm in aerodynamic diameter throughout the respiratory tract, with an analysis carried out region by region. An example is presented in Fig. 2.2 [5].

From this predictive model (see Fig. 2.2), it can be concluded that, the smaller the particles, the more likely they are to be deposited in the respiratory tract, with a specific distribution in each region depending on the size of the nanoparticles. Indeed, while 5 nm nanoparticles can be deposited at similar levels in each of the three regions, deposition in the nasopharyngeal region tends to dominate for particles smaller than 5 nm, whereas particles bigger than 5 nm tend to be deposited preferentially in the alveolar region. The maximal alveolar deposition (50–60%) is predicted for 20 nm particles, the total deposition probability for this class of nanoparticles being 80%. These differences in deposition (total and regional) can be expected to influence the subsequent biological effects of the nanoparticles. Moreover, it should be noted that this model was made considering breathing through the mouth and at rest. It is easy to understand that the deposition parameters are likely to be modified under rapid breathing conditions, e.g., when making a physical effort, since the volume of air taken in will be greater per unit time, and with greater perturbations in the flow. Another situation where one would expect modifications is in the presence of a respiratory pathology such as asthma, bronchitis, and so on, which also modulates the air flow in the airways. Indeed, mathematical models predict increased deposition of nanoparticles in pathological or constricted airways. This is confirmed by the fact that airways affected by an obstructive pulmonary disease or asthma display higher pulmonary retention of nanoparticles [6–9]. Finally, it should be noted that this model was made for spherical nanoparticles, and its relevance for deposition of nanoparticles with other shapes remains to be established.

Clearance Mechanisms

Clearance of a given substance is defined by the capacity of an organ to eliminate it totally. In the lungs, clearance of deposited particles is governed by two types of mechanism: chemical clearance and physical translocation of particles.

Chemical clearance involves processes which dissolve either the particle or its soluble components, lixiviation which consists in removing certain chemical elements from the particle matrix, or absorption or binding to proteins, allowing the particles to pass into the bloodstream or lymph system. These chemical clearance processes can occur throughout the respiratory apparatus, but with different levels of efficiency depending on the intra- and extracellular environment and in particular the pH [10].

In contrast to chemical clearance, physical translocation mechanisms are more specific to the region of the respiratory apparatus. Two main clearance mechanisms are conventionally considered in the respiratory system:

- *The Mucociliary Escalator.* The nasal mucous membrane and the tracheobronchial region are endowed with a highly effective clearance mechanism, jointly accomplished by the ciliated epithelial cells and the mucus secreting cells (see Fig. 2.1B). These cells form a mucociliary escalator, allowing the migration of a lining of mucus toward the pharynx. This is a very fast clearance mechanism for solid particles, which are eliminated from the tracheobronchial region in just 24 hours [10].

- *Phagocytosis by Alveolar Macrophages.* Another absolutely classic pulmonary clearance mechanism, effective in the alveolar region this time, involves the alveolar macrophages (see Fig. 2.1C) and their ability to internalise particles by phagocytosis. This mechanism is all the more efficient in that the macrophages are guided by chemical attraction to the point of deposition of the particles. The macrophages recruited in this way quickly internalise the particles, then migrate toward the mucociliary escalator. The whole process takes several days. The efficiency of this clearance system seems to depend on the size of the particles, becoming less efficient for nanoparticles than for micrometric particles [10–13]. However, since all particles are phagocytized within 6 to 12 hours of their deposition [10], this implies that other clearance mechanisms come into play for nanoparticles. Among these mechanisms are epithelial translocation, transit into the blood or lymph systems, and translocation via sensorial neurons [10]. These different mechanisms are explained later in the chapter.

2.2.2 Cutaneous Route

The skin constitutes an important barrier, protecting against all forms of environmental agression, and hence potentially against nanoparticles. The skin has an area in the range 1.5–2 m^2, and it is structured in three layers: the epidermis, the dermis, and the hypodermis (see Fig. 2.3).

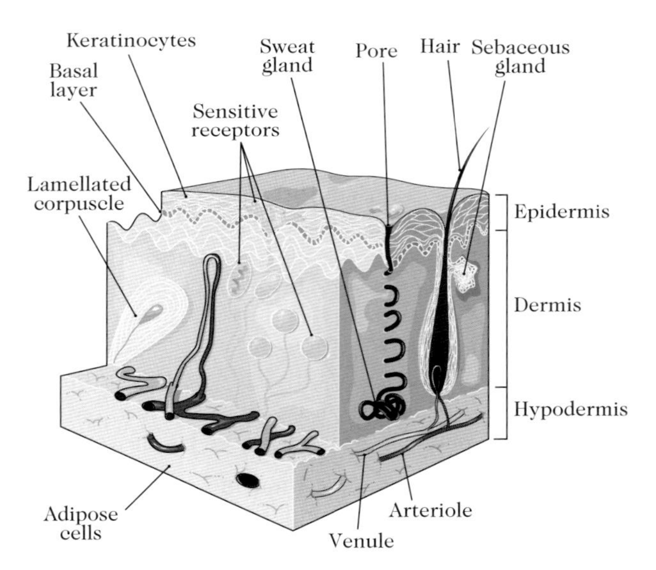

Fig. 2.3. Structure of the skin: epidermis, dermis, and hypodermis (see colour plate). Illustration produced using Servier Medical Art, www.servier.fr. The epidermis is a stratified non-vascularized epithelium, separated from the dermis by the basal layer. The vascularized dermis contains the cutaneous appendages. The lowermost layer is the hypodermis, comprising mainly adipocytes

Epidermis

The epidermis is the outermost layer of the skin and thus forms the first physical barrier to environmental assault. It is a stratified epithelium which is continually renewed. It is made up mainly of keratinocytes (90–95% of the total cell population), but also melanocytes, Langerhans cells, and Merkel cells. It is not irrigated by any blood vessels, but contains many nerve endings. The epidermis is itself composed of four sublayers: the *stratum corneum*, the *stratum granulosom*, the Malpighian layer, and the basal layer.

The *stratum corneum* is the outermost layer of the epidermis. It is very thin over most of the body (of order μm), except at certain specific locations, such as the soles of the feet and the palms of the hands. It contains only dead cells, without nuclei and highly keratinized, forming an impermeable and flexible layer. There is a desquamation process of the keratinocytes making up the *stratum corneum*, leading to complete renewal of this layer roughly once a month.

The *stratum granulosom* is made up of keratinocytes which are beginning to lose their nucleus, releasing a kind of lipid-based 'cement' which strengthens intercellular cohesion and helps the epidermis to fulfill its role of protective barrier.

The main feature of the Malpighian layer, or underlying mucous layer, apart from being the thickest layer of the epidermis, is that it contains Langerhans cells. These are cells from the bone marrow, which intercalate between the keratinocytes and fulfill an immunological role.

Finally, the deepest layer of the epidermis, the basal layer, is also made up of keratinocytes, joined together and joined to the underlying dermis by desmosomes. These are cells that proliferate and ensure renewal of the epidermis. Melanocytes are also found in this layer. The function of these cells is to synthesise a pigment called melanin. And then there are Merkel cells, with neuroendocrine and epithelial functions.

Dermis

Located beneath the epidermis, this is 10 to 40 times thicker than the latter. It is the thickest layer of the skin. Its resident cells are fibroblasts, essential for synthesising the constituents of the conjunctive tissue, dendrocytes (dendritic mesenchymal cells), and mastocytes (mononucleated medullary cells). The dermis is also the place where most skin structures are found, including the sweat and fat glands, the hair follicles, the nerve endings, the blood vessels which carry oxygen and nutrients to the skin, and the lymphatic vessels which contain immune cells for fighting infections. This layer also contains two elements that are essential for its cohesion and flexibility, namely collagen and elastin fibres, respectively. Finally, it constitutes an important stock of water for the organism as a whole.

Hypodermis

The lowermost layer of the skin is the hypodermis. It is attached to the overlying dermis by elastin and collagen fibres. In this layer, the main cells are the adipocytes, which serve to store fats. For this reason, and by virtue of its many blood vessels, the hypodermis serves as an energy store through the adipocytes which supply fats to the organism, but also as a protection against temperature variations, since fat is a good thermal insulator, and against mechanical assault.

2.2.3 Digestive Route

The digestive system is a potential entry route for nanoparticles when the organism ingests contaminated foodstuffs or water, or consumes processed foods into which nanoparticles have been introduced intentionally during fabrication. Oral exposure may also occur by hand-to-mouth transfer. Finally, inhaled nanoparticles which have been eliminated from the respiratory system by mucociliary clearance are subsequently ingested if they are not expectorated.

The intestine has a surface area of around $240\,\mathrm{m}^2$ and contributes significantly to the total area of the digestive tract. This high surface area is designed for efficient absorption of nutrients. It is achieved by the length of the organ and by the formation of finger-shaped folds called villi, which cover the intestinal wall. It is further increased by the presence of microvilli at the apical pole of the enterocytes which constitute the most abundant cell type in the intestinal mucosa (see Fig. 2.4A). This mucosa is covered with a monostratified epithelium made up of enterocytes and goblet cells joined together by tight junctions which guarantee the cohesion of the tissue and its role as a barrier (see Fig. 2.4B). The main functions of the enterocytes are to control the transfer of macromolecules and micro-organisms and at the same time to allow the absorption of nutrients. The goblet cells secrete a protective viscous fluid, mucus, made up of glycoproteins (mucins). Mucus defends the mucosa against the adhesion or penetration of toxins, bacteria, and antigens.

Dispersed through the intestinal mucosa are regions known as Peyer's patches which are not involved in digestive activities, but which play a role in local immunity (see Fig. 2.4C). These patches are bounded on the luminal side by a specialised epithelium containing so-called M cells, used by micro-organisms to cross the intestinal mucosa and reach the underlying lymphoid follicles. These M cells represent only 10–20% of the cells in the Peyer's patches, and barely 10^7 cells of the intestinal epithelium. They represent a potential entry route for nanoparticles, because they have a great capacity for transcytosis and can transport a wide range of materials, including nanoparticles. It is generally accepted that particles of size less than $1\,\mu\mathrm{m}$ are phagocytized by M cells and transported to the basal region, whereas those with sizes greater than $5\,\mu\mathrm{m}$ are also phagocytized by the M cells, but remain trapped in the Peyer's patches [14].

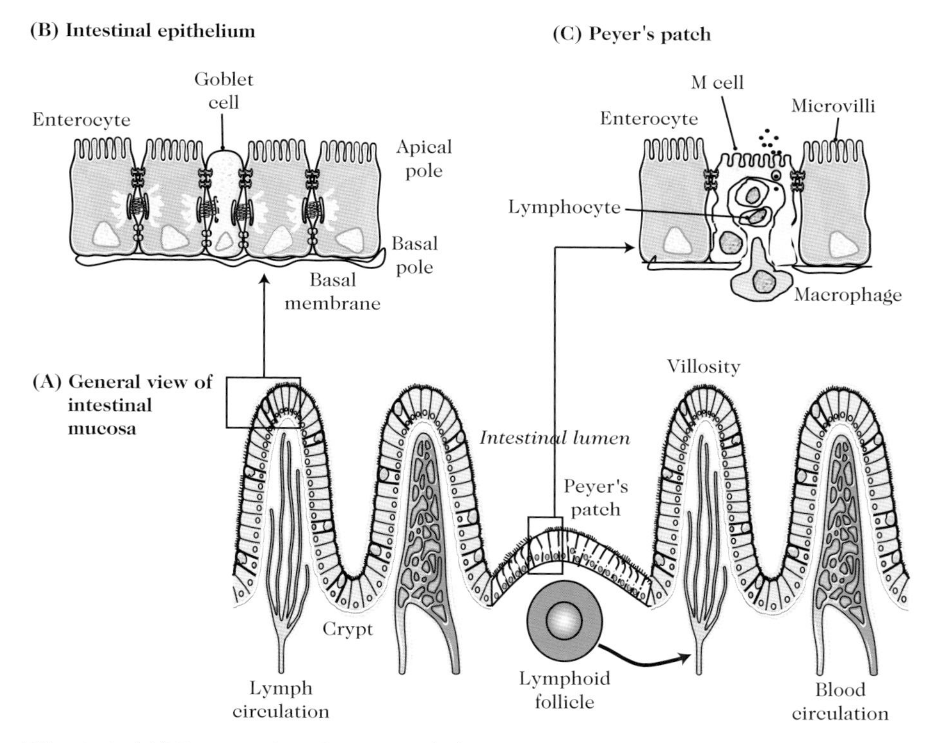

Fig. 2.4. (**A**) Intestinal wall showing (**B**) epithelial villi covered by a monostratified epithelium made up of enterocytes and goblet cells in contact via different types of junction, including tight junctions preventing paracellular transfer and resting on a basal membrane, and (**C**) a Peyer's patch with specialised epithelium containing M cells among enterocytes above a lymphoid follicle. Illustration produced using Servier Medical Art, www.servier.fr (see colour plate)

2.3 Barrier Crossing

Whatever epithelium one may consider (respiratory, epidermic, or intestinal), it constitutes a physical barrier to penetration by foreign substances by virtue of impermeable junctions around the cells. Excluding pathological situations or exposure to toxic substances affecting the permeability of the epithelia, the paracellular route, i.e., between cells, for nanoparticle uptake is highly unlikely. Nanoparticle penetration across epithelial barriers thus implies transcellular transfer.

2.3.1 Internalisation Mechanisms

The plasma membrane around a cell regulates and coordinates the entry and exit of molecules in order to maintain an inner medium that differs from the one around the cell. It comprises a lipid bilayer and proteins. The lipidic aspect of the membrane allows it to block the transfer of ions and large polar molecules. The presence of pumps, carriers, and protein channels allows selective transfer of ions and solutes. As far as macromolecules and particles are concerned, they can only be captured in the extracellular medium by a mechanism known as endocytosis, which results in their being internalised in vesicles from the plasma membrane.

Cell biologists distinguish two main types of endocytosis: phagocytosis, which consists in the internalisation of large particles, and pinocytosis, which involves the capture of fluid substances and solutes (see Fig. 2.5) [15].

Phagocytosis

Phagocytosis is normally carried out by professional phagocytes, i.e., macrophages, monocytes, and polynuclear neutrophils, to protect the organism against invasion by pathogens. During phagocytosis, bacteria, yeasts, cell debris, or large particles are internalised in phagosomes of diameter 0.1–10 μm. The fusion of these phagosomes with lysosomes destroys the pathogen through

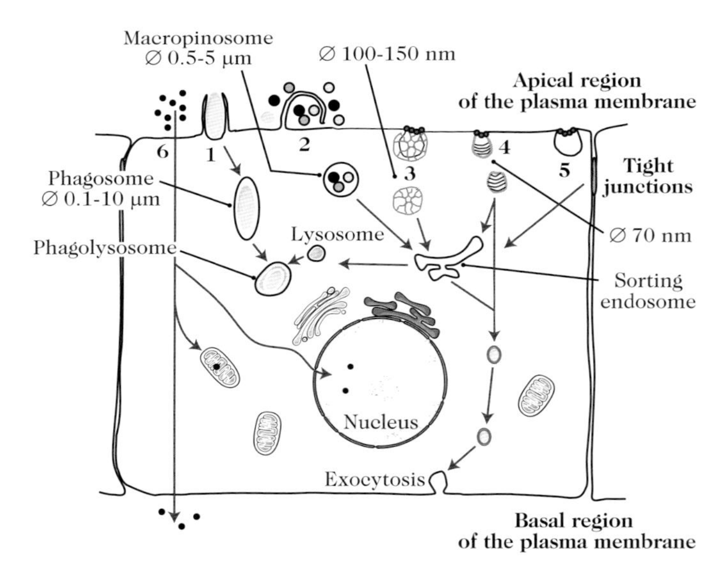

Fig. 2.5. Different types of internalisation by cells. The endocytosis routes differ by the size of the resulting vesicles, the nature of the endocytosed compounds, and the vesicle formation mechanism. Illustration adapted from [15] and produced using Servier Medical Art, www.servier.fr. (1) Phagocytosis, mediated by receptors and leading to the formation of a phagosome in which the ingested material is degraded after fusion with lysosomes. (2) Macropinocytosis. (3) Clathrin-dependent endocytosis. (4) Caveolae-dependent endocytosis. (5) Endocytosis not mediated by clathrins or caveolae. (6) Diffusion

the combined action of oxidants, proteases, and hydrolases acting in an acidic environment. Compounds resisting this degradation perdure in the phagolysosomes and constitute residual bodies.

The process of phagocytosis is triggered when the particle is fixed on receptors present in the membrane of the phagocyte. These may be specific receptors for molecules which attach themselves to the foreign body, such as opsonins, or class A scavenger receptors specialised in the phagocytosis of unopsonised environmental particles [16]. When the particle fixes onto the receptor, it triggers the assembly of actin in the cell. Actin is a protein of the cytoskeleton which allows the plasma membrane to stretch around the particle and enclose it, as shown in Fig. 2.5(1).

Depending on the shape and size of the particle, the cell may not manage to completely enclose it, and this leads to what is known as frustrated phagocytosis, likely to trigger an inflammatory reaction. This phenomenon has been observed in mice exposed to carbon nanotubes [17]. Furthermore, several studies have shown that, in the breathing system, phagocytosis by alveolar macrophages is less effective against nanoparticles than against larger particles, leading to persistence of nanoparticles in the alveolar compartment and subsequent deposition on the alveolar epithelium [10].

Pinocytosis

In contrast to phagocytosis, which is induced by the binding of a particle to the membrane surface, pinocytosis is a constitutive process occurring all the time. There are four different mechanisms: macropinocytosis, clathrin-dependent endocytosis, caveolae-dependent endocytosis, and endocytosis that is independent of both clathrins and caveolae.

Macropinocytosis consists in the internalisation of large amounts of extracellular fluid thanks to membrane protrusions, using a mechanism involving actin from the cytoskeleton. However, some bacteria use this route to penetrate cells. It leads to the formation of macropinosomes with diameters in the range 0.5–5 μm, as shown in Fig. 2.5(2).

Clathrin-dependent endocytosis involves the receptors of the plasma membrane. It provides a way of recycling them and concentrating the ligands of the surrounding extracellular medium if they bind to the receptors. This type of endocytosis occurs in specialised areas of the membrane lined on the intracellular side by a protein network made up of clathrin. Invagination of the plasma membrane detaches a vesicle coated with clathrin, with diameter in the range 100–150 nm, as shown in Fig. 2.5(3).

Caveolae-dependent endocytosis leads to the formation of small vesicles called caveolae, with diameter around 70 nm. It occurs in microdomains of the plasma membrane that are rich in glycosphingolipids and in cholesterol which interacts with caveolin, a membrane protein. In most cells, this internalisation mechanism is considered to be secondary compared with clathrin-mediated internalisation, with the exception of the endothelial cells. In these cells, it allows transcytosis, i.e., the transfer of serum proteins from the blood via the caveolae to tissues juxtaposing the endothelial cells, as shown in Fig. 2.5(4).

The last form of endocytosis, involving neither clathrin nor caveolae, has been observed to occur in lipid microdomains or rafts. The mechanisms involved here remain poorly understood at the present time [see Fig. 2.5(5)].

With the exception of macropinocytosis, most other internalisation mechanisms involve membrane receptors. The adsorption of ligands present in the biological fluids on nanoparticles in a way that depends on their physicochemical characteristics might favour their recognition by membrane receptors, and hence their internalisation.

Following pinocytosis, the vesicles fuse with the sorting endosome and are then degraded by lysosomes or addressed to the opposite membrane domain where transcytosis can occur (see Fig. 2.5).

Other Mechanisms

In all the endocytosis processes mentioned above, internalised material is enclosed in a vesicle and thereby separated from the cytoplasm by a membrane. However, free nanoparticles have been observed in the cytoplasm, and even in the nucleus and mitochondria, where they may then interact directly with macromolecules. For example, the inhalation of small doses of TiO_2 nanoparticles by rats has resulted in the observation of nanoparticles that are not enclosed by a membrane in epithelial and endothelial cells, in the conjunctive tissue, and even in red blood cells [18].

This therefore suggests that nanoparticles can enter cells by different mechanisms to the usual ones, in particular, those involving actin. Indeed, in the presence of a substance depolymerising actin, neutral and charged polystyrene nanoparticles are nevertheless internalised by macrophages, in contrast to 1 μm particles. For example, some studies suggest that, by adhesive interactions, nanoparticles may be able to diffuse passively through the plasma membrane by virtue of temporary pore creation, as shown in Fig. 2.5(6) [18]. This has been observed with neutral and charged fluorescent polystyrene nanoparticles, gold nanoparticles, and TiO_2 nanoparticles found in red blood cells, which are cells with no standard endocytosis process. According to the authors of these investigations, the surface properties and chemical composition of the nanoparticles may not be relevant to nanoparticle uptake by these mechanisms.

If nanoparticles can enter cells by this diffusion mechanism, they can also leave by the same route, thereby favouring their transfer through to the other side of the epithelial barrier, and thus achieving transcytosis.

Many studies have demonstrated the rapid internalisation of a wide range of nanoparticles by different cell types, whether or not they are specialised in phagocytosis. However, the mechanisms coming into play have not yet been carefully investigated. Moreover, most of these studies have been carried out in vitro under conditions where contact between the nanoparticles and the culture medium causes proteins to adsorb onto them which by their very nature will facilitate receptor-mediated internalisation, not to mention the aggregation of the particles. For example, a study made with cerium oxide nanoparticles of different sizes (20–500 nm), used in very low concentrations (0.02–0.2 $\mu g/cm^2$) has shown that their internalisation by fibroblasts, probably macropinocytosis, increases with particle size [19]. The sedimentation of the largest nanoparticles favours their contact with the cells, in contrast with what happens for smaller nanoparticles, since these gain access to the fibroblasts only by diffusion, even though they may aggregate.

2.3.2 Particle Translocation

Studies carried out so far suggest that nanoparticles can cross epithelial barriers, thereby gaining access to the blood compartment, whereupon they may be distributed throughout the organism.

Air–Blood Translocation

Several studies have been carried out on humans and animals to estimate the capacities of inhaled particles to translocate through the alveolo-capillary barrier [20]. According to these studies, it seems that translocation is variable, which could be explained by the type of particle used and the mode of administration of the nanoparticles (inhalation, instillation).

For example, a one hour inhalation of radioactive iridium nanoparticles (15 and 80 nm) by rats led to a low level of translocation since, 7 days later, less than 1% of the radioactivity was observed in secondary organs, such as the liver, spleen, heart, and brain. The nanoparticles were mainly eliminated in the feces after pulmonary clearance and ingestion [21]. Monitoring over 6 months, it was shown that the nanoparticles are first trapped in the interstitium, but can return to the alveolar lumen to be eliminated by alveolar macrophages [22].

In contrast, intratracheal instillation of TiO_2 particles in rats results in significant translocation (50%) for 12 nm particles, whereas it reaches only 4% for 220 nm particles [10]. Still in rats, intratracheal instillation of 22 nm ferric oxide nanoparticles ($^{59}Fe_2O_3$) results in rapid transfer to the blood (10 min) through the alveolo-capillary barrier, with ensuing distribution in the liver, spleen, kidneys, and testicles. The plasma half-life is estimated here at 22.8 days [23].

Following intratracheal instillation of 20 nm colloidal gold nanoparticles in mice, these particles are then observed in the basal membrane between the alveolar cells and the endothelial cells, but also on the surface of endothelial cells in the blood vessels. However, the amount of nanoparticles ending up in the bloodstream is very low [24]. On the other hand, alveolar macrophages containing particles migrate into the blood flow and hence to extrapulmonary organs, suggesting that translocation may be not only direct, but also indirect via the macrophages. Another study comparing 1.4 and 18 nm gold nanoparticles demonstrated significant translocation of the 1.4 nm nanoparticles, while the 18 nm particles were retained in the lungs [25].

Finally, in humans, the administration of carbon-containing particles coupled with technetium produced contradictory results, still under discussion [26, 27].

In vitro investigations of alveolar barrier crossing using primary cultures of alveolar epithelial cells from rats have clarified the role of particle characteristics on translocation mechanisms. The phenomenon is more significant for smaller polystyrene nanoparticles (20 as compared with 100 nm) and for positively charged particles [28]. Furthermore, it does not occur at 4°C, reminding us that this is an energy-consuming mechanism.

To sum up, these studies show that nanoparticle translocation through the alveolo-capilliary barrier is possible, but that the extent depends significantly on the nanoparticles themselves, and in particular their physicochemical properties. In addition, these studies have all involved a single exposure, and we may imagine that a situation of repeated exposure will lead to higher levels of translocation into the bloodstream. In the same way, it may be increased in a pathological context. For example, polystyrene nanoparticles with diameters 56 and 202 nm administered to rats by intratracheal instillation lead to higher levels of systemic transfer if the rats are first treated with lipopolysaccharide, i.e., if they display inflammation [29]. Likewise, the exposure of perfused rat lungs to iridium nanoparticles only results in detection of nanoparticles in the perfusate if the lungs are first treated with hydrogen peroxide, which simulates oxidative stress conditions occurring during inflammation, or in the presence of histamine, which increases vascular permeability [30].

Neuronal Translocation

As mentioned earlier (see the discussion of deposition on p. 40), studies using models for the deposition of non-aggregated particles in the breathing apparatus have shown that the smaller particles are efficiently deposited in the nose. The olfactory epithelium in the nasal cavities represents a potential route for the uptake of nanoparticles by the central nervous system (see Fig. 2.6). Indeed, 50 nm gold nanoparticles have been observed to translocate in monkeys, while 35 nm carbon nanoparticles and 30 nm manganese oxide nanoparticles have been observed to translocate in rats, entering the olfactory bulb by following the axon of the olfactory nerve [10]. In particular, experiments carried out with MnO on rats lead to the estimate that 11% of deposited particles end up in the olfactory bulb, while some reach an even more distal location in the brain. In addition, inflammatory effects have been noted in the olfactory bulb [31]. Moreover, a study in which mice were exposed to TiO_2 nanoparticles with diameters 80 and 155 nm for one month confirmed the transfer of nanoparticles to the brain via the olfactory bulb. Evidence has been found of accumulation in the hippocampus using synchrotron radiation

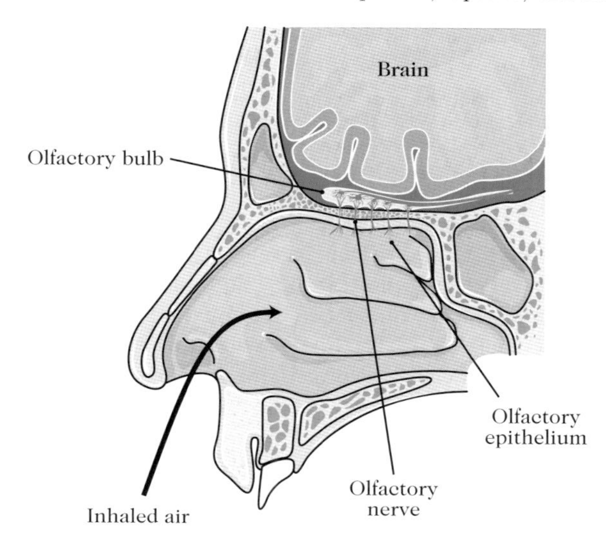

Fig. 2.6. Olfactory epithelium. Located on the roof of the nasal cavity, it comprises several cell types, including olfactory nerve cells. Their axon communicates with the olfactory bulb. Illustration produced using Servier Medical Art, www.servier.fr

X-ray fluorescence (SRXRF) analysis. This accumulation is associated with lesions, notably oxidative lesions [32].

These experiments were carried out on rodents with a more developed olfactory mucosa ($8\,\mathrm{cm}^2$ in rats, or 50% of the nasal mucosa) than in humans ($5\,\mathrm{cm}^2$, or 5% of the nasal mucosa), and with entirely nasal respiration. But the olfactory epithelium is nevertheless a non-negligible translocation region, and it will be important to assess the implications for neurodegenerative disorders.

Another way nanoparticles might get into the central nervous system is by somehow crossing the blood–brain barrier. This is characterised by the existence of very effective junctions between the endothelial cells, so as to avoid all penetration of particles by the paracellular route. At the present time, there is no proof that nanoparticles can transfer via this route. For example, for intravenous or intraperitoneal injections of 40 nm gold particles, no particles were subsequently observed in the brain [33].

However, there is currently a great deal of research to develop nanoparticles that could deliver drugs directly to the brain, and some of these have already been designed to cross the blood–brain barrier. The binding of these nanoparticles to serum apolipoproteins seems to be the factor favouring internalisation. The latter is mediated by the low density lipoprotein receptors present on the membranes of endothelial cells [34].

Cutaneous Translocation

Transcutaneous transfer of nanoparticles has been very carefully investigated for titanium dioxide (TiO_2) and zinc oxide (ZnO) nanoparticles, since they are used in sun creams to block out ultraviolet radiation.

Several studies have shown that, with repeated application of TiO_2 on healthy human skin, the nanoparticles remain on the skin or in the upper layers of the *stratum corneum* of the epidermis, and neither cross nor even penetrate the living part of the epidermis [35]. Similar results have been obtained with ZnO. It should also be noted that nanoparticles can accumulate in the hair follicles. They then constitute a reservoir in which nanoparticles can persist, until eliminated by the flow of sebum. At the present time, there is no evidence of nanoparticles transferring from hair follicles to the dermis.

Applying quantum dots of different sizes (14–45 nm), shapes (spherical and ellipsoidal), and electrical charges on the skin of a healthy pig, there is significant absorption of these nanoparticles, whatever their characteristics. However, with 8 hours of exposure, no transfer has been observed to the perfusate, i.e., the liquid around the basal part of the skin fragment exposed to the nanoparticles [36].

While penetration through healthy skin seems limited, and dependent on the type of nanoparticle, doubts remain with regard to damaged skin (injuries, erythema, eczema, etc.) and flexion zones. For example, using an in vitro method on human skin subjected to mechanical flexion (20 flexions of 45° per minute), epidermic and dermic penetration of fluorescent particles (0.5 and 1 μm) was observed after 60 min of exposure and flexions, whereas larger particles (2 and 4 μm) remained on the *stratum corneum* [37]. This penetration is not systematic and only concerns 50% of the skin samples tested and a small percentage of the particles applied to the skin. In the same way, on a model of pig skin subjected to mechanical flexions, fullerene nanoparticles functionalised with an amino acid have been shown to penetrate, and also to accumulate in the lipid-rich intercellular spaces of the *stratum granulosom* of the epidermis [38]. On a rat skin model, it transpired that only an abrasion of the skin would allow quantum dots to reach the dermis [39]. Finally, in vivo experiments with mice exposed to ultraviolet radiation favoured the penetration of quantum dots [40].

In the current state of knowledge and for the tested nanoparticles, it seems that nanoparticles can in fact penetrate the epidermis, or even the dermis, but that transcutaneous transfer is not possible with undamaged skin.

Digestive Translocation

It has been reported that, when rats were exposed to 50 or 3 000 nm polystyrene beads by daily force feeding for 10 days, their intestines absorbed about 34% of the 50 nm nanoparticles and 26% of the 100 nm nanoparticles [41]. Absorption occurred in the Peyer's patches, followed by transfer to the mesenteric lymph. It is easier for smaller nanoparticles to cross the layer of mucus lining the intestinal epithelium, and easier also for nanoparticles that do not carry positive electrical charge [42]. TiO_2 particles are absorbed and end up in the blood [14]. On the other hand, administration of 18 nm iridium 192 nanoparticles by force feeding rats did not result in gastro-intestinal absorption [21]. The data available at the present time thus suggest that particle size and composition do influence their ability to cross the intestinal barrier.

The physiological state of the subject can affect intestinal permeability. For example, bacterial invasion results in overexpression of transport proteins in the epithelial cells of the Peyer's patches. In addition, using an in vitro approach with epithelial cells of the Caco-2 cell line [42], transcytosis of fluorescent nanoparticles was observed during joint exposure in the presence of *Yersinia* bacteria expressing invasin, a bacterial adhesion molecule.

Placental Translocation

The possibility of nanoparticles passing into the bloodstream raises the question of whether they could cross the placental barrier and hence exhibit fetotoxicity. Intravenous or intraperitoneal injection of 2 and 40 nm gold nanoparticles in gravid rats does not result in transfer to the placenta [33].

2.4 Nanoparticle Biodistribution in the Organism. Elimination

Nanoparticles present systemically in the organism can be eliminated in two ways: by the urine, after filtering in the kidneys, or by the feces, after transfer to the bile in the liver. The latter route concerns nanoparticles that cannot be eliminated via the renal route.

Renal clearance involves glomerular filtration, tubular secretion, then elimination via the urine. Filtration of molecules through the glomerular capillary wall depends on their size. Those with diameters less than 5.5 nm can be filtered, since they correspond to the diameter of pores in the vascular endothelial cells. Those larger than 8 nm remain in the bloodstream and are dealt with by the reticulo-endothelial system. Between these two diameters, electrical charge is relevant, since it can result in adsorption of molecules, thereby increasing the hydrodynamic diameter of the nanoparticle, combined with the fact that the capillary wall carries negative charges [43]. Glomerular filtration is therefore favourable for nanoparticles in the range 6–8 nm which are neutral or positively charged. Once in the tubule, the filtered nanoparticles can nevertheless be reabsorbed by the tubular epithelium, but at the present time there is no data to either support or contradict the occurrence of such reabsorption.

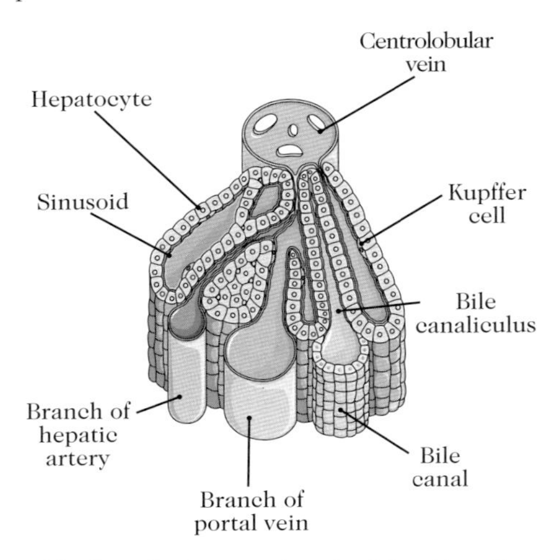

Fig. 2.7. The liver. Toxic substances absorbed by the digestive route arrive in the liver by the portal vein, while those absorbed by other routes arrive by the hepatic artery. Blood reaches the centrolobular vein via the sinusoidal capillaries. Hepatocytes form a monolayer around each capillary and produce bile. The Kupffer cells are macrophages located in the sinusoidal lumen. Illustration produced using Servier Medical Art, www.servier.fr

One of the physiological functions of the liver is to efficiently capture and remove particles with sizes in the range 10–20 nm (see Fig. 2.7). Two cell types are involved here:

- Hepatic epithelial cells or hepatocytes (60% of cells), capable of endocytosis and responsible for enzymatic degradation of particles and their removal via the bile.
- Kupffer cells, belonging to the reticulo-endothelial system (40% of cells). These are macrophages able to engage in endocytosis thanks to the many receptors for opsonised particles which they carry at their surface. Particle removal by these cells is based solely on intracellular degradation. If the latter is not accomplished, nanoparticles are retained in these cells, and hence remain in the organism. Nanoparticles developed for therapeutic purposes and whose half-life thus needs to be increased, are given a hydrophilic coating, e.g., polyethylene glycol, to prevent them from being opsonised and subsequently captured by the reticulo-endothelial system.

There have not yet been studies of nanoparticle metabolisation. It seems unlikely that nanoparticles of gold, silver, titanium dioxide, or fullerenes could be metabolised by hepatic enzymes. However, functionalised nanoparticles may lose their functional groups, or the latter may be modified [44].

Many studies have been carried out to study what happens to nanoparticles after intravenous administration. Although this route is not relevant to occupational or environmental exposure, it still provides data on the biodistribution of nanoparticles and their elimination from the organism.

Quantum dots are rapidly removed from the bloodstream after intravenous injection. They are then deposited in the liver, the skin, the bone marrow, and lymph nodules, depending on their surface coating [45]. Twenty-eight day toxicokinetic studies of mice have confirmed the short blood half-life of 13 nm QD705 quantum dots, viz., 18.5 hr, which accumulate in the liver, the lungs, the kidneys, and the spleen, while no appreciable excretion has ever been observed [46]. However, one 5 day study with another type of quantum dot led to urinary and fecal elimination of these nanoparticles, with only 8% remaining trapped in the liver [47]. The characteristics of surface coatings and the size of quantum dots determine whether they are eliminated from the kidney or captured by the reticulo-endothelial system and hence trapped [48, 49].

In contrast, after intravenous administration, functionalised single-wall carbon nanotubes do not end up in the liver or the spleen, but are rapidly eliminated from systemic circulation by the renal excretion route [50]. Intravenous administration of TiO_2 in rats results in an accumulation in the liver and spleen over 28 days [51]. Smaller levels are found in the lungs and kidneys, and these return to the control level.

Gold nanoparticles injected intravenously in mice [52] and rats [53] distribute themselves differently depending on their size. The smallest (10–15 nm) have a wider distribution, but particularly in the liver, followed by the lungs, the kidneys, and spleen, while an accumulation has been observed in the brain in the case of mice. In contrast, the bigger the nanoparticle size, the fewer organs are concerned, while it is still the liver that retains the most nanoparticles. Another study, using even smaller gold nanoparticles (1.4 nm, compared with 18 nm nanoparticles) has shown that their biodistribution in rats depends on the administration route (intravenous or intratracheal) [25].

2.5 Conclusion

The present and future dissemination of nanoparticles makes increased unintentional exposure to them quite unavoidable. There are already some experimental results showing that nanoparticles can cross the epithelial barriers, whence they may enter the bloodstream and get distributed throughout the organism. However, it seems that the uptake mechanisms, the level of transfer, and the biopersistence of nanoparticles remain poorly understood, even though it is fairly clear that these things will depend significantly on the physicochemical characteristics of the nanoparticles. Future research must identify the main factors determining the absorption of nanoparticles and what happens to them subsequently. The characterisation of exposure and understanding of nanoparticle toxicokinetics will be essential if we are to correctly assess human health risks.

References

1. S.J. Klaine, P.J. Alvarez, G.E. Batley, T.F. Fernandes, R.D. Handy, D.Y. Lyon, S. Mahendra, M.J. McLaughlin, J.R. Lead: Nanomaterials in the environment: Behavior, fate, bioavailability, and effects. Environ. Toxicol. Chem. **27**, 1825–1851 (2008)
2. G. Nichols, S. Byard, M.J. Bloxham, J. Botterill, N.J. Dawson, A. Dennis, V. Diart, N.C. North, J.D. Sherwood: A review of the terms agglomerate and aggregate with a recommendation for nomenclature used in powder and particle characterization. J. Pharm. Sci. **91**, 2103–2109 (2002)
3. www.nano.gov/html/res/faqs.html
4. www.nanotechproject.org/inventories/consumer/
5. O. Witschger, J.F. Fabries: Particules ultrafines et santé au travail. 2. Sources et caractérisation de l'exposition. Hygiène et Sécurité au Travail, ND 2227, **199**, 37–54 (2005)
6. P.J. Anderson, J.D. Wilson, F.C. Hiller: Respiratory tract deposition of ultrafine particles in subjects with obstructive or restrictive lung disease. Chest **97**, 1115–1120 (1990)
7. J.W. Card, D.C. Zeldin, J.C. Bonner, E.R. Nestmann: Pulmonary applications and toxicity of engineered nanoparticles. Am. J. Physiol. Lung Cell. Mol. Physiol. **295**, L400–411 (2008)
8. D.C. Chalupa, P.E. Morrow, G. Oberdorster, M.J. Utell, M.W. Frampton: Ultrafine particle deposition in subjects with asthma. Environ. Health Perspect. **112**, 879–882 (2004)
9. A. Farkas, I. Balashazy, K. Szocs: Characterization of regional and local deposition of inhaled aerosol drugs in the respiratory system by computational fluid and particle dynamics methods. J. Aerosol. Med. **19**, 329–343 (2006)
10. G. Oberdörster, E. Oberdörster, J. Oberdörster: Nanotoxicology: An emerging discipline evolving from studies of ultrafine particles. Environ. Health Perspect. **113**, 823–839 (2005)
11. W.G. Kreyling, M. Semmler, F. Erbe, P. Mayer, S. Takenaka, H. Schulz, G. Oberdörster, A. Ziesenis: Translocation of ultrafine insoluble iridium particles from lung epithelium to extrapulmonary organs is size dependent but very low. J. Toxicol. Environ. Health A **65**, 1513–1530 (2002)
12. G. Oberdörster, J. Ferin, P.E. Morrow: Volumetric loading of alveolar macrophages (AM): A possible basis for diminished AM-mediated particle clearance. Exp. Lung Res. **18**, 87–104 (1992)
13. M. Geiser, M. Casaulta, B. Kupferschmid, H. Schulz, M. Semmler-Behnke, W. Kreyling: The role of macrophages in the clearance of inhaled ultrafine titanium dioxide particles. Am. J. Respir. Cell. Mol. Biol. **38**, 371–376 (2008)
14. A. des Rieux, V. Fievez, M. Garinot, Y.J. Schneider, V. Préat: Nanoparticles as potential oral delivery systems of proteins and vaccines: A mechanistic approach. J. Control. Release **116**, 1–27 (2006)
15. S.D. Conner, S.L. Schmid: Regulated portals of entry into the cell. Nature **422**, 37–44 (2003)
16. M.S. Arredouani, A. Palecanda, H. Koziel, Y.C. Huang, A. Imrich, T.H. Sulahian, Y.Y. Ning, Z. Yang, T. Pikkarainen, M. Sankala, S.O. Vargas, M. Takeya, K. Tryggvason, L. Kobzik: MARCO is the major binding receptor for unopsonized particles and bacteria on human alveolar macrophages. J. Immunol. **175**, 6058–6064 (2005)

17. C.A. Poland, R. Duffin, I. Kinloch, A. Maynard, W.A. Wallace, A. Seaton, V. Stone, S. Brown, W. Macnee, K. Donaldson: Carbon nanotubes introduced into the abdominal cavity of mice show asbestos-like pathogenicity in a pilot study. Nat. Nanotechnol. **3**, 423–428 (2008)
18. B. Rothen-Rutishauser, S. Schürch, P. Gehr: Interactions of particles with membranes. In K. Donaldson, P. Borm, *Particle Toxicology*, pp. 139–160. Informa Healthcare (2007)
19. L.K. Limbach, Y. Li, R.N. Grass, T.J. Brunner, M.A. Hintermann, M. Muller, D. Gunther, W.J. Stark: Oxide nanoparticle uptake in human lung fibroblasts: Effects of particle size, agglomeration, and diffusion at low concentrations. Environ. Sci. Technol. **39**, 9370–9376 (2005)
20. C. Mühlfeld, B. Rothen-Rutishauser, F. Blank, D. Vanhecke, M. Ochs, P. Gehr: Interactions of nanoparticles with pulmonary structures and cellular responses. Am. J. Physiol. Lung Cell. Mol. Physiol. **294**, L817–829 (2008)
21. W.G. Kreyling, M. Semmler, F. Erbe, P. Mayer, S. Takenaka, H. Schulz, G. Oberdörster, A. Ziesenis: Translocation of ultrafine insoluble iridium particles from lung epithelium to extrapulmonary organs is size dependent but very low. J. Toxicol. Environ. Health A **65**, 1513–1530 (2002)
22. M. Semmler-Behnke, S. Takenaka, S. Fertsch, A. Wenk, J. Seitz, P. Mayer, G. Oberdörster, W.G. Kreyling: Efficient elimination of inhaled nanoparticles from the alveolar region: Evidence for interstitial uptake and subsequent reentrainment onto airways epithelium. Environ. Health Perspect. **115**, 728–733 (2007)
23. M.T. Zhu, W.Y. Feng, Y. Wang, B. Wang, M. Wang, H. Ouyang, Y.L. Zhao, Z.F. Chai: Particokinetics and extrapulmonary translocation of intratracheally instilled ferric oxide nanoparticles in rats and the potential health risk assessment. Toxicol. Sci. **107**, 342–351 (2009)
24. A. Furuyama, S. Kanno, T. Kobayashi, S. Hirano: Extrapulmonary translocation of intratracheally instilled fine and ultrafine particles via direct and alveolar macrophage-associated routes. Arch. Toxicol. **83**, 429–437 (2008)
25. M. Semmler-Behnke, W.G. Kreyling, J. Lipka, S. Fertsch, A. Wenk, S. Takenaka, G. Schmid, W. Brandau: Biodistribution of 1.4- and 18-nm gold particles in rats. Small **4**, 2108–2111 (2008)
26. A. Nemmar, P.H.M. Hoet, B. Vanquickenborne, D. Dinsdale, M. Thomeer, M.F. Hoylaerts, H. Vanbilloen, L. Morelmans, B. Nemery: Passage of inhaled particles into the blood circulation in humans. Circulation **105**, 411–414 (2002)
27. N.L. Mills, N. Amin, S.D. Robinson, A. Anand, J. Davies, D. Patel, J.M. de la Fuente, F.R. Cassee, N.A. Boon, W. Macnee, A.M. Millar, K. Donaldson, D.E. Newby: Do inhaled carbon nanoparticles translocate directly into the circulation in humans? Am. J. Respir. Crit. Care Med. **173**, 426–431 (2006)
28. N.R. Yacobi, L. Demaio, J. Xie, S.F. Hamm-Alvarez, Z. Borok, K.J. Kim, E.D. Crandall: Polystyrene nanoparticle trafficking across alveolar epithelium. Nanomedicine **4**, 139–145 (2008)
29. J. Chen, M. Tan, A. Nemmar, W. Song, M. Dong, G. Zhang, Y. Li: Quantification of extrapulmonary translocation of intratracheal-instilled particles in vivo in rats: Effect of lipopolysaccharide. Toxicology **222**, 195–201 (2006)
30. J.J. Meiring, P.J. Borm, K. Bagate, M. Semmler, J. Seitz, S. Takenaka, W.G. Kreyling: The influence of hydrogen peroxide and histamine on lung permeability and translocation of iridium nanoparticles in the isolated perfused rat lung. Part. Fibre Toxicol. **2**, 3 (2005)

31. A. Elder, R. Gelein, V. Silva, T. Feikert, L. Opanashuk, J. Carter, R. Potter, A. Maynard, Y. Ito, J. Finkelstein, G. Oberdörster: Translocation of inhaled ultrafine manganese oxide particles to the central nervous system. Environ. Health Perspect. **114**, 1172–1178 (2006)
32. J. Wang, Y. Liu, F. Jiao, F. Lao, W. Li, Y. Gu, Y. Li, C. Ge, G. Zhou, B. Li, Y. Zhao, Z. Chai, C. Chen: Time-dependent translocation and potential impairment on central nervous system by intranasally instilled TiO_2 nanoparticles. Toxicology **254**, 82–90 (2008)
33. E. Sadauskas, H. Wallin, M. Stoltenberg, U. Vogel, P. Doering, A. Larsen, G. Danscher: Kupffer cells are central in the removal of nanoparticles from the organism. Part. Fibre Toxicol. **4**, 10 (2007)
34. H.R. Kim, K. Andrieux, S. Gil, M. Taverna, H. Chacun, D. Desmaële, F. Taran, D. Georgin, P. Couvreur: Translocation of poly(ethylene glycol-co-hexadecyl)cyanoacrylate nanoparticles into rat brain endothelial cells: Role of apolipoproteins in receptor-mediated endocytosis. Biomacromolecules **8**, 793–799 (2007)
35. G.J. Nohynek, J. Lademann, C. Ribaud, M.S. Roberts: Grey goo on the skin? Nanotechnology, cosmetic and sunscreen safety. Crit. Rev. Toxicol. **37**, 251–277 (2007)
36. J.P. Ryman-Rasmussen, J.E. Riviere, N.A. Monteiro-Riviere: Penetration of intact skin by quantum dots with diverse physicochemical properties. Toxicol. Sci. **91**, 159–165 (2006)
37. S.S. Tinkle, J.M. Antonini, B.A. Rich, J.R. Roberts, R. Salmen, K. DePree, E.J. Adkins: Skin as a route of exposure and sensitization in chronic beryllium disease. Environ. Health Perspect. **111**, 1202–1208 (2003)
38. J.G. Rouse, J. Yang, J.P. Ryman-Rasmussen, A.R. Barron, N.A. Monteiro-Riviere: Effects of mechanical flexion on the penetration of fullerene amino acid-derivatized peptide nanoparticles through skin. Nano Lett. **7**, 155–160 (2007)
39. L.W. Zhang, N.A. Monteiro-Riviere: Assessment of quantum dot penetration into intact, tape-stripped, abraded and flexed rat skin. Skin Pharmacol. Physiol. **21**, 166–180 (2008)
40. L.J. Mortensen, G. Oberdörster, A.P. Pentland, L.A. Delouise: In vivo skin penetration of quantum dot nanoparticles in the murine model: The effect of UVR. Nano Lett. **8**, 2779–2787 (2008)
41. P.H. Hoet, I. Brüske-Hohlfeld, O.V. Salata: Nanoparticles – known and unknown health risks. J. Nanobiotechnology **2**, 12 (2004)
42. E.G. Ragnarsson, I. Schoultz, E. Gullberg, A.H. Carlsson, F. Tafazoli, M. Lerm, K.E. Magnusson, J.D. Söderholm, P. Artursson: *Yersinia pseudotuberculosis* induces transcytosis of nanoparticles across human intestinal villus epithelium via invasin-dependent macropinocytosis. Lab. Invest. **88**, 1215–1226 (2008)
43. M. Longmire, P.L. Choyke, H. Kobayashi: Clearance properties of nano-sized particles and molecules as imaging agents: Considerations and caveats. Nanomed. **3**, 703–717 (2008)
44. W.I. Hagens, A.G. Oomen, W.H. de Jong, F.R. Cassee, A.J. Sips: What do we (need to) know about the kinetic properties of nanoparticles in the body? Regul. Toxicol. Pharmacol. **49**, 217–229 (2007)
45. B. Ballou, B.C. Lagerholm, L.A. Ernst, M.P. Bruchez, A.S. Waggoner: Noninvasive imaging of quantum dots in mice. Bioconjug. Chem. **15**, 79–86 (2004)

46. R.S. Yang, L.W. Chang, J.P. Wu, M.H. Tsai, H.J. Wang, Y.C. Kuo, T.K. Yeh, C.S. Yang, P. Lin: Persistent tissue kinetics and redistribution of nanoparticles, quantum dot 705, in mice: ICP-MS quantitative assessment. Environ. Health Perspect. **115**, 1339–1343 (2007)
47. Z. Chen, H. Chen, H. Meng, G. Xing, X. Gao, B. Sun, X. Shi, H. Yuan, C. Zhang, R. Liu, F. Zhao, Y. Zhao, X. Fang: Biodistribution and metabolic paths of silica coated CdSeS quantum dots. Toxicol. Appl. Pharmacol. **230**, 364–371 (2008)
48. H.S. Choi, W. Liu, P. Misra, E. Tanaka, J.P. Zimmer, B. Itty Ipe, M.G. Bawendi, J.V. Frangioni: Renal clearance of quantum dots. Nat. Biotechnol. **25**, 1165–1170 (2007)
49. M.L. Schipper, G. Iyer, A.L. Koh, Z. Cheng, Y. Ebenstein, A. Aharoni, S. Keren, L.A. Bentolila, J. Li, J. Rao, X. Chen, U. Banin, A.M. Wu, R. Sinclair, S. Weiss, S.S. Gambhir: Particle size, surface coating, and PEGylation influence the biodistribution of quantum dots in living mice. Small **5**, 126–134 (2009)
50. R. Singh, D. Pantarotto, L. Lacerda, G. Pastorin, C. Klumpp, M. Prato, A. Bianco, K. Kostarelos: Tissue biodistribution and blood clearance rates of intravenously administered carbon nanotube radiotracers. Proc. Natl. Acad. Sci. USA **103**, 3357–3362 (2006)
51. E. Fabian, R. Landsiedel, L. Ma-Hock, K. Wiench, W. Wohlleben, B. van Ravenzwaay: Tissue distribution and toxicity of intravenously administered titanium dioxide nanoparticles in rats. Arch. Toxicol. **82**, 151–157 (2008)
52. G. Sonavane, K. Tomoda, K. Makino: Biodistribution of colloidal gold nanoparticles after intravenous administration: Effect of particle size. Colloids Surf. B. Biointerfaces **66**, 274–280 (2008)
53. W.H. De Jong, W.I. Hagens, P. Krystek, M.C. Burger, A.J. Sips, R.E. Geertsma: Particle size-dependent organ distribution of gold nanoparticles after intravenous administration. Biomaterials **29**, 1912–1919 (2008)

3

Experimental Models in Nanotoxicology

Armelle Baeza-Squiban, Ghislaine Lacroix, and Frédéric Y. Bois

The aim of toxicology is to characterise the potentially harmful effects of solid, liquid, or gaseous substances for humans. Having evaluated the hazards, and given the level of exposure to the substance, we can then assess the risks.

The term 'nanotoxicology' was first used in the editorial of a scientific review in 2004 [1]. The authors explicitly recommended the creation of a new branch of toxicology called nanotoxicology, which would focus on the specific problems that might be raised by nanoparticles. Even then, it was expected that the particular physicochemical properties of nanomaterials might lead to novel toxic effects requiring special investigative methods.

In experimental toxicology, the underlying principle is always the same. Individuals, tissues, or cells are exposed to the substance under investigation, and the resulting response is compared with a control group treated under the same conditions, but without the exposure to the substance. As far as this approach is concerned, nanoparticles are no different from other more conventional substances like chemical products.

In some cases, humans are deliberately exposed. We then speak of controlled exposure. But for obvious ethical reasons, such experimentation is limited, and restricted to parameters accessible by non-invasive techniques.

The vast majority of toxicological studies appeal to animal models (in vivo toxicology) or cell models (in vitro toxicology). When extrapolating results to humans, these models clearly require some reflection. At this point, mathematical modelling (in silico toxicology) can sometimes be of use.

3.1 In Vivo Models

3.1.1 Different Animal Species Used

In France, toxicological studies use around 11% of the animals supplied for scientific activities (see Fig. 3.1 left). The animals most frequently used in toxicology are mammals (see Fig. 3.1 right). In principle, any mammal could be

P. Houdy et al. (eds.), *Nanoethics and Nanotoxicology*,
DOI 10.1007/978-3-642-20177-6_3, © Springer-Verlag Berlin Heidelberg 2011

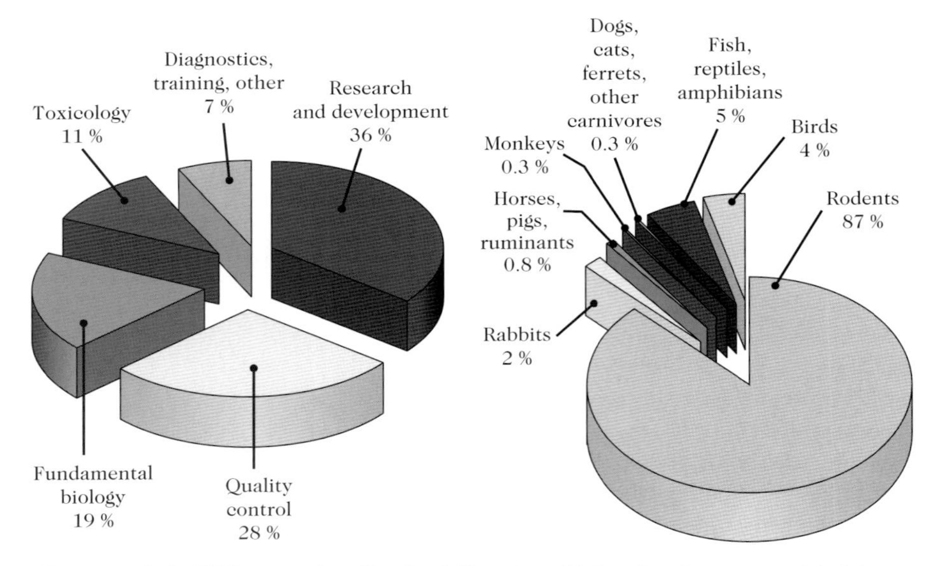

Fig. 3.1. *Left*: Different scientific disciplines in which animals were used in France in 2001 [2]. *Right*: Animal species used for scientific work in France in 2001 [2]

employed for toxicological assessment of a substance, but certain species are more widely solicited, or indeed more commonly permitted from a regulatory point of view. The main species are mice, rats, guinea pigs, rabbits, dogs, pigs, small ruminants, and primates. Rodents (mice and rats) nevertheless constitute the majority of species used in experimental toxicology, and almost all the species used in nanotoxicology. We shall therefore focus more specifically on these in the rest of the chapter.

3.1.2 Types of Animal Model

Animal models used in toxicology are either healthy animals, animals with some spontaneously occurring or deliberately introduced preexisting pathology, or genetically modified animals.

In the latter case, the genotype of the animal has been altered from the wild type, making it more sensitive to some kind of disease, e.g., cancer, or deficient in some molecule of particular biological significance, e.g., a protein. This type of model is used to study the impact of exposure to a toxic substance or a given pathogenic process or to assess the role of a given molecule in the biological response obtained after exposure to the toxic substance. Transgenic animals are still rarely used in nanotoxicology, which begins by studying the responses obtained from non-genetically modified models. However, mice deficient in apolipoprotein E have been used to study the role played by this protein in nanoparticle translocation across the blood–brain barrier [3]. More recently, mice deficient in the protein p53 (p53+/−), hence susceptible to cancer, have been used to study the capacity of carbon nanotubes to induce pleural mesotheliomas [4].

Animals models with preexisting pathologies have been used in toxicology for decades to evaluate the impact of toxic substances on vulnerable populations, e.g., people with cardiac conditions or suffering from bronchitis, emphysemas, or allergies. Various models have been used to assess in particular the impact of particles in air pollution that have been accused of exacerbating certain respiratory pathologies. These models are (non-exhaustive list):

- *Pulmonary Infection.* The aim here is to imitate a form of pneumonia. The healthy animal is inoculated with bacteria such as *Streptococcus pneumoniae*, *Haemophilus influenzae*, or *Pseudomonas aeruginosa*, or viruses such as *Influenza* or respiratory syncytial virus, and often treated with immunosuppressants to avoid rapid elimination of the pathogens [5].
- *Emphysemas.* These pathologies lead to destruction of the pulmonary parenchyma. The most widespread method for imitating this infection in animals is to instil a proteolytic enzyme such as elastase, but a similar result can be obtained by exposing animals to cigarette smoke [6].
- *Pulmonary Fibrosis.* This pathology is characterised by excessive deposition of collagen in the pulmonary parenchyma. The best characterised animal model is obtained by intratracheal instillation of bleomycin, but silica can also be used, or the animals irradiated [7].
- *Chronic Bronchitis.* This pathology is characterised by an increase in the size of the mucous glands in the airways. It is related to chronic irritation of the airways by inhaled substances, such as cigarette smoke, pollution, or occupational exposure. Experimentally, this disease is obtained with animals after exposing them to SO_2, cigarette smoke, endotoxins, enzymes, or adrenergic or cholinergic substances [8].
- *Asthma.* This is a chronic inflammatory disease affecting the whole of the lungs. This typically human pathology displays several characteristics such as bronchoconstriction due to sensitivity to the IgE-mediated antigen, an increase in airway resistance, inflammation (eosinophilia), accumulation of mucus, alteration of mucociliary clearance, and so on. No single animal model can simulate the whole complexity of this affliction. The mouse, which is the most widely used model, is a good tool for studying humoral and inflammatory reactions, but exhibits very little bronchoconstriction, unlike the guinea pig. In rats, the Brown Norway is most commonly used because it develops significant eosinophilia [9].

At the present time, most studies in nanotoxicology concern the respiratory system. As mentioned earlier, rats and mice are the two most commonly used species. Some of the models described above are beginning to be used. For example, several studies have considered the capacity of carbon-containing nanoparticles to induce an adjuvant effect on the respiratory allergy in ovalbumin-sensitized mice [10–12]. Other groups have used bleomycin-treated rats to induce pulmonary fibrosis and thereby investigate the impact of manufactured or combustion-derived nanoparticles [13].

3.1.3 Types of Exposure

A toxic substance can enter the organism by the three conventional routes, namely, oral, respiratory, and cutaneous. As far as nanoparticles are concerned, a fourth route should be added, namely, the parenteral route, i.e., by injection. The latter is rather particular, since it is effectively a form of voluntary exposure, with medical objectives (nanomedicine). It is not therefore part of conventional environmental toxicology, where only unintentional exposure is considered. However, this parenteral route can be implemented for mechanistic studies to simulate pulmonary or intestinal barrier crossing, for example.

Exposure by Respiratory Route

This is particularly relevant for nanoparticles since, owing to their very small dimensions, they enter easily into suspension and can penetrate deep into the lungs. At the present time, this is one of the exposure routes that has received the most attention in the case of nanoparticles.

Exposure by inhalation best imitates naturally occurring pulmonary exposure [14]. It is nevertheless a delicate matter to implement. Nanoparticles usually occur in the form of a dry powder. Two techniques can generally be applied in this case: mechanical dispersion and encapsulation. In the first, the powder is mechanically set in suspension in an air flow, e.g., using small rotating brushes. This technique has the advantage of generating large mass concentrations. However, it is often hard to avoid the formation of aggregates comprising several elementary particles. Encapsulation consists in generating a polydisperse aerosol from a liquid, whose physical properties such as viscosity and surface tension are chosen so as to suitably adjust the average size of the liquid droplets. One advantage with this process lies in the fact that these particles are relatively easy to produce. If the solid particles that have to form the aerosol are insoluble in the chosen liquid and if the size of the droplets is suitably matched to the size of the solid particles, this technique can be used to form an aerosol of solid particles once the liquid has evaporated. In some cases, encapsulation can produce suspensions in which the particles agglomerate only slightly, or not at all. The resulting concentrations are generally lower than with direct dispersion. Most work published so far has used mechanical dispersion.

In every case, the material to be tested must be available in sufficient amounts to generate a high enough concentration for a long enough time to evaluate its toxicity. Furthermore, the process used to generate the aerosol must not significantly alter the physicochemical characteristics of the substance under assessment. In any case, the respiratory system of the animal subject must be monitored by careful measurement (metrology). This is particularly problematic for nanoparticles, since a lot of the equipment currently available commercially for analysing the granulometry, surface properties, etc., has only been validated for particles with dimensions greater than 100 nm. Nanometric particles often exhibit high surface reactivity and are particularly sensitive to electrostatic effects and agglomeration. Any alterations like to occur during generation must be carefully recorded and monitored, if there is no way of controlling them, otherwise it will be impossible to reach a correct interpretation of the observed results, or make comparisons with results obtained elsewhere.

Apart from these purely technical considerations, the safety of all those handling these materials must be treated with the utmost caution. The toxicity of nanoparticles is still largely unknown, and the precautionary principle must apply. In the absence of data, they must be treated as highly toxic, so the experimenter must be protected from all exposure. At the present time, there is no standardised system for inhalation exposure to nanoparticles. Some systems are commercially available, but most are more or less custom built using whatever techniques happen to lie to hand. As in conventional toxicology, the animals can be exposed in nose-only or whole-body systems. In the first case, only the nose of the animal comes into contact with the aerosol. This technique is more stressful for the animal, which is immobilised in a tube during the whole exposure period. However, it has the advantage that there is no exposure to the toxic substance by any other route than inhalation [14, 15]. In a whole-body system, the particles can be adsorbed onto the fur and subsequently swallowed by licking.

Intratracheal instillation is a simpler method to implement. It imitates inhalation exposure [16]. In this case, a known amount of the substance in question is deposited directly in the airways by means of a nozzle inserted in the trachea. With this technique, the amount of substance actually introduced into the lungs is perfectly controlled, in contrast with the inhalation methods just described. There is also a lesser risk of exposure for the experimenter. On the other hand, the method is less physiological than the last and requires the nanoparticles to be suspended in a liquid. Depending on the kind of nanoparticle, and in particular its hydrophobicity, this suspension may or may not be homogeneous, and this reduces the reproducibility and repeatability of the experiment. With hydrophobic nanoparticles, such as carbon nanotubes, a substance is often added to improve the uniformity of the suspensions, e.g., surfactant, Tween 80, serum, albumin, etc. [17, 18]. But then one must make absolutely sure that this additive itself has negligible toxicity, and that its presence does not significantly modify the toxicity of the nanoparticles under investigation, e.g., by coating.

It is very important to take into account the phenomenon of agglomeration when intratracheal instillation is used, perhaps even more so than for inhalation, where natural filtration phenomena come into play to block the larger particles. (Those measuring more than 10 μm are arrested by the nasal airways.) Indeed, instillation forces even non-respirable particles or clusters of particles, i.e., bigger than 10 μm, to enter the lungs. But according to certain studies, especially concerning carbon nanotubes which easily form agglomerates, it turns out that the biological response may depend on the form of administration (inhalation or instillation) and hence on the size of the agglomerates [19].

Exposure by Oral Route

This route is even easier to exploit. For nanoparticles, it is mainly implemented by force feeding through a catheter. Once again, the nanoparticles must be in suspension, so the limitations mentioned above for intratracheal exposure are relevant here, even though tube-feeding is closer to the reality than intratracheal instillation is to the reality of inhalation.

Exposure by Cutaneous Route

This is done by applying the nanoparticles to the skin. Since certain cosmetic products contain nanoparticles, e.g., some sunscreen creams contain titanium dioxide nanoparticles, studies have focused on the cutaneous effect of nanoparticles in a formulation (emulsion, cream). This also makes it easier to implement. However, occupational exposure must also be considered, where workers may be exposed to 'dry' nanoparticles.

3.1.4 Targets

Once the nanoparticles have entered the organism, two issues come under investigation:

1. The fate of the nanoparticles in the organism, particularly with regard to their biodistribution, translocation (transfers from one biological compartment to another), and elimination.
2. Their toxicity.

These two aspects are of course complementary and often assessed simultaneously.

The best way to ascertain whether the particles have actually entered the organism is to visualise them using imaging techniques like optical, confocal, or electron microscopy, for example. However, this can be difficult to implement, especially when there are not many particles, because the probability of not seeing them is quite high. By definition, microscopy techniques can visualise fields of varying extent, but only in two dimensions. (This is less true for confocal microscopy, which can explore the thickness of an organ.) It is indeed rather like looking for a needle in a haystack! If one has to increase the field to improve the accuracy of the search, time soon becomes a limiting factor.

Some nanoparticles have specific properties making them easy to detect. This is the case for nanoparticles possessing magnetic susceptibility, such as magnetite (Fe_3O_4), which can be detected by nuclear magnetic resonance (NMR) [20]. Some groups have used the autofluorescence property of single-wall carbon nanotubes to monitor their progress through the organism [21]. Others have used Raman spectroscopy to detect carbon nanotubes in the organs [22, 23]. Another common option is to assay the majority component of the nanoparticle, e.g., titanium for TiO_2, but one has to ensure that the particle is insoluble and that one does not assay the solubilised molecular constituent.

External tracers, e.g., fluorescent, radioactive, etc., can be used for nanoparticles that are difficult to detect. In this case, one must ensure that the presence of the tracer does not significantly alter the intrinsic behaviour of the nanoparticle, and that it does indeed remain firmly fixed on the particles when it has entered the organism. The distribution of carbon nanoparticles radio-tagged with technetium has been studied in humans after inhalation [24]. The authors concluded that the nanoparticles pass quickly into the blood circulation, but their results were subsequently questioned on the grounds that the tracer had at least in part detached itself from the nanoparticle [25]. An interesting alternative is to use an impurity of the nanoparticle. This has been done with carbon nanotubes, for example [26]. Since these require metal catalysts for their fabrication, they are contaminated, e.g., by iron or nickel. Whenever these contaminants are not labile, they provide a good way of detecting the nanotubes in the organism. The nanoparticle or its tracer are detected by conventional chemical assay methods such as ICP mass spectrometry or optical ICP (inductively coupled plasma) [26], or by imaging techniques such as NMR [27].

The toxic effects of nanoparticles are currently assessed in conventional ways (inflammation, oxidative stress, effects on the genome, etc.), and will be discussed in later chapters.

3.2 In Vitro Models

There has been a considerable effort to develop in vitro models in toxicology in the context of the 3 R's campaign, i.e., replacement, reduction, refinement, which results from an ethical imperative to find alternatives to animal experimentation. While they allow large scale screening of the toxicity of molecules prior to animal experimentation, their strong point is that they are choice models for studying the action mechanism of the toxic substance directly on its target. The absence of humoral, metabolic, and neural interference facilitates the analysis of molecular and cellular effects. In vitro methods are simpler, quicker to implement, and less expensive than animal studies, but they cannot reproduce the full complexity of the organism, and do not take into account the toxicokinetic phase. In addition, they are mainly developed to carry out short term studies. A fully satisfactory characterisation of the effects of a toxic substance must therefore combine in vitro studies, which identify action mechanisms, with in vivo studies, which check the relevance of data acquired in vitro and establish dose–response relationships.

The basic principle of any in vitro culture is to keep alive, outside the organism, an organ, a tissue, or cells not organised as part of a tissue but able to divide and express a metabolism and specific functions in vitro. So there are in vitro models on each level of biological organisation, from organs (perfused isolated organ, e.g., a lung), to tissues (e.g., tracheal rings, organ sections, excised skin fragments, etc.), right down to a single cell, with a concomitant reduction in complexity with regard to the diversity of the cell populations, an improvement in reproducibility, and better control of environmental conditions. These different systems are selected depending on the question being investigated and the analytical methods used, which require different amounts of biological material.

3.2.1 Cell Cultures

Different Types of Culture

Cell cultures are the most widely used in vitro methods. There are two main types of culture:

- *Primary Cultures.* These derive from a tissue sample taken from a human or animal source. The advantage with this type of culture is that it provides the best representation of the original tissue. However, depending on the type of cell, the culture methods used cannot always maintain a satisfactory state of differentiation among these cells, and they then lose their characteristics. Moreover, their lifetime in culture is limited. Finally, the supply of tissues, notably human tissues, may be restricted and it may raise ethical problems.
- *Cell Lines.* These have the special feature of being immortal. They derive from samples of tumour cells, or they are obtained by transfection of a gene allowing them to divide indefinitely. The advantage with these cell lines is their permanent and unlimited availability, although the acquisition of such proliferative properties is often accompanied by the loss of certain specific functions and a modified karyotype.

The specific features of the response from a given organ, cell type, or species can be investigated using cells from different organs, from different sources (epithelial, conjunctive, muscular, neural, etc.), and from different species, and in particular, humans.

Thanks to progress in molecular biology, cells can be manipulated so as to cause gene extinction, or indeed gene overexpression, and hence investigate the role played by the target gene in the action mechanism of a toxic substance.

Culture Methods

The simplest cell culture methods involve growing cells in a sterile plastic box, supplying them with a suitable culture medium, and maintaining them under suitable conditions of temperature and humidity. The conditions in this form of culture are sometimes quite different from the real living conditions of cells in the organism, and more elaborate systems have been developed in order to simulate in vivo conditions and favour the expression of differentiated characters. For example, many epithelial cells located at the interface with the external medium can be cultivated in two-compartment chambers in such a way as to expose either their luminal side (facing outward) or their basal side (facing inward) with the toxic substance, depending on the uptake route of the substance.

Furthermore, these setups can be further refined by making co-cultures, i.e., by associating several types of cells in order to investigate the role of their interactions in toxic effects. For example, a triculture model has been used to study the translocation of nanoparticles. Alveolar epithelial cells were grown at confluence on a porous membrane lining the bottom of a culture insert to reconstitute an epithelial barrier. Antigen presenting dendritic cells were placed on the other face of the porous membrane and alveolar macrophages were added on the epithelial cells. This model was used to investigate the relative abilities of these different cell types to phagocytize fluorescent nanoparticles, revealing the extensions produced by the dendritic cells to capture the nanoparticles on the luminal side of the epithelium [28].

3.2.2 In Vitro Methods in Regulatory Toxicology

While cell cultures are widely used in mechanistic toxicology, their use for risk assessment in the regulatory context is still rather limited. Among the in vitro methods that have now been validated [29, 30], many concern the skin, since safety assessments of cosmetic products in Europe can no longer appeal to animal experimentation. These methods aim to evaluate the general toxicity, e.g., absorption, phototoxicity, irritation, corrosivity, and genotoxicity.

These tests can only assess acute toxicity, and there is as yet no validated method for evaluating long term toxicity effects. In addition, the validation of these methods did not include particulate toxic substances, and it is unlikely that they could be directly transposed to nanoparticles (see Sect. 3.2.4).

3.2.3 In Vitro Methods for Assessing Nanoparticle Toxicity

In the emerging field of nanotoxicology, in vitro methods will probably help us to improve our understanding of several issues:

- Uptake and transfer of nanoparticles across physiological barriers.
- Cytotoxicity and cellular effects.

- Induction of oxidative stress, considered a key feature in the toxicity mechanisms of nanoparticles [31].
- Mutagenicity and genotoxicity of nanoparticles.

Internalisation and Translocation of Nanoparticles

The ability of nanoparticles to cross physiological barriers is a crucial issue (see Chap. 2), and in vitro models of the epithelial barrier should help us to understand the mechanisms involved and also to make quantitative assessments of this barrier crossing, depending on the characteristics of the nanoparticle.

Concerning skin absorption, there are models consisting of excised human skin maintained in a diffusion chamber (OECD test guideline 428) with which one can make quantitative assessments of the nanoparticles retained by the skin and those entering the perfusate in which the basal part of the skin fragment is bathed. Commercially reconstituted human skin models (Episkin^{TM}) can also be used.

Concerning the lungs, the bronchial and alveolar epithelial barrier can be reproduced by performing cultures in two-compartment chambers. Human bronchial (16HBE, BEAS-2B, Calu-3) and alveolar (A549) cell lines exist, but they do not exhibit all the features of in vivo cells, in particular, their ability to form a perfect junctional epithelium in vitro. With regard to bronchial cells, primary cultures can be made in which it is also possible to modulate the state of differentiation in such a way as to imitate normal or pathological conditions existing in vivo [32].

Concerning the intestine, the most widely used cells are from the human Caco-2 cell line derived from an adenocarcinoma of the colon.

The discovery that nanoparticles can enter the bloodstream requires the development of models for the blood–brain barrier [33] and blood–placenta barrier, to assess the risk of nanoparticles translocating from the blood to the brain or the placenta.

Whatever the model, assessment of nanoparticle uptake and translocation also requires microscopy techniques (electron or confocal microscopy if the nanoparticles are fluorescent) to locate the nanoparticles, and sensitive analytical (spectroscopic) techniques to quantify nanoparticles used in small doses. On the level of the cell, these microscopy techniques are a determining factor when undertaking studies of nanoparticle internalisation mechanisms (see Chap. 2) according to the cell type and the physicochemical properties of the particles.

Cytotoxicity and Cell Damage

There is a wide range of tests available to assess cytotoxicity, and they can be applied to both cell lines and primary cultures in which one chooses the origin and species according to the toxic substance under scrutiny. Cytotoxicity can be investigated by examining the integrity of the membrane, metabolic activity, and apoptosis. The most widely used methods are:

- Membrane damage measurements based on:
  - Exclusion of dyes by living cells, e.g., trypan blue or propidium iodide, which can only enter dead cells and which can be counted under the microscope or by flow cytometry, respectively.
  - Release of cytosolic enzymes, e.g., lactate dehydrogenase (LDH), whose activity is then measured in the culture medium of the damaged cells.
  - Inclusion of dyes by living cells, e.g., neutral red, which is retained in the lysosomes of healthy cells, or calcein AM, which is cleaved to yield a fluorescent product within living cells.
- Observation of metabolic changes evaluated by:
  - Measuring mitochondrial activity, e.g., using MTT (a tetrazolium salt), which is reduced to coloured formazan in the mitochondria of viable cells, or Alamar blue, resazurin reduced to fluorescent resorufin.
  - Measuring the level of adenosine triphosphate (ATP).
- Evaluating apoptosis, a multistage process of controlled death, by:
  - Measuring caspase activity.
  - Labelling phosphatidylserine residues on the extracellular side of apoptotic cells with annexin V.
  - The TUNEL assay which assesses DNA fragmentation.

Induction of Oxidative Stress

Depending on their extent, modifications in the intracellular redox state are involved in modulating the expression of genes for antioxidant defence and pro-inflammatory response, but they can lead to cell death. A lot of research has already shown that the toxicity of nanoparticles is largely due to their ability to generate oxidative stress [31].

Oxidative stress can be assessed by measuring the production of reactive oxygen species (ROS) using electron paramagnetic resonance or fluorescent probes more or less specific to some ROS, but which inform us about the intracellular redox status of exposed cells [34].

It can also be assessed indirectly by evaluating its molecular and cellular consequences. For example, one can measure the ratio of reduced glutathione to oxidised glutathione, the activity of antioxidant enzymes, or the level of lipid peroxidation, or one can look for oxidative DNA lesions (8-OH-deoxyguanosine).

Mutagenicity and Genotoxicity

There are a certain number of in vitro tests on mammalian cells which can assess genotoxicity, such as the chromosomal aberration assay (OECD 473), the gene mutation assay (OECD 476), the micronucleus assay (OECD 487), the sister chromatid exchange assay (OECD 479), and the DNA repair (UDS) assay (OECD 482). However, these protocols must be adapted to the nanoparticle problem situation, especially with regard to kinetics (exposure time), leaving the nanoparticles sufficient time to reach the nucleus.

Cells that actively proliferate in vivo are the most sensitive to genotoxic effects associated with carcinogenic processes, since mutations only become established in proliferating cells. Cells with these characteristics in the main organs exposed to nanoparticles are type II pneumocytes for the lungs, keratinocytes in the basal layer of the epidermis for the skin, and intestinal epithelial cells. Existing tests should be applied to these cell types.

3.2.4 Specific Problems for Assessing in Vitro Toxicity of Nanoparticles

Interaction with the Culture Medium

Cells are exposed to nanoparticles by suspending them in a culture medium, possibly adding serum or a serum substitute. The complex composition of these media results in adsorption of molecules, especially proteins, at their surface. One study carried out with polystyrene nanoparticles [35], differing through the presence of different molecular surface groups, showed that serum proteins adsorb very quickly (in a few seconds), but that the nature of the adsorbed proteins can then evolve by the Vroman effect, i.e., proteins initially adsorbed because they have high diffusion rates or because they establish simple interactions are subsequently replaced by others with stronger affinity for the nanoparticle. This adsorption leads to an increase in the size of the nanoparticle and a modification of its zeta potential. The amount of adsorbed proteins can vary, depending on the molecular groups carried by the nanoparticles, but their identity cannot. Thus the adsorption of serum proteins decreases when the nanoparticles carry neutral groups (CH_3 and polyethylene glycol), suggesting the intervention of electrostatic interactions [35]. This protein corona which forms around nanoparticles can alter the cell response. It has been shown that, adding serum to the culture medium forestalls the cytotoxicity of carbon and TiO_2 nanoparticles with regard to bronchial epithelial cells [36]. In addition, if growth factors are adsorbed by the nanoparticles, this can result in indirect cytotoxicity due to depletion of nutrients in the medium [37].

When nanoparticles are suspended in biological media, this also causes them to aggregate, suggesting that this may modify their toxicity. The aggregation problem is not restricted to in vitro studies, but also occurs in vivo, depending on the form of administration (see above). However, a certain number of in vitro studies have shown that, despite their aggregation, the nanoparticles have different effects to those caused by micrometric particles of the same kind [38].

Interference When Assessing Biological Effects

The surface properties of nanoparticles may also interfere with methods for assessing induced biological effects. For example, some reagents used to evaluate cell viability, e.g., MTT,[1] the substrate used to measure LDH activity, neutral red, etc., adsorb onto nanoparticles, resulting in erroneous cytotoxicity assessments [39]. The presence of nanoparticles in the medium in which the absorbance measurement is made can attenuate the signal [40]. It is important to check that the optical properties of the nanoparticles do not interfere with the detection system being used, e.g., absorbance, fluorescence, diffraction of light, etc.

Proteins released by cells when they are exposed to nanoparticles, e.g., cytokines, cannot be correctly quantified in the culture medium, because they adsorb onto the nanoparticles, thereby masking the effect under investigation [36].

The importance of interference, whether it be the behaviour of the nanoparticles in the culture medium of the exposed cells or their interaction with the substrates or parameters measured to assess cytotoxicity, is directly related to the physicochemical characteristics of the nanoparticles, i.e., size, electrical charge, hydrophobicity, etc.

Whatever toxicological assay is intended, it is important to understand the physicochemical characteristics of the nanoparticles, to check that there is no interference between the nanoparticles and the measured parameter by carrying out a suitable series of controls, to use several methods to evaluate each effect, and to include standard reference particles which are not yet available.

3.3 Predicting Penetration and Fate of Nanoparticles in the Body

It is already known that some, and maybe all, nanoparticles can enter our body if we are exposed to them (see Chap. 2). This is clear for inhalation exposure: in all experiments carried out by inhalation, some of the inhaled particles are deposited in the lungs. The problem here is just to find out what happens to them subsequently. It is less obvious for exposure by ingestion or on the skin. In the latter case, the epithelial barrier seems relatively effective, unless it is damaged, or the nanoparticle in question has been specifically designed to cross this barrier, e.g., in medical applications. But even when we succeed in measuring the penetration of any given nanoparticle, its subsequent fate is another important matter. Once deposited in the lungs, it may be that some fraction of the nanoparticles comes back up toward the gastrointestinal tract. Would that be good news? When we observe a very slight diffusion toward

[1] 3-(4,5-dimethylthiazol-2-yl)-2,5-diphenyltetrazolium bromide.

other organs, does this augur well for the health of the pulmonary tissue? If some of the inhaled or ingested nanoparticles should reach the blood or the lymph, will they be quickly eliminated from the body, or will they accumulate in certain organs and damage them?

Questions like these are not specific to nanomaterials, since they are already raised by all the many substances to which we are exposed, and in particular, drugs. There are tools, and even a whole scientific discipline, devoted to solving just this kind of problem: pharmacokinetics.

3.3.1 Pharmacokinetics

Pharmacokinetics, or toxicokinetics in the case of non-medical substances, studies the fate of products in the body. The corresponding process is described as the pharmacokinetic process, and it includes four simultaneous phases: absorption, distribution, metabolism, and excretion (ADME for short). Each of these phases may exhibit specificities intrinsic to nanomaterials.

Absorption

The nanomaterial may cross the biological membranes separating the absorption site, e.g., the lungs in the case of inhalation, from the blood. In this case, it enters what is known as the systemic circulation and from there can be distributed throughout the whole body. Otherwise, the product may accumulate at the absorption site, which may raise problems of toxicity at some point. For inhaled nanoparticles, a particular absorption process, poorly understood from a quantitative standpoint, is phagocytosis by macrophages, themselves able to migrate within body tissue, carrying with them the particles they have internalised.

Distribution

Chemical substances that reach the blood circulation may bind more or less strongly, and reversibly, with the plasma proteins, e.g., lipoproteins, albumin, globulins, etc. The prevalence of this phenomenon for nanomaterials remains almost unknown at the present time.

Depending on their physicochemical and biochemical properties, nanomaterials may then accumulate in certain organs or tissues, where the blood, the lymph, or macrophages carry them. Lipophilic substances thus accumulate in fats. In the same way, the affinity of the nanomaterial for the different tissues is probably a determining factor in establishing their distribution, but such affinities are poorly understood, and in any case specific to each material.

Metabolism

Many chemical substances can be transform by enzymes in the organism, especially in the liver, an organ that specialises in such transformations (especially of nutrients). The metabolites produced in this way may subsequently exhibit no toxic or other activity (metabolism is in this case an elimination route), but they may also be more toxic than the initial product. We then speak of metabolic activation. The possible metabolic transformation of nanomaterials, together with its consequences for their toxicity, is poorly understood and deserves to be given more attention.

Excretion

The substances absorbed by the organism or their metabolites are used either as sources of energy or as structural components for the body, otherwise eliminated from the organism by excretion. There are several excretion routes, the main ones being the urinary route and the biliary route (which leads to feces). Many other organs can contribute to elimination: the lungs (by exhalation), the skin (by perspiration, desquamation, and accumulation in the integumentary system, e.g., hair, nails, etc.), the salivary glands, lachrymal glands, mammary glands, and so on.

The kidney is the main organ concerned with direct excretion of substances via the blood. The products are excreted by simple glomerular filtration or by active tubular secretion (for cationic forms). There may also be a phenomenon of tubular reabsorption of previously excreted substances. While the distribution and metabolic transformation of nanomaterials are poorly understood, their excretion is likewise, and a great deal of work remains to be done in this area.

In short, the toxicokinetics of nanomaterials is still poorly understood at the time of publication of this book. It must be said that the same is true for many other chemical substances, but the concern inspired by the possible toxicity of nanomaterials is unlikely to be allayed by such a lack of knowledge.

3.3.2 Pharmacokinetic Models

The experimental methods described in the remainder of this chapter aim to identify, among other things, the temporal evolution of the concentrations of nanomaterials in different parts of the body. Mathematical models of the same phenomena are complementary, since they provide ways to improve the interpretation of experimental results and extrapolate them to unobserved, or even unobservable conditions, e.g., pregnant women.

A distinction is made between conventional compartmental models and physiologically-based pharmacokinetic models.

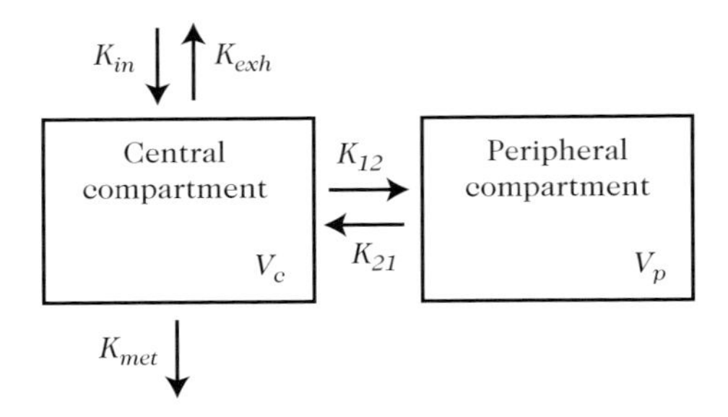

Fig. 3.2. Schematic of a compartmental pharmacokinetic model. Parameters are the volumes V and the transfer rate constants K

Conventional Multi-Compartmental Models

A multi-compartmental model is a mathematical model to describe the transport of materials between the compartments of a system. Each compartment is assumed to represent a homogeneous region of space. For example, in a pharmacokinetic model, the compartments may represent the different parts of the body within which the concentrations of a chemical substance are assumed to be equal. It is also required that the distribution of the substance within each compartment can be treated as instantaneous. Typically, the amount of material in each compartment is a state variable, characterising the state of the system at each moment of time, whose temporal evolution is governed by an explicit differential equation. Figure 3.2 shows a two-compartment model that has been successfully used to describe the kinetics of butadiene elimination in humans [41].

Physiologically-Based Pharmacokinetic Models

One class of multi-compartmental models is particularly interesting when describing complex phenomena, extrapolating, or making predictions applicable to risk assessment. These are the physiologically-based pharmacokinetic models (PBPK).

PBPK models are mechanistic mathematical descriptions of anatomical, physiological, physical, and chemical phenomena involved in the absorption, distribution, and so on, of any kind of substance, and hence by extension, nanomaterials. These models are nevertheless always a simplification of the real situation, often involving a certain level of empirical input, but their range of validity is usually much broader than that of the conventional pharmacokinetic models.

PBPK models attempt to reproduce the anatomical and physiological structure of the body. The compartments correspond to well defined organs or tissues, interconnected by flows of blood or lymph (and in some cases, diffusion phenomena). A set of differential equations can always be specified.

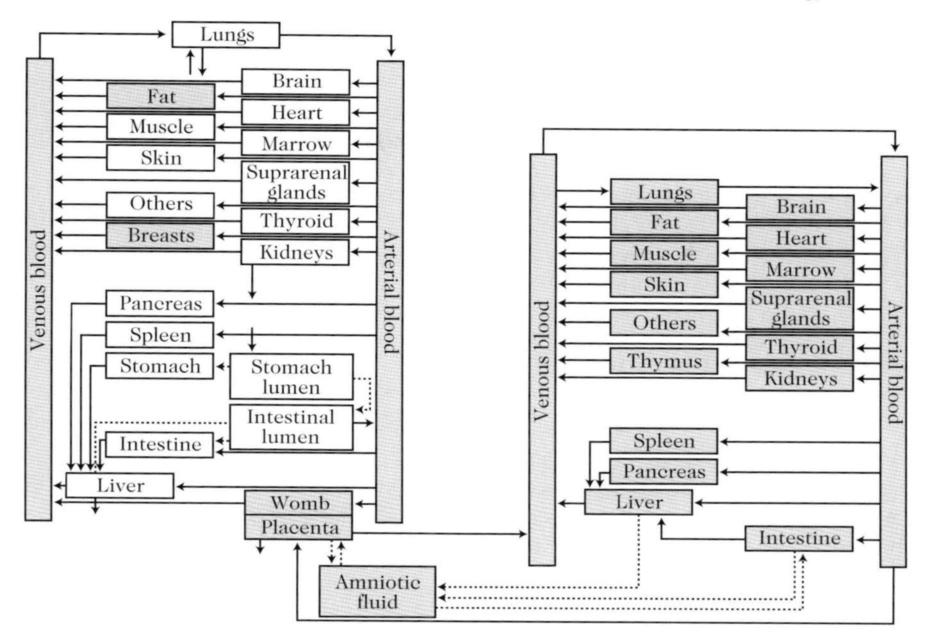

Fig. 3.3. Schematic of a physiologically-based pharmacokinetic model applicable to a pregnant woman and fetus

Their parameters then represent the blood flow rates, the pulmonary ventilation rate, the volumes of the organs, etc. Information about each of these parameters is available in the scientific literature, so it is easier to fix their values and hence give the model greater predictivity and a particular capacity for extrapolation. An example of a generic physiological model that can be applied to many chemical substances is shown in Fig. 3.3.

It is useful to remember that the first pharmacokinetic model described in the scientific literature was in fact a PBPK model [42]. However, it led to calculations that could not be carried out at the time. This is why simpler models were set up, now called conventional models, for which analytic solutions were available. With the development of fast calculators and numerical integration algorithms in the 1970s, PBPK models came back into the fore [43].

These models can be used for purely predictive applications. In the first place, they provide a way of synthesising what may appear to be disparate forms of data, obtained by physicochemical or biochemical experiments, toxicological or pharmacological studies in vitro or in vivo, and so on. They can also be used to determine internal concentrations of administered products and their metabolites, in particular, at their site of action, whether their effects are therapeutic or toxic. Finally, they can also help to interpolate or extrapolate data acquired in different contexts:

- *Dose.* For example, from high concentrations often used in the lab to the low levels encountered in the environment.
- *Administration Route.* For example, from inhalation to ingestion.
- *Exposure Time.* For example, from discontinuous exposure to continuous exposure, or from single to repeated exposure.
- *Species.* For example, to transpose from rodent to human, just before administering a new drug to volunteers for the first time, in the framework of clinical trials, or when experimentation on humans is unacceptable.
- *Individuals.* For example, from men to women, from adults to children, from non-gestating to gestating women, etc.

Some of these extrapolations are parametric, in the sense that only the input values or the parameters of the model need to be modified to make the extrapolation. This is usually the case when extrapolating dose or duration. Others are non-parametric, in the sense that the very structure of the model must be changed, e.g., to transpose to a pregnant woman, one must include equations describing the fetus. Other uses such as statistical inference are also possible with the development of Bayesian methods [44, 45], as described in the example below.

3.3.3 Examples of Applications to Nanomaterials

The real meaning, in terms of risk to human health, of the many studies so far published or in progress on the in vitro effects of nanomaterials remains an open question. To answer it, we need to understand the pharmacokinetics of these nanomaterials, to determine in particular the extent to which they are able to cross the barriers in the body. In the absence of a general rule, each nanomaterial has to treated as a special case. And it will probably only be by investigating a large enough number of different cases that general rules will eventually emerge.

As an example, we have used a PBPK model to analyse data obtained in Louvain [24] from volunteers exposed to carbon nanoparticles tagged with ^{99m}technetium (Tc), using the Technegas process, which serves to explore the respiratory function in a clinical context. Nemmar et al. [24] concluded from their data that the nanoparticles were able to transfer from the lungs to the blood. However, other groups contested these results, on the grounds that their own observations could be explained by the presence of free technetium, not bound to nanoparticles [25]. It seemed to us useful to reexamine the data of Nemmar et al. using a physiologically-based model giving a finer description of the phenomena coming into play [46].

Nemmar et al. obtained data on the Technegas distribution in five healthy volunteers aged 24–47 years. Technegas is an aerosol of carbon nanoparticles tagged with ^{99m}Tc. Particle sizes are in the range 5–10 nm. The volunteers were exposed to about 100 MBq of Technegas. The radioactivity in the blood was measured 1, 5, 10, 20, 30, 45, and 60 min after inhalation of the Technegas.

Images of the radioactivity distribution in the body were obtained using a gamma camera after 5, 10, 20, 30, and 45 min. The relative intensity of the radioactivity (compared with the intensity measured in the liver after 5 min) is specified in three regions of interest, namely, the liver, stomach contents, and urine in the bladder, indicated on the images.

The basic structure of our PBPK model for humans is shown in Fig. 3.3. The model attempts a realistic, although simplified, description of the mechanisms underlying absorption, distribution, and elimination of technetium-tagged nanoparticles and free technetium in the body. It subdivides the human body into 24 compartments. Initial absorption is by inhalation. The particles are supposed to deposit themselves in the upper airways and the lungs. One part is quickly transferred to the stomach by deglutition. We considered three ^{99m}Tc fractions: the first bound to small particles able to transfer from the lungs to the blood, the second bound to large particles, unable to cross the aveolo-capillary barrier, and the third free, i.e., not bound to nanoparticles. The simultaneous distribution of the three fractions was then modelled. Once in the blood, the small particles and free ^{99m}Tc diffuse into the various compartments, but not the brain, which is protected by the blood–brain barrier, as can be seen from the data of Nemmar et al. The free ^{99m}Tc is assumed to be eliminated by filtration in the kidney. During the time of the experiment (60 min), renal elimination of ^{99m}Tc bound to particles is assumed negligible, since only free ^{99m}Tc was found in the urine by Nemmar et al. We also assumed that the affinity of the particles and of the free ^{99m}Tc was the same for all organs [47]. Physiological parameters such as volumes of the organs and blood flow rate were fixed at their average values for an adult human (see Table 3.1). Other parameters, and in particular those specific to the Technegas, were treated as random variables using Bayesian statistics [44].

Once fitted to the data, the model is consistent with the hypothesis that a small proportion (about 10%) of the Technegas nanoparticles is able to reach the blood, more slowly than free technetium. The free technetium fraction is estimated at about 5%, which is consistent with data in the literature.

A PBPK model was developed for a quantum dot containing cadmium (QD 705), in mice, after intravenous injection [49]. The authors show that QD 705 accumulates in the spleen, the liver, and the kidneys, with a low level of elimination. But that is typical of cadmium, even in ionised form, and the relevance of the nanoparticulate form is not clear.

Other work has used PBPK models for QD 705 [50]. The model correctly predicts the persistence of quantum dots in the tissues (of rodents), but only poorly reproduces the initial kinetics of the products. The authors conclude that more sophisticated models need to be developed, and more specific to nanomaterials, if these are to be used for risk assessment.

In conclusion, the mechanisms underlying possible barrier crossing by nanomaterials in the body, and the exact permeability of the barriers, remain poorly understood at the present time. In such a context, models can only evolve hand in hand with experimental studies. There thus remains much to

Table 3.1. Volumes and flow rates (blood, but not the urinary flow rate) used in the Technegas kinetic model [47, 48]

Tissue or organ	Volume [l]	Flow rate [l/min]
Fats	18.8	0.564
Suprarenal glands	0.014	0.02
Arterial blood	1.40	–
Venous blood	4.20	–
Bones	2.75	–
Brain	1.45	0.78
Mammary glands	0.025	0.002
Intestine	1.02	0.98
Intestinal lumen	0.65	–
Heart	0.33	0.35
Kidney	0.31	1.23
Liver	1.80	0.45
Lungs	0.50	6.72
Bone marrow	3.65	0.29
Muscles	29.0	1.11
Pancreas	0.14	0.065
Skin	3.30	0.33
Spleen	0.15	0.19
Stomach	0.15	0.065
Stomach lumen	0.25	–
Testicles	0.056	0.004
Thyroid	0.019	0.094
Others	7.06	0.19
Urinary flow rate	–	0.001

be done before we can devise models reliable enough to extrapolate experimental results on nanomaterials from animals to humans, or indeed from one nanomaterial to another, or even between different sizes of a given material.

3.4 Conclusion

Up to now, nanotoxicology has progressed on the basis of existing experimental models, the main features of which have been discussed in this chapter. Although nanotoxicological research is only in its infancy, a certain number of problems have already arisen with regard to the realisation and/or exploitation of these studies. The limitations of conventional experimental methods are due to the specificities of nanoparticles, and were never so crucial for the toxicology of larger particles.

While progress can be expected in the implementation and interpretation of data resulting from conventional approaches to toxicity, nanotoxicology may well prove to be a driving force in the development of new ways of assessing toxicity. A good example is predictive toxicology, integrating mathematical modelling (methods of predictive chemistry, PBPK models, systemic biological modelling, etc.) and experimentation (high throughput methods, omics data, etc.) to determine risks and decide upon the necessary safety measures, even before exposure, and hence also its consequences, have had a chance to occur.

References

1. K. Donaldson, V. Stone, C.L. Tran, et al.: Nanotoxicology. Occup. Environ. Med. **61**, 727–728 (2004)
2. European Commission: Statistical data on the use of laboratory animals in France in 2001. Working party for the preparation of the multilateral consultation of parties to the European convention for the protection of vertebrate animals used for experimental and other scientific purposes. Council of Europe, GT 123 (2003)
3. J. Kreuter, D. Shamenkov, V. Petrov, et al.: Apolipoprotein-mediated transport of nanoparticle-bound drugs across the blood–brain barrier. J. Drug Target **10**, 317–325 (2002)
4. A. Takagi, A. Hirose, T. Nishimura, et al.: Induction of mesothelioma in p53+/− mouse by intraperitoneal application of multi-wall carbon nanotube. J. Toxicol. Sci. **33**, 105–116 (2008)
5. C.A. Conn, F.H.Y. Green, K.J. Nikula: Animal models of pulmonary infection in the compromised host: Potential usefulness for studying health effects of inhaled particles. Inhal. Toxicol. **12**, 783–827 (2000)
6. T.H. March, F.H.Y. Green, F.F. Hahn, et al.: Animal models of emphysema and their relevance to studies of particle-induced disease. Inhal. Toxicol. **12**, 155–187 (2000)
7. B.B. Moore, C.M. Hogaboam: Murine models of pulmonary fibrosis. Am. J. Physiol. Lung Cell. Mol. Physiol. **294**, L152–160 (2008)
8. K.J. Nikula, F.H.Y. Green: Animal models of chronic bronchitis and their relevance to studies of particle-induced disease. Inhal. Toxicol. **12**, 123–153 (2000)
9. D.E. Bice, J. Seagrave, F.H.Y. Green: Animal models of asthma: Potential usefulness for studying health effects of inhaled particles. Inhal. Toxicol. **12**, 829–862 (2000)
10. K. Inoue, H. Takano, R. Yanagisawa, et al.: Effects of nanoparticles on antigen-related airway inflammation in mice. Respir. Res. **6**, 106 (2005)
11. K. Inoue, H. Takano, R. Yanagisawa, et al.: Effects of nanoparticles on cytokine expression in murine lung in the absence or presence of allergen. Arch. Toxicol. **80**, 614–619 (2006)
12. F. Alessandrini, H. Schulz, S. Takenaka, et al.: Effects of ultrafine carbon particle inhalation on allergic inflammation of the lung. J. Allergy Clin. Immunol. **117**, 824–830 (2006)
13. S.A. Evans, A. Al-Mosawi, R.A. Adams, et al.: Inflammation, edema, and peripheral blood changes in lung-compromised rats after instillation with combustion-derived and manufactured nanoparticles. Exp. Lung. Res. **32**, 363–378 (2006)

14. J. Pauluhn: Overview of testing methods used in inhalation toxicity: From facts to artifacts. Toxicol. Lett. **140–141**, 183–193 (2003)
15. J. Pauluhn: Overview of inhalation exposure techniques: Strengths and weaknesses. Exp. Toxicol. Pathol. **57** (Suppl. 1), 111–128 (2005)
16. K.E. Driscoll, D.L. Costa, G. Hatch, et al.: Intratracheal instillation as an exposure technique for the evaluation of respiratory tract toxicity: Uses and limitations. Toxicol. Sci. **55**, 24–35 (2000)
17. D. Elgrabli, S. Abella-Gallart, O. Aguerre-Chariol, et al.: Effect of BSA on carbon nanotube dispersion for in vivo and in vitro studies. Nanotoxicology **1**, 266–278 (2007)
18. K. Donaldson, R. Aitken, L. Tran, et al.: Carbon nanotubes: A review of their properties in relation to pulmonary toxicology and workplace safety. Toxicol. Sci. **92**, 5–22 (2006)
19. J.G. Li, W.X. Li, J.Y. Xu, et al.: Comparative study of pathological lesions induced by multiwalled carbon nanotubes in lungs of mice by intratracheal instillation and inhalation. Environ. Toxicol. **22**, 415–421 (2007)
20. A. Al Faraj, G. Lacroix, H. Alsaid, et al.: Longitudinal ^{3}He and proton imaging of magnetite biodistribution in a rat model of instilled nanoparticles. Magn. Reson. Med. **59**, 1298–1303 (2008)
21. P. Cherukuri, C.J. Gannon, T.K. Leeuw, et al.: Mammalian pharmacokinetics of carbon nanotubes using intrinsic near-infrared fluorescence. Proc. Natl. Acad. Sci. USA **103**, 18882–18886 (2006)
22. M.L. Schipper, N. Nakayama-Ratchford, C.R. Davis, et al.: A pilot toxicology study of single-walled carbon nanotubes in a small sample of mice. Nat. Nanotechnol. **3**, 216–221 (2008)
23. Z. Liu, C. Davis, W. Cai, et al.: Circulation and long-term fate of functionalized, biocompatible single-walled carbon nanotubes in mice probed by Raman spectroscopy. Proc. Natl. Acad. Sci. USA **105**, 1410–1415 (2008)
24. A. Nemmar, P.H. Hoet, B. Vanquickenborne, et al.: Passage of inhaled particles into the blood circulation in humans. Circulation **105**, 411–414 (2002)
25. N.L. Mills, N. Amin, S.D. Robinson, et al.: Do inhaled carbon nanoparticles translocate directly into the circulation in humans? Am. J. Respir. Crit. Care Med. **173**, 426–431 (2006)
26. D. Elgrabli, M. Floriani, S. Abella-Gallart, et al.: Biodistribution and clearance of instilled carbon nanotubes in rat lung. Part. Fibre Toxicol. **5**, 20 (2008)
27. A. Al Faraj, K. Cieslar, G. Lacroix, et al.: In vivo imaging of carbon nanotube biodistribution using magnetic resonance imaging. Nano Lett. **9**, 1023–1027 (2009)
28. F. Blank, B. Rothen-Rutishauser, P. Gehr: Dendritic cells and macrophages form a transepithelial network against foreign particulate antigens. Am. J. Respir. Cell. Mol. Biol. **36**, 669–677 (2007)
29. W. Lilienblum, W. Dekant, H. Foth, et al.: Alternative methods to safety studies in experimental animals: Role in the risk assessment of chemicals under the new European Chemicals Legislation (REACH). Arch. Toxicol. **82**, 211–236 (2008)
30. SCCP: Safety of nanomaterials in cosmetic products. Scientific Commitee on Consumer Products, European Commission (2007)
31. A. Nel, T. Xia, L. Madler, et al.: Toxic potential of materials at the nanolevel. Science **311**, 622–627 (2006)

32. K. Million, F. Tournier, O. Houcine, et al.: Effects of retinoic acid receptor-selective agonists on human nasal epithelial cell differentiation. Am. J. Respir. Cell. Mol. Biol. **25**, 744–750 (2001)
33. M. Vastag, G.M. Keseru: Current in vitro and in silico models of blood–brain barrier penetration: A practical view. Curr. Opin. Drug Discov. Dev. **12**, 115–124 (2009)
34. J.G. Ayres, P. Borm, F.R. Cassee, et al.: Evaluating the toxicity of airborne particulate matter and nanoparticles by measuring oxidative stress potential: A workshop report and consensus statement. Inhal. Toxicol. **20**, 75–99 (2008)
35. M.S. Ehrenberg, A.E. Friedman, J.N. Finkelstein, et al.: The influence of protein adsorption on nanoparticle association with cultured endothelial cells. Biomaterials **30**, 603–610 (2009)
36. S. Val, S. Hussain, S. Boland, et al.: Carbon black and titanium dioxide nanoparticles induce pro-inflammatory response in bronchial epithelial cells: Need of multiparametric evaluation due to adsorption artefacts. Inhal. Toxicol. **21** (Suppl. 1), 115–122 (2009)
37. L. Guo, A. Von Dem Bussche, M. Buechner, et al.: Adsorption of essential micronutrients by carbon nanotubes and the implications for nanotoxicity testing. Small **4**, 721–727 (2008)
38. C. Monteiller, L. Tran, W. MacNee, et al.: The pro-inflammatory effects of low-toxicity low-solubility particles, nanoparticles and fine particles, on epithelial cells in vitro: The role of surface area. Occup. Environ. Med. **64**, 609–615 (2007)
39. N.A. Monteiro-Riviere, A.O. Inman, L.W. Zhang: Limitations and relative utility of screening assays to assess engineered nanoparticle toxicity in a human cell line. Toxicol. Appl. Pharmacol. **234**, 222–235 (2009)
40. A. Kroll, M.H. Pillukat, D. Hahn, J. Schnekenburger: Current in vitro methods in nanoparticle risk assessment: Limitations and challenges. Eur. J. Pharm. Biopharm. **72**, 370–377 (2009)
41. F.Y. Bois, T. Smith, A. Gelman, et al.: Optimal design for a study of butadiene toxicokinetics in humans. Toxicol. Sci. **49**, 213–224 (1999)
42. T. Teorell: Kinetics of distribution of substances administered to the body. Archives Internationales de Pharmacodynamie et de Thérapie **57**, 205–240 (1937)
43. K.B. Bischoff, R.L. Dedrick, D.S. Zaharko, et al.: Methotrexate pharmacokinetics. J. Pharmaceutical Sci. **60**, 1128–1133 (1971)
44. A. Gelman, F.Y. Bois, J. Jiang: Physiological pharmacokinetic analysis using population modeling and informative prior distributions. J. Am. Statistical Assoc. **91**, 1400–1412 (1996)
45. S. Micallef, C. Brochot, F.Y. Bois: L'analyse statistique bayésienne de données toxicocinétiques. Environnement, Risque et Santé **4**, 21–34 (2005)
46. A. Péry, C. Brochot, P. Hoet, et al.: Development of a physiologically-based kinetic model for ^{99m}technetium labelled carbon nanoparticles inhaled by humans. Inhal. Toxicol. **21**, 1099–1107 (2009)
47. ICRP Publication 80: Radiation dose to patients from radiopharmaceuticals. International Commission on Radiological Protection. Annals of the ICRP 2 (3). Oxford, Pergamon Press (1999)
48. L.R. Williams, R.W. Leggett: Reference values for resting blood flow to organs of man. Clinical Physics and Physiological Measurement **10**, 187–217 (1989)

49. P. Lin, J.W. Chen, L.W. Chang, et al.: Computational and ultrastructural toxicology of a nanoparticle, Quantum Dot 705, in mice. Environ. Sci. Technol. **42**, 6264–6270 (2008)
50. H.A. Lee, T.L. Leavens, S.E. Mason, et al.: Comparison of quantum dot biodistribution with a blood-flow-limited physiologically based pharmacokinetic model. Nano Lett. **9**, 794–799 (2009)

4

Nanoparticle Toxicity Mechanisms: Oxidative Stress and Inflammation

Béatrice L'Azou and Francelyne Marano

4.1 Introduction

4.1.1 From Particulate Toxicology to Nanotoxicology

Toxicology plays a key role in understanding the potentially harmful biological effects of nanoparticles, since epidemiological studies are still difficult to implement given the lack of data concerning exposure. For this reason, in 2005, Günter Oberdörster coined the term 'nanotoxicology' to specify the emerging discipline that dealt with ultrafine particles (UFP). It involves in vivo or in vitro studies under controlled conditions to establish the dose–response relationship, so difficult to expose by epidemiological studies. It also aims to determine the thresholds below which biological effects are no longer observed. It is concerned with the role played by properties specific to nanoparticles in the biological response: size, surface reactivity, chemical composition, solubility, etc. Nanotoxicology is also the study of interactions with biological molecules such as proteins, lipids, or nucleic acids, which can modify the retention and translocation properties of nanoparticles in the organism. Finally, it is essential for understanding the action mechanisms that may be responsible for physiopathological responses in exposed individuals. It does have its limitations, however, insofar as the complexity of the human environment cannot be perfectly reconstituted in the laboratory. But it remains one of the essential building blocks when undertaking risk assessment.

Data has accumulated over the last 15 years about the consequences for human health of fine particles ($PM_{2.5}$ and PM_1, i.e., aerodynamic diameters less than or equal to 2.5 and 1 µm, respectively) and ultrafine particles (UFP, $PM_{0.1}$, i.e., aerodynamic diameters less than 100 nm) which end up in the atmosphere as a result of combustion processes, or which form in a secondary manner as a result of nucleation reactions, and this has raised concern over the toxicity of nanoparticles [1–3]. Some of the data concerns diesel particles (DiP), and some concerns experimental studies comparing the biological effects and toxicology of various fine and ultrafine manufactured particles, in

P. Houdy et al. (eds.), *Nanoethics and Nanotoxicology*,
DOI 10.1007/978-3-642-20177-6_4,

particular, carbon, silica, TiO_2, and ZnO nanoparticles. The results of recent epidemiological studies which relate the amounts of UFPs in the atmosphere and the increase in cardio-respiratory morbidity and mortality show that this concern is justified [4]. One of the problems raised over the last few years is that fine and ultrafine atmospheric particles may have systemic effects on organs such as the heart, which are not themselves direct targets. Similar questions arise for nanoparticles (NP). But these exhibit significant differences with UFPs. The latter have variable sizes and complex chemical makeup, whereas NPs are more uniform and have well-defined chemical composition. Furthermore, the number of manufactured NPs is continually increasing, with a very wide range of properties and uses, and this implies a number of different uptake routes. The main route is the respiratory system, but the digestive and cutaneous routes are also relevant (see Chap. 2).

Toxicological studies of NPs were thus developed on the understanding that they would cause the same type of pathologies as fine and ultrafine particulate matter (PM), i.e., pathologies associated with the oxidative stress they induce in tissues and which leads to an inflammatory response. If exposure is continuous, even at low doses, and if the particles persist in the body, this can lead to chronic pathologies, such as fibrosis and cancers, or to an exacerbation of other pathologies, such as asthma and chronic obstructive bronchopneumopathy.

But what link can we now establish between NPs and oxidative stress, and what are the molecular, cellular, and tissular mechanisms whereby this initial stress could induce such pathologies?

4.1.2 Nanoparticles, Oxidative Stress, and Inflammation

Many studies attest to the fact that oxidative stress is induced by fine and ultrafine atmospheric particles, and also by some manufactured nanoparticles. This stress sets off a series of molecular and cellular events which themselves have a range of consequences: inflammatory response, modulation of cell proliferation and differentiation, or even cell death. This hypothesis is supported by the review articles [2,3,5,6], with the further suggestion in the case of manufactured nanoparticles that oxidative stress may be a central mechanism in their toxicological effects. The harmful effects seem to be exercised either directly on the target tissues due to the toxicity of the reactive oxygen derivatives, or indirectly as a result of the effects of certain reactive oxygen derivatives on the production of inflammatory and immune system mediators, mainly pro-inflammatory cytokines. This oxidative stress can deactivate antiproteases and at the same time activate metalloproteases, thus favouring proteolysis and uncontrolled cell destruction. By activating transcription factors sensitive to oxidative stress, the transcription of genes for pro-inflammatory factors is stimulated, and this results in the release of many inflammatory mediators.

Some authors have suggested that inflammation is a primary response, while oxidative stress is just a consequence of that. Indeed, particles are recognised by the organism as foreign bodies that must be eliminated by means of the inflammatory reaction. The interactions between nanoparticles and proteins in biological fluids play a decisive role in their ability to be recognised by cells of the immune system responsible for their elimination, and also by cells in the covering tissues, the first target of these NPs, also able to emit pro-inflammatory signals. Inflammation may then accelerate the production of reactive oxygen species (ROS) and reduce the antioxidant defence capacity, favouring the appearance of oxidative stress and associated tissue damage. Today these ROS are in fact considered to be secondary messengers through which inflammation exercises its main actions. The inflammatory reaction favours and maintains oxidative stress, which in return accelerates the recruitment and activation of inflammatory cells.

Whatever the situation, oxidative stress and inflammation go hand in hand. This is why it seemed important to specify the general mechanisms of oxidative stress, then describe the present state of our understanding of how nanoparticles generate ROS. Finally, we discuss the role played by inflammation in nanoparticle toxicity and in the development of acute or chronic pathologies.

To understand oxidative stress, one needs to examine the effects of the various ROS. An excess of free radicals not neutralised by the organism's defence system is very harmful for biological macromolecules, resulting in genetic and functional perturbations which may lead to a loss of proliferation control and even cell death.

However, the physicochemical mechanisms occurring at the nanoparticle–cell interface and responsible for this oxidative stress remain poorly understood at the present time. Many studies seeking to establish a dose–response relationship, or an exposure–effect relationship, have revealed the importance of the size and/or surface area of the particle. However, it remains to find out whether ultrafine particles of comparable sizes pose the same threat and whether the ability to generate free radicals can be taken as a useful biomarker for the effects of these particles.

4.1.3 Acute Inflammatory Reaction and Inflammatory Defence Against Chronic Pathologies

The inflammatory response is a defence mechanism of the organism used by the higher animals to fight attack of any kind, be it biological, chemical, or physical, in order to maintain its integrity [7]. This 'exogenous' inflammatory response is said to be non-specific. It may be associated with damaged tissues or cells emitting signals initiating a series of responses, in particular, in the blood vessels and circulating cells. It is therefore a beneficial adaptive response which tends to reestablish the integrity of the organism. However, it uses destructive methods directed against the attacker which may have harmful consequences if they are not properly controlled.

A conventional inflammatory reaction occurs in five stages:

1. Recognition of the attack which triggered the reaction.
2. A vascular response leading to vasodilation.
3. Activation of endothelial cells and circulating cells in the blood: polynuclear, neutrophils, then monocytes, which migrate through the endothelium to the tissue where the foreign body is located.
4. Release of mediators favouring the elimination of the foreign body, in particular, phagocytosis.
5. Repair of the damaged tissue.

The transition to chronic inflammation occurs when it has not been possible to eliminate the foreign body and it then results in more or less serious lesions, while cells involved in the inflammation remain more or less activated.

Rapid changes in the cell redox potential are considered to be among the initial inflammatory signals. They are related to excessive ROS production in so-called sentinel cells, such as monocytes. However, various epithelial cells can emit such signals. ROS production can be direct and NPs, like UFPs and many physical and chemical environmental factors, are able to produce these in biological media. It can also be indirect, mediated by inflammatory cells, in particular macrophages. The modification of the intracellular redox potential is generally associated with activation of ubiquitous transcription factors such as NF-κB. The latter plays a key role in the response to many agents, in particular physical ones like UV irradiation or chemical ones, e.g., metals such as nickel or cobalt. NF-κB is also activated by asbestos fibres or particles, and in particular atmospheric particles. This cytoplasmic factor comprises two subunits, P50 and P65, deactivated by IκB. The activation of NF-κB occurs when IκB detaches from the NF-κB complex, which then migrates to the nucleus and binds to the promoters of many genes then activated for transcription. Among these are the genes of many cytokines and other inflammatory factors: IL-1β, IL-6, IL-8, M-CSF, GM-CSF, TNF-α, iNOS, etc. (see Fig. 4.1). By taking part in the activation of this set of genes, ROS play not only an initiating role in the inflammatory response, but also an amplifying role, insofar as feedback activation mechanisms are subsequently set up. For example, by binding with these membrane receptors, TNF-α is responsible for the production of superoxide anions by the mitochondria of the target cell, and this may result in death by necrosis, i.e., accidental cell death, or apoptosis, i.e., programmed cell death. Apart from ROS, other signals are responsible for initiating inflammation, in particular, the kinases which contribute to raising the level of phosphorylation.

4.2 Interactions Between Nanoparticles and Biological Media, Including Proteins

When the NPs come into contact with the biological fluids in the respiratory apparatus, the digestive system, or the blood, these fluids enter the pores of the NPs, whether they are isolated or occur in aggregates. Proteins, either

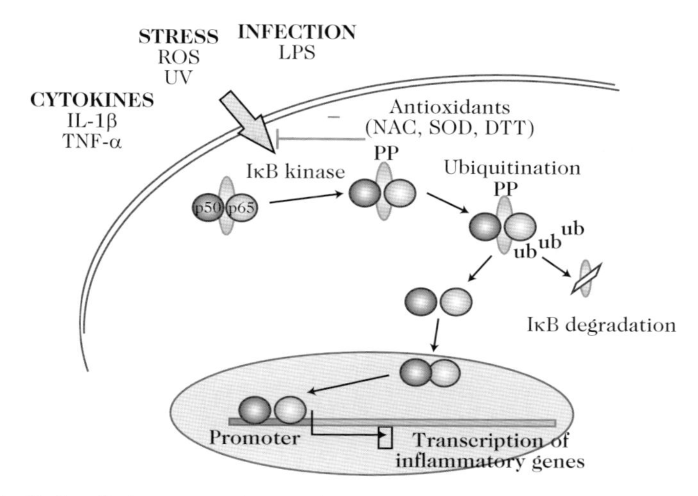

Fig. 4.1. Role of the transcription factor NF-κB in the inflammatory response following cell stress. Stress can be induced by various factors: cytokines, UV, reactive oxygen species (ROS), bacterial infection (LPS). The transcription factor NF-κB is in a deactivated form in the cell cytoplasm associated with its inhibiting protein IκB. The different stress factors activate an enzyme, IκB kinase, and this induces degradation of IκB (following phosphorylation and ubiquitination). Once released, NF-κB migrates to the nucleus and binds to different gene promoters, including genes associated with the inflammatory response

alone or associated with lipids, e.g., the pulmonary surfactant, can then coat the particle surface, forming a corona [8], which will modify the ability of the particle to interfere with the tissues and influence any biological responses (see Fig. 4.2). Many proteins form transient complexes with the NPs, depending on their physical and chemical characteristics. Among the protein interactions that have been studied, albumin and fibrinogen have a strong affinity and a large dissociation constant, higher than what is observed, for example, with the apolipoprotein A1. Such differences of affinity can determine the protein constitution of the corona. The resulting coating of the NPs will subsequently play a decisive role in the chances of capture by tissues and the inflammatory and immunological response to the NPs.

Depending on the uptake route, the NPs will be recognised by the organism as foreign bodies. The coating molecules may be opsonins. These are proteins involved in the phagocytosis of foreign bodies by macrophages. After binding with the NPs, they will be recognised by cells of the immune system carrying receptors for them. For example, in the respiratory system, the MARCO receptor, part of the respiratory antibacterial defence system, is employed in anti-particle defences, too [9]. Proteins in the complement can also attach to particles. These interactions have a knock-on effect until inflammation is induced. Other interactions also play an important role, depending on the

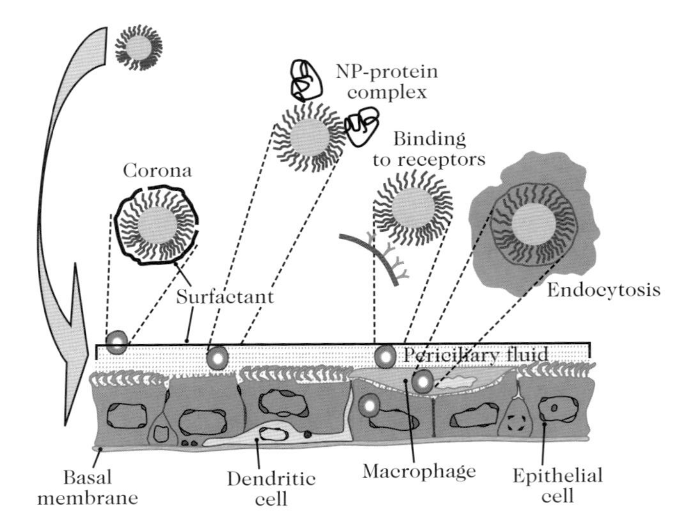

Fig. 4.2. Interactions between proteins, surfactant, and NPs in biological fluids of the respiratory system, leading to endocytosis by macrophages and epithelial cells (see colour plate). Adapted from Kreyling 2007. The complex between proteins of the pulmonary fluids and the NP results in the formation of a corona which can be coated with surfactant. Once coated in this way, the NP, alone or in an aggregate, can interact more easily with membrane receptors of the epithelial cells and hence be phagocytosed

uptake route. For example, for the respiratory route, fine atmospheric particles sequester the surfactant and various compounds present in the bronchoalveolar fluid.

At the present time, the exact role of these interactions in the biological responses is still poorly understood, but it seems likely to be important in the inflammatory response. The cytokines, molecular signals secreted during the inflammatory reaction, can interact with the NPs, modifying this response downstream [10, 11]. The interactions depend on the type of NP and the type of cytokine. For any interpretation of the inflammatory effects of NPs, it is therefore essential to take into account these interactions with proteins. In a recent review, Lynch and Dawson [12] analyse current data on NP–protein interactions. These are particularly complex, since they are not static, and depend on the evolution of the protein environment in the organism, noting that highly abundant proteins can be gradually replaced by less abundant proteins with a stronger affinity for the NPs. These modifications may influence not only the inflammatory response, but also the accumulation and translocation of these particles.

Finally, NPs can interact with proteins and, in doing so, modify their properties. They induce in vitro the assembly of proteins and peptides to form amyloid fibrils [13]. It has been shown recently that different sorts of

NPs, e.g., quantum dots, carbon nanotubes, etc., can induce the nucleation of β2-microglobulin fibrils in vitro. This is an abundant protein in the central nervous system. This observation may have very important consequences insofar as NPs may migrate to the brain along the olfactory nerve following deposition in the olfactory mucous membrane. However, no in vivo study has yet demonstrated such a phenomenon.

4.3 Nanoparticles and Oxidative Stress

The involvement of oxidative stress in particle toxicity mechanisms was originally demonstrated in the context of occupational exposure to coal particles, glass fibres, quartz particles, and asbestos fibres [14–16]. Studies carried out in vitro have shown that ultrafine particles generally generate more reactive oxygen species than fine particles. This increased production, greater than the elimination capacities of the antioxidant systems of the organism, will be responsible for a lot of molecular damage and alteration of biological functions.

4.3.1 Reactive Oxygen Species (ROS)

The concept of oxidative stress has existed in human biology for many years now to explain dysfunctions that lead to pathologies, e.g., following ischemia (where a tissue is deprived of oxygen after an infarction, for example), or in age-related illnesses. However, the physiological role of the ROS and nitrogen has already been demonstrated when they are released in a controlled way. The ROS contribute to cell homeostasis, affecting signal transduction, and regulating the expression of redox-sensitive genes. As a consequence, physiopathological mechanisms appear when there is overproduction of radical species or when the organism is unable to defend itself due to a deficiency in the antioxidant systems, e.g., superoxide dismutase (SOD), catalase, glutathione peroxidase (GPx). The result is an imbalance between ROS production and antioxidant defence capacity [17].

Free radicals are molecules or atoms with one or more unpaired electrons in their outer shell. This state confers upon them a thermodynamic instability and reaction kinetics which explain their high level of reactivity. ROS production can be generated naturally in each cell of the organism and is essentially of enzymatic origin in the mitochondrial complex of the respiratory chain or membrane NADPH oxidase (see Fig. 4.3). Other sources, cytosolic or within different cell organelles (smooth endoplasmic reticulum, peroxisomes), can also play a role in signal modulation (xanthine oxidase, enzymes of the arachidonic acid pathway, lipoxygenases, cyclooxygenases) [18].

The leading reactive oxygen species is the superoxide anion $O_2^{\bullet -}$.

Organic matter is composed of atoms in which all electrons are paired and occur in the singlet state, while a molecule with one unpaired electron is called a free radical

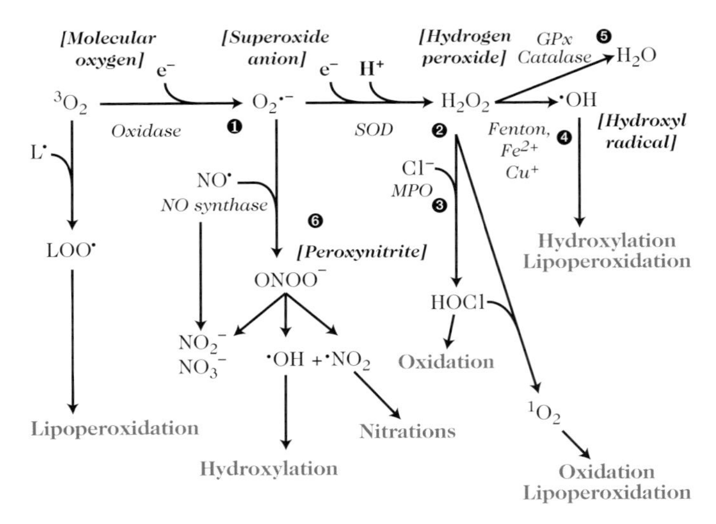

Fig. 4.3. Production of reactive oxygen species from molecular oxygen. From [19–22]

and occurs in the doublet state. In the presence of radiation, metals, pollutants, etc., oxygen gives rise to reactive oxygen species (ROS), either during symmetric breaking of a covalent bond (homolytic fission or homolysis) in which each atom keeps its electron, or during a redox reaction with electron loss or gain from a non-radical compound. This electron transfer is tightly controlled by enzymes (biological catalysts), namely, oxidases and oxygenases. The role of these enzymes is to transform O_2 and the organic molecule in such a way that one of them becomes a doublet (free radical). For example, the oxidases, e.g., NADPH oxidase, transform triplet oxygen to the doublet state and can lead to the formation radicals such as the superoxide anion $O_2^{\bullet -}$ (1), while oxygenases trans form organic molecules into free radicals. The superoxide radical $O_2^{\bullet -}$ is transformed under the action of SOD into hydrogen peroxide H_2O_2 (2). Chemically speaking, the latter is not an oxygenated free radical like most other reactive oxygen species, but from a biological standpoint, it does behave as such. Several things may happen to the resulting H_2O_2. By the HOCL pathway, it can synthesis new unstable derivatives. H_2O_2 also gives rise to the hydroxyl radical $^{\bullet}OH$, if there are metal ions in the medium, such as iron Fe^{2+} complexed with an activating ligand. This reaction is called the Fenton reaction (4). $^{\bullet}OH$ is an oxidation agent, in particular for aromatic rings. H_2O_2 can also undergo detoxification reactions by catalase or glutathione peroxidase, or interact with vitamins E and C to prevent their accumulation (5). The oxygen and nitrogen metabolisms intersect. Starting with $O_2^{\bullet -}$, another possible product is peroxynitrite (6) (non-radical $ONOO^-$) by reaction with nitrogen monoxide (radical $NO^{\bullet}$) produced by NO synthase. The unstable peroxynitrite is highly oxidising and forms new active species, some of which are radical, e.g., $^{\bullet}OH$ and $^{\bullet}NO_2$, the basis for nitrations and hydroxylations.

The H_2O_2 concentration is regulated by so-called antioxidant enzymes such as catalase (present in peroxisomes) and the glutathione peroxidases (mainly found in the cytosol). Catalase catalyses the dismutation of H_2O_2 into oxygen and water, while glutathione peroxidase catalyses the oxidation of glutathione into the oxidised form of glutathione.

The use of oxygen thus involves specialised enzymes which may not be present in sufficient amounts, depending on the enzyme resources of the cell. This imbalance can lead to deficiencies in ROS metabolism, and the resulting biological consequences of oxidative stress can vary enormously, depending on the ROS excess and cell type.

4.3.2 Reactive Oxygen Species and Their Effects

An increased concentration of reactive forms of oxygen, exceeding the system's antioxidant capacity, can thus lead, either directly or indirectly, to oxidative damage on the molecular level and considerably affect cell mechanisms (see Fig. 4.4). These reactive products can attack most macromolecules, e.g., sugars, proteins, nucleic acids, lipids, disorganising their chemical structure and altering their biological functions.

The ROS act non-specifically, so all molecules may be affected. However, some are more sensitive than others, such as the unsaturated lipids, certain amino acides, and aromatic compounds.

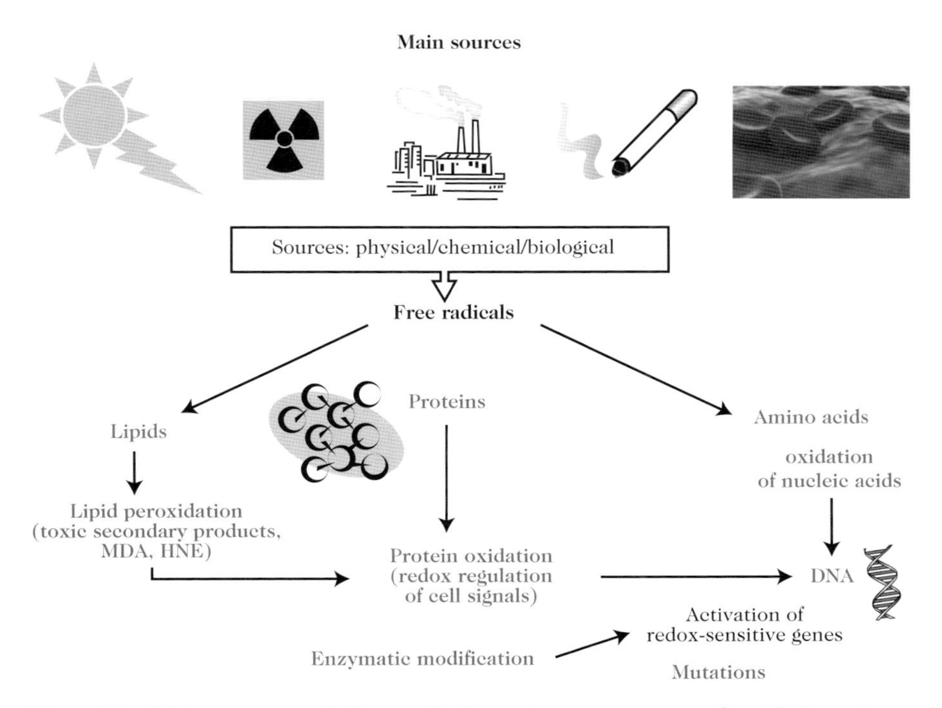

Fig. 4.4. Main sources of free radicals and consequences of oxidative stress. From [23]

For example, owing to their chemical composition, membranes are choice targets for attack by radicals. When membrane lipid double bonds are attacked, cascade peroxidation processes will result (rearrangement of the double bonds, leading to conjugated dienes, followed by the formation of lipid peroxides $ROO^{\bullet}$), ending up in the complete disorganisation of the membrane, and thereby altering its exchange, barrier, and information functions [24].

The most sensitive proteins to attack by free radicals are undoubtedly those carrying an amino acid with a sulfur atom (methionine, cysteine) or a sulfhydryl (SH) group. This is the case for many cellular enzymes and transport proteins which will thus be oxidised and deactivated. With regard to aromatic amino acids, the addition of hydroxyl radicals on double bonds gives rise to specific oxidation reactions. Apart from these oxidative lesions, fragmentation of polypeptide chains is also observed, causing irreversible lesions. The toxicity of ROS, which also acts on proteins, can result in modifications of the cell signalling mechanisms. Indeed, the ROS can act on receptors, nuclear transcription factors, and certain protein kinase cascades. They can modify the enzyme activity of the tyrosine kinases and serine/threonine kinases (such as the mitogen-activated protein kinases or MAPK), thereby activating transcription factors that initiate the expression of redox-sensitive genes. When this domain is altered, phosphorylation is perturbed and signal transduction modified.

For example, free radicals can deactivate or degrade the NF-κB inhibitor IκB by activating phosphorylation cascades favouring proteolysis [9]. Other transcription factors, such as AP-1, are also partly under the control of reactive oxygen derivatives. AP-1 comprises two proteins, c-fos and c-jun, and participates in the cell differentiation process, and in the modulation of the expression of cytokines and other mediators with an immunological role [25, 26].

Oxidative damage induced by $^{\bullet}OH$ can also affect DNA bases, generating intrachain adducts, strand breakage, and DNA–protein crosslinks [23, 27]. Oxidative stress can attack the bond between the base and the deoxyribose, creating an abasic site, or attack the sugar itself, creating a single-strand break. Indirectly, damage can result from effects on lipids whose peroxidation generates mutagenic aldehydes such as malondialdehyde (MDA), creating adducts of the form MDA–guanine on the DNA bases or etheno derivatives. Free radical attack can also affect structural chromatin proteins such as histones, as well as replication and transcription factors and enzymes.

Oxidative stress can be measured directly by electron paramagnetic resonance. It can also be done indirectly by measuring the metabolites resulting from radical reactions, e.g., lipid peroxidation, protein oxidation, DNA oxidation (see Fig. 4.5). However, the end products of oxidation formed for each biomolecule are many and complex.

In addition, there are inherent difficulties due to the fugacity of radical species. Free radical status is investigated by measuring the production of radicals (pro-oxidant status), but also by measuring the specific biochemical dis-

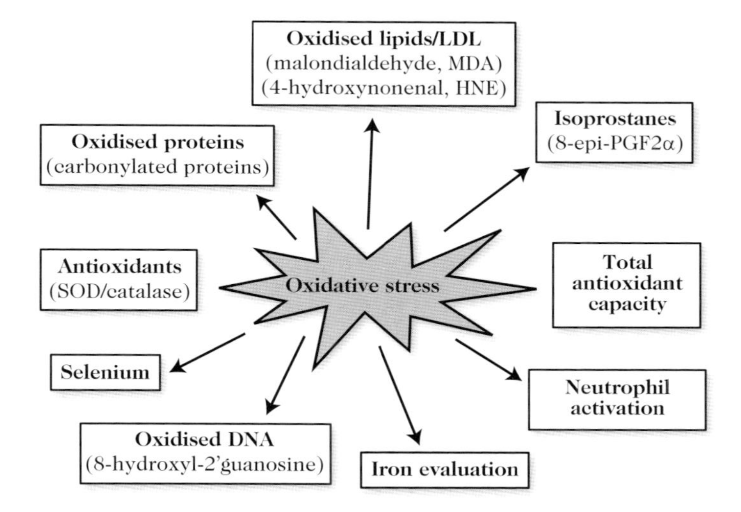

Fig. 4.5. Main methods for assessing the state of oxidative stress in humans. From [28]

orders resulting from an antioxidant/pro-oxidant imbalance. The involvement of oxidative stress can be measured by various methods, notably chemoluminescence techniques, or using fluorescent molecules such as dichlorofluorescein diacetate (H_2-DCFDA). Another method uses a fluorescent lipophilic compound called C11-BODIPY to quantify lipid oxidation of cell membranes induced by NP exposure [29, 30].

Evaluation of the defensive capacity of the organism by measuring the antioxidant status is considered as an indirect proof of ROS production. The concentration of antioxidants such as vitamin C, glutathione, SOD, and GPx, is easily measured by spectrophotometry, as described in the presence of fullerenes C_{60} [29], silver nanoparticles [31], and titanium nanoparticles [32].

4.3.3 Nanoparticles, ROS production, and Oxidative Stress

Oxidation properties and biological responses depend on the environmental particles tested. Atmospheric particles are difficult to study owing to the extreme complexity and heterogeneity of their action mechanisms. Indeed, it has been shown that inhaled atmospheric particles are sources of free radicals, on the one hand because they exacerbate phagocytosis by macrophages and on the other because their surfaces are covered with highly reactive elements, such as organic compounds, and in particular the polycyclic aromatic hydrocarbons (PAH) and quinones, but also transition metals [33, 34]. Li et al. [35] observed a linear relation between the redox activity of particles, the level of PAHs, and the ability to induce an antioxidant enzyme, namely heme oxygenase (HO-1). A higher level of free radicals is produced by UFP samples if they are compared with cruder samples of the same atmospheric particles [36,37]. These results thus emphasise the fact that the surface properties of particles modulate their ability to induce oxidative stress, and that, the smaller they are, the higher will be their specific surface area per unit mass and hence the greater will be their ability to transport toxic substances and to produce free radicals [36–38]. Indeed, for the same mass, the number of surface atoms available to be oxidised or reduced is greater for NPs than for larger particles. Some groups have observed much higher toxicity with, for example, CeO_2 or TiO_2 NPs compared in vitro with particles of the same chemical composition but much larger in size, with an increase in ROS production, activation of caspase 3, and chromatin fragmentation resulting in cell death by apoptosis [39].

With ultrafine TiO_2, carbon black, or cobalt particles in vivo, more serious and more persistent pulmonary lesions have been observed than with fine particles [40, 41]. These biological alterations are likely to be due to the increase in the number of particles for a given mass and the reduction in size, as has already been observed for UFPs [40, 42, 43].

At the present time, there is not enough experimental evidence to select a measurement criterion that would serve to compare the effects of every particle type: area rather than size, number, or mass, which is the criterion currently preferred. However, it is very important to determine the best way of expressing the data if we are to standardise assays for NP toxicity assessment.

Most available studies thus show a close link between nanoparticles and oxidative stress, although this relationship is complex and depends on many structural aspects of the nanoparticles, making it difficult to generalise results.

Different types of NPs have been studied on a range of biological targets to determine the respective roles of the chemical makeup and size of the NPs in relation with the cell type. For the epidermis, which is an important target for NPs through their use in cosmetic products, the epidermic cell lines HT1080 and A431 were treated with silver NPs. This induced ROS production, and lipid peroxidation with consequent dose-dependent cell death by apoptosis, revealed by DNA fragmentation and increased caspase 3 activity [31]. Other experimental studies on pulmonary cell cultures demonstrate just as clearly the generation of oxidative stress leading to DNA damage (micronuclei, comet test) [44].

Cytotoxicity with associated ROS generation has been confirmed on cultures of many other cell types, e.g., human bronchial cells 16HBE, fibroblasts NIH3T3, alveolar epithelial cell lines SV40T2, alveolar macrophages, and renal epithelial cells LLC-PK_1, and with other nanoparticles (TiO_2, carbon black, ZnO, CeO_2, etc.) [45–49]. Cell cultures provide a good way of investigating the mechanisms involved, and it may be possible in the near future to develop standardised assays for high throughput screening of NP cytotoxicity and oxidative stress.

In parallel, in vivo studies, mainly on rats, also reveal the existence of NP-induced oxidative stress. Different biomarkers have been used, such as the expression of messenger RNA coding for manganese-dependent superoxide dismutase (MnSOD), an antioxidant enzyme involved in pulmonary defence against ROS [50]. In contrast to fine particles with diameter 1 μm, 20 nm TiO NPs caused a significant increase in the expression of mRNA coding for MnSOD. The increases were more moderate for other antioxidants such as catalase, glutathione peroxidase, and CuZnSOD. MnSOD induction may therefore provide a predictive indicator for pulmonary oxidative stress, along with the reduction in intracellular glutathione levels and cell metabolic activity [32, 51].

4.4 Nanoparticles and Inflammatory Response

As we have seen, ROS production is generally considered to be one of the key mechanisms in the toxicity of fine and ultrafine atmospheric particles, as well as nanoparticles. It leads to an inflammatory response, as has been revealed essentially in the respiratory system, since this has so far been the most widely studied exposure route. This reaction is above all related to the defence of the organism, and is in principle beneficial. However, many human pathologies are linked with a chronic inflammatory state. It has been clearly demonstrated that environmental particles, such as silica, asbestos, or carbon black, when they accumulate in the tissues in a persistent way (overloading), result in serious pathologies like fibrosis and cancer. This therefore raises the same question rather urgently for manufactured nanoparticles, since some are already widely used, while the occupational, consumer, and environmental risks have not yet been properly evaluated. However, experimental data clearly demonstrates that some of them give rise to an inflammatory response, but it would be difficult to conclude for the moment as to whether it could lead to the development of chronic pathologies in humans.

4.4.1 Fine and Ultrafine Atmospheric Particles, Man-Made Nanoparticles, and Inflammation: A Clearly Established Relationship

Over the last 20 years or so, the large amount of epidemiological data regarding long and short term effects of atmospheric particles has encouraged active experimental work to provide a causal explanation for the effects observed either on the general public or on specific populations [52, 53]. These studies were mainly carried out with model particles, diesel particles (DiP), fine and ultrafine particles of carbon black and metal oxides, and in some cases fine and ultrafine fractions of atmospheric particles. They show a clear, general relationship between particle exposure and inflammatory response in the organ in which the particles are deposited, viz., the lungs, but also remote effects, notably on the cardiovascular system.

Controlled Human Exposure and Inflammatory Effects of Diesel Particles

Several controlled exposure studies have been carried out on humans with diesel particles. There are two approaches: nasal instillation and inhalation via an exposure chamber. Exposure of healthy volunteers to DiP by nasal instillation induces an increase in the number of inflammatory cells, cytokines, chemokines (signals responsible for chemotactism), and specific immunoglobulin E (IgE) of the allergic response [54]. The result of an inflammatory process is also observed in bronchoalveolar lavages (BAL) carried out on healthy volunteers after exposure to dilute diesel exhaust fumes, representative of environmental exposure [29]. The inflammatory infiltrate shows an increase in neutrophils, cells playing a major role in chronic bronchitis, but also in asthma and allergic rhinitis, B lymphocytes, mastocytes, T lymphocytes (CD4+ and CD8+), and histamine, a molecule involved in the allergic response [55].

These different results on humans, following controlled exposure, all point toward an inflammatory response induced by DiPs, while no study on manufactured NPs has yet been published. However, this inflammatory response is complex, and animal and in vitro studies have led to an understanding of the action mechanisms.

Pulmonary Exposure of Animals to Fine Particles and Nanoparticles, and Associated Response

Fine and Ultrafine Atmospheric Particles, Nanoparticles, and Pulmonary Pathologies

Models used here are rodents, mainly rats, which accumulate in a very different way to what is observed in humans. In addition, exposure doses used in animal experimentation are often significantly higher than those in the surrounding atmosphere, or those used for human exposure. Care is therefore needed in extrapolating from animals to humans. Notwithstanding, short term exposure also causes an inflammatory response. For example, rat studies have shown that exposure by instillation or inhalation to DiPs or atmospheric particles induces oxidative stress and pulmonary inflammation, characterised in particular by an influx of polynuclear neutrophils, an increase in the amounts of proteins in the BAL fluid, and an increase in the expression of pro-inflammatory cytokines, i.e., a response that is quite comparable with the one observed for controlled human exposure to DiPs [56–59]. The inflammatory reaction is reduced in the presence of antioxidants like SOD and catalase [60], thus revealing the role of ROS.

One major advantage with animal studies is the use of pathological models, e.g., asthma, emphysema, cardiovascular pathologies, which imitate human pathologies. Age and sex can also be taken into account, to assess possible differences in sensitivity. These pathological animal models have revealed the role played by DiP and PM (particulate matter) exposure in asthma, allergic rhinitis, and chronic bronchitis. Particles can play the role of an adjuvant, i.e., an amplifying cofactor, in association with an allergen. This is particularly true of fine and ultrafine atmospheric particles which can adsorb biological molecules from pollens, fungal spores, or bacterial cell walls, e.g., bacterial lipopolysaccharide (LPS). When they penetrate the lungs as far as the alveoles, inhalable particles carry these molecules with them, exposing the subject to a possible allergic reaction. The presence of bacterial LPS on certain particles causes an inflammatory response that is independent of ROS production by the particles, inducing chemotactic signals of foreign body recognition. These observations lead to the idea of a Trojan horse. The UFPs serve as a carrier for molecules causing the biological response. Although such ideas have not been published in the context of man-made NPs, it seems likely, depending on their reactivity, that they too could serve as carriers for adsorbed biological molecules, thereby allowing uncontrolled uptake of these molecules during the phagocytosis process (see Sect. 4.2).

Nanoparticles and Cardiovascular Effects

Over the past few years, several epidemiological studies have describe the short term effects of exposure to atmospheric particles on cardiovascular pathologies. For example, Peters et al. [61] have shown that 2 h exposure increases the risk of myocardial infarction. Furthermore, experimental data on rats suggests that particles in the ultrafine fraction (UFP of diameter less than 100 nm) are the ringleaders [62].

Current hypotheses regarding the mechanisms whereby inhaled particles might have extrapulmonary effects tend in two directions, although probably not mutually exclusive. The first seeks out the possible consequences of pulmonary inflammation for the heart and other systems, such as blood coagulation and the cardiovascular system [38, 63]. Increased heart rate and rate abnormalities, arterial vasoconstriction, and an increase in neutrophils and platelets in the peripheral blood have been linked to periods of particulate pollution.

With regard to the other hypothesis, the fact that NPs translocate, even in small amounts, from the lungs to the systemic circulation in rats suggests possible direct effects on the vessels. However, the toxicological mechanisms whereby these particles exercise their harmful effects on the extrapulmonary compartments remain poorly understood at the present time.

Carbon Nanotubes and the Fibre Effect

Among the NPs studied in animals, carbon nanotubes raise specific problems. Two recent publications have stimulated great concern over the risks associated with exposure to carbon nanotubes. The first [64] shows that nanotubes injected into the mouse abdomen can induce inflammation and the formation of granulomas (fibrous cell clusters), similar to those induced by asbestos fibres. This response only arises when the nanotubes have certain length and shape characteristics (lengths of a few μm and straight) which make them similar to asbestos fibres. This non-physiological administration route was chosen after comparison with studies carried out on asbestos and fibres used as substitutes for it. It gives fast biological responses on the peritoneum, a tissue similar to the pulmonary mesothelioma. The suspected similarities between asbestos fibres and carbon nanotubes have been corroborated by another study [65], where intraperitoneal injection of a mouse line selected for its susceptibility to develop tumours shows that carbon nanotubes induce mesotheliomas at a higher rate than asbestos for equivalent doses.

This therefore raises a question of carcinogenic power through a fibre effect for certain carbon nanotubes. It looks as though they may behave like fibres from the point of view of the macrophages, cells responsible for eliminating particles by phagocytosis. If the fibre or nanotube were too long, this would result in frustrated phagocytosis. The macrophage would be unable to eliminate it, and this might result in persistence and accumulation in certain tissues. Although this picture remains to be confirmed, it shows that it would be advisable to enforce total confinement for any use of carbon nanotubes that might result in aerosol formation.

4.4.2 Comparability of Cellular and Molecular Toxicity Mechanisms for Fine and Ultrafine Atmospheric Particles and Nanoparticles

As we have seen, currently available data on the inflammatory response associated with nanoparticles deals mainly with the lungs, but that does not rule out potential effects on other organs, depending on the uptake route and accumulation points within the organism. In the lungs, the bronchial and alveolar epithelia function as dynamical barriers by participating in the inflammatory process induced by oxidative stress. ROS production by NPs may activate signalling pathways in the cell and nuclear transcription factors which regulate the expression of genes involved in a range of biological processes, including growth, apoptosis, inflammation, responses to stress (see Sects. 4.3.1 and 4.3.2). The resulting ROS can induce pro-inflammatory mediators through the activation of signalling pathways, in particular, the one for proteins in the kinase family (MAPK) which are heavily involved in transduction of signals and transcription factors sensitive to the redox status of the cell, in particular, AP-1 and NF-κB [66, 67]. Furthermore, it has been observed on a human monocytic cell line, and also on cells in rat bronchoalveolar lavage fluid, that an increased concentration of calcium ions in the cytosol could be induced by ultrafine carbon black particles [66, 68]. These carbon black nanoparticles might activate the opening of calcium channels via a mechanism inducing ROS production. The rise in intracytosolic calcium concentrations might then result in activation of genes controlling inflammation with increased production of TNF-α, IL-2, IL-6, IL-8, ICAM-1, and E-selectin. This in turn would lead to functional changes by paracrine action of epithelial secretions (cytokines and growth factors) on surrounding tissues. Notwithstanding, finer particles may have a direct effect on neighbouring tissues by transcytosis through the epithelia (see Fig. 4.6).

The accumulation and persistence of particles in tissues may cause chronic inflammation. In the lungs, it can lead to bronchial remodelling, characterised by thickening of the smooth muscle tissue, mucous metaplasia, and peribronchial fibrosis [53]. These modifications are also observed in patients suffering from chronic obstructive bronchopneumopathy and asthma. If the effects are too great, adaptive mechanisms cannot be implemented and the cell moves towards death by apoptosis, or even necrosis. Induction of apoptosis has been observed in alveolar macrophages and bronchial epithelial cells in culture in response to oxidative stress induced by ultrafine particles and

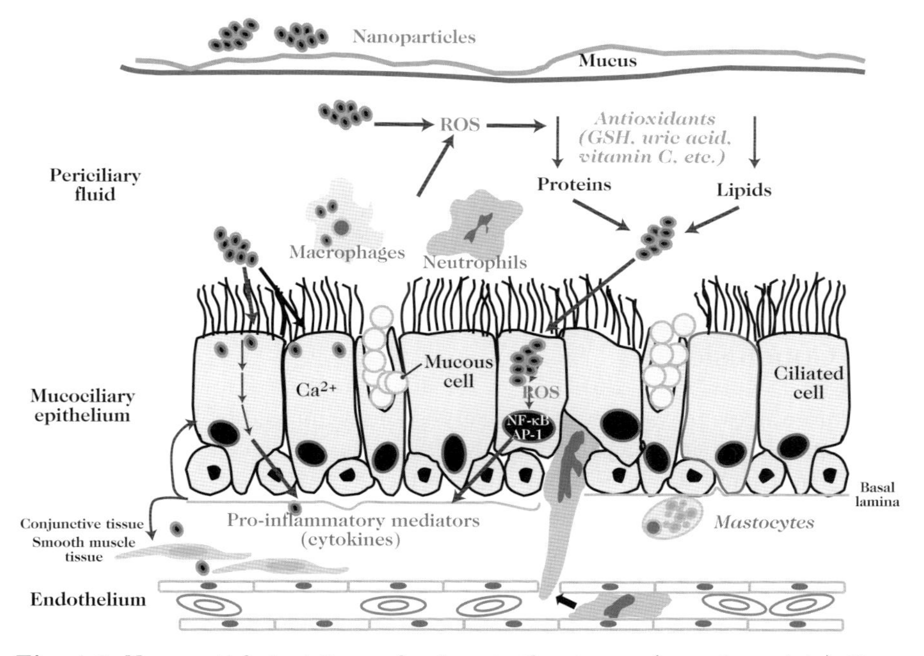

Fig. 4.6. Nanoparticle toxicity mechanisms in the airways (see colour plate). From [34]. Atmospheric particles and nanoparticles act by similar cellular and molecular mechanisms. Mucociliary transport and phagocytosis by macrophages are the main protection mechanisms against PM and NPs. However, their surface reactivity and the presence of transition metals may initiate extracellular ROS production. Upon contact with the epithelium, they may be phagocytosed and, possibly, cross the barrier by transcytosis. In the cell, the signalling pathways involved in the pro-inflammatory response are the same. One specificity of particulate stress concerns activation of the epidermal growth factor receptor (EGFR) and the synthesis of its ligands underlying paracrine action on neighbouring tissues

nanoparticles. Reactive oxygen species may cause mitochondrial damage and initiate pro-apoptotic cascade through a drop in the transmembrane potential of the mitochondrion and cytosolic release of cytochrome C. Doubtless due to their greater capacity for phagocytosis, macrophages are more sensitive to induction of apoptosis than epithelial cells. This observation is important to explain the persistence of particles in the deep lung associated with chronic inflammation and remodelling.

The EGF receptor (EGFR) and its ligands might play a decisive role in these responses, as has been shown for asbestos, diesel particles, and atmospheric particles [34]. Indeed, they regulate the growth and differentiation of epithelial and conjunctive cells in the lungs and are strongly expressed in asthmatic patients, leading to an overproduction of mucus and thickening of the basal membrane. Asbestos, $PM_{2.5}$, and DiPs cause cytokine secretion following EGFR activation. This might occur in the absence of specific ligands by a mechanism known as transactivation, related to ROS. These particles also induce secretion of different EGFR ligands, such as amphiregulin, heavily involved in bronchial remodelling. This essentially basolateral secretion can explain a paracrine effect on neighbouring conjunctive and smooth muscle tissues [34]. These mechanisms have not yet been demonstrated with nanoparticles, but it seems likely that there is an interaction between the affected epithelium and neighbouring tissues via molecular signals emitted by the epithelial cells and/or macrophages, thereby initiating the physiopathological processes.

4.5 Conclusion

Oxidative stress plays a key role in cell responses induced by fine and ultrafine atmospheric particles, and very likely also in those induced by nanoparticles. The size, chemical composition, and surface reactivity of nanoparticles are important features here. One criterion in the context of regulatory procedures for the protection of exposed individuals might be obtained by determining the oxidative potential of these particles. Finally, individuals are subject to repeated exposure and it seems likely that nanoparticles can act in association with other molecules such as gases in the atmosphere.

Appendix: Table of Acronyms

AP-1	Activating protein 1
BAL	Bronchoalveolar lavage
CD4+, CD8+	Clusters of differentiation. Glycoproteins at the surface of T lymphocytes
DCF	Dichlorofluorescein
DCFDA	Dichlorofluorescein diacetate

DiP	Diesel particles
DNA	Deoxyribonucleic acid
DTT	Dithiothreitol
EGF	Epidermal growth factor
GM-CSF	Granulocyte macrophage colony stimulating factor
GPx	Glutathione peroxidase
HNE	Hydroxynonenal
HO	Heme oxygenase
ICAM	Intercelluar adhesion molecule
IgE	Immunoglobulin E
IL-1β, IL-6, IL-8	Interleukins 1, 6, and 8
iNOS	Inducible nitric oxide synthase
IκB	Inhibitor of kappa B
LDL	Low density lipoprotein
LPS	Lipopolysaccharide
MAPK	Mitogen-activated protein kinase
MARCO	Membrane receptor
M-CSF	Macrophage colony stimulating factor
MDA	Malondialdehyde
MnSOD	Manganese superoxide dismutase
NAC	N-acetylcysteine
NADPH	Nicotinamide adenine dinucleotide phosphate peroxidase
NF-κB	Nuclear factor κ B
NP	Nanoparticle
PAH	Polycyclic aromatic hydrocarbon
PG	Prostaglandins
PM	Particulate matter
ROS	Reactive oxygen species
SOD	Superoxide dismutase
TNF-α	Tumour necrosis factor α
UFP	Ultrafine particles
UV	Ultraviolet

References

1. G. Oberdörster, E. Oberdörster, J. Oberdörster: Nanotoxicology: An emerging discipline evolving from studies of ultrafine particles. Environ. Health Perspect. **113**, 823–839 (2005)
2. K. Donaldson, V. Stone, C.L. Tran, W. Krieling, P.J. Borm: Nanotoxicology. Occup. Environ. Med. **61**, 727–728 (2004)
3. A. Nel, T. Xia, L. Mädler, N. Li: Toxic potential of materials at the nano level. Science **311**, 622–627 (2006)
4. Extrapol: Pollution atmosphérique: Particules ultrafines et santé, no. 3 (2007)

5. P.H. Hoet, A. Nemmar, B. Naemery: Health impact of nanomaterials? Nat. Biotechnol. **22** (2004)
6. S. Lanone, J. Boczkowski: Biomedical applications and potential health risks of nanomaterials: Molecular mechanisms. Curr. Mol. Med. **6**, 651–663 (2006)
7. F. Russo-Marie, A. Peltier, B. Polla: *L'inflammation*. Paris, John Libbey Eurotext, 565 pages (1998)
8. T. Cedervall, I. Lynch, S. Lindman, T. Berggard, E. Thulin, H. Nilsson, K.A. Dawson, S. Linse: Understanding the nanoparticle–protein corona using methods to quantify exchange rates and affinities of proteins for nanoparticles. PNAS USA **104**, 2050–2055 (2007)
9. M. Arredouani, Z. Yang, Y. Ning, et al.: The scavenger receptor MARCO is required for lung defense against pneumoccocal pneumonia and inhaled particles. J. Exp. Med. **200**, 267–272 (2004)
10. A. Kocbach, A.I. Totlandsda, M. Lag, M. Refsnes, P.E. Schwarze: Differential binding of cytokines to environmentally relevant particles: A possible source for misinterpretation of in vitro results? Toxicol. Lett. **176**, 131–137 (2008)
11. S. Val, S. Hussain, S. Boland, R. Hamel, A. Baeza-Squin, F. Marano: Carbon black and titanium dioxide nanoparticles induce pro-inflammatory response in bronchial epithelial cells: Need for multiparametric evaluation due to adsorption artefacts. Inhal. Toxicol. **21**, 115–122 (2009)
12. I. Lynch, K.A. Dawson: Protein–nanoparticle interaction. Nano Today **3**, 40–47 (2008)
13. S. Linse, C. Cabaleiro-Lago, W.F. Wue, et al.: Nucleation of protein fibrillation by nanoparticles. Proc. Natl. Acad. Sci. USA **104**, 8691–8696 (2007)
14. M.I. Gilmour: Interaction of air pollutants and pulmonary allergic responses in experimental animals. Toxicology **105**, 335–342 (1995)
15. V. Castranova, V. Vallyathan, D.M. Ramsey, J.L. McLaurin, D. Pack, S. Leonard, M.W. Barger, J.Y. Ma, N.S. Dalal, A. Teass: Augmentation of pulmonary reactions to quartz inhalation by trace amounts of iron-containing particles. Environ. Health Perspect. **105**, 1319–1324 (1997)
16. M.C. Jaurand, F. Lévy: Effets cellulaires et moléculaires de l'amiante. Med. Sci. **15**, 1370–1378 (1999)
17. J.M. Reimund: Stress oxydant au cours des syndromes inflammatoires chroniques. Nutrition clinique et métabolisme **16**, 275–284 (2002)
18. J.L. Beaudeux, J. Peynet, D. Bonnefont-Rousselot, P. Therond, J. Delattre, A. Legrand: Stress oxydant: Sources cellulaires des espèces réactives de l'oxygène et de l'azote. Ann. Pharm. Fr. **64**, 373–381 (2006)
19. W.A. Pryor, G.L. Squadrito: The chemistry of peroxynitrite. Am. J. Physiol. **268**, L622–L699 (1995)
20. G. Deby-Dupont, C. Deby, M. Lamy: Données actuelles sur la toxicité de l'oxygène. Réanimation **11**, 28–39 (2002)
21. P. Masion, J.C. Presier, J.L. Balligrand: Les espèces réactives de l'azote: Bénéfiques ou délétères? Nutrition clinique et métabolisme **16**, 248–252 (2002)
22. M. Gardès-Albet: Stress oxydant: Aspects physico-chimiques des espèces réactives de l'oxygène. Ann. Pharm. Fr. **64**, 365–372 (2006)
23. A. Favier: Le stress oxydant: Mécanismes biochimiques. L'Actualité chimique November–December, 108–115 (2003)
24. Y.H. Huang, T.C. Zhang: Effects of dissolved and oxygen on formation of corrosion products and concomitant oxygen and nitrate reduction in zero-valent iron systems with or without aqueous Fe(II). Water Res. **39**, 1751–1760 (2005)

25. J. Pincemail, R. Limet, J.O. Defraigne: Stress oxydant et transmission cellulaire: Implication dans le développement du cancer. Medi. Sphere **134**, 1–4 (2001)
26. S.S. Leonard, G.K. Harris, X. Shi: Metal-induced oxidative stress and signal transduction. Free Radical Biol. Med. **37**, 1921–1942 (2004)
27. J.J. Marnett, J.N. Riggins, J.D. West: Endogenous generation of reactive oxidants and electrophiles and their reactions with DNA and protein. J. Clin. Invest. **111**, 583–593 (2003)
28. J. Pincemail, M. Meurisse, R. Limet, J.O. Defraigne: Méthodes d'évaluation du stress oxydatif chez l'homme: Importance en matière de prévention. Cancérologie **95**, 1–4 (1999)
29. C.M. Sayes, A.M. Gobin, K.D. Ausman, J. Mendez, J.L. West, V.L. Colvin: Nano-C60 cytotoxicity is due to lipid peroxidation. Biomaterials **26**, 7587–7595 (2005)
30. B.J. Marquis, S.A. Lave, K.L. Braun, C. Haynes: Analytical methods to assess nanoparticle toxicity. Analyst **134**, 425–439 (2009)
31. S. Arora, J. Jain, J.M. Rajwade, K.M. Paknikar: Cellular response induced by silver nanoparticles: In vitro studies. Toxicol. Letters **179**, 93–100 (2008)
32. E.J. Park, J. Yi, R.H. Chung, D.Y. Ryu, J. Choi, K. Park: Oxidative stress and apoptosis induced by titanium dioxide nanoparticles in cultured BEAS-2B cells. Toxicol. Letters. **180**, 222–229 (2008)
33. T. Xia, P. Korge, J.N. Weiss, N. Li, M.I. Venkatesen, C. Sioutas, A. Nel: Quinones and aromatic chemical compounds in particulate matter induce mitochondrial dysfunction: Implications of ultrafine particle toxicity. Environ. Health Perspect. **112**, 1347–1358 (2004)
34. A. Baeza, F. Marano: Pollution atmosphérique et maladies respiratoires: Un rôle central pour le stress oxydant. Med. Sci. **23**, 497–501 (2007)
35. N. Li, C.Sioutas, A. Cho, D. Schmitz, C. Misra, J. Sempf, M. Wang, T. Oberley: Ultrafine particulate pollutants induce oxidative stress and mitochondrial damage. Environ. Health Perspect. **11**, 709–731 (2003)
36. K. Donalson, X.Y. Li, W. MacNee: Ultrafine (nanometer) particle mediated lung injury. J. Aerosol. Sci. **29**, 553–560 (1998)
37. Q. Zhang, Y. Kusaka, K. Sato, K. Nakakuki, N. Kohyama, K. Donalson: Differences in the extent of inflammation caused by intratracheal exposure to three ultrafine metals: Role of free radicals. J. Toxicol. Environ. Health A **53**, 423–438 (1998)
38. S. Salvi, S.T. Holgate: Mechanism of particulate matter toxicity. Clin. Exp. Allergy **29**, 1187–1194 (1999)
39. E.J. Park, J. Choi, Y.K. Park, K. Park: Oxidative stress induced by cerium oxide nanoparticles in cultured BEAS-2B cells. Toxicology **245**, 90–100 (2008)
40. G. Oberdörster, J. Ferin, B.E. Lehnert: Correlation between particle size, in vivo particle persistence and lung injury. Environ. Health Perspect. **102**, 173–179 (1994)
41. X.Y. Li, D. Brown, S. Smith, W. MacNee, K. Donalson: Short-term inflammatory responses following intratracheal instillation of fine and ultrafine carbon black in rats. Inhal. Toxicol. Environ. Health Perspect. **105**, 1279–1283 (1999)
42. K.E. Driscoll, J.K. Maurer: Cytokine and growth factor release by alveolar macrophages: Potential biomarkers of pulmonary toxicity. Toxicol. Pathol. **19**, 398–405 (1991)

43. D. Höhr, Y. Steinfartz, R.P. Schins, A.M. Knaapen, G. Martza, B. Fubini, P.J. Borm: The surface area rather then the surface coating determines the acute inflammatory response after instillation of fine and ultrafine TiO_2 in the rat. Int. J. Hyg. Environ. Health. **205**, 239–244 (2002)
44. P.V. Asharani, G.L.K. Mun, S. Valiyaveettil: Cytotoxicity and genotoxicity of silver nanoparticles in human cells. ACS Nano. **3**, 279–290 (2009)
45. V. Stone, J. Shaw, D. Brown, W. MacNee, S.P. Faux, K. Donalson: The role of oxidative stress in the prolonged inhibitory effect of ultrafine carbon black on epithelial cell function. Toxicol. In Vitro **12**, 649–659 (1998)
46. E. Koike, T. Kobayashi: Chemical and biological oxidative effects of carbon black nanoparticles. Chemosphere **65**, 946–951 (2006)
47. Y.H. Hsin, C.F. Chen, S. Huang, T.S. Shih, P.S. Lai, P.J. Chueh: The apoptotic effect of nanosilver is mediated by a ROS- and JNK-dependent mechanism involving the mitochondrial pathway in NIH3T3 cells. Toxicol. Letters **179**, 130–139 (2008)
48. B. L'Azou, J. Jorly, D. On, E. Sellier, F. Moisan, J. Fleury-Feith, J. Cambar, P. Brochard, C. Ohayon: In vitro effects of nanoparticles on renal cells. Particle and Fibre Toxicology **5**, 22 (2008)
49. H. Hussain, S. Boland, A. Baeza-Squiban, R. Hamel, L.C. Thomassen, J.A. Martens, M.A. Billon-Galland, J. Fleury-Feith, F. Moisan, J.C. Pairon, F. Marano: Oxidative stress and proinflammatory effects of carbon black and titanium dioxide nanoparticles: Role of particle surface area and internalized amount. Toxicology **260**, 142–149 (2009)
50. Y.M.W. Jassen, N.H. Heintz, J.P. Marsh, P. Borm, B.T. Mossman: Induction of c-fos et c-jun proto-oncogenes in target cells of the lung and pleura by carcinogenic fibers. Am. J. Physiol. Lung Cell. Mol. Biol. **11**, 522–530 (1997)
51. W. Lin, Y. Huang, X.D. Zhou, Y. Ma: In vitro toxicity of silica nanoparticles in human lung cancer cells. Toxicol. Appl. Pharmacol. **217**, 252–259 (2006)
52. I. Annessi-Maesano, U. Ackermann, C. Boudet, L. Filleul, S. Medina, R. Slama, G. Viegi: Effets des particules atmosphériques sur la santé: Revues des études épidémiologiques. Environnement, Risques et Santé **3**, 97–110 (2004)
53. F. Marano, M. Aubier, P. Brochart, F. De Blay, R. Marthan, B. Nemery, A. Nemmar, B. Wallaert: Impacts des particules atmosphériques sur la santé: Aspects toxicologiques. Environnement, Risques et Santé **3**, 87–96 (2004)
54. D. Diaz-Sanchez, A. Tsien, J. Fleming, A. Saxon: Combined diesel exhaust particulate and ragweed allergen challenge markedly enhances human in vivo nasal ragweed-specific IgE and skews cytokine production to a T helper cell 2-type pattern. J. Immunol. **158**, 2406–2413 (1997)
55. J.A. Nightingale, R. Maggs, P. Cullinan, L.E. Donnely, D.F. Rogers, R. Kinnersley, K. Fan Chung, P.J. Barnes, M. Ashmore, A. Newman-Taylor: Airway inflammation after controlled exposure to diesel exhaust particulates. Am. J. Respir. Crit. Care Med. **162**, 161–166 (2000)
56. N. Li, C. Sioutas, A. Cho, D. Schmitz, C. Misra, J. Sempf, M. Wang, T. Oberley, J. Froines, A. Nel: Ultrafine particulate pollutants induce oxidative stress and mitochondrial damage. Environ. Health Perspect. **111**, 455–460 (2003)
57. P.H.N. Salvida, R.W. Clarke, B.A. Coull, R.C. Stearns, J. Lawrence, G.C.K. Murthy, E. Diaz, P. Koutrakis, H. Suh, A. Tsuda, J.J. Godleski: Lung inflammation induced by concentrated ambient air particles is related to particle composition. Am. J. Respir. Crit. Care Med. **165**, 1610–1617 (2002)

58. I.Y. Adamson, R. Vincent, J. Bakowska: Differential production of metalloproteinases after instilling various urban particle samples to rat lung. Exp. Lung Res. **29**, 375–388 (2003)
59. X. Liu, Z. Meng: Effects of airborne fine particulate matter on antioxidant capacity and lipid peroxidation in multiple organs of rats. Inhal. Toxicol. **17**, 467–473 (2005)
60. M. Sagai, H. Saito, H., T. Ichinose, M. Kodama, Y. Mori: Biological effects of diesel exhaust particles. I. In vitro production of superoxide and in vivo toxicity in mouse. Free Radic. Biol. Med. **14**, 37–47 (1993)
61. A. Peters, S. Perz, A. Doring, J. Stieber, W. Koenig, H.E. Wichmann: Increases in heart rate during an air pollution episode. Am. J. Epidemiol. **150**, 1094–1098 (1999)
62. D.B. Yeats, J.L. Mauderly: Inhaled environmental/occupational irritants and allergens: Mechanisms of cardiovascular and systemic responses. Introduction. Environ. Health Perspect. **109** (Suppl. 4), 479–481 (2001)
63. K. Donaldson, V. Stone, A. Seaton, W. MacNee: Ambient particle inhalation and the cardiovascular system: Potential mechanisms. Environ. Health Perspect. **109** (Suppl. 4), 523–527 (2001)
64. C.A. Poland, R. Duffin, R.I. Kinloch, A. Maynard, W.A.H. Wallace, A. Seaton, V. Stone, S. Brown, W. MacNee, K. Donaldson: Carbon nanotubes introduced into the abdominal cavity of mice show asbestos-like pathogenicity in a pilot study. Nature Nanotech. **10**, 1038 (2008)
65. A. Tagaki, A. Hirose, T. Nshimura, N. Fukumori, A. Ogata, N. Ohashi, S. Kitajima, J. Kanno: Induction of mesothelioma in p53+/– mouse by intraperitoneal application of multi-wall carbon nanotube. J. Toxicol. Sci. **33**, 105–116 (2008)
66. V. Stone, M. Tuinman, J.E. Vamvakpoulos, J. Shaw, D. Brown, S. Petterson, S.P. Faux, P. Borm, W. MacNee, F. Michnaelangeli, K. Donalson: Increased calcium influx in a monocytic cell line on exposure to ultrafine carbon black. Eur. Respir. J. **15**, 297–303 (2000)
67. K. Donalson, V. Stone, P. Borm, A. Jimenez, P. Gilmour, R.P. Schins, A.M. Knaapen, I. Rahman, S.P. Faux, D.M. Brown, W. MacNee: Oxidative stress and calcium signalling in the adverse effects of environmental particles PM10. Free Radical Biol. Med. **34**, 1369–1382 (2003)
68. V. Stone, D. Brown, N. Watt, M. Wilson, K. Donalson, H. Ritchie, W. MacNee: Ultrafine particle-mediated activation of macrophages : Intracellular calcium signalling and oxidative stress. Inhal. Toxicol. **12**, 345–351 (2000)

5

Nanoparticle Toxicity Mechanisms: Genotoxicity

Alain Botta and Laïla Benameur

Despite the relatively small amount of convincing experimental data, the potentially genotoxic nature of certain nanoparticles seems plausible, owing in particular to the presence of reactive oxygen species (ROS) such as the superoxide anion $O_2^{\bullet-}$, the hydroxyl radical $^{\bullet}OH$, and singlet oxygen $^1O_2$, and reactive nitrogen species (RNS) such as nitrogen monoxide NO, the peroxynitrite anion $ONOO^-$, the peroxynitrite radical $ONOO^{\bullet}$, and dinitrogen trioxide N_2O_3, a powerful nitration agent.

These species turn up in many studies of tissular and cellular nanoparticle toxicity. The genotoxic potential of these nanocompounds would thus appear to be closely linked to oxidative stress resulting from hyperproduction of radical species.

Note. The appendix at the end of this chapter contains a table of acronyms and a lexicon, among other things.

5.1 Mechanisms for Radical Species Production

Nanoparticle-mediated ROS and RNS production mechanisms have been thoroughly investigated and can be classified into three groups: intrinsic production, production by interaction with cell targets, and production mediated by the inflammatory reaction. The three groups share responsibility for most of the genotoxic effects so far observed with nanoparticles.

5.1.1 Intrinsic Production

This encompasses the following cases:

- Reactivity of transition metals at the surface of nanoparticles, but also in the presence of oxidising groups (e.g., quinones, silicon dioxide SiO_2), or free radicals.

P. Houdy et al. (eds.), *Nanoethics and Nanotoxicology*,
DOI 10.1007/978-3-642-20177-6_5, © Springer-Verlag Berlin Heidelberg 2011

- Reactivity of certain particles in aqueous solution, leading to the production of radical species.
- Electron transfer mechanisms in relation with the semiconducting properties of certain nanocompounds.
- Bioactivation of polycyclic aromatic hydrocarbons (PAH) and their nitrated derivatives (nitro-PAH, much more genotoxic, mutagenic, and carcinogenic than their unsubstituted counterparts) adsorbed at the surface of nanoparticles. For the latter, nitroreductase involvement (futile cycle) leads to direct ROS production.

5.1.2 Production by Interaction with Cell Targets

This concerns mitochondrial alteration (after distribution and accumulation in mitochondria of nanoparticles resorbed by cells) involving interactions with the respiratory electron transport chain and alteration of enzyme mechanisms underlying antioxidant defence.

5.1.3 Production Mediated by Inflammatory Reaction

This appears to be the main process whereby nanoparticles generate oxidative stress. This has featured in almost all studies so far described. It is a complex defence process against all forms of endogenous and exogenous attack on cell or tissue, in which macrophages and polynuclear neutrophils (PNN) are activated, and chemical mediators (e.g., histamine, etc.), cytokines (e.g., interleukines, interferons, TNF), prostaglandins, and leukotrienes are brought into play. One of the main consequences of the inflammatory process, related notably to PNN activation, is ROS and RNS production, the latter by induction of iNOS (inducible nitric oxide synthase), myeloperoxidases, and NAD(P)H-dependent oxidases.

5.2 General Genotoxicity Mechanisms

Genotoxicity resulting from the ROS and RNS produced by the three groups of mechanisms described above will be referred to as primary (direct or indirect) or secondary (see Fig. 5.1). Primary genotoxicity is generally thought to be without threshold, while it seems that thresholds can be specified for secondary genotoxicity effects.

The genotoxic potentials of nanoparticles thus appear to be directly related to oxidative damage to DNA and proteins caused by ROS and RNS. The mechanism may be clastogenic, direct or indirect (damage to the genetic material itself during the interphase or during the mitotic process), or it may be aneugenic (alteration of proteins making up the mitotic apparatus, in particular the spindle and nucleoli). Another consequence is DNA adducts, generated by

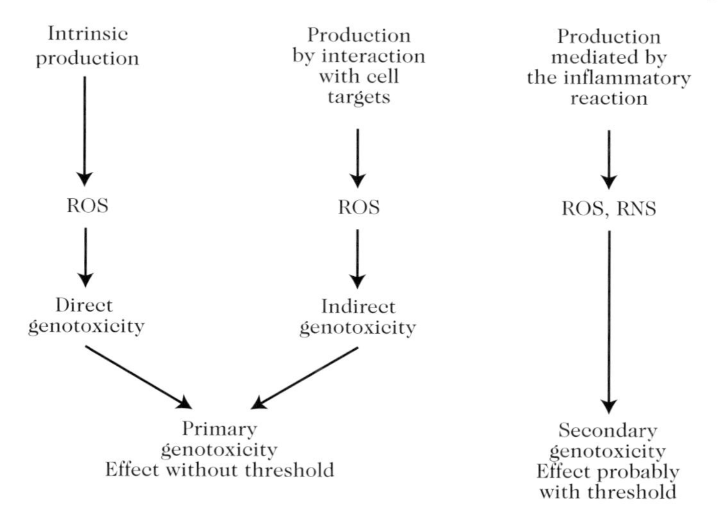

Fig. 5.1. ROS and RNS production and nanoparticle genotoxicity

electrophilic metabolites resulting from bioactivation of PAH adsorbed onto the nanoparticles and thereby delivered to the cytosol. Figure 5.2 summarises these mechanisms.

5.2.1 Direct Clastogenic Mechanisms

These underlie many DNA lesions, such as base oxidisation to produce 8-hydroxy,2′-deoxyguanosine (8-OHdG), for example, a lesion usually repaired by base excision repair (BER), or base nitration by RNS, methylation, oxidative deamination, depurination generating abasic sites, ring opening, and finally, strand breakages, especially single strand breakages (SSB), but also double strand breakages (DSB) by deoxyribose ring opening and breakage. This mechanism may have carcinogenic consequences, since the mutations resulting from oxidative DNA lesions, e.g., base pair mutations, deletions, and insertions, often turn up in oncogenes and tumour suppressing genes silenced in cancers [1].

5.2.2 Indirect Clastogenic Mechanism

This is relayed by the preliminary lipid peroxidation due to ROS which generates electrophilic unsaturated α and β aldehydes such as malondialdehyde (MDA) and 4-hydroxynonenal (4-HN) underlying the production of exocyclic DNA adducts (etheno and propano adducts).

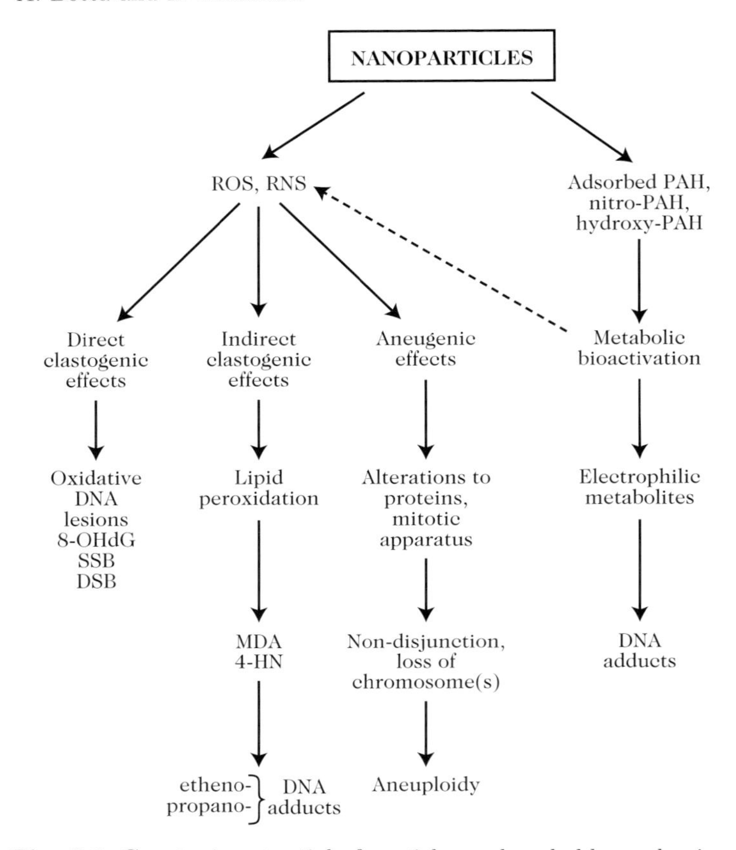

Fig. 5.2. Genotoxic potential of particles and probable mechanisms

5.2.3 Aneugenic Mechanism

This involves protein oxidative lesions, e.g., oxidation of cysteines or nitration by RNS and their derivatives such as dinitrogen trioxide N_2O_3 which results in particular in nitrotyrosination. When lesions affect components of the mitotic apparatus (achromatic spindle, microtubule organising centers, kinetochores), this can lead to dysfunction of chromosome segregation and migration during mitosis. The result may be non-disjunction in the anaphase and loss of the chromosome.

5.2.4 Production of DNA Adducts

The possibility that polycyclic aromatic hydrocarbons (PAH) can be adsorbed onto nanoparticles and delivered to the cytosol represents a particularly worrying phenomenon, since it may produce an intracellular concentration of mutagenic and carcinogenic genotoxic substances. Indeed, many PAH are bioactivated as electrophilic metabolites generating bulky DNA and protein adducts. It has been shown that certain nanoparticles, such as diesel or carbon black particles, adsorb, transport, deliver, and salt out PAH like benzo[a]pyrene (b[a]p), but also nitrated and hydroxylated PAH derivatives that are much more genotoxic than their unsubstituted counterparts. For example, one finds benzo[a]pyrenediol-epoxide–DNA (BPDE-DNA) adducts, electrophilic metabolites of b[a]p. These adducts are normally repaired by nucleotide excision repair (NER), but the main worry is that the ROS and RNS produced by PNN activation might themselves be enzymatic inducers of electrophilic PAH bioactivation. This may in turn raise their biologically effective dose (BED), as shown by the significant increase in the production of BPDE-DNA adducts in the presence of activated PNN [2]. Furthermore, the ROS and RNS produced by PNN activation might also inhibit nucleotide excision repair (NER), a process demonstrated on the human alveolar epithelial cell line A549 [3].

5.3 Detection and Characterisation of Genotoxicity

Genotoxicity detection and characterisation appeals to short term assays that fall into three main families depending on the type of abnormality to be detected: primary DNA alterations, gene mutations, or chromosome mutations.

5.3.1 Detecting Primary DNA Alterations

The tests conventionally used are the determination of sister chromatid exchanges, the unscheduled DNA synthesis (UDS) test, and the comet test. Here we shall discuss only the UDS and comet tests.

The UDS test reveals genotoxic lesions by measuring the intensity of DNA synthesis required for repairs to the genotype. Although it still applies the principle of base complementarity, this unscheduled synthesis differs from the programmed synthesis carried out during DNA replication prior to mitosis. The UDS test can be used in vitro, e.g., on primary cultured hepatocytes, and in vivo, e.g., on rodent hepatocytes.

The comet test or single cell gel electrophoresis assay is a fast, reproducible, and sensitive microelectrophoretic technique (4 h), able to visualise and evaluate single and double strand breakages, alkali-labile sites, crosslinks, and sites not yet fully repaired (excision phase of the BER and NER systems) in single prokaryote and eukaryote cells. It can be used in vitro on cell cultures, ex vivo on human lymphocytes or epithelial cells, and in vivo on the whole animal to pinpoint any organ specificity of the genotoxic substance. But it can also be used to identify apoptotic cells and to test the cell's capacity to repair DNA lesions. In addition, the comet test can be carried out in the presence of bacterial endonucleases, such as formamidopyrimidine DNA glycosylase (Fpg) or endonuclease III, which handle specific excision repair of oxidative DNA lesions (with ensuing resynthesis of the eliminated strand), and which thus give rise to SSB following the excision stage. This provides indisputable evidence of oxidative lesions in the DNA bases created by ROS.

5.3.2 Detecting Gene Mutations

Gene mutations are either substitution of a base pair (point mutations) or alterations (addition or deletion) of several base pairs (frameshift mutations).

The classic assays are the Ames test, the mouse lymphoma test, the HPRT test, and the use of transgenic mouse strains.

The Ames test detects gene mutations in strains of *Salmonella typhimurium* carrying a mutation in the operon governing the synthesis of the amino acid histidine, making them unsuited to develop in a histidine-deficient culture medium (auxotrophy). In the presence of a genotoxic substance, the reverse mutation gives the bacteria the ability to synthesise histidine once again (prototrophy), whereupon they can then develop in a medium deficient in this amino acid. Various strains have been developed with different sensitivities to genotoxic substances. For oxidative lesions, the strain TA102 is the most appropriate. The Ames test is a good tool for detecting gene mutations, but it does not detect clastogenic or aneugenic chromosome mutations. One major advantage of this test is that it lends itself just as well to detection of gene mutations induced by directly genotoxic substances as to detection of those induced by indirectly genotoxic substances, i.e., requiring some previous bioactivation which gives rise to electrophilic metabolites, the true agents of genotoxicity. To distinguish these two mechanisms, the test is carried out with or without an in vitro metabolising mixture, S9 Mix, which is a rat liver fraction induced by Aroclor (an enzyme inducer of CYP450 mono-oxygenases), combined with NADP(H) generating cofactors.

The mouse lymphoma assay is carried out on the L5178Y mouse lymphoma cell line, heterozygous at the thymidine kinase locus (tk+/−). Deactivation of the tk+ allele induces resistance to trifluorothymidine, allowing selection of the tk−/− mutants within the tk+/− cell population. This test reveals both gene mutations and clastogenic and aneugenic chromosome mutations.

The HPRT assay detects gene mutations at the hypoxanthine guanine phosphoribosyl transferase (HPRT) locus in the V79 cell lines of Chinese hamster pulmonary fibroblasts or Chinese hamster ovary (CHO) cell lines. The basis of the test is the catalysis by the enzyme HPRT of the phosphoribosylation of 6-thioguanine to produce a cytotoxic monophosphate derivative. This property can be used to assess mutations at the HPRT locus by counting clones resistant to 6-thioguanine.

The use of transgenic mouse strains has become a classic method for assessing genotoxicity in vivo. The BigBlue model contains the gene *Lac I* as target for the genotoxic substance and the gene *Lac Z* as reporter. *Lac I* represses the activity of β-galactosidase which normally hydrolyses the substrate X-Gal to produce galactose, resulting in the appearance of blue lysis plaques. *Lac Z* yields a functional β-galactosidase. Any gene mutation of *Lac I* will result in a non-functional Lac*1 protein repressor, thereby allowing hydrolysis of X-Gal by β-galactosidase and the appearance of blue plaques. Another model has been developed, called Muta Mouse, which uses the gene *Lac Z* directly as target. Based on the same principle as BigBlue, this method uses the toxicity of galactose for the bacterium *Escherichia coli* Cgal/E$^-$, which cannot develop in the presence of galactose. Any gene mutation of *Lac Z* prevents the production of galactose and thus allows the bacteria to develop.

5.3.3 Detecting Chromosome Mutations

Chromosome mutations involve several tens of kilobases, or even whole chromosomes. There are two cases:

- Structural (or qualitative) abnormalities generated by clastogenic genotoxic substances. These result from double strand DNA breakages.
- Numerical (or quantitative) abnormalities which consist in changes in the number of chromosomes, induced by aneugenic genotoxic substances creating lesions in the proteins of the mitotic apparatus.

To detect and evaluate chromosome structure and number damage, the micronucleus test is the most widely used. Micronuclei (MN) are nuclear entities independent of the main nucleus, numbering anywhere between 1 and 6 per cell, with diameters between 1/3 and 1/16 of the diameter of the main nucleus. These micronuclei are formed during cell division and comprise either acentric chromosome fragments which, not having a centromere, cannot position themselves at the equator of the achromatic spindle (clastogenic effect), or whole chromosomes that have been lost during the anaphase due to lesions of the spindle proteins (aneugenic effect).

To distinguish these two types of occurrence and hence specify whether the genotoxic substance induces a clastogenic and/or aneugenic effect, the micronucleus test is combined with fluorescent in situ hybridization (FISH) using pancentromeric DNA probes, which provide a precise fluorescent visualisation of the presence (aneugenic) or absence (clastogenic) of centromeres within the micronucleus. In certain cell types, the FISH technique can be usefully replaced by immunocytochemistry, using a monoclonal antibody to immunomark the constitutive centromere protein CENPA. The test can be carried out in vivo, ex vivo, and in vitro on Chinese hamster ovary (CHO) cell lines, mouse lymphoma L5175Y cell lines, V79 cell lines of Chinese hamster pulmonary fibroblasts, and primary cultured human cells (lymphocytes, fibroblasts, keratinocytes, melanocytes, enterocytes, etc.).

Figure 5.3 summarises the advantages of a conventional methodology associating the comet test, micronucleus test, and FISH/CENPA.

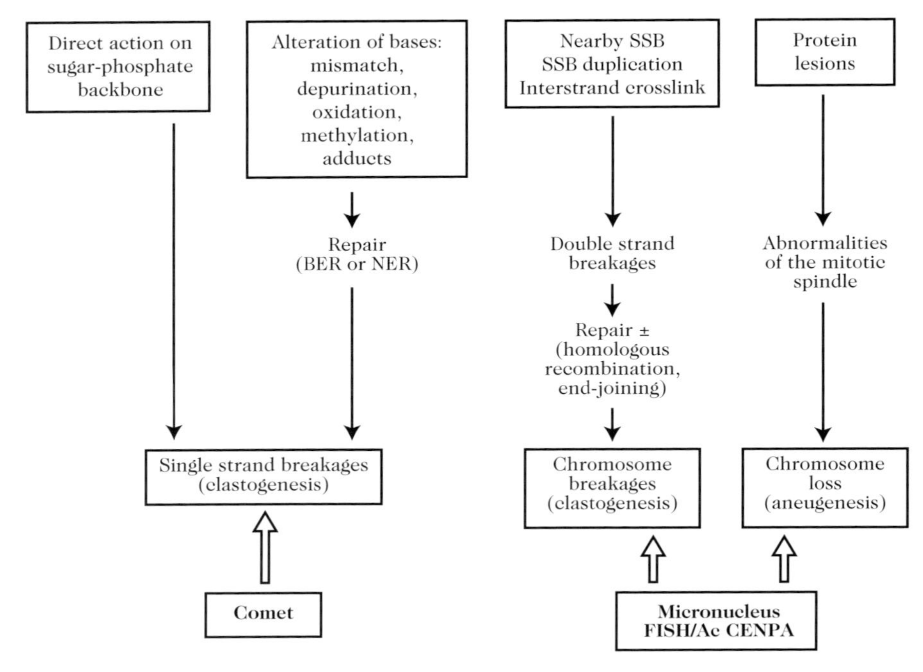

Fig. 5.3. Nanoparticle genotoxicity. Comet, micronucleus, and FISH assay protocols

5.4 Nanoparticle Genotoxic Action Mechanisms: Current Data from the Main Scientific Studies

Most studies investigate ROS production by nanoparticles in abiotic and biotic conditions, along with their harmful effects on proteins, lipids, and genotype [4]. The affinity of nanoparticles for DNA is illustrated by the ability of single wall carbon nanotubes (SWCNT) to direct the self-assembly of DNA using gold nanoparticles as binder [5], while fullerenes can bind to nucleotides and deform the double helix, suggesting the potentially negative impact of fullerenes on the structure, stability, and biological functions of DNA [6]. In addition, the strong and non-specific adsorption of oligonucleotides onto metal nanoparticles can inhibit hybridization of complementary DNA sequences [7]. This affinity of certain nanoparticles for DNA underpins certain therapeutic strategies which exploit their genotoxic aspects. For example, functionalised cationic carbon nanotubes are often used as specific cell vectors of functional DNA or siRNA (silencer RNA or micro-RNA, short nucleotide sequences taking part in post-transcriptional regulation of gene expression) to specifically modify the expression of a target gene [8].

While DNA lesions, caused either directly or indirectly by oxidative stress, are now well documented, the precise mechanism by which nanoparticles produce ROS is still under investigation to assess the relative importance of the direct mechanism and the indirect mechanisms, particularly when mediated by the inflammatory cell response to the presence of nanoparticles. Furthermore, it seems possible that cells containing a high concentration of antioxidants, e.g., reduced glutathione, or antioxidant enzymes, e.g., catalase, peroxidase, superoxide dismutase, might be more resistant to the toxic action of nanoparticles. Finally, our understanding of nanoparticle surface properties suggests that very small particles might be more toxic than their larger counterparts, owing to their greater specific surface area, entailing a greater bioavailability [9].

The main mechanisms whereby cells interact with nanoparticles, generators of oxidative stress, fall into three categories, according to the most widely accepted hypothesis:

- Direct involvement of the surface effect.
- Involvement of redox mechanisms due to transition metals salted out by nanoparticles.
- Activation of membrane receptors such as epidermal growth factor receptor (EGFR, a gene coding for a cell surface protein inducing proliferation, hyperexpressed in many cancers), by transition metals following their intracellular diffusion.

In the first two cases, oxidative stress is accompanied by increased cytosolic calcium concentrations and activation of signalling pathways inducing activation of transcription factors, notably, nuclear factor-$\kappa\beta$ (NF-$\kappa\beta$, a regulatory pathway for genes important for the survival of the cell), involved in the transcription of key genes.

In the third case, the activation of membrane receptors leads to the mitochondrial distribution of nanoparticles and concomitant production of oxidative stress.

All in all, the involvement of oxidative stress discovered in in vivo and in vitro studies for most nanoparticles would appear to be a key stage in their genotoxicity mechanism. For example, ROS production has been demonstrated for fullerenes, single wall carbon nanotubes, multiwall carbon nanotubes, quantum dots, and metal-containing nanoparticles. This ROS production is sometimes affected by simultaneous exposure to visible light or UV radiation [10].

5.4.1 Carbon-Containing Nanoparticles

Single Wall Carbon Nanotubes

Genotoxic effects mediated by oxidative stress have been observed in in vitro and in vivo experimental studies, but the exact mechanism remains to be identified.

Single wall carbon nanotubes in cultures of immortalised human epidermal keratinocytes (HaCat cell line) induce ROS production, lipid peroxidation, and antioxidant depletion [11]. In cultures of human embryo kidney cells

(HEK293 cell line), single wall carbon nanotubes inhibited cell proliferation and adhesion, arrested the cell cycle in the G1 phase, and induced apoptosis. Internucleosomal fragmentation and overexpression of proapoptotic genes such as *p53* and *bax* are observed [12]. In Chinese hamster lung fibroblasts (V79 cell line), single wall carbon nanotubes purified for 3 h and at 96 μg/cm^2 induced single and double strand DNA breakages, as revealed by the comet test, which also identified alkali-labile sites. However, for the same exposure, the micronucleus test did not reveal any significant increase in the number of micronucleated cells, and the Ames test (strains YG1024 and YG1029) was negative. The genotoxicity of single wall carbon nanotubes is thus confirmed by the comet test, which assesses primary DNA lesions, while the negative micronucleus test suggests that these primary lesions were effectively repaired [13].

In contradiction with these results, a study comparing samples of single wall carbon nanotubes with different purities showed that the most highly purified nanotubes of amorphous carbon lost their ability to induce acute toxicity and oxidative stress. The latter result, reported as significant but not supported by experimental data, led the authors to conclude that oxidative stress, along with the induction of inflammation, seemed to be directly related to the presence of metal impurities, including nickel, iron, and other persistent heavy metals [14].

Finally, an in vivo study was carried out on C57BL/6 mice after administration of single wall carbon nanotubes of diameter 0.8–1.2 nm and length 100–1 000 nm, using two different forms of exposure: closed-chamber inhalation (5 mg/m^3 of single wall carbon nanotubes, 5 h per day for 4 days) and aspiration through the lungs of a particle suspension deposited in the pharynx (5, 10, or 20 μg per mouse). Both exposure routes generated an immediate inflammatory reaction, oxidative stress, fibrosis, and hyperplasia of bronchial epithelium cells, but only inhalation exposure caused genotoxicity as evaluated by analysing mutations of the gene *k-ras*, persisting after 28 days of exposure [15].

Multiwall Carbon Nanotubes

Multiwall carbon nanotubes at concentrations of 5 and 100 μg/ml induced apoptosis of mouse embryo stem cells (ES cell line) by activating the protein p53. This protein can arrest the cell cycle in the case of DNA lesions, thereby allowing the cell to implement DNA repair systems, but it can also induce apoptosis if these repair systems are insufficient or if the lesions cannot be repaired. In addition, on the same cell type, multiwall carbon nanotubes produced hyperexpression of two isoforms of OGG1 (8-oxo-guanine DNA glycosylase 1, a key enzyme in the base excision repair system for oxidative DNA lesions), thus suggesting that multiwall carbon nanotubes may induce oxidative DNA damage, notably on guanine, via ROS. Moreover, in the same experiments, the multiwall carbon nanotubes used seemed to play a role in the hyperexpression of Rad51 and XRCC4, proteins involved in the repair of double strand DNA breakages, but also in the increased phosphorylation of the histone H2AX, a protein participating in the organisation of chromatin and the repair of double strand DNA breakages, and the sumoylation of XRCC4. [Protein sumoylation is a SUMO-type transcriptional modification (small ubiquitin-like modifier), involved among other things in transcriptional regulation and promotion of the cell cycle.] Such results suggest that these nanoparticles induce double strand DNA breakages.

Finally, mutagenesis studies using the endogenous molecular marker Aprt (adenine phosphoribosyl transferase) show that multiwall carbon nanotubes significantly increase the mutation rate compared with the spontaneous mutation rate in mouse embryonic stem cells [16].

Applying the micronucleus test to rat lung epithelial cell cultures exposed to multiwall carbon nanotubes at concentrations of 10, 25, and 50 μg/ml, it was shown that this exposure leads to a significant increase in the number of micronucleated cells and the number of micronuclei per cell. In addition, a simultaneously clastogenic and aneugenic genotoxic mechanism was identified using in situ hybridization of fluorescent pancentromeric probes on human epithelial cell lines (MCF-7) exposed to multiwall carbon nanotubes. Moreover, intratracheal instillation of multiwall carbon nanotubes (0.5 or 2 mg) in rats over 3 days led to a dose-dependent increase in micronuclei in type II pneumocytes [17]. Results obtained in vitro and in vivo thus seem sufficiently concordant to conclude the genotoxicity of multiwall carbon nanotubes with both clastogenic and aneugenic mechanisms of genomic mutation.

Fullerenes

Fullerenes (C_{60}) are responsible for various dysfunctions in human cells, e.g., dermal fibroblasts, hepatocarcinoma cells, and neuronal astrocytes, mediated by excess ROS production causing lipid peroxidation. When an antioxidant (L-ascorbic acid) is added to the culture medium, oxidative damage resulting from the presence of C_{60} is completely averted [18].

Exposure of FE1-Muta mouse lung epithelial cell lines to C_{60}, single wall carbon nanotubes, and carbon black nanoparticles induced ROS production. Applying the comet test to cells exposed to compact aggregates of C_{60} (n-C_{60}) and to single wall carbon nanotubes did not reveal any increase in the rate of DNA strand breakages, but combining the comet test with a preliminary treatment by the endonuclease FPG demonstrated that there were oxidative DNA lesions. Furthermore, n-C_{60} and single wall carbon nanotubes did not induce mutagenic effects as evaluated by measuring the mutation rate at the cII locus in the FE1-Muta mouse [19].

It has been demonstrated that photoactivation is relevant in the initial stages of the oxidative stress production at the root of the genotoxic mechanism. In the presence of UV or visible irradiation, C_{60} molecules (complexed with cyclodextrin) induced oxidative damage in rat liver microsomes [20]. In contrast, without light irradiation, the C_{60} molecules behave as antioxidants [21]. The mutagenicity of pure C_{60} ($> 99.9\%$) dissolved in polyvinylpyrrolidone has been investigated using the Ames test with and without S9 Mix on the strains TA102, TA104, and YG3003 (mutation of TA102). No mutagenic effect was observed on the strain TA102 (with or without metabolic activation) except when the preparation was first irradiated by visible light. The level of mutagenicity was then dose dependent and varied with the time of irradiation. The same results were obtained with the strain YG3003 and to a lesser extent with the strain TA104 [22].

A different result was obtained with a mixture of C_{60} and C_{70}. Indeed, the Ames test on strains TA100, TA1535, TA98, and TA1537, with and without metabolic activation, and a test carried out on *Escherichia coli* (WP2uvrA/pKM101) revealed no mutagenic effect, even at fullerite concentrations above 5 mg per dish. In addition, a chromosome aberration assay on Chinese hamster lung cell lines (CHL/IV) showed no abnormality of number or structure, and no clastogenic effect, even at concentrations of 5 mg/ml [23].

The difficulty in interpreting experimental results and the need to standardise methodologies are well illustrated by the following two studies. The first involved exposure of fish to colloidal suspensions of C_{60} dispersed in tetrahydrofurane (THF). A significant level of lipid peroxidation was induced in the brain, together with a depletion of reduced glutathione (GSH) in the liver and gills but no oxidative lesions, notably in proteins [24]. Today this study is the subject of some controversy, since it seems that the effects attributed to C_{60} may have been caused by a decomposition product of the THF [25]. Another study prepared stable aqueous colloidal suspensions of C_{60} using two methods, first by dispersing C_{60} in ethanol and then redispersing it in water, the other by directly dissolving it in water. A primary culture of human lymphocytes was exposed to these two preparations and primary DNA lesions were then identified by the comet test. Both types of suspension induced a dose-dependent increase in DNA damage, but the suspension prepared by direct dispersion in water turned out to be more genotoxic than the one first prepared in ethanol. The reason given was the formation of an ethanol–C_{60} complex, together with hydroxylation of the C_{60}, thereby limiting its reactivity and hence its toxicity [26].

Also in the literature are studies demonstrating the antioxidant potential of the fullerenes, comparable with those of vitamins C and E, which might therefore by relevant in the prevention of cellular oxidative stress! This dichotomy highlights the need to perfect research methods for dealing with the interactions between nanomaterials and their cell targets, especially with regard to standardising the doses used, exposure conditions, and the biological effects that need to be identified [9].

Carbon Black

It has been demonstrated that oxidative stress plays a key role in the genotoxic potential of carbon black. Subchronic inhalation of nanometric carbon black (16 and 70 nm) results in increased levels of 8-oxo-deoxyguanosine (8-oxodG) in rat lungs [27]. Carbon black nanoparticles of diameter 14 nm, co-administered to mice with bacterial endotoxins caused a pulmonary edema and worsened the inflammatory state of the animals. This state, associated with an increased level of pro-inflammatory markers and 8-OHdG in the lungs independently of the effects of the endotoxins, suggests that carbon black nanoparticles may also facilitate the effects of other environmental stimuli [4]. Primary genotoxic lesions have been identified in lung epithelial cells of transgenic FE1-Muta mice, where carbon black nanoparticles increased rates of DNA SSB, notably in the presence of Fpg. On the other hand, long term exposure to low concentrations of these nanoparticles only resulted in a slightly increased mutation rate for the genes *cII* and *Lac Z*.

Carbon black nanoparticles would thus appear to be genotoxic for eukaryotic cell lines, the mechanism being relayed by oxidative stress [28]. In addition, carbon black nanoparticles non-functionalised and functionalised by benzo[a]pyrene induced SSB, identified by the comet test, in human lung epithelial cells, as well as activating p53 proteins and inducing the production of the transcription factors NF-$\kappa\beta$ and AP-1 (regulation factor for expression of the target gene, comprising a combination of the proteins Jun and Fos) [29]. Finally, comparing carbon black nanoparticles of the same size and composition, but with different specific surface areas (300 vs. 37 m^2/g), it was found that the resulting biological effects, e.g., inflammation, genotoxicity, depend on the specific surface area rather than the mass of the particles. Furthermore, similar work on the carcinogenic effects of inhaled particles has shown that tumour incidence is more closely correlated with specific surface area than with particle mass [30].

5.4.2 Metal-Containing Nanoparticles

Cobalt

The genotoxic effects of cobalt-containing nanoparticles have been investigated on pure cobalt, but also cobalt–chromium and cobalt–iron mixtures. These studies have aimed to assess the influence of particle size on the nature and strength of the effects. A comparison between the genotoxic effects of 100 nm cobalt nanoparticles and cobalt ions Co^{2+} in human leukocytes from voluntary donors has shown a very significant level of intracellular internalisation of cobalt nanoparticles. The micronucleus test revealed a significant increase in the number of micronucleated cells under exposure to Co^{2+} ions, whereas the comet test shows that cobalt nanoparticles induce instead primary DNA lesions. This suggests that genotoxic effects leading to SSB are probably well repaired, but modulated by the salting out of Co^{2+} ions. Moreover, the genetic polymorphism of the donors, especially in the repair gene hOGG1 (a gene coding for a protein that excises the oxidised base 8-oxo-guanine), may also modulate the genotoxic response by more or less completely repairing all the ROS-induced SSB [31].

The influence of size has been studied by comparing the cytotoxic and genotoxic effects of nanoparticles (30 nm) and microparticles (2.9 μm) of cobalt–chromium alloys in human fibroblast cultures. The nanoparticles were internalised by the fibroblasts and induced more DNA damage, as assessed by the comet test, than the microparticles. However, the micronucleus test did not reveal any significant difference between the micronucleated cells produced by the two types of particle, whereas the nanoparticles induced more aneuploid lesions than the microparticles [32]. One can hypothesise that the two types of particle induce different genotoxic action mechanisms, even though they have the same chemical composition: clastogenic CSB could be fully repaired, leaving only aneugenic effects.

In addition, analysis of the potential cytotoxic and genotoxic effects of cobalt–ferrite nanoparticles ($CoFe_2O_4$) of nanometric (5.6 nm) and micrometric (10 and 120 μm) sizes has been carried out in cultures of human peripheral lymphocytes. The nanoparticles significantly reduced the cell proliferation index (attesting to diminished cell viability) and a significant increase in the frequency of micronucleated binucleated lymphocytes (MBNL), while the 10 μm microparticles only increased the frequency of MBNL. In addition, to see whether the genotoxicity might not have been caused by salted out ions, the nanoparticles were coated, precisely to block this effect. The results showed that there were micronucleated cells, but that they were four times less common than for exposure to bare nanoparticles. This tends to corroborate the hypothesis that Co^{2+} ions play a role in the genotoxicity of cobalt–ferrite nanoparticles [33].

Silver

The genotoxicity of two types of silver nanoparticle, functionalised, i.e., coated, by surface polysaccharides, and non-functionalised, i.e., uncoated, was investigated on two types of mouse embryo cells: stem cells (mES) and fibroblasts (MEF). The two types of nanoparticle induced the expression, in both cell types, of Rad51, a protein involved in the repair of DNA DSB. The hypothesis that they cause DSB was confirmed by immunofluorescence and immunoblot. Furthermore, the two types of nanoparticle raised the level of expression of *p53*. Finally, the different surface chemistries led to different alterations in the DNA. The functionalised nanoparticles had more effect on the DNA than the non-functionalised ones. It thus seems logical to assume that the functionalised nanoparticles were barely agglomerated, and hence well distributed, while the non-functionalised nanoparticles were highly agglomerated, leading to lower availability and limited access to organelles [34].

Cerium

Cerium nanoparticles seem to display redox cycles, especially in the presence of hydrogen peroxide. This might result in ROS production through Fenton-like reactions [35]. In addition, exposure of primary cultured human fibroblasts to cerium dioxide (CeO_2) nanoparticles revealed a dose-dependent production of DSB (evaluated by the comet test), significantly reduced in the presence of the antioxidant L-ergothioneine, but also a significant induction of micronuclei. Furthermore, when nanoparticles were incorporated into the cell culture medium, 25–30% of the surface atoms were reduced from Ce^{4+} to Ce^{3+}. This reduction could generate ion and electron transfers underlying the oxidative stress and genotoxicity observed in human fibroblasts [36]. On the other hand, the Ames test applied to 9 nm CeO_2 nanoparticles gave negative results both with and without metabolic activation, at all tested concentrations [37].

Titanium

Titanium dioxide (TiO_2) nanoparticles, coated or uncoated, have been the subject of much experimental work, on both cultured animal cells and cultured human cells. Studies of the influence of photoactivation of these nanocompounds on the strength of the genotoxic response have led to some contradictory results, and no formal conclusions are yet possible.

After a 1 h inhalation of an aerosol of 22 nm TiO_2 nanoparticles, on average 24% of these nanocompounds ended up within the epithelial barrier, but also in the main compartments of the pulmonary tissue, in the cell cytoplasm and nucleus [38]. TiO_2 nanoparticles in anatase form, coated with vanadium pentoxide V_2O_5, induced a higher level of cytogenotoxicity than the same particles when they were not coated, as evaluated by the micronucleus test on V79 cells, Chinese hamster lung fibroblasts. Furthermore, the coated particles produced more ROS and unsaturated α,β aldehydes (resulting from lipid peroxidation) than the uncoated ones. Anatase nanoparticles coated with V_2O_5 are thus more genotoxic than bare nanoparticles, and their genotoxicity is mediated by oxidative stress [39]. After UVA irradiation, these nanoparticles induce DNA strand breakages and reduce the integrity of lysosomal membranes in fish cell lines [40]. Likewise, in goldfish skin cells exposed to concentrations of 1, 10, and 100 μg/ml of TiO_2 nanoparticles (anatase form, 5 nm), with and without UVA radiation, much greater genotoxic damage was observed after irradiation, including reduced cell viability and increased number of DNA oxidation sites (comet test in the presence of Fpg and endonuclease III). Moreover, using the analytic technique known as electron spin resonance (ESR), it was shown that ROS production (including the hydroxyl radical $^\bullet OH$) was greater after UVA irradiation [41].

The genotoxicity mechanism was investigated by studying Syrian hamster embryo (SHE) fibroblasts in which TiO_2 nanoparticles with diameters smaller than 20 nm and concentrations in the range 0.5–10 μg/cm^2 produced micronuclei by a clastogenic mechanism demonstrated by analysing kinetochores in the micronuclei using CREST antibodies (centromeric antiprotein antibodies common in certain autoimmune disorders). Furthermore, these nanoparticles induced apoptosis via internucleosomal cleavage and chromatin compaction. The mechanism put forward appeals to interactions between the nanoparticles and the fibroblast cell membranes. These would induce ROS production, which would in turn induce lipid peroxidation, disturb intracellular Ca^{2+} homeostasis, and alter the metabolic pathways. Disturbing Ca^{2+} activates endonucleases which may in turn initiate chromatin fragmentation, a key feature of apoptosis [42].

The importance of form has also been demonstrated by studying the photoclastogenic effects of three types of titanium dioxide nanoparticle, viz., anatase, rutile, and a mixture of both, in coated, doped, and uncoated forms in Chinese hamster ovary cells (CHO-WBL), with and without UV irradiation. Results show that not all forms of titanium dioxide induce an increased rate of chromosome aberrations (photochemical genotoxicity) with and without UV irradiation [43].

Contradicting all previous data, a study carried out in vivo by instilling rats with different doses (0.15, 0.3, 0.6, and 1.2 mg) of two types of 20 nm TiO_2 nanoparticle (hydrophilic, with untreated surface, and hydrophobic, with surface silanised by trimethoxyoctylsilane) for 90 days did not reveal any evidence of 8-oxoguanine production in the DNA of alveolar epithelial cells [44].

Studies on cultured human cells have brought some progress in understanding the genotoxicity mechanisms of TiO_2 nanoparticles. For example, in human lymphoblastoid cell lines, they reduce cell viability, induce DNA damage as measured by the micronucleus and comet tests, and increase the mutation rate in the HPRT test [45]. In addition, TiO_2 nanoparticles doped with cerium IV have been found on human hepatoma cell membranes and resorbed in the cytosol by phagocytosis. If these nanoparticles are first irradiated by visible light, they induce micronuclei and internucleosomal fragmentation of DNA, probably by activation of endogenous endonucleases bringing about apoptosis [46].

In 2008, a genotoxicity study of 25 nm TiO_2 nanoparticles (Degussa P25: 70–85% anatase/30–15% rutile) was carried out on cultured peripheral human lymphocytes, using the comet and micronucleus tests. The lymphocytes treated with nanoparticles exhibited dose-dependent production of micronuclei and DNA strand breakage, but also increased ROS production which could cause breakage or loss of genetic material in the lymphocytes. In addition, pretreating the lymphocytes with N-acetylcysteine (NAC), an antioxidant reduced glutathione precursor, itself a free radical scavenger, significantly inhibited ROS production and oxidative DNA damage.

Apart from this, TiO_2 nanoparticles induced the accumulation and activation of P53 proteins, but without affecting expression of the molecular targets of P53, viz., P21 and BAX. The results of this study demonstrate the genotoxicity of TiO_2 nanoparticles in human lymphocytes mediated by oxidative stress and the activation of P53 but without concomitant stimulation of its transactivation activity, required to arrest the cell cycle and produce apoptosis [47]. Finally, another study has shown that, without photoactivation, 10–20 nm anatase induces oxidative DNA lesions, lipid peroxidation, and formation of micronuclei in human bronchial epithelial cell lines (BEAS-2B). On the other hand, larger particles (anatase 200 nm) did not induce oxidative stress under the same conditions. This seems to confirm that particle size reduction is the sole factor leading to oxidative damage [48].

Iron

Nanoparticles of maghemite (γFe_2O_3) coated with dimercaptosuccinic acid (DMSA) have been found adsorbed onto the outer membranes of fibroblasts, then internalised via endocytosis vesicles. In this case, no cytotoxicity or genotoxicity were observed. The stability of the DMSA coating was monitored during contact between the nanoparticles and the fibroblasts. The DMSA remained chemically adsorbed to the surface of the maghemite nanoparticles, thus forming a stable organic layer which protects the cells from direct contact with the surface of these nanoparticles [49].

Iron–platinum (FePt) nanoparticles coated with tetraethylammonium hydroxide (9 nm) were subjected to the Ames test on strains TA98, TA100, TA1535, and TA1537, and also on *Escherichia coli* WP2uvrA/pKM101 strains, with and without metabolic activation. All tested concentrations gave negative results, except the strain TA100 [50].

Magnetoliposomes (14 nm nanoparticles made from magnetite Fe_3O_4 and coated with a lipid bilayer) were administered intravenously to SWISS mice. The micronucleus test applied after 12, 24, and 48 h to (anucleated) polychromatic erythrocytes showed a tendency for micronucleus induction, but only at 24 h [51]. The same test applied to magnetite nanoparticles coated with polyaspartic acid (8.5 nm), administered from 1 to 30 days, showed micronucleus induction at 1 to 7 days [52]. The presence of magnetic nanoparticles located in hematopoietic stem cells may constitute a cancer risk and increase the frequency of leukaemias caused by prolonged exposure to electromagnetic fields [53].

Zinc

The genotoxicity of zinc oxide nanoparticles coated with tetraethylammonium hydroxide has been assessed using the Ames test on the strains TA98, TA100, TA1535, and TA1537, and on the *Escherichia coli* strains WP2uvrA(−), both with and without metabolic activation. The results were negative at all tested concentrations [54].

5.4.3 Quantum Dots

There have not been many genotoxicity studies here. Some investigate the possibility of using quantum dots for targeted therapy, especially against cancers. Human breast carcinoma cells (MCF-7) treated with cadmium telluride quantum dots revealed the following phenomena in the nucleus: nuclear reorganisation, hypoacetylation of histone H3, reduced expression of genes involved in preventing cell death [heat shock protein 70 (Hsp70) and the apoptosis inhibitor cIAP-1], totally inhibited expression of genes for glutathione peroxidase, and overexpression of apoptotic genes controlled by P53 [p53-upregulated modifier of apoptosis (PUMA) genes and NADPH oxidase activator 1 (NOXA) genes]. In mitochondria, these quantum dots induce membrane damage and increased intracytosolic ROS production, causing accumulation of Bcl-2 associated X (BAX) proteins involved in salting out pro-apoptotic factors. These quantum dots thus induce both a genotoxic response via P53 and a global epigenetic response, which will lead in the long term to genetic reprogramming [55]. Likewise, photoactivated CdSe–ZnS quantum dots produced strand breakage and nucleobase (purine and pyrimidine) damage in a plasmid. This damage to plasmid DNA (pDNA) is correlated with ROS production under photoactivation conditions. The authors of this study show how these quantum dots might be used to target cancer cell nuclei in photodynamic therapies [56].

5.4.4 Other Types of Nanoparticle

Cationic Polystyrene Nanospheres

Cationic polystyrene nanospheres are inert and do not produce ROS in abiotic media. However, when these nanoparticles enter a biological medium, e.g., when they are placed in contact with murine macrophages, they induce superoxide anions $O_2^{\bullet -}$, mitochondrial damage, lysosome loss, and in some cases apoptosis [4].

Silica-Containing (SiO_2) Nanoparticles

Results of studies on both colloidal and crystalline silica are not always fully consistent, but on the whole it looks as though these nanocompounds do have genotoxic potential.

Low-dose instillation of ultrafine colloidal silica in ICR mice produces a moderate to severe inflammation in lung epithelial cells and macrophages, together with apoptosis and tissue lesions which seem to be related to induction of 8-OHdG, an oxidative stress marker [4]. These silica nanoparticles can also enter into human and rodent neuronal and epithelial (nasal, pulmonary) cell nuclei, where they cause an alteration of the nuclear structures, inducing the formation of aberrant clusters of nucleoplasmic topoisomerase II, and aggregation of proteins such as ubiquitin, huntingtin, and proteasomes.

One consequence of the formation of intranuclear protein aggregates would be inhibition of replication, transcription, and cell proliferation, without significantly altering proteasomal activity or cell viability [57].

Administered to transgenic Lac I rats, crystalline silica particles (crocidolite) induced transversion mutations G→T in the gene *Lac I*, correlated with 8-OHdG production [58]. Genotoxicity was investigated by exposing human lymphoblastoid cells (WIL2-NS) to SiO_2 nanoparticles (< 100 nm) at concentrations of 0, 30, 60, and 120 μg/ml for periods of 6, 24, and 48 h. The micronucleus test pinpointed two dose-dependent effects: an increased level of micronucleated cells and a reduced cell proliferation index, while the comet test gave no significant results, making it very difficult to understand the genotoxic mechanism. However, at a concentration of 120 μg/ml, these nanoparticles induced a significant increase in the number of mutants detected by the HPRT test [59]. On the other hand, the genotoxic effects did not turn up in all studies. For example, commercial laboratory-synthesised silica nanoparticles (Glantreo, 30 and 80 nm) in contact with 3T3-L1 fibroblasts for periods of 3, 6, and 24 h at concentrations of 4 and 40 μg/ml did not generate a detectable genotoxic effect under the comet test.

These apparently surprising results have nevertheless been independently validated in two different laboratories [60]. Moreover, a genotoxicity study of light-emitting silica nanoparticles in human lung epithelial cells (A549) showed no genotoxic effects below a concentration of 0.1 mg/ml [61]. Finally, in a study that demonstrated ROS and RNS induction by bare crystalline silica particles in rat lung cells, at much higher levels than for the same particles coated with polyvinylpyridine-*N*-oxide, it was shown that this production of radical species was not associated with higher 8-OHdG levels. On the other hand, hyperexpression of the genes APE/Ref1 coding for a protein mediating base excision repair (BER) of DNA lesions was detected. The inflammatory process may therefore be accompanied by a compensatory induction of APE/Ref1 translation, and hence effective repair of oxidative DNA lesions [62]. Finally, with regard to the possibility of nanoparticles carrying and delivering pollutants, a Trojan horse effect has been suggested for the way silica nanoparticles can facilitate cell penetration by associated heavy metals. The resulting oxidative stress was some eight times greater than for the heavy metals alone in an aqueous solution [63].

Diesel Particles

When BigBlue Lac I transgenic rats are exposed to diesel particles containing various concentrations of nanoparticles, it causes transversion mutations G→T in the gene *Lac I*, correlated with 8-OHdG production and induction of mRNA for CYP450 1A1 (isoform of CYP450 monooxygenases) [64].

5.4.5 Comparative Studies Between Different Nanoparticles

Crystalline silica, carbon black, and titanium dioxide particles induced mutations of the gene *HPRT* in rat alveolar cells for doses at which they also produce an inflammatory reaction through accumulation of polynuclear neutrophils [65].

Single wall carbon nanotubes and carbon black, ZnO, and SiO_2 nanoparticles in contact with primary mouse fibroblast cultures induced a significant depletion of GSH, inhibition of superoxide dismutase activity, and dose-dependent production of ROS and MDA. Comparative analysis shows that the nanoparticle composition probably plays a key role in cytotoxic effects, while genotoxic potential seems to be more closely related to form [66].

Human lung epithelial cell lines (A549) were exposed to metal oxide nanoparticles (CuO, TiO_2, $CuZnFe_2O_4$, Fe_3O_4, and Fe_2O_3) and multiwall carbon nanotubes. The CuO nanoparticles were the most active, causing cytotoxicity, DNA damage, oxidation lesions (assessed by the comet test), and intracellular ROS production. The ZnO nanoparticles reduced cell viability and caused DNA damage. The TiO_2 (rutile and anatase) and $CuZnFe_2O_4$ nanoparticles caused DNA damage. Iron oxide (Fe_2O_3, Fe_3O_4) nanoparticles exhibited little or no toxic effect. Finally, the multiwall carbon nanotubes induced DNA damage at the lowest tested doses [67].

5.4.6 Review of Genotoxicity Mechanisms

Figure 5.4 summarises the possible ROS production mechanisms in relation with nanoparticle genotoxicity, while Fig. 5.5 reviews our present state of understanding of genotoxic mechanisms. Tables 5.1, 5.2, and 5.3 summarise all genotoxicity mechanisms so far identified for the various nanoparticles that have been investigated.

5.5 Conclusion

Investigations carried out to date on the genotoxic potentials of nanoparticles seem to justify concern. Observed effects are probably mediated by oxidative stress, although direct genotoxic effects should not be excluded, and nor should combined effects involving the delivery of adsorbed pollutants. The inflammatory reaction seems to be an important relay mechanism for genotoxicity, but it is certainly not the only one that needs to be considered. Furthermore, the disagreement between certain results raises some doubts about the validity

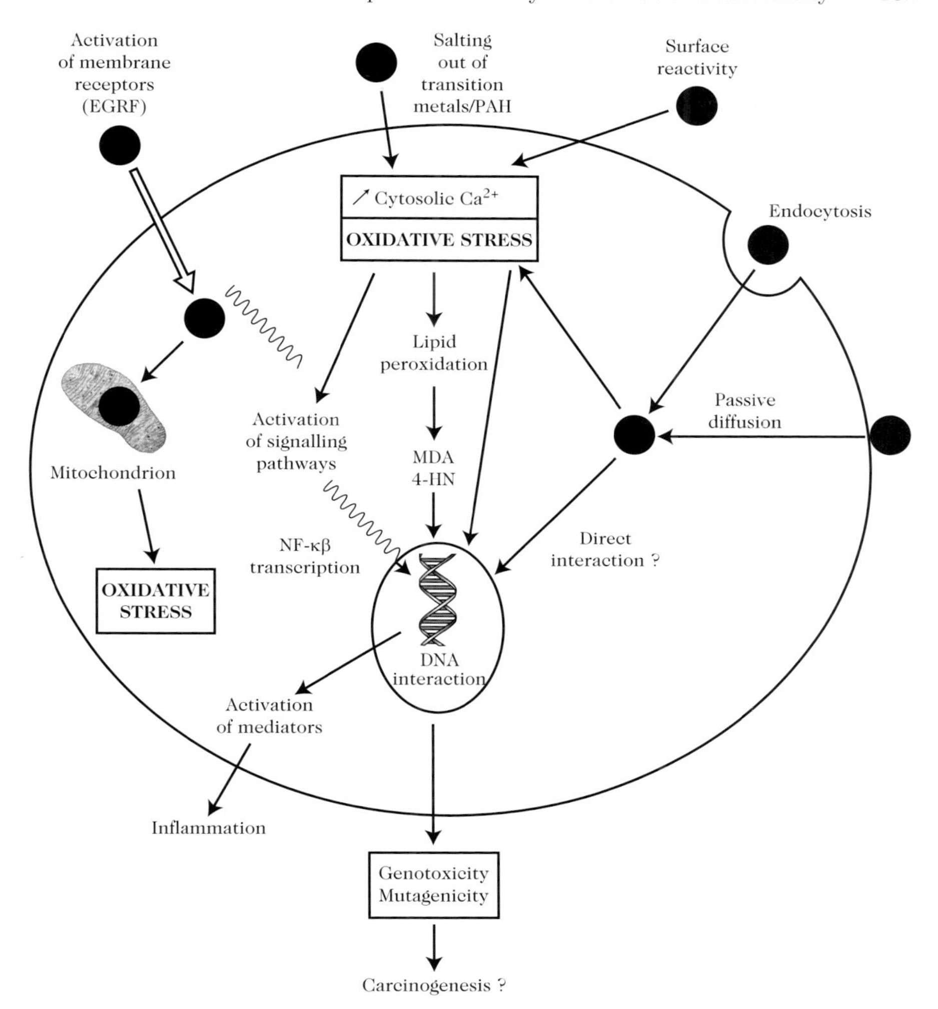

Fig. 5.4. Hypotheses regarding ROS production mechanisms in relation with nanoparticle genotoxicity. Adapted from [10]

of conventional toxicological methods where nanoparticles are concerned. The relevance of current assays needs to be reassessed, and more suitable methods are likely to be developed, not only to shed light on biological barrier crossing mechanisms, but also to understand processes leading to accumulation in target tissues and the exact nature of the interactions with biological macromolecules. Likewise, cell models, sample preparations, doses, and contact times must all be reconsidered, taking into account on the one hand variations in the compositions and impurity levels of the given nanocompounds,

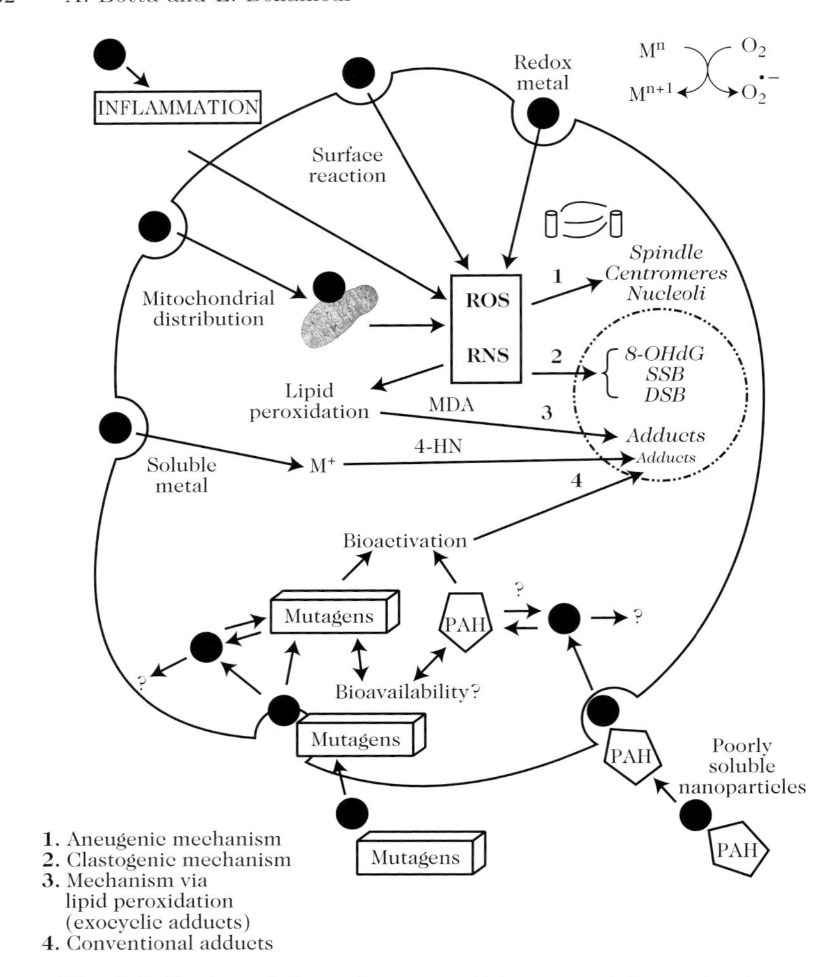

Fig. 5.5. Review of the main nanoparticle genotoxicity mechanisms

and on the other the influence of the new length scale relevant to nanoparticles, not to mention the two-phase results sometimes found as a function of the concentration. A significant example is provided by the fullerenes. These prove to be antioxidant at low doses, but generate oxidative stress, relayed by inflammation, at high doses. Finally, genotoxic mechanisms are unlikely to be the only ones to carry the cell toward mutagenesis and carcinogenesis. Certain epigenetic mechanisms such as the impact on gene expression regulators like histones or micro-RNA also deserve further investigation.

Table 5.1. Genotoxic effects of carbon-containing nanoparticles and nanomaterials

Nanoparticle/ nanomaterial	System investigated	Identified effects	Ref.
Single-wall carbon nanotubes (SWCNT)			
SWCNT	Human epidermal keratinocytes (HaCat)	ROS production Lipid peroxidation Antioxidant depletion	[11]
SWCNT	Human embryo kidney cells (HEK293)	Cell cycle arrest in G1 Apoptosis	[12]
Purified SWCNT	Chinese hamster pulmonary fibroblasts (V79) Strains YG1024 and YG1029 of *Salmonella typhimurium*	Comet test: primary DNA lesions No mutagenic effect or MN induction (Ames and MN tests)	[13]
SWCNT with traces of metals Fe, Cu, Cr, Ni 0.8–1.2 nm diameter 100–1 000 nm length	C57BL/6 mouse lung	Inflammation and oxidative stress Mutagenic effect (gene *k-ras*)	[15]
Multiwall carbon nanotubes (MWCNT)			
Purified MWCNT	Mouse embryo stem cells (ES)	P53 activation and apoptosis Induction of repair systems (double strand breakage and base excision) Mutagenic effect (gene *Aprt*)	[16]
Purified MWCNT (98% C, traces of Co and Fe)	In vivo: intratracheal administration to female Wistar mice and recovery of type II pneumocytes (AT-II) In vitro: Rat epithelial cells (RLE) and human epithelial cell line (MCF-7)	MN test: increased MN induction in vivo and in vitro FISH: clastogenic and aneugenic effects in cells (MCF-7)	[17]
Fullerenes			
C_{60}	Human cells: dermal fibroblasts, neuronal astrocytes, and hepatocarcinoma cells	ROS production Lipid pexoxidation	[18]
nC_{60} SWCNT Suspended in water	Mouse lung epithelial cell line (Muta-FE1)	ROS production Comet test: oxidative DNA lesions No mutagenic effect (gene *cII*)	[19]

(*Continued*)

Table 5.1. (Continued)

Nanoparticle/ nanomaterial	System investigated	Identified effects	Ref.
C_{60} complexed by cyclodextrin	Rat liver microsomes	Oxidative membrane damage	[20]
Pure nC_{60} suspended in polyvinylpyrrolidone	*Salmonella typhimurium* strains (TA102, 104, and YG3003)	Ames test: mutagenicity depends on visible irradiation, dose, and duration	[22]
Fullerites (mixture of C_{60} and C_{70})	*Salmonella typhimurium* strains (TA100, TA1535, TA98, TA1537) *Escherichia coli* (WP2uvrA/pKM101) Chinese hamster lung cell lines (CHL/IV)	No mutagenic or genotoxic effect (Ames test and chromosomal aberrations)	[23]
nC_{60} suspended in THF	Juvenile largemouth bass	Brain: lipid peroxidation Gills, liver: GSH depletion	[24]
nC_{60} prepared in water nC_{60} prepared in ethanol	Primary human lymphocyte culture	Comet test: primary DNA lesions	[26]
Carbon black			
Carbon black (16 and 70 nm)	Rat lung (subchronic inhalation)	Increased 8-OHdG	[27]
Carbon black (14 nm)	Mouse lung epithelial cell line (Muta-FE1)	Breakage of DNA strands and FPG sites (oxidised purines) measured by comet test Mutagenic effect (genes *cII* and *Lac Z*)	[28]
Carbon black functionalised or not by benzo[a]pyrene (14 nm)	Human lung epithelial cell lines (A549)	DNA strand breakages measured by comet test p53 activation Increased NF-κB and AP-1	[29]

Table 5.2. Genotoxic effects of metal and non-metal nanoparticles

Nanoparticle/nanomaterial	System investigated	Identified effects	Ref.
Cobalt-containing nanoparticles			
Co nanoparticles (100 nm)	Human leukocytes from voluntary donors	Comet test: primary DNA lesions MN test negative	[31]
Co–Cr nanoparticles (30 nm)	Human fibroblasts	Comet test: primary DNA lesions MN test negative	[32]
$CoFe_2O_4$	Human peripheral lymphocytes	MN induction measured by MN test	[33]
Silver-containing nanoparticles			
Ag nanoparticles	Mouse embryo stem cells (mES) Mouse embryo fibroblasts (MEF)	Increased expression of proteins involved in repairing DNA double strand breakages (Rad51) Increased P53 expression	[34]
Cerium-containing nanoparticles			
CeO_2 nanoparticles	Primary human fibroblast culture	Comet test: primary DNA lesions MN induction Genotoxicity may be caused by reduction of a fraction of surface atoms	[36]
CeO_2 nanoparticles (9 nm)	*Salmonella typhimurium* strains (TA98, TA100, TA1535, and TA1537)	Ames test negative	[37]
Titanium-containing nanoparticles			
TiO_2 nanoparticles (anatase) coated with V_2O_5 or not	Chinese hamster pulmonary fibroblasts (V79)	MN induction Lipid peroxidation ROS production Coated nanoparticles have higher genotoxicity	[39]

(Continued)

Table 5.2. (Continued)

Nanoparticle/nanomaterial	System investigated	Identified effects	Ref.
TiO_2 nanoparticles with UVA radiation	Fish cell lines	DNA strand breakages Reduced integrity of lysosomal membranes	[40]
TiO_2 nanoparticles (anatase, 5 nm) with and without UVA radiation	Goldfish skin cells	After irradiation: increased ROS production and oxidative DNA lesions (comet test)	[41]
TiO_2 nanoparticles (< 20 nm)	Syrian hamster embryo fibroblasts (SHE)	MN induction Clastogenic effects Apoptosis	[42]
TiO_2 nanoparticles (20 nm) coated with trimethoxy-octylsilane or uncoated	Female Wistar rats (intratracheal instillation)	No genotoxic effects measured by 8-oxo-guanine	[44]
Different forms of TiO_2 nanoparticle, with and without UV radiation	Chinese hamster ovary cells (CHO-WBL)	No photoclastogenic effects (chromosome aberration test)	[43]
TiO_2 nanoparticles	Human lymphoblastoid cell lines	Increased DNA damage (comet and MN tests) Mutagenic effect (HPRT test)	[45]
TiO_2 nanoparticles doped with cerium IV (visible irradiation)	Human hepatoma cell lines (Bel 7402)	MN induction and apoptosis	[46]
TiO_2 nanoparticles (25 nm, Degussa P25: 70–85% anatase/30–15% rutile)	Peripheral human lymphocytes	Increased DNA damage (comet and MN tests) ROS production p53 activation	[47]

TiO_2 nanoparticles (anatase, 10–20 nm)	Human bronchial epithelial cell lines (BEAS 2B)	Oxidative DNA lesions Lipid peroxidation MN induction	[48]
Iron-containing nanoparticles			
Fe_2O_3 nanoparticles coated with DMSA (20 nm)	Primary human fibroblast culture	No genotoxicity	[49]
Magnetic liposomes (14 nm)	In vivo: intravenous administration to SWISS mouse with recovery of polychromatic erythrocytes	MN induction after 24 h	[51]
Magnetite nanoparticles coated with polyaspartic acid (8.5 nm)	In vivo: endovenous administration to SWISS mouse with recovery of polychromatic erythrocytes	MN induction after 1–7 days	[52]
Fe–Pt nanoparticles coated with tetraethylammonium hydroxide (9 nm)	*Salmonella typhimurium* strains (TA98, TA100, TA1535, and TA1537) *Escherichia coli* strain [WP2uvrA(−)]	Ames test negative except for TA100	[50]
Zinc-containing nanoparticles			
ZnO nanoparticles coated with tetraethylammonium hydroxide	*Salmonella typhimurium* strains (TA98, TA100, TA1535, and TA1537) *Escherichia coli* strain [WP2uvrA(−)]	Ames test negative	[54]
Quantum dots (QD)			
Cadmium telluride QD	Human breast carcinoma cell lines (MCF-7)	Genotoxic response via p53 Global epigenetic response ROS production Damaged mitochondrial membranes	[55]
Photoactivated CdSe–ZnS QD	Plasmid DNA	Strand breakage and nucleobase damage correlated with ROS production	[56]

(*Continued*)

Table 5.2. (Continued)

Nanoparticle/nanomaterial	System investigated	Identified effects	Ref.
Silica-containing nanoparticles			
Crystalline silica particles (crocidolite)	Lac 1 transgenic rats	Mutagenic effect: G → T transversion in gene *Lac 1*	[58]
Crystalline silica particles coated or not coated by aluminium lactate or polyvinylpyridine-*N*-oxide	Rat pulmonary cells	ROS and RNS production Increased APE/Ref1: mediator of DNA lesion repair by BER	[62]
SiO_2 nanoparticles (30, 80 nm)	Fibroblasts 3T3-L1	No genotoxicity (comet test)	[60]
Silica nanoparticles	Human and rodent epithelial and neuronal cells	Alteration of nuclear structures and nucleoplasmic topoisomerase II clusters Protein aggregates	[57]
SiO_2 nanoparticles (< 100 nm)	Human lymphoblastoid cell lines (WL2-NS)	Comet test negative MN induction Mutagenic effect (HPRT test, concentration 120 µg/ml)	[59]
Light-emitting silica nanoparticles	Human lung epithelial cell lines (A549)	No genotoxic effects below concentration 0.1 mg/ml	[61]
Diesel particles	BigBlue transgenic rat (Lac 1)	Mutagenic effect: G → T transversion in gene *Lac 1* correlated with 8-oxodG production Induction of CYP450 1A1	[64]

Table 5.3. Comparative genotoxicity studies on different nanoparticles

Nanoparticle/nanomaterial	System investigated	Identified effects	Ref.
SWCNT, carbon black, ZnO and SiO_2 nanoparticles	Primary mouse fibroblast cultures	GSH depletion and SOD inhibition ROS production and lipid peroxidation Genotoxicity attributed to nanoparticle shape	[66]
Crystalline silica, carbon black, and TiO_2 nanoparticles	Rat alveolar cells	Mutagenic effect (HPRT test)	[65]
Metal oxide nanoparticles (CuO, TiO_2, $CuZnFe_2O_4$, Fe_3O_4, Fe_2O_3)	Human lung epithelial cell lines (A549)	DNA strand breakages and oxidative damage measured by comet test (except Fe_2O_3 nanoparticles)	[67]
Multiwall carbon nanotubes			

Appendices

A. Table of Acronyms

BED	Biologically effective dose
BER	Base excision repair
DSB	Double strand breakage
FISH	Fluorescent in situ hybridization
HPRT	Hypoxanthine phosphoribosyl transferase
iNOS	Inducible nitric oxide synthase
MDA	Malondialdehyde
MN	Micronucleus
MWCNT	Multi-wall carbon nanotube
NER	Nucleotide excision repair
NF-$\kappa\beta$	Nuclear factor $\kappa\beta$
8-OHdG	8-hydroxy-2′-deoxyguanosine
PAH	Polycyclic aromatic hydrocarbon
PNN	Polynuclear neutrophil
ROS	Reactive oxygen species
SSB	Single strand breakage
SWCNT	Single-wall carbon nanotube
THF	Tetrahydrofuran
TNF	Tumour necrosis factor (cytokine inducing cell necrosis)
UV	Ultraviolet

B. Mutagenesis Assays

Unscheduled DNA Synthesis (UDS) Assay. This reveals lesions in the genome by identifying unprogrammed DNA repairs.

Comet or Single Cell Gel Electrophoresis Assay (SCGEA). Electrophoretic technique revealing single and double strand breakages in isolated cells. Applicable in vitro, ex vivo, and in vivo, there are several versions of this test depending on pH conditions during DNA melting and electrophoresis, since these modulate the sensitivity of the test. For example, the neutral version reveals double strand breakages, whereas the alkaline version detects single and double strand breakages, alkali-labile sites, and gaps induced by repairs. Adding stages of digestion by specific glycosylases increases the sensitivity and specificity of the test.

Ames Test. This detects point mutations and frameshift (shifted read frame) mutations induced by genotoxic substances.

Mutation Assay sur Escherichia coli. Variant of the Ames test using the *Escherichia coli* strain WP2/uvrA/pKM101, designed to detect the defective SOS response.

Hypoxanthine Phosphoribosyl Transferase (HPRT) Assay. Assesses mutations by counting cell clones resistant to 6-thioguanine.

Micronucleus Assay. Usually the cytokinesis blocked micronucleus assay (CBMN). This assesses damage to chromosome structure and number in culture cells observed during mitosis after blocking cytokinesis, in order to consider only binucleated cells, i.e., cells that have just divided their genetic material. Associated with fluorescent in situ hybridization (FISH), this method can identify the clastogenic or aneugenic mechanism underlying micronucleus production.

Fluorescent in Situ Hybridization (FISH) Technique. Hybridization of alphoid probes, represented by complementary DNA fragments, with the relevant DNA sequences. For example, when pancentromeric DNA probes, i.e., present in all centromeres of all chromosomes, are applied to micronuclei, fluorescent spots reveal the loss of one or more whole chromosomes, thus signalling an aneugenic event.

C. Strains Used for the in Vitro Ames Test

Classic strains derive from *Salmonella typhimurium* LT2. They are auxotrophic for the amino acid histidine (his), i.e., they cannot grow in the absence of histidine. They carry a specific mutation (His−) in one of the genes of the operon governing the synthesis of this amino acid, and in the presence of mutagenic agents, this mutation (His−) can revert to the wild type (His+), which is prototrophic for histidine. In addition, to increase their sensitivity, further mutations have been introduced into the His operon: rfa mutation increasing permeability of the bacterial membrane and ΔuvrB mutation neutralising excision repair and thus favouring the error-prone SOS repair. Likewise, introducing the plasmid pKM 101 which carries the genes *mucA* and *mucB* in the strains TA97, TA98, TA100, and TA102 increases the error-prone SOS response and thus favours errors in the DNA repair. Depending on the genetic characters of the strains, a certain level of specificity can be defined with regard to genotoxic substances. For example, the strains TA1535, TA1538, TA97, and TA98 preferentially detect mutagens causing frameshift mutations (inserting or deleting pairs of bases), whereas the strains TA1535, TA100, and TA102 tend to detect mutagens inducing base pair substitutions (transversion or translation). The strain TA102 which carries the specific mutation hisG428 and the rfa mutation mainly detects oxidising mutagens, such as the radical species responsible for cell oxidative stress. Finally, specialised strains have been created to detect certain environmental genotoxic substances, such as nitrated PAH derivatives and aromatic amines carcinogenic for the bladder. For example, the strains YG1021 and YG1026 contain the plasmid pYG216, which carries the gene for nitroreductase, the strains YG1024 and YG1029 contain the plasmid pYG219, which carries the gene for O-acetyltransferase, and the strains YG1041 and YG1042 contain the plasmid pYG233, which carries both genes.

D. Lexicon

Anaphase. Third stage of cell division, following the prophase and metaphase. During the anaphase, the chromosomes split, separate, and move towards the cell poles.

Aneugenesis. Chromosome number mutations resulting from alteration of protein structures involved in the migration of the chromosomes during mitosis (centromeres, mitotic spindle, kinetochore, nuclear membrane) by aneugenic agents.

Apoptosis. Set of cell phenomena during which the cell dies under physiological conditions. This really is a programmed cell death, involving in particular the proteins P53 and BAX. This active programme of self-destruction should be contrasted with death by necrosis, which occurs when a cell finds itself in extreme non-physiological conditions.

Clastogenesis. Induction of chromosome structure mutations, also called chromosome rearrangements, resulting from double strand breakages in the DNA molecule caused by clastogenic agents.

Fibroblasts. Cells of the conjunctive tissues responsible for producing the collagen fibres that make up the muscles and skin.

p53. Tumour suppressor gene coding for the protein P53 involved in arresting the cell cycle, apoptosis, and repairing DNA lesions.

Oxidative Stress. Physiological and pathological effects induced by the cellular and molecular consequences of accumulating reactive oxygen and nitrogen species.

DNA Repair Systems. Efficient and faithful repair systems protecting the integrity and stability of the genome. The main system is excision repair. There are two variants, base excision repair (BER), which deals with damage that does not lead to significant modifications in the spatial conformation of the DNA double helix, and nucleotide excision repair (NER), which repairs bulky adducts causing significant distortion of the double helix.

References

1. H. Wiseman, B. Halliwell: Damage to DNA by reactive oxygen and nitrogen species: Role in inflammatory disease and progression to cancer. Biochem. J. **313**, 17–29 (1996)
2. P.J. Borm, A.M. Knaapen, et al.: Neutrophils amplify the formation of DNA adducts by benzo[a]pyrene in lung target cells. Environ. Health Perspect. **105**, 1089–1093 (1997)
3. N. Güngör, R.W. Godschalk, et al.: Activated neutrophils inhibit nucleotide excision repair in human pulmonary epithelial cells: Role of myeloperoxidase. FASEB J. **21**, 2359–2367 (2007)

4. N. Li, T. Xia, et al.: The role of oxidative stress in ambient particulate matter-induced lung diseases and its implications in the toxicity of engineered nanoparticles. Free Radic. Biol. Med. **44**, 1689–1699 (2008)
5. S. Li, P. He, et al.: DNA-directed self-assembling of carbon nanotubes. J. Am. Chem. Soc. **127**, 14–15 (2005)
6. X. Zhao, A. Striolo, et al.: C_{60} binds to and deforms nucleotides. Biophys. J. **89**, 3856–3862 (2005)
7. J. Yang, J.Y. Lee, et al.: Inhibition of DNA hybridization by small metal nanoparticles. Biophys. Chem. **120**, 87–95 (2006)
8. X.J. Liang, C. Chen, et al.: Biopharmaceutics and therapeutic potential of engineered nanomaterials. Curr. Drug Metab. **9**, 697–709 (2008)
9. M.R. Wiesner, G.V. Lowry, et al.: Assessing the risks of manufactured nanomaterials. Environ. Sci. Technol. **40**, 4336–4345 (2006)
10. G. Oberdörster, E. Oberdörster, et al.: Nanotoxicology: An emerging discipline evolving from studies of ultrafine particles. Environ. Health Perspect. **113**, 823–839 (2005)
11. A.A. Shvedova, V. Castranova, et al.: Exposure to carbon nanotube material: Assessment of nanotube cytotoxicity using human keratinocyte cells. J. Toxicol. Environ. Health A **66**, 1909–1926 (2003)
12. D. Cui, F. Tian, et al.: Effect of single wall carbon nanotubes on human HEK293 cells. Toxicol. Lett. **155**, 73–85 (2005)
13. E.R. Kisin, A.R. Murray, et al.: Single-walled carbon nanotubes: Geno- and cytotoxic effects in lung fibroblast V79 cells. J. Toxicol. Environ. Health A **70**, 2071–2079 (2007)
14. J.M. Worle-Knirsch, K. Pulskamp, et al.: Oops they did it again! Carbon nanotubes hoax scientists in viability assays. Nano Lett. **6**, 1261–1268 (2006)
15. A.A. Shvedova, E. Kisin, et al.: Inhalation vs. aspiration of single-walled carbon nanotubes in C57BL/6 mice: Inflammation, fibrosis, oxidative stress, and mutagenesis. Am. J. Physiol. Lung Cell. Mol. Physiol. **295**, L552–565 (2008)
16. L. Zhu, D.W. Chang, et al.: DNA damage induced by multiwalled carbon nanotubes in mouse embryonic stem cells. Nano Lett. **7**, 3592–7 (2007)
17. J. Muller, I. Decordier, et al.: Clastogenic and aneugenic effects of multi-wall carbon nanotubes in epithelial cells. Carcinogenesis **29**, 427–433 (2008)
18. C.M. Sayes, A.M. Gobin, et al.: Nano-C_{60} cytotoxicity is due to lipid peroxidation. Biomaterials **26**, 7587–7595 (2005)
19. N.R. Jacobsen, G. Pojana, et al.: Genotoxicity, cytotoxicity, and reactive oxygen species induced by single-walled carbon nanotubes and C_{60} fullerenes in the FE1-Muta mouse lung epithelial cells. Environ. Mol. Mutagen. **49**, 476–487 (2008)
20. J.P. Kamat, T.P. Devasagayam, et al.: Oxidative damage induced by the fullerene C_{60} on photosensitization in rat liver microsomes. Chem. Biol. Interact. **114**, 145–159 (1998)
21. I.C. Wang, L.A. Tai, et al.: C_{60} and water-soluble fullerene derivatives as antioxidants against radical-initiated lipid peroxidation. J. Med. Chem. **42**, 4614–4620 (1999)
22. N. Sera, H. Tokiwa, et al.: Mutagenicity of the fullerene C_{60}-generated singlet oxygen dependent formation of lipid peroxides. Carcinogenesis **17**, 2163–2169 (1996)

23. T. Mori, H. Takada, et al.: Preclinical studies on safety of fullerene upon acute oral administration and evaluation for no mutagenesis. Toxicology **225**, 48–54 (2006)
24. E. Oberdorster: Manufactured nanomaterials (fullerenes, C_{60}) induce oxidative stress in the brain of juvenile largemouth bass. Environ. Health Perspect. **112**, 1058–1062 (2004)
25. G.D. Nielsen, M. Roursgaard, et al.: In vivo biology and toxicology of fullerenes and their derivatives. Basic Clin. Pharmacol. Toxicol. **103**, 197–208 (2008)
26. A. Dhawan, J.S. Taurozzi, et al.: Stable colloidal dispersions of C_{60} fullerenes in water: Evidence for genotoxicity. Environ. Sci. Technol. **40**, 7394–7401 (2006)
27. J. Gallagher, R. Sams, et al.: Formation of 8-oxo-7,8-dihydro-2′-deoxyguanosine in rat lung DNA following subchronic inhalation of carbon black. Toxicol. Appl. Pharmacol. **190**, 224–231 (2003)
28. N.R. Jacobsen, A.T. Saber, et al.: Increased mutant frequency by carbon black, but not quartz, in the lacZ and cII transgenes of muta mouse lung epithelial cells. Environ. Mol. Mutagen. **48**, 451–461 (2007)
29. R.M. Mroz, R.P. Schins, et al.: Nanoparticle carbon black driven DNA damage induces growth arrest and AP-1 and NFkappaB DNA binding in lung epithelial A549 cell line. J. Physiol. Pharmacol. **58** (Suppl. 5) (Pt. 2), 461–470 (2007)
30. P.H. Hoet, I. Bruske-Hohlfeld, et al.: Nanoparticles: Known and unknown health risks. J. Nanobiotechnol. **2**, 12 (2004)
31. R. Colognato, A. Bonelli, et al.: Comparative genotoxicity of cobalt nanoparticles and ions on human peripheral leukocytes in vitro. Mutagenesis **23**, 377–382 (2008)
32. I. Papageorgiou, C. Brown, et al.: The effect of nano- and micron-sized particles of cobalt–chromium alloy on human fibroblasts in vitro. Biomaterials **28**, 2946–2958 (2007)
33. R. Colognato, A. Bonelli, et al.: Analysis of cobalt ferrite nanoparticles induced genotoxicity on human peripheral lymphocytes: Comparison of size and organic grafting-dependent effects. Nanotoxicology **1**, 301–308 (2007)
34. M. Ahamed, M. Karns, et al.: DNA damage response to different surface chemistry of silver nanoparticles in mammalian cells. Toxicol. Appl. Pharmacol. **233**, 404–410 (2008)
35. E.G. Heckert, S. Seal, et al.: Fenton-like reaction catalyzed by the rare earth inner transition metal cerium. Environ. Sci. Technol. **42**, 5014–5019 (2008)
36. M. Auffan, J. Rose, et al.: CeO_2 nanoparticles induce DNA damage towards human dermal fibroblasts in vitro. Nanotoxicology **1**, 11 (2009)
37. B. Park, P. Martin, et al.: Initial in vitro screening approach to investigate the potential health and environmental hazards of Enviroxtrade mark: A nanoparticulate cerium oxide diesel fuel additive. Part. Fibre Toxicol. **4**, 12 (2007)
38. M. Geiser, B. Rothen-Rutishauser, et al.: Ultrafine particles cross cellular membranes by nonphagocytic mechanisms in lungs and in cultured cells. Environ. Health Perspect. **113**, 1555–1560 (2005)
39. K. Bhattacharya, H. Cramer, et al.: Vanadium pentoxide-coated ultrafine titanium dioxide particles induce cellular damage and micronucleus formation in V79 cells. J. Toxicol. Environ. Health A **71**, 976–980 (2008)
40. W.F. Vevers, A.N. Jha: Genotoxic and cytotoxic potential of titanium dioxide (TiO_2) nanoparticles on fish cells in vitro. Ecotoxicology **17**, 410–420 (2008)

41. J.F. Reeves, S.J. Davies, et al.: Hydroxyl radicals (•OH) are associated with titanium dioxide (TiO_2) nanoparticle-induced cytotoxicity and oxidative DNA damage in fish cells. Mutat. Res. **640**, 113–122 (2008)
42. Q. Rahman, M. Lohani, et al.: Evidence that ultrafine titanium dioxide induces micronuclei and apoptosis in Syrian hamster embryo fibroblasts. Environ. Health Perspect. **110**, 797–800 (2002)
43. E. Theogaraj, S. Riley, et al.: An investigation of the photo-clastogenic potential of ultrafine titanium dioxide particles. Mutat. Res. **634**, 205–219 (2007)
44. B. Rehn, F. Seiler, et al.: Investigations on the inflammatory and genotoxic lung effects of two types of titanium dioxide: Untreated and surface treated. Toxicol. Appl. Pharmacol. **189**, 84–95 (2003)
45. J.J. Wang, B.J. Sanderson, et al.: Cyto- and genotoxicity of ultrafine TiO_2 particles in cultured human lymphoblastoid cells. Mutat. Res. **628**, 99–106 (2007)
46. L. Wang, J. Mao, et al.: Nano-cerium-element-doped titanium dioxide induces apoptosis of Bel 7402 human hepatoma cells in the presence of visible light. World J. Gastroenterol. **13**, 4011–4014 (2007)
47. S.J. Kang, B.M. Kim, et al.: Titanium dioxide nanoparticles trigger p53-mediated damage response in peripheral blood lymphocytes. Environ. Mol. Mutagen. **49**, 399–405 (2008)
48. J.R. Gurr, A.S. Wang, et al.: Ultrafine titanium dioxide particles in the absence of photo-activation can induce oxidative damage to human bronchial epithelial cells. Toxicology **213**, 66–73 (2005)
49. M. Auffan, L. Decome, et al.: In vitro interactions between DMSA-coated maghemite nanoparticles and human fibroblasts: A physicochemical and cyto-genotoxical study. Environ. Sci. Technol. **40**, 4367–4373 (2006)
50. S. Maenosono, T. Suzuki, et al.: Mutagenicity of water-soluble FePt nanoparticles in Ames test. J. Toxicol. Sci. **32**, 575–579 (2007)
51. V.A.P. Garcia, L.M. Lacava, et al.: Magnetoliposome evaluation using cytometry and micronucleus test. Eur. Cells Materials **3** (Suppl. 2), 154–155 (2002)
52. N. Sadeghiani, L.S. Barbosa, et al.: Genotoxicity and inflammatory investigation in mice treated with magnetite nanoparticles surface coated with polyaspartic acid. Journal of Magnetism and Magnetic Materials **289**, 466–468 (2005)
53. V. Binhi: Do naturally occurring magnetic nanoparticles in the human body mediate increased risk of childhood leukaemia with EMF exposure? Int. J. Radiat. Biol. **84**, 569–579 (2008)
54. R. Yoshida, D. Kitamura, et al.: Mutagenicity of water-soluble ZnO nanoparticles in Ames test. J. Toxicol. Sci. **34**, 119–122 (2009)
55. A.O. Choi, S.E. Brown, et al.: Quantum dot-induced epigenetic and genotoxic changes in human breast cancer cells. J. Mol. Med. **86**, 291–302 (2008)
56. A. Anas, H. Akita, et al.: Photosensitized breakage and damage of DNA by CdSe–ZnS quantum dots. J. Phys. Chem. B **112**, 10005–10011 (2008)
57. M. Chen, A. von Mikecz: Formation of nucleoplasmic protein aggregates impairs nuclear function in response to SiO_2 nanoparticles. Exp. Cell Res. **305**, 51–62 (2005)
58. K. Unfried, C. Schurkes, et al.: Distinct spectrum of mutations induced by crocidolite asbestos: Clue for 8-hydroxideoxyguanosine-dependent mutagenesis in vivo. Cancer Res. **62**, 99–104 (2002)
59. J.J. Wang, B.J. Sanderson, et al.: Cytotoxicity and genotoxicity of ultrafine crystalline SiO_2 particulate in cultured human lymphoblastoid cells. Environ. Mol. Mutagen. **48**, 151–157 (2007)

60. C.A. Barnes, A. Elsaesser, et al.: Reproducible comet assay of amorphous silica nanoparticles detects no genotoxicity. Nano Lett. **8**, 3069–3074 (2008)
61. Y. Jin, S. Kannan, et al.: Toxicity of luminescent silica nanoparticles to living cells. Chem. Res. Toxicol. **20**, 1126–1133 (2007)
62. C. Albrecht, A.M. Knaapen, et al.: The crucial role of particle surface reactivity in respirable quartz-induced reactive oxygen/nitrogen species formation and APE/Ref-1 induction in rat lung. Respir. Res. **6**, 129 (2005)
63. L.K. Limbach, P. Wick, et al.: Exposure of engineered nanoparticles to human lung epithelial cells: Influence of chemical composition and catalytic activity on oxidative stress. Environ. Sci. Technol. **41**, 4158–4163 (2007)
64. H. Sato, H. Sone, et al.: Increase in mutation frequency in lung of BigBlue rat by exposure to diesel exhaust. Carcinogenesis **21**, 653–661 (2000)
65. K.E. Driscoll, L.C. Deyo, et al.: Effects of particle exposure and particle-elicited inflammatory cells on mutation in rat alveolar epithelial cells. Carcinogenesis **18**, 423–430 (1997)
66. H. Yang, C. Liu, et al.: Comparative study of cytotoxicity, oxidative stress and genotoxicity induced by four typical nanomaterials: The role of particle size, shape and composition. J. Appl. Toxicol. **29**, 69–78 (2009)
67. H.L. Karlsson, P. Cronholm, et al.: Copper oxide nanoparticles are highly toxic: A comparison between metal oxide nanoparticles and carbon nanotubes. Chem. Res. Toxicol. **21**, 1726–1732 (2008)

6

Elements of Epidemiology

Agnès Lefranc and Sophie Larrieu

6.1 Generalities

Epidemiology is defined as the study of the distribution of diseases and their determining factors [1,2]. In the field of environmental health, it thus investigates the relationship between different aspects of environmental exposure and human health. Epidemiology does not consider individuals, but rather groups of individuals specified by some common characteristic, e.g., exposure to some given substance, a pathology, etc. It then compares these groups of individuals, for example, to answer a question like: when individuals are exposed to a given substance, are they more often affected by a certain pathology?

While this approach provides important information about the relationship between environmental exposure and health, it does not of course inform as to the mechanisms that may underlie those relationships on the individual level. So epidemiological and experimental studies are complementary, and mutually supportive when they lead to concordant results.

In addition, the observation of significant correlation between exposure and health in the context of epidemiological studies alone is not sufficient to draw conclusions about the causal nature of the observed relation. This question has been widely debated, and lists of criteria put forward to define situations in which causality can be reasonably inferred. The best known are those specified by Hill in 1965 [3], which include the following:

- Constancy of observed associations: the findings of an epidemiological study must be confirmed by other epidemiological studies, if possible using different methods.
- Temporality: exposure must precede appearance of the effect.
- Biological plausibility: known or plausible biological mechanisms should explain the ways exposure affects health. Toxicology is in this respect a major source of support or refutation for any causal hypothesis.
- Consistency of results: Findings obtained should be consistent with those available from other sources referring to the same subject.

P. Houdy et al. (eds.), *Nanoethics and Nanotoxicology*,

DOI 10.1007/978-3-642-20177-6_6,

Consequently, to avoid risk of overinterpretation, it is crucial to consider the findings of epidemiological studies in the light of this kind of criterion.

6.2 Studies of Ultrafine Particles and Lack of Data for Nanoparticles

At the present time, there have been no epidemiological studies of exposure to nanomaterials. To illustrate this, it is instructive to submit the request *(Nanoparticles OR Nanostructures OR Nanotubes OR Nanomaterials) AND Epidemiology* to the Medline data base. At the time of writing (31 December 2009) only 53 publications corresponded to these criteria, but closer examination revealed that none of them constituted a genuine epidemiological study of exposure to nanomaterials.

In contrast, the health consequences of exposure to airborne particles have been investigated in a great many epidemiological studies. These particles suspended in the atmosphere form a heterogeneous mix, not only in terms of size and chemical composition, but also in terms of their sources. There are several coexisting classifications, but the most commonly used is based on the mean aerodynamic diameter of these particles, disregarding their nature and origin. For example, a distinction is generally made between different sizes of particulate matter: PM_{10} with diameters less than 10 μm, coarse particles $PM_{2.5–10}$ with diameters in the range 2.5–10 μm, fine particles $PM_{2.5}$, smaller than 2.5 μm, and ultrafine particles (UFP) $PM_{0.1}$ with diameters less than 0.1 μm. Owing to their nanometric dimensions, the latter correspond to the definition of nanoparticles and thus exhibit some similar properties to the products of the nanotechnologies. However, they arise from a wide range of both natural and anthropic sources, and their chemical compositions thus vary enormously, in contrast to the specifically engineered materials in nanotechnology.

The findings of many epidemiological studies [4], together with the conclusions of experimental studies on animals and humans, tend to suggest a causal relation between exposure to fine particles ($PM_{2.5}$) and short or long term health effects, mainly of a cardiorespiratory nature. Ultrafine particles, with aerodynamic diameters less than 0.1 μm, i.e., 100 nm, are suspected of playing an important role in effects observed with $PM_{2.5}$, bearing in mind that UFPs fall within the $PM_{2.5}$ category. There are several reasons for this suspicion:

- They pass through the nasopharyngeal region, where some are deposited [5]. They then enter the deepest confines of the respiratory system, where they may once again be deposited, and all the more so if the subject suffers from some preexisting respiratory pathology, e.g., asthma [6] or obstructive pulmonary disease [7], or when the subject is taking physical exercise [6].
- They can also very quickly cross the epithelial wall to the pulmonary interstitium, thanks to their small size [8], and enter the blood circulation,

whereupon they will be distributed to target organs like the heart, the liver, or even the brain, where they may have toxic effects [9, 10].
- For the same mass, their number and global specific surface area are much higher than those of larger particles. This increases their capacity to induce pro-inflammatory and allergic phenomena [11].
- They may contain many toxic substances, such as oxidising gases, organic compounds, metals, and so on. These may be included within the particle itself, or adsorbed onto the particle surface. These substances may modify the properties of the particles, and at the same time the particles may assist in the uptake of the toxic substance by the organism.

Ultrafine particles have many sources. They may be directly emitted (so-called primary particles), in particular by combustion phenomena occurring in industrial processes or domestic activities, or in vehicle engines. But they may also form in a secondary manner by condensation, at the surface of existing particles or by homogeneous nucleation, of compounds emitted in gaseous form at a higher temperature than the surrounding atmosphere, leading to the formation of compounds with low saturated vapour pressure, which are thus likely to condense. The diversity of UFP sources is what results in the wide range of concentrations and chemical compositions, both in time and in space [12].

6.3 Review of Epidemiological Studies of Ultrafine Particles Suspended in the Surrounding Atmosphere

Considering the special case of nanoparticles suspended in the ambient air, a search on *(ultrafine particles OR UFP) AND Epidemiology* picked up 79 publications as of 31 December 2009, once again carefully inspecting the results in order to select only those publications which actually presented original epidemiological studies of the health effects of UFPs. This systematic search was complemented by using the references cited in the selected papers. The few available epidemiological studies thus identified all postdate 1997 (see Table 6.1). The vast majority were carried out in towns taking part in two multicenter research programmes:

- The *Exposure and Risk Assessment for Fine and Ultrafine Particles in Ambient Air* study (ULTRA), carried out in three northern European towns, viz., Erfurt in Germany, Helsinki in Finland, and Amsterdam in Holland, investigated the associations between UFP exposure and many health parameters [13–31].
- The *Health Effects of Air Pollution on Susceptible Subpopulations* study (HEAPSS), carried out in five European towns, viz., Augsburg in Germany, Barcelona in Spain, Helsinki in Finland, Rome in Italy, and Stockholm in Sweden, investigated mortality and hospital admissions [32–34].

Table 6.1. Published epidemiological studies investigating the relationship between UFPs and health

Assessment of UFP levels	Health indicators	Ref.
UFP count at a site in the town of Erfurt	Respiratory symptoms and respiratory function parameters in asthmatic adults	[13]
	Mortality (all non-accidental causes, for respiratory and cardiovascular causes)	[14]
	Respiratory symptoms and use of anti-asthmatic medication in asthmatic adults	[15]
	Change in ventricular repolarisation in patients suffering from ischemic cardiopathy	[16]
	Blood markers of inflammation and coagulation in patients suffering from ischemic cardiopathy	[17]
	Platelet activation markers in patients suffering from ischemic cardiopathy	[18]
	Mortality (all non-accidental causes, for respiratory and cardiovascular causes)	[19]
	Blood markers of inflammation and coagulation in patients suffering from chronic obstructive bronchopneumopathy	[20]
	Mortality (all non-accidental causes)	[21]
UFP count at a site in the town of Helsinki	Respiratory function parameters in asthmatic children	[22]
	Respiratory function parameters, respiratory functions, and use of anti-asthmatic medication in asthmatic adults	[23]
	Respiratory function parameters in asthmatic adults	[24]
	Electrocardiogram parameters in patients with coronary heart disease	[25]
	Death by stroke of persons aged 65 or over	[26]
	Electrocardiographic parameters in patients with coronary heart disease	[27]
UFP counts at one site in each town featuring in the ULTRA study	Cardiorespiratory symptoms in patients suffering from cardiovascular disease	[28]
	Urinary concentration of the lung Clara cell protein CC16 in patients suffering from cardiovascular disease	[29]
	Blood pressure, heart rate, and heart rate variability in patients suffering from cardiovascular disease	[30]
	Heart rate and heart rate variability in patients suffering from cardiovascular disease	[31]

UFP count at a site in Rome over the period April 2001 to June 2002, after the health data collection period (1998–2000). Retrospective assessment of the UFP count using a linear regression model, integrating the levels of other atmospheric pollutants and meteorological conditions	Out-of-hospital coronary deaths	[32]
UFP counts at one site in each of the five towns featuring in the HEAPSS study over a period of about one year, after the health data collection period. Retrospective assessment of the UFP count using a linear regression model, integrating the levels of other atmospheric pollutants and meteorological conditions	Hospital cardiac readmissions of myocardial infarction survivors	[33]
	Hospital admission for first myocardial infarction	[34]
UFP count at a site in Copenhagen over the whole period of the study (15 May 2001 to 31 December 2004). Comparison between levels at this site and those measured close to traffic and in a rural area over limited periods (105 and 47 days, respectively)	Hospital admissions of elderly people (> 65 yr) for cardiovascular and respiratory reasons, and admissions of children (5–18 yr) for asthma	[35]
	Wheezing symptoms in infants (0–1 yr)	[36]

6.3.1 Assessing Exposure

The main difficulty in epidemiological studies is to characterise the extent to which individuals are exposed to UFPs [12], especially given the spatial and temporal variability already mentioned. Moreover, the methods used for routine surveillance of the mass concentrations of fine or coarse particles in the atmosphere [37] cannot be directly applied to UFPs, in particular due to their very low mass. Since UFP concentrations in the surrounding air are currently subject to no form of regulation, they are not measured on a routine basis. As a consequence, the identified epidemiological studies are all based on specifically implemented UFP measurements.

In every case covered in Table 6.1, a single UFP measurement site is available per town, while the measured levels are used to evaluate the exposure of all subjects within a certain zone, usually the town or corresponding conurbation, which may extend to some distance from this measurement site.

It should be noted that, while UFPs constitute only a very small fraction of the total mass of atmospheric particles, they are extremely numerous among them. In addition, they possess a much higher reactive surface than coarser particles for the same mass. Since UFP mass measurements with relatively high temporal resolution, e.g., 24 hours, are particularly problematic from a methodological standpoint, owing to the very small masses coming into play, ultrafine particles have thus been assessed in terms of number concentration in all the studies listed here, while PM_{10} and $PM_{2.5}$ are generally assessed in terms of mass concentration.

In all the studies in Table 6.1, the measurement station is described as being implanted in a background situation, i.e., away from any immediate influence of a source of pollution. The underlying idea of most of these studies is to seek a link between daily variations in the level of some exposure indicator and the incidence of some kind of health effect. In this context, an exposure indicator is acceptable, in the sense of not introducing bias, if the daily variations are reasonably well correlated with daily variations in the average individual exposure (even if the two variables have rather different values). So the use of a single measurement station to describe UFP exposure levels for a population living in a given zone, which may be relatively extensive (a town or conurbation), is likely to introduce bias when assessing the exposure of subjects included in the study if the temporal variations of the levels it measures are not correlated with those of the average individual exposure levels. This may happen, in particular, due to the spatial variability of pollution levels [38]. For UFPs, only one recent epidemiological study makes a formal comparison between the levels measured at different sites, i.e., in a background situation, near road traffic, and rural, and it concludes that a central measurement site is not in fact very representative [35,36]. However, the rare evidence available also shows that the levels measured at different background sites in the same conurbation exhibit rather well correlated temporal variations from one to the other (in Barcelona, Rome, and Stockholm for a measurement campaign in the HEAPSS study [39], and also in Helsinki [40]). Other parameters may nevertheless contribute to reducing the correlation between UFP levels measured in a background station and the average of individual exposures. For example, sources inside buildings affect individual exposure, and the diffusion of such particles from the outside in is relatively low [12].

Finally, in some studies, UFP measurements are not available for the period over which the health indicators were gathered [32–34]. Linear regression models are then constructed for the period over which measurements of UFPs and other pollutants (but also climatic parameters) are simultaneously available, and these models are used to infer UFP levels retrospectively over the period of health data acquisition. The use of such models may of course introduce further uncertainties, on top of those mentioned above. These studies also investigate links between the levels of particles of different sizes, and they also show that UFP number concentrations were weakly correlated to $PM_{2.5}$ and/or PM_{10} mass concentrations (≤ 0.5) [26,32–34].

6.3.2 Health Indicators

The epidemiological studies listed in Table 6.1 were concerned with short term effects of UFP exposure. In the light of accumulated knowledge of the health effects of fine particles [4], almost all the studies target respiratory or cardiovascular effects. In addition, a large majority of them are concerned with these effects in populations that are likely to be more sensitive to the effects of particulate atmospheric pollution, notably because they suffer from preexisting chronic respiratory or cardiovascular pathologies.

The health indicators studied cover events with different levels of seriousness, from changes in subclinical[1] markers [16–18,20–22,25,27,30,31], through symptoms [13,15,23,24,28,29,36], use of medication [15,23,24], and hospitalisation [33–35], to death [14,19,21,26,32]. Analyses using morbidity indicators in the ULTRA study [15–18, 20–25, 28–31] are based on panels of patients suffering from cardiovascular or respiratory disease, from whom much data is collected by means of questionnaires (symptoms or use of medication) or during medical examinations (electrocardiogram, respiratory function parameters). In addition, analysis of blood samples provides levels of inflammatory and coagulation markers.

In the HEAPSS study, the aim was to target populations considered to be particularly at risk of being affected by UFP exposure. Subjects with pathological histories liable to increase their sensitivity were identified and monitored using preexisting registers and administrative medical data bases [32, 33].

[1] Subclinical means not clinically manifest, but which can be brought to light by laboratory tests or imaging.

6.3.3 Different Types of Analyses Carried out

Results concerning the links between UFP exposure and health come from three types of study, as shown schematically in Fig. 6.1: time series studies, panel studies, and case cross-over studies.

Time series studies investigate correlations between ambient UFP levels in a zone within which pollution background levels can be considered uniform, usually a town or its conurbation, and daily variations in the number of health

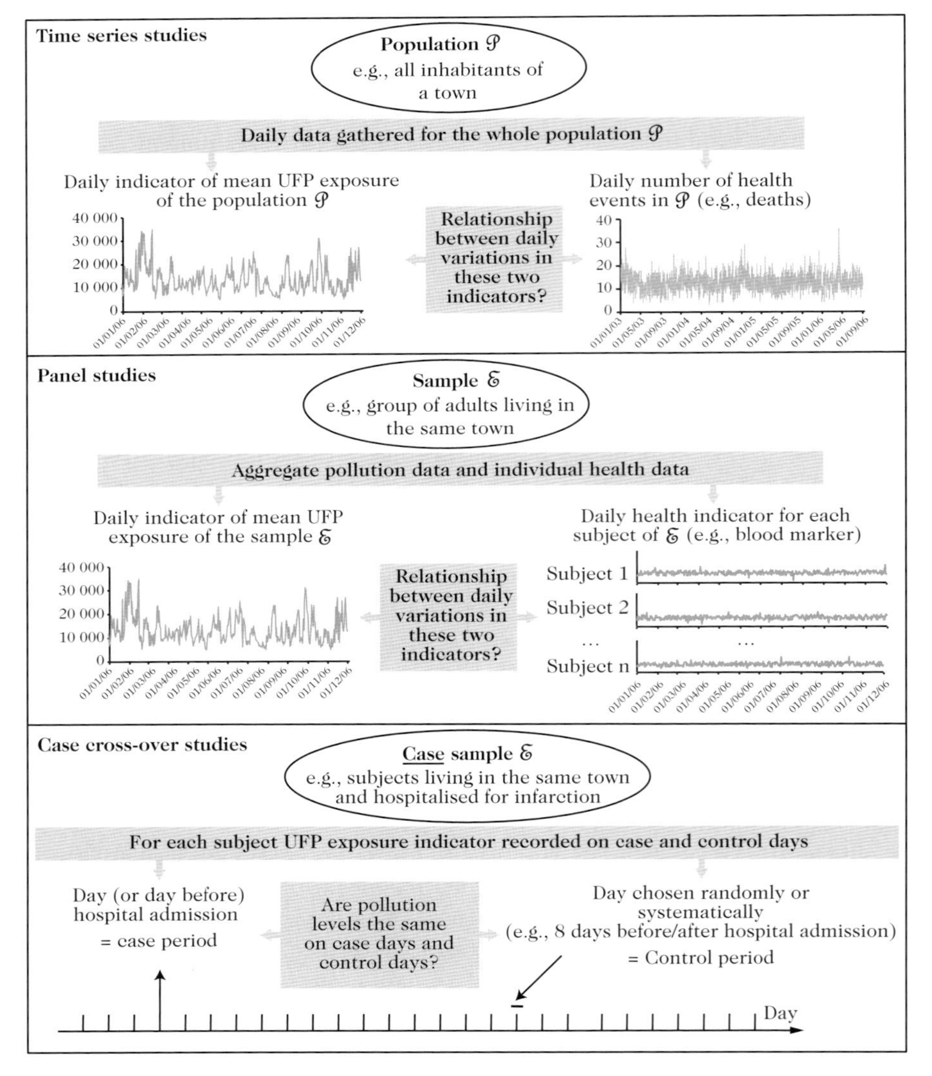

Fig. 6.1. General principles of the different epidemiological studies

events, e.g., deaths, emergency admissions, hospital admissions, etc., among the population living in this same zone. So the daily number of health events is modelled as a function of the daily UFP exposure indicator, taking into account potential confounding factors, i.e., parameters which may be linked both to the health indicator and to ambient UFP levels, e.g., seasonal and long term variations, day of the week, climatic factors, etc. It thus becomes possible to estimate the increase in the risk of suffering the given health event following an increase in the UFP levels in the study zone [14,19,21,26,33–36].

Panel studies follow a cohort of subjects over time for a given period and collect individual health data repeatedly, at regular intervals of time, from the members of this cohort. In parallel, UFP exposure is estimated with the same time interval, but collectively, in the same way as in time series studies. In this way, one can estimate correlations between individual variations of the chosen health indicator and those of the UFP exposure indicator [13, 15–18, 20, 22–25, 27–31].

Finally, case cross-over studies only involve subjects who have suffered the chosen health event. This method, developed in the 1990s [41], compares UFP exposure levels observed over the period immediately preceding the studied event, the so-called case period, with exposure levels observed over one or more periods prior to or subsequent to the event, the so-called control period(s) [32]. These control periods can be chosen randomly or predetermined, e.g., 7 days before and/or after the case period, inserting at least a few days between the two (usually one week in the case of atmospheric pollution). If an increased UFP level is indeed a trigger for the health event, a significantly higher level should be recorded on case days than on control days. This hypothesis is tested using a conditional logistic regression model which delivers an odds ratio[2] (OR) fitted to the potential confounding factors (notably meteorological conditions). On the other hand, since each subject acts as their own control, individual characteristics not depending on time do not play the role of confounding factor and do not therefore need to be taken into account. This method is particularly well suited to identifying the possible effects of time-variable exposure, such as the effects of atmospheric pollution, on occurrence of an acute health event.

6.3.4 Results

Findings vary rather significantly from one study to another. Regarding mortality, significant correlations, or at the limit of being significant, are rather consistently observed between high UFP levels on a given day and an increased risk of death on the following days. Investigations into the delay that may occur between exposure and the occurrence of death have observed effects that persist for 4–5 days, sometimes with greater effects for a delay of a few days [14, 19], sometimes with greater effects immediately after exposure [32].

[2] The odds ratio is a measure of the correlation between exposure and risk.

Finally, taking into account cumulative exposure over 6 and 15 days, greater effects are demonstrated [21]. The observed relationships are generally slightly stronger for mortality due to respiratory or cardiovascular causes than when all non-accidental causes of death are taken into account [14, 19]. It seems that the presence of high blood pressure or an obstructive pulmonary disease increases the risk of coronary death after exposure to UFPs [32]. Likewise, the season seems to modify the correlation between UFP levels and mortality, but the observed variations would not appear to be consistent from one study to the next [14, 26]. Finally, it should be noted that, although the study carried out over the first period at Erfurt [14] showed higher relative risk of comparable death in relation with interquartile increases[3] in the levels of $PM_{2.5}$ (by mass) and UFPs (by number), the continuation of the study over a longer period [19] revealed much higher risks associated with UFPs than with $PM_{2.5}$. These quantitative variations in the relation between UFP levels and the relative risk of death over one decade have also been found in a subsequent study of data gathered at Erfurt [21]. Finally, regarding coronary deaths occurring outside hospital, the higher relative risk observed in relation with UFPs is much greater than with PM_{10} [32].

Results concerning the relationship between UFP exposure and hospital admissions for cardiovascular reasons are more varied. The two HEAPSS studies of admissions for a first myocardial infarction and readmissions for this same pathology reveal significant correlations [33, 34]. In contrast, Andersen et al. [35] observed no link between UFP number concentrations and cardiovascular hospital admissions in the elderly (65 yr or more), whereas significant links were observed with PM_{10} mass concentrations. When indicators concerning cardiorespiratory symptoms are studied (questionnaire to evaluate the quality of life of patients suffering from angina pectoris [28]), $PM_{2.5}$ mass levels are associated with many symptoms, and also with avoidance of physical exercise. But UFP levels are significantly correlated only with the latter.

Contrasting results are also observed for inflammatory and coagulation markers, blood pressure, and parameters describing cardiac activity (heart rate and its fluctuations, ventricular repolarisation, etc.). For example, Ibald-Mulli et al. [30] observed no significant link between UFP number concentration and blood pressure or heart rate, whereas, among the same subjects, Timonen et al. [31] found significant correlations between this concentration and reduced heart rate variability, a factor that is strongly correlated with cardiac mortality in predisposed patients. A significant association is also observed between UFP number concentrations over the 2 days prior to taking

[3] The interquartile range is the interval between the first and third quartiles (percentiles 25 and 75), the three quartiles dividing a statistical distribution into four equal groups of observations.

the electrocardiogram and the risk of ST segment depression,[4] a marker for myocardial infarction [25]. However, this association does not show up when the levels taken into account are limited to just a few hours immediately prior to taking the electrocardiogram [27]. Still other relationships have been noted between UFP number levels and blood concentrations of C-reactive protein, a systemic inflammatory marker [17], the protein sCD40L, released during platelet activation [18], and fibrinogen, a protein involved in coagulation mechanisms [20]. Such relationships are not observed systematically, however. Finally, for other biological markers, e.g., markers of endothelial dysfunction or coagulation, no significant correlation has been identified [17].

Regarding respiratory effects of UFP exposure, a single study deals with hospital admissions, and it is only concerned with asthma admissions for children (5–18 yr) [35], where associations lie just within the significance limit. Furthermore, panel studies of asthmatic adults converge globally, suggesting correlations between UFP number concentrations and various respiratory health indicators, such as peak expiratory flow rate (PEFR) [13], use of asthma treatments [15,23], and respiratory symptoms [23]. However, in certain asthmatic panels (children [22] and adults [24]), opposite but non-significant relationships were observed between UFP levels and PEFR, whereas significant correlations were observed in relation with PM_{10} mass concentrations or number concentrations of particles with diameters in the range 0.1–1 μm.

6.3.5 Interpretation of Findings

All things said, the results obtained from epidemiological studies reveal a certain variability. However, the following points should be noted:

- In all these studies, exposure levels are evaluated from measurements carried out at only one measurement station per town. The small amount of data available about UFP measurements made at different background sites across the same town show that temporal concentration variations at these different sites are relatively well correlated [39, 40]. However, given the high spatial variability of UFP levels, their low levels of penetration into living areas, and the significance of internal sources, some authors suggest that, in the context of time series studies, using measurements all of which were made at the same station would introduce a more significant bias in the case of UFP number measurements than is the case, for example, for $PM_{2.5}$ mass measurements [42]. On the other hand, since this bias is in principle non-directional, it would lead one to underestimate the relationship between ambient UFP levels and health indicators, and this might in part explain the absence of any significant correlation in some studies.

[4] The ST segment is part of the electrocardiogram plot corresponding to a ventricular repolarisation phase, in which a modification may reflect a cardiovascular pathology.

- The exact nature of the UFPs may vary with the relative importance of their different sources. This variability in their chemical composition might in part be responsible for the differences observed in the findings of epidemiological studies, especially when there are differences in time and place. Indeed, the intrinsic toxicity of UFPs may vary with their chemical composition, and hence also with the time of year and location.
- Biological markers generally exhibit a high level of interindividual variability. In this context, it may be difficult to identify significant effects of UFP exposure on these parameters in the framework of an epidemiological study, even assuming these effects exist. In particular, given the complexity of the methods of data collection (questionnaires, samples, and medical examinations) implemented in most of the studies discussed here, the number of subjects is extremely small, and this limits the statistical power of such studies.

While the results of epidemiological studies are sometimes inconsistent for some health indicators, they do emphasise the significance of the health risk due to UFP exposure, since several studies have demonstrated an effect on mortality. In addition, comparison with experimental findings tends to corroborate the hypothesis, especially the ability of UFPs to cross epithelial barriers (see, for example, [5, 42] for reviews).

In short, the findings of epidemiological studies remain somewhat incomplete, especially due to the relative rarity of ambient UFP measurements that are suitable for use in the framework of these studies, i.e., recorded over long enough periods, under conditions allowing a reasonable chance of inferring exposure levels from the measured values. In addition, they come from studies carried out exclusively in European conurbations, so they cannot be extrapolated to the whole population. Indeed, a better understanding of the size dependence of particle effects can only be obtained by studying them in different geographical contexts, with different sources and different levels of pollution. However, even though these findings highlight the difficulties involved in such studies, they must nevertheless be taken as an encouragement to pursue research in this area. They show that there is a need for close collaboration with experts in the fields of atmospheric pollution and metrology to develop appropriate methods for measuring UFPs in the air and to describe the spatial and temporal variability in their levels. In addition, a permanent exchange must always be maintained with toxicology, whose results can be used to clarify and orient epidemiological work, in particular, the selection of suitable exposure indicators (granulometric range, chemical species, etc.) and health indicators. For example, some experimental studies stress the potential action of inhaled UFPs, apart from respiratory and cardiovascular effects. Indeed, it is suspected that they may be able to enter the central nervous system and deposit themselves there [10]. In the face of such evidence, it seems particularly important to set up epidemiological studies to study the connections between chronic exposure to UFPs in the surrounding air and certain neurodegenerative pathologies.

6.4 Drawing Conclusions about Intrinsic Nanoparticle Effects

The first epidemiological findings about health risks due to UFP exposure suggest a significant effect in terms of morbidity and mortality. These conclusions support the findings of a great many epidemiological studies on fine particles [4], which provide robust and clear evidence of a strong link between exposure to particles in the ambient air and both short and long term health effects, mainly of a cardiorespiratory nature. These results, added to the conclusions of experimental studies on animals and humans, strongly suggest an effect that increases as the particle size diminishes. Furthermore, a panel of experts has recently concluded that the likelihood of a causal relation between UFP exposure and short term health effects, such as increased mortality and hospital admissions for respiratory and cardiovascular pathologies, is medium to high [43].

As a consequence, even though no epidemiological study has yet been specifically carried out on nanotechnological products and byproducts, all currently available evidence regarding the effects of particles tends to suggest that these entities could indeed induce cardiorespiratory effects in exposed populations, just like the ultrafine particles they so closely resemble through their nanometric dimensions.

On the other hand, the findings of these studies on ambient UFPs are not yet sufficient to fully understand the whole range of potential effects of nanoparticles, since they correspond to quite different situations to those expected in the development and use of nanotechnologies:

- Ultrafine particles are mainly produced unintentionally, and their chemical composition may for this reason differ significantly from certain nanoparticles manufactured for some precisely defined application. So some manufactured nanoparticles with a specific chemical composition may be totally absent from UFPs suspended in the ambient atmosphere, whence they will not be taken into account in epidemiological studies of the latter.
- Occupational exposure of people working in industries using nanoprocesses may reach much higher levels than those measured in the ambient air, and these exposure levels would not then be represented in studies concerning the general population.
- Exposure to UFPs suspended in the ambient air occurs mainly by inhalation. Other exposure routes, such as ingestion and cutaneous contact, which may be relevant depending on the context for manufactured nanoparticles, e.g., during the fabrication process or the use of products in which the materials have been integrated, are not therefore properly accounted for in epidemiological studies on atmospheric UFPs so

far available. But these other exposure routes may not only contribute to increasing global exposure levels. They may in fact generate specific effects. In particular, studies have shown that orally ingested nanoparticles may be able to enter the blood circulation in significant amounts, and thereby attack target organs such as the liver, kidneys, or spleen [44].

For these reasons, even though the currently available epidemiological studies tend to confirm a significant health risk associated with exposure to nanoparticles present in the ambient air, they are not sufficient either to quantify or to conclude as to the exact nature of the resulting health effects. Potential effects on organs like the kidneys or liver, as revealed by toxicological studies, have never yet been considered by epidemiological studies, since these have all focused on cardiorespiratory effects up to now.

In parallel with implementing the necessary safety measures, it is essential to accompany the current expansion in the manufacture and voluntary use of nanoparticles by epidemiological studies monitoring the consequences for the most exposed individuals. Only this kind of study could watch over the whole range of potential health consequences for these populations and evaluate the risks associated with exposure to manufactured nanoparticles. To achieve this, even though many measures have been planned or implemented to prevent or at least reduce their exposure, those working with these products probably constitute the most suitable population, at least to begin with, not only to study the more acute effects of nanoparticle exposure, but also to monitor its longer term consequences.

References

1. J. Bouyer et al.: *Epidémiologie. Principes et méthodes quantitatives.* INSERM, Paris, 498 p. (1995)
2. K. Rothman, S. Greenland: *Modern Epidemiology.* Lippincott Williams & Wilkins, Philadelphia, 752 p. (1998)
3. F.C. Hill: The environment and disease: Association or causation? Proc. R. Soc. Med. **58**, 295–300 (1965)
4. C.A. Pope, D.W. Dockery: Health effects of fine particulate air pollution: Lines that connect. J. Air Waste Manag. Assoc. **56**, 709–742 (2006)
5. G. Oberdorster, E. Oberdorster, J. Oberdorster: Nanotoxicology: An emerging discipline evolving from studies of ultrafine particles. Environ. Health Perspect. **113**, 823–839 (2005)
6. D.C. Chalupa et al.: Ultrafine particle deposition in subjects with asthma. Environ. Health Perspect. **112**, 879–882 (2004)
7. P.J. Anderson, J.D. Wilson, F.C. Hiller: Respiratory tract deposition of ultrafine particles in subjects with obstructive or restrictive lung disease. Chest **97**, 1115–1120 (1990)
8. A. Elder et al. Translocation of inhaled ultrafine manganese oxide particles to the central nervous system. Environ. Health Perspect. **114**, 1172–1178 (2006)
9. W.G. Kreyling, M. Semmler-Behnke, W. Möller: Ultrafine particle–lung interactions: Does size matter? J. Aerosol Med. **19**, 74–83 (2006)

10. A. Peters et al. Translocation and potential neurological effects of fine and ultrafine particles: A critical update. Part. Fibre Toxicol. **3**, 13 (2006)
11. C. de Haar et al.: Ultrafine but not fine particulate matter causes airway inflammation and allergic airway sensitization to co-administered antigen in mice. Clin. Exp. Allergy **36**, 1469–1479 (2006)
12. C. Sioutas, R.J. Delfino, M. Singh: Exposure assessment for atmospheric ultrafine particles (UFPs) and implications in epidemiologic research. Environ. Health Perspect. **113**, 947–955 (2005)
13. A. Peters et al.: Respiratory effects are associated with the number of ultrafine particles. Am. J. Respir. Crit. Care Med. **155**, 1376–1383 (1997)
14. H.E. Wichmann et al.: Daily mortality and fine and ultrafine particles in Erfurt, Germany. Part I: Role of particle number and particle mass. Res. Rep. Health Eff. Inst. **98**, 5–86 (2000)
15. S. von Klot et al.: Increased asthma medication use in association with ambient fine and ultrafine particles. Eur. Respir. J. **20**, 691–702 (2002)
16. A. Henneberger et al.: Repolarization changes induced by air pollution in ischemic heart disease patients. Environ. Health Perspect. **113**, 440–446 (2005)
17. R. Ruckerl et al.: Air pollution and markers of inflammation and coagulation in patients with coronary heart disease. Am. J. Respir. Crit. Care Med. **173**, 432–441 (2006)
18. R. Ruckerl et al.: Ultrafine particles and platelet activation in patients with coronary heart disease: Results from a prospective panel study. Part. Fibre Toxicol. **4**, 1 (2007)
19. M. Stolzel et al.: Daily mortality and particulate matter in different size classes in Erfurt, Germany. J. Expo. Sci. Environ. Epidemiol. **17**, 458–467 (2007)
20. K. Hildebrandt et al.: Short-term effects of air pollution: A panel study of blood markers in patients with chronic pulmonary disease. Part. Fibre Toxicol. **6**, 25 (2009)
21. S. Breitner et al.: Short-term mortality rates during a decade of improved air quality in Erfurt, Germany. Environ. Health Perspect. **117**, 448–454 (2009)
22. J. Pekkanen et al.: Effects of ultrafine and fine particles in urban air on peak expiratory flow among children with asthmatic symptoms. Environ. Res. **74**, 24–33 (1997)
23. P. Penttinen et al.: Ultrafine particles in urban air and respiratory health among adult asthmatics. Eur. Respir. J. **17**, 428–435 (2001)
24. P. Penttinen et al.: Number concentration and size of particles in urban air: Effects on spirometric lung function in adult asthmatic subjects. Environ. Health Perspect. **109**, 319–323 (2001)
25. J. Pekkanen et al.: Particulate air pollution and risk of ST-segment depression during repeated submaximal exercise tests among subjects with coronary heart disease: The Exposure and Risk Assessment for Fine and Ultrafine Particles in Ambient Air (ULTRA) study. Circulation **106**, 933–938 (2002)
26. J. Kettunen et al.: Associations of fine and ultrafine particulate air pollution with stroke mortality in an area of low air pollution levels. Stroke **38**, 918–922 (2007)
27. T. Lanki et al.: Hourly variation in fine particle exposure is associated with transiently increased risk of ST segment depression. Occup. Environ. Med. **65**, 782–786 (2008)

28. J.J. De Hartog et al.: Effects of fine and ultrafine particles on cardiorespiratory symptoms in elderly subjects with coronary heart disease: The ULTRA study. Am. J. Epidemiol. **157**, 613–623 (2003)
29. K.L. Timonen et al.: Daily variation in fine and ultrafine particulate air pollution and urinary concentrations of lung Clara cell protein CC16. Occup. Environ. Med. **61**, 908–914 (2004)
30. A. Ibald-Mulli et al.: Effects of particulate air pollution on blood pressure and heart rate in subjects with cardiovascular disease: A multicenter approach. Environ. Health Perspect. **112**, 369–377 (2004)
31. K.L. Timonen et al.: Effects of ultrafine and fine particulate and gaseous air pollution on cardiac autonomic control in subjects with coronary artery disease: The ULTRA study. J. Expo. Sci. Environ. Epidemiol. **16**, 332–341 (2006)
32. F. Forastiere et al.: A case-crossover analysis of out-of-hospital coronary deaths and air pollution in Rome, Italy. Am. J. Respir. Crit. Care Med. **172**, 1549–1555 (2005)
33. S. von Klot et al.: Ambient air pollution is associated with increased risk of hospital cardiac readmissions of myocardial infarction survivors in five European cities. Circulation **112**, 3073–3079 (2005)
34. T. Lanki et al.: Associations of traffic-related air pollutants with hospitalisation for first acute myocardial infarction. The HEAPSS study. Occup. Environ. Med. **63**, 844–851 (2006)
35. Z.J. Andersen et al.: Size distribution and total number concentration of ultrafine and accumulation mode particles and hospital admissions in children and the elderly in Copenhagen, Denmark. Occup. Environ. Med. **65**, 458–466 (2007)
36. Z.J. Andersen et al.: Ambient air pollution triggers wheezing symptoms in infants. Thorax **63**, 710–716 (2008)
37. F. Mathé et al.: La mesure des particules en suspension dans l'air ambiant: Applications dans les réseaux français de surveillance de la qualité de l'air. Analusis Magazine **26**, 27–33 (1998)
38. J. Pekkanen, M. Kulmala: Exposure assessment of ultrafine particles in epidemiologic time-series studies. Scand. J. Work Environ. Health. **30**, 9–18 (2004)
39. P. Aalto et al.: Aerosol particle number concentration measurements in five European cities using TSI-3022 condensation particle counter over a three-year period during health effects of air pollution on susceptible subpopulations. J. Air Waste Manag. Assoc. **55**, 1064–1076 (2005)
40. G. Buzorius et al.: Spatial variation of aerosol number concentration in Helsinki city. Atmospheric Environment **33**, 553–565 (1999)
41. M. Maclure: The case-crossover design: A method for studying transient effects on the risk of acute events. Am. J. Epidemiol. **133**, 144–153 (1991)
42. F. Marano et al.: Impacts des particules atmosphériques sur la santé: Aspects toxicologiques. Environnement, Risques et Santé **3**, 87–96 (2004)
43. A.B. Knol et al.: Expert elicitation on ultrafine particles: Likelihood of health effects and causal pathways. Part. Fibre Toxicol. **6**, 19 (2009)
44. Z. Chen et al.: Acute toxicological effects of copper nanoparticles in vivo. Toxicol. Lett. **163**, 109–120 (2006)

7

Monitoring Nanoaerosols and Occupational Exposure

Olivier Witschger

As for any new form of technology, it is essential to assess the potential risks involved in the nanotechnologies, and more exactly, those raised by nanoparticles and nanomaterials. The very chemical and/or physical properties, sometimes unprecedented, on which nanomaterials and resulting products are based, some of them extremely interesting, may lead to new risks for the environment and for human health [1].

The risk for human health refers to the probability, low or high, of a person being affected by (exposed to) a hazard [2]. The hazard due to a chemical compound is the set of all its properties with the potential to cause toxic effects that are harmful to health. An intrinsic property of the compound, its toxicity is only one aspect of the risk. Indeed, the risk derives from a combination of toxicity and exposure. In the presence of a dangerous compound, the risk is nevertheless zero if there is no exposure. For this reason, a good appreciation of the exposure, and that includes the available collective and personal means of protection, is essential to controlling exposure and managing risk, as well as setting up suitable preventive measures [3,4]. Quantitative assessment of exposure is also a crucial aspect of epidemiological studies looking for links between a given form of particulate air pollution and its effects on health [5].

Regarding the risk to human health, there are three possible exposure routes: ingestion, the percutaneous route, and inhalation. The latter is considered to be the main form of exposure, especially in the workplace, and will thus be the subject of the present chapter. Exposure of the general population to nanoparticles and the consequent question of public health will be discussed in Chap. 8.

The general body of knowledge produced internationally over the past 15 years by studies of toxicology and effects on humans contains evidence attesting to certain harmful consequences of the specific properties of some nanoparticles. The resulting hypotheses regarding respiratory and cardiovascular effects, and consequences for the central nervous system and immune system, suggest that caution is in order when handling nanoparticles [6–11].

P. Houdy et al. (eds.), *Nanoethics and Nanotoxicology*,
DOI 10.1007/978-3-642-20177-6_7, © Springer-Verlag Berlin Heidelberg 2011

Published evidence also throws doubt on toxicological notions developed over a long period of time now for known substances, e.g., titanium dioxide.

Regarding exposure to aerosols of nanoparticles, referred to as nano-aerosols, our knowledge remains scant. One reason for this is the lack of agreement over measurement criteria, but this is compounded by a panoply of largely inappropriate instrumentation and non-standardised measurement strategies. Apart from this, the production and uses of nanoparticles and nanomaterials throughout the world of research and industry remain largely unknown, cooperation between specialists measuring exposure levels and industrial installations or research laboratories can be a delicate matter to set up, and results are sometimes difficult to publish and hence remain poorly advertised among the scientific community.

The aim of this chapter is to provide an overview of the main points regarding exposure to nanoaerosols, and in particular to summarise the possibilities provided by nanoaerosol measurement tools and strategies for characterising exposure. We also examine different ways of establishing reference exposure levels. Finally, we review research requirements for the years to come.

It is no easy matter to describe the whole issue of exposure to aerosols of nanoparticles today, given the many facets of the problem, each with its own potential significance in a context of scientific uncertainty. This chapter cannot therefore claim to be exhaustive. For instance, the instruments and their performance will not be described in full detail, and the reader is referred to various papers or books (or chapters in books) which go into greater depth, particularly regarding the measurement of aerosols [12–18].

7.1 Terminology and Definitions

Despite the many committees and reports set up worldwide, the question of definitions remains controversial, and disagreement is still mainly over the frontiers of the fields covered by the terms 'nanoparticle' and 'nanomaterial'. The meaning of these terms is different depending on who is using them, viz., researchers, institutions, or companies, but also the area of science, viz., physics, chemistry, or biology, and the technology being considered [19]. The list of documents published by various institutions over the past 5 years [20–25] gives a false impression of consensus, something not yet achieved. The aim in this section will thus be to set out explicitly what is meant by the terms 'nanoparticle' and 'nanomaterial' in the context of occupational health, and more precisely, the characterisation of exposure.

7.1.1 Nanoparticles

In a document published recently [25], the technical committee ISO/TC 229 of the International Standards Organisation (ISO) devoted to nanotechnologies proposed to adopt the term 'nano-object' as the generic term for any material

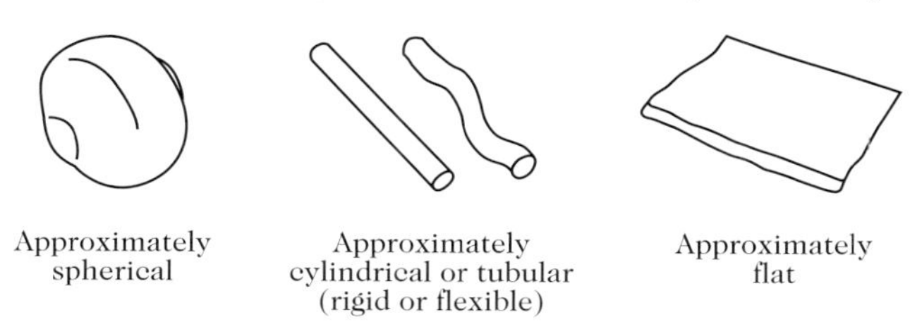

Fig. 7.1. Three possible shapes of nano-object

(functionalised matter) in which at least one external dimension is nanometric, and whose physicochemical properties are specific to this field. Here the nanometric length scale is roughly 1–100 nm, and there are three families of nano-objects depending on their approximate shape, as shown in Fig. 7.1:

- For spherical nano-objects, the three dimensions must be nanoscale.
- For cylindrical or tubular nano-objects, i.e., nanotubes with hollow interior in the latter case, the longest external dimension (the length) must be greater by a factor of at least 3 compared with the other two dimensions, and the length itself can be greater than 100 nm.
- For flat nano-objects, i.e., nanoplatelets formed by certain clays, only the thickness, the shortest dimension, need be nanometric.

For many reasons related to their behaviour and environment, e.g., during their fabrication [26, 27], nano-objects rarely occur in free form, i.e., isolated from one another. They tend to group together into more or less stable but disordered clusters, some dimensions of which may be significantly longer than 100 nm. The term 'primary particle' is also used to designate the elements making up a cluster [22]. There are two types of cluster:

- If the sum of the surface areas of the nano-objects making up the cluster, i.e., the primary particles, is close to the outer surface area of the cluster, this means that the nano-objects adhere to one another by weak physical bonds, e.g., Van der Waals forces, or else are merely tangled up, e.g., as happens with nanotubes. This kind of cluster is called an agglomerate.
- If the cluster comprises nano-objects (primary particles) connected by strong chemical bonds (covalent bonds), or if indeed they have partially coalesced, the resulting external surface area of the cluster may be significantly less than the sum of the surface areas of the nano-objects taken individually. In this case, the cluster is referred to as an aggregate.

While low energy processes such as shaking and ultrasound can dislocate agglomerates (deagglomeration), disaggregation requires higher energy processes, when it is possible at all [28, 29].

Another point, not specified in the ISO document, is that a nano-object may comprise different chemical elements or compounds. This compositional heterogeneity may occur in a range of different ways, e.g., core–shell, inclusions, and so on.

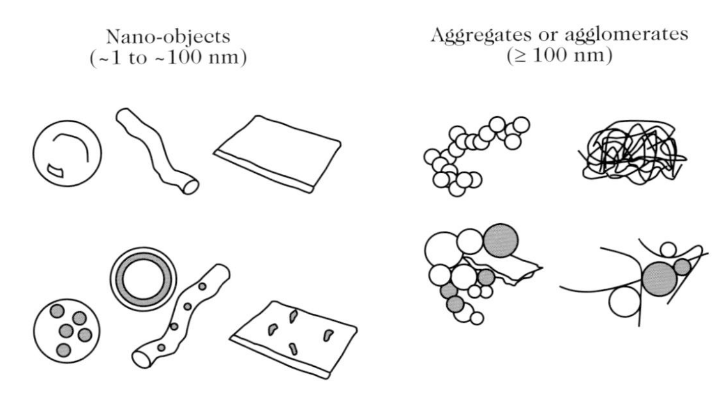

Fig. 7.2. Typology of nanostructured particles for the purpose of assessing occupational exposure. Adapted from [30]. Colour differences illustrate a difference of composition

The term 'nanostructured particle' is used to refer to the nano-object ensemble, aggregate or agglomerate. It indicates a particle whose structural features (or primary particles) have at least one nanometric dimension, i.e., less than 100 nm, and can influence its chemical, physical, or biological properties [30,31]. Nanostructured particles can have varying degrees of complexity and one dimension significantly greater than 100 nm. Figure 7.2 illustrates the typology of nanostructured particles in their individual or cluster forms.

To indicate that an aerosol is made up of nanostructured particles, the term 'nanoaerosol' is used [22].

7.1.2 Nanomaterials

The term 'nanomaterial' refers to a material, i.e., functionally specific matter, which, owing to its nanometric structure, has a modified chemical or physical property (or combination of properties) that is improved, adapted, or new compared with the bulk material of the same composition [32].

In a recently published document [23], the British Standards Institution (BSI) makes the more precise definition of a nanomaterial as being either a nanoparticle (in the sense of a nano-object), or a nanostructured material whose dimensions exceed the nanometric scale. The latter is said to be nanostructured either because it has some intrinsic nanometric structure, e.g., a nanoporous material, or because it contains nano-objects. In both cases, this nanostructure can be uniformly distributed throughout the piece of matter, or localised at the surface of it, for example. To refer to a nanomaterial containing nano-objects like nanotubes or metal nanoparticles in polymers, the generic term is 'nanocomposite' [24]. Figure 7.3 illustrates the different categories subsumed under the term 'nanomaterial'.

To simplify the discussion, we shall hereafter include nanostructured particles under the generic term 'nanoparticle', unless otherwise specified,

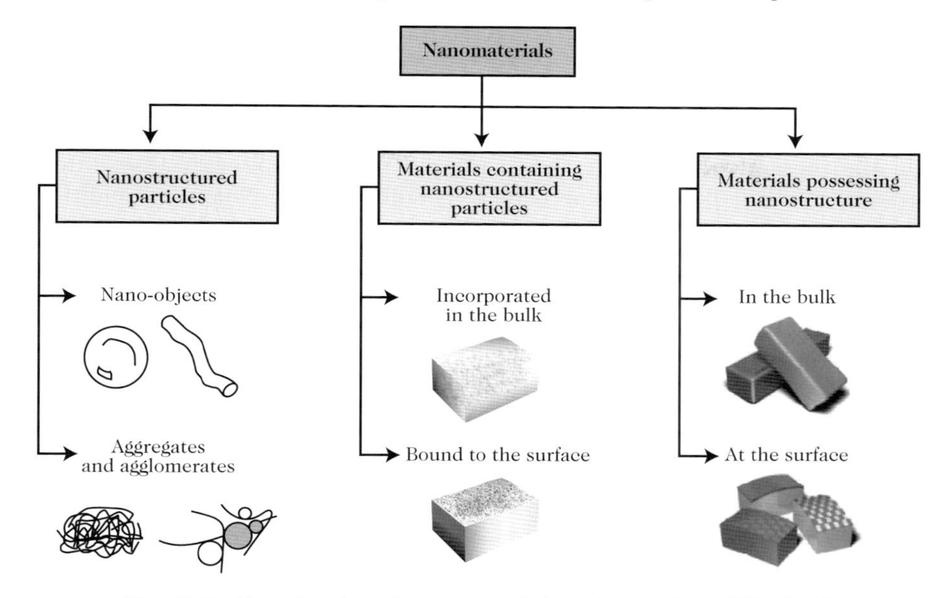

Fig. 7.3. Classification of nanomaterials. Adapted from [23,30,33]

e.g., when discussing nanotubes or nanofibres. Likewise, we shall use the term 'primary particle' to refer to the elements making up an agglomerate or aggregate.

Finally, it should be borne in mind that the appellation 'nanoparticle' is reserved for all intentionally produced nanostructured particles manipulated for industrial or commercial purposes. When referring to particles of nanometric size, the aggregates and agglomerates naturally present in the environment or else in industrial sources such as solder fumes, diesel emissions, etc., we shall use the term 'ultrafine particle'.

7.2 Characterising Occupational Exposure

Although the ultimate aim of aerosol measurements in occupational health is to inform about worker safety, there are nevertheless several other objectives:

- To evaluate personal exposure for comparison with some regulatory occupational exposure limit (OEL).
- To obtain exposure data for studies assessing exposure–effect or dose–effect relationships in humans.
- To identify and characterise emission sources in the context of a general assessment.
- To evaluate the effectiveness of existing or new means of controlling exposure, e.g., extraction systems, fume hoods, and so on.

For each of these objectives, there is generally a type of instrument, several methods, and a strategy.

Exposure to a substance can be defined as the amount of particles of the given substance likely to be inhaled and to reach a target organ or tissue. In the present case, the target is the walls of the respiratory tract. Many parameters will be important here, depending on the exposure conditions (duration, frequency, air flow, etc.), the respiration (inhalation, deposition, etc.), and the characteristics of the aerosol (granulometry, concentration, etc.).

The concentration is the amount of particles in a given volume of air. There are several possible metrics for this: mass per unit volume of air, with units mg/m^3 or $\mu g/m^3$, or number per unit volume, with units $1/cm^3$, or again surface area per unit volume, with units $\mu m^2/m^3$. In almost all occupational contexts, the aerosol is polydisperse, i.e., involves several different particle sizes, and the concentration will vary with particle size.

Exposure is a more precise notion than simply the contact concentration that may occur between the airways of the worker and the aerosol particles in which the worker is working and breathing. In principle, any measure of occupational exposure should produce a result that can be interpreted in terms of the level of occupational health risk.

7.2.1 Conventional Approach to Aerosols

For more than half a century, occupational exposure has been characterised quantitatively by the mass concentration, in units of mg/m^3 or $\mu g/m^3$, associated with the size ranges of particles entering different regions of the respiratory system (inhalable, thoracic, and alveolar fractions [34]), with the exception of fibres, where the concentration criterion is based on the number of fibres per unit volume of air. This approach applies to any chemical substance in the form of an aerosol and whatever the size of the particles that originally made up the aerosol.

The reason for this choice of criteria, viz., chemical composition, mass concentration in the air, and specific fraction of the ambient aerosol, is just that positive correlations could be established between them and toxic effects in animals (inhalation toxicology studies), or indeed harmful effects in humans (epidemiological studies).

Exposure assessment is also based as far as possible on a so-called personal measurement, i.e., using a portable instrument able to make the measurement as close as possible to the airways of the individual and right through the working day.

All the methods and arrangements made regarding chemical risk assessment and control are based on these fundamental principles [35]. In particular, in France and in many other countries, all occupational exposure limits (OEL) are based on the measurement of one of these health-related fractions. In the vast majority of them, it is in fact the inhalable or alveolar fraction. It is important to note that there are OELs even for poorly soluble

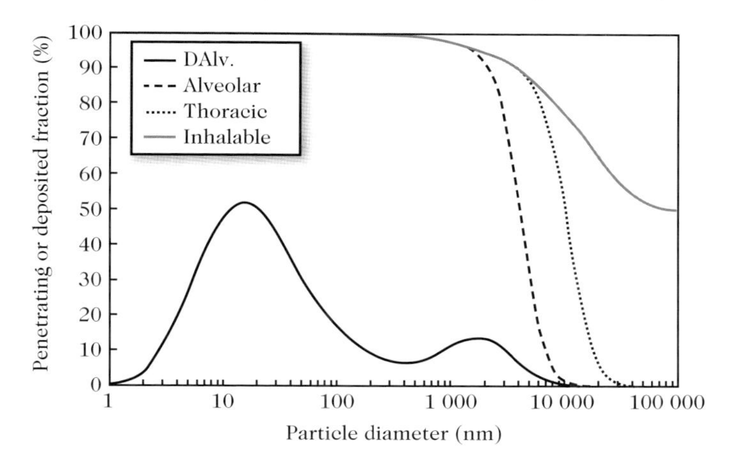

Fig. 7.4. Conventional inhalable, thoracic, and alveolar fraction curves [34], and deposition curve in the alveolar region. Calculations assume a spherical particle of density $\rho = 1\,\mathrm{g/cm^3}$ and a standard worker. From [43]

solid inorganic aerosols, said to be without specific effects, since inhalation of excessive amounts of particles likely to enter the airways, deposit themselves in the lungs, and remain there for a certain length of time creates a pulmonary overload that can weaken the organism's defences.

There is a range of instruments which meet the requirements for measuring occupational exposure to aerosols [36, 37]. These instruments must have certain levels of sampling efficiency depending on the particle size (expressed by the equivalent aerodynamic diameter) as close as possible to one of the three curves describing the so called conventional fractions (see Fig. 7.4).

In the panoply of portable commercial instruments, some are designed to sample the inhalable fraction, e.g., the IOM Sampler, the Button Sampler, or the CIP 10-I, while others sample the alveolar fraction, e.g., the Dorr-Oliver cyclone, or the thoracic fraction (CIP 10-T) [36]. There are also portable instruments able to measure several fractions simultaneously, like the Respicon [38]. In addition, instruments have been designed to make measurements of the conventional fractions at a fixed position, e.g., the CATHIA device [39]. The performance of all these instruments has been widely studied in different configurations over the last few years [40, 41]. Our understanding has thus reached a relatively stable level, even though there are still some questions, e.g., effects of the electrical charge of the particles on sampling, and there are still some improvements to be made with a view to reducing detection limits, miniaturising instruments, acquiring data in real time rather than averaging over the measurement period, and establishing simplified test protocols for the instruments [42].

All occupational exposure data acquired in France, Europe, and North America has been based for many years on the inhalable, thoracic, and alveolar conventions. While these conventions were set up to be linked to health through experimental data concerning particle inhalation, penetration, and deposition in the airways [15], this is not so for those used in the public health field under the appellations PM_{10} or $PM_{2.5}$, which refer to particles of equivalent aerodynamic diameter less than 10 and 2.5 μm, respectively. Initially set up to characterise particulate pollution sources in the general environment, the PM_{10} and $PM_{2.5}$ curves do not resemble any of the conventional inhalable, thoracic, or alveolar fraction curves. Occupational exposure to aerosols and nanoaerosols should not therefore be assessed on the basis of these PM_x indicators. Indeed, data produced on the basis of different standards could in future result in interpretive problems that would be hard to solve.

7.2.2 Measurement Criteria for Exposure to Nanoaerosols

Even though the subject of nanoparticle toxicity is far from being exhausted, the current body of knowledge in this area throws doubt on the conventional approach to assessing exposure. In addition, given the increasing number of people expected to be exposed in their workplace, a critical evaluation of the conventional approach is justified, together with a reassessment of criteria that were original treated as of secondary importance, but which may now predominate, in view of certain properties specific to nanoparticles. The question as to which measurement criteria should be chosen to assess exposure to nanoaerosols inevitably requires us to identify the parameters of nanoaerosols that are relevant when assessing effects on health, and which of these are measurable.

Many factors are potentially relevant to the toxic effects of nanoparticles [6, 8, 9, 11]: chemical composition, size distribution, shape, porosity and density of the particles, level and stability of agglomeration or aggregation, total surface area and surface reactivity, crystal structure, solubility and electric charge in biological media, and so on. Some of these factors were already identified for particles of micrometric size, e.g., chemical composition, size, crystal structure, but others take on much greater importance, and still others are quite novel for nanoparticles, such as surface area, level and stability of agglomeration, and so on.

Furthermore, it is useful to bear in mind that at least two conditions must be satisfied before we can consider that there is a risk due to nanoparticle inhalation [31]:

- The nanoparticle must be able to interact with the body in such a way that its nanostructure becomes biological accessible.
- The particle must have the potential to produce a biological response associated with its nanostructure.

While the second condition relates to toxicity considerations, the first indicates that, to a first approximation, any inhaled nanostructured particle should be taken into account as soon as contact can be set up between the particle and some deposition area in the respiratory tract.

Finally, three main criteria should be retained for exposure measurements: they relate to the particle size range, the aerosol fraction, and the concentration in the air, in particular, its metric.

Nanoparticle Size Range

As we saw in Sect. 7.1.1, in most institutional publications, the upper limit of the standard size range for nanoparticles is $\sim 100\,\mathrm{nm}$, since it is below this value that the specific physical and chemical properties of the nanometric scale tend to appear [20–22, 25]. However, this upper limit needs to be reconsidered, since one must:

- specify the relevant equivalent diameter,
- integrate biological considerations into the specification of this limit,
- take into account agglomerates and aggregates.

Quite generally, the result of a diameter measurement depends on the method used. In the context of aerosol monitoring and exposure characterisation, several equivalent diameters can be used: Stokes, aerodynamic, electrical mobility, diffusion, projected area, etc. [13, 14]. At the present time, there is no consensus regarding the choice of standard equivalent diameter. Such a consensus could only exist if there were validated instrumentation, designed to suit the constraints of an exposure measurement, which is not yet the case (see Sect. 7.2.3).

Since in the vast majority of cases the particles making up nanoaerosols are neither spherical nor of density $1\,\mathrm{g/cm^3}$, the various equivalent diameters do not result in the same value. As an example, Fig. 7.5 shows the theoretical ratio between two standard equivalent diameters used in aerosol measurements, viz., the equivalent aerodynamic diameter (AD) and the equivalent electrical mobility diameter (MD), for three combinations of particle density (ρ) and dynamic shape factor (χ). Note that the deviations between the equivalent diameters are significant and not monotonic. For example, $\mathrm{MD} \approx 100\,\mathrm{nm}$ corresponds to $\mathrm{AD} \approx 310\,\mathrm{nm}$ for a spherical particle of density $\rho = 5\,\mathrm{g/cm^3}$.

A first biological argument for specifying the upper size limit is that particle deposition must be possible in the respiratory tract, and particularly in the alveolar region (deep lung). As shown in Fig. 7.4, there is a minimum on the alveolar deposition curve, lying between ~ 300 and $\sim 500\,\mathrm{nm}$. This deposition minimum also corresponds to the minimum of the total deposition curve [43]. A second biological argument is the fact that inhaled particles deposited in

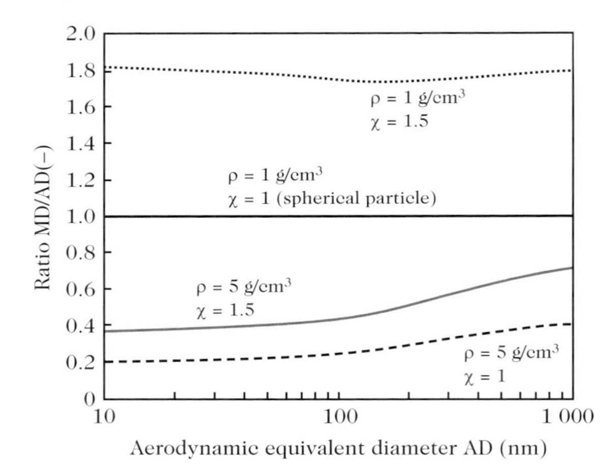

Fig. 7.5. Theoretical ratio of the equivalent electrical mobility diameter (MD) and the equivalent aerodynamic diameter (AD) for three combinations of particle mass density (ρ) and dynamic shape factor (χ)

the alveolar region would not be recognised by alveolar macrophages if they have one dimension greater than $\sim 500\,\text{nm}$ [44].

It is also essential to integrate the question of aggregates and agglomerates into the specification of the limit, since the biological impact is likely to be very different depending on their ability to decompose into smaller objects [5]. We may thus assume that, even if the aggregates and agglomerates are stable, they can induce a specific effect due to the nanostructuring of the ramifications entering into contact with the pulmonary tissue. Likewise, if the biological response is associated with the surface, it is crucial to take aggregates and agglomerates into account, and all the more so in that their morphology remains open (3D fractal dimension less than 2). In any case, it is clear that aggregates and agglomerates need to be taken into account if they manage to reach the alveolar region.

The presence of nanoparticles (and hence of nanoscale features) must also be envisaged in a size range that differs from the one in which they were originally emitted. Indeed, microscale particles – not nanostructured – already present in the work atmosphere and emitted from different sources, can play the role of attractor for nanoparticles under certain conditions (heterogeneous coagulation). This point, recently demonstrated experimentally under laboratory conditions representative of those encountered in the work environment [45], tells us that the upper limit of the nanoparticle size range should be placed well above the value usually cited, viz., 100 nm, when assessing occupational exposure.

Given the points just discussed, in a context of uncertainty and lack of consensus, one must apply the principle of precaution. As a consequence, when the problem is to characterise occupational exposure, one must integrate the whole length interval specified by the alveolar region, viz., approximately 1–5 000 nm. However, two situations can be distinguished:

- When the nanoparticles are in their free form, the upper limit will be roughly 500 nm.
- For aggregates and agglomerates, the upper limit is extended to around 5 000 nm.

This proposal agrees with one recently published (see Sect. 7.3.2).

Aerosol Fraction

The question of the aerosol fraction is fundamental since, quite generally, not all particles present in the respiratory tract of an individual are inhaled, and not all inhaled particles are deposited. The particles that are not inhaled or those that are exhaled do not interact with the respiratory tract and thus should not be included in the exposure.

While the specification of the inhalable, thoracic, and alveolar conventions was indeed an improvement in the field of exposure assessment, these conventions are still not fully satisfactory, since the existence of differences between penetrating and deposited fractions for the same region of the respiratory tract leads to biases of varying degrees when evaluating doses [43]. In nanoaerosol exposure assessment, it is thus necessary to integrate a deposition criterion rather than a penetration criterion for a given respiratory compartment (the principle underlying the conventions). In practice, this can be done through suitable measurements, e.g., granulometry, concentration, and a deposition calculation using a model, such as the one proposed by the International Commission on Radiological Protection (ICRP) [46], on the understanding that the model does indeed correspond to experimental data obtained recently for humans [47].

Metric (Concentration in the Air)

As discussed in Sect. 7.2.1, concentration measurements in units of mass per unit volume of air, i.e., mg/m^3 or $\mu g/m^3$, are the norm for occupational exposure assessment, with the exception of fibres, where the number concentration per unit volume of air is preferred.

Given our current understanding based on epidemiological and toxicological studies, it is becoming ever clearer that, for insoluble or poorly soluble substances like titanium dioxide, exposure to nanoaerosols cannot be assessed purely on the basis of the two indicators provided by mass and chemical composition [9]. But specifying just how it should be assessed in these cases remains an ambitious objective today, because the list of determining factors is long, and the number of substances studied in the nanoparticle state is still limited. However, it seems fairly clear that [30]:

- It is appropriate to measure the concentration in terms of surface area per unit volume (i.e., in units of $\mu m^2/m^3$) in many circumstances, but it cannot be generalised to all.
- It is appropriate to measure the number concentration (i.e., in units of cm^{-3}) when surface area is not the main factor underlying toxicity. In addition, since this measure brings out the finer fraction of any polydisperse aerosol, it is useful for identification purposes.
- It is still useful to measure the mass concentration in certain situations, provided that a suitable granulometric selection has been carried out upstream. In addition, since this measure remains the norm for aerosols, it provides a modicum of continuity with historical exposure data.

At the present time there is still no certainty about which concentration (surface, number, or mass) or which parameter (particle shape, surface reactivity, solubility, charge, etc.) to measure apart from the size, composition, and chemical structure. Research is ongoing for many nanoparticles to determine the relative importance of these different factors. Eventually, only a few of them should be retained in the framework of a methodology tailored to measure occupational nanoaerosol exposure.

In this context, the consensual approach here is therefore to adopt as far as possible a measurement strategy that can characterise several complementary nanoaerosol parameters. The aim is to obtain results that could be interpreted in their entirety in the light of future knowledge of toxicity and health consequences. From a practical standpoint, this involves determining the area, number, and mass concentrations of the nanoaerosols, but also if possible their size distribution, particle shape, chemical composition, crystal structure, and so on.

7.2.3 Instrumentation and Methods

In the field of aerosol monitoring, there currently exist instruments and methods to achieve the following objectives:

- To measure particles in the length range between the nanometer and a few tens of micrometers. Different equivalent diameters can be measured, e.g., Stokes, aerodynamic, electrical mobility, diffusion, etc.
- To measure directly or indirectly the area, number, and mass concentrations.
- To obtain information concerning parameters that cannot be directly measured in the aerosol phase, such as the particle density, shape (through the fractal dimension), and electric charge.
- To obtain samples by granulometric class or otherwise for gravimetric, electron microscope, or physicochemical analyses, the latter including X-ray diffraction, inductively coupled plasma mass spectrometry (ICPMS), specific surface area measurements using the Brunauer–Emmett–Teller (BET) method, etc.

Some of these instruments can make measurements in real time while others require post-processing analysis, i.e., weighing, chemical analysis, etc., before obtaining the result.

Most of these instruments and methods were developed for research applications to the physics of aerosols and nanoparticles, atmospheric aerosols, and so on, or to meet specific industrial requirements, such as engine emissions, processes for synthesising nanopowders, etc., with the consequence that none of them is designed to match the constraints of occupational exposure assessment. An ideal instrument for such measurements would be one with the following attributes:

- Able to provide information in real time for a range of parameters, including surface, number, and mass concentrations, particle size distribution, and particle charge, while distinguishing the relevant nanoparticles from other nanoscale particles in the ambient air of the workplace (either of natural origin or produced by engine emissions).
- Small and portable, so that measurements can be made as close as possible to the worker's airways.
- Qualified for use in an industrial environment, i.e., autonomous and authorised in potentially explosive environments.
- Relatively cheap for use by a large number of companies and research centers for routine measurements.

Such an instrument does not yet exist. It will be one of the major challenges in the field of nanoaerosol measurement over the next few years [7]. On the European level, the research programme NANODEVICE, a collaboration between 26 partners, has just been launched to try to meet this challenge [48].

For each of these metrics (number, area, and mass), Table 7.1 lists the instruments or methods available to characterise exposure to nanoaerosols. The table is an updated version of several published sources [22,49,50]. In the table, coupling refers to whether or not coupling is incorporated for chemical or microscopic analysis.

Area concentration measurement is currently an important subject of research, since it may well turn out to be a relevant exposure indicator for certain exposure situations. Just as there are several definitions of particle surface area, there are several methods for measuring it. The best known approach is based on the specific surface area obtained using the BET method. While this robust technique remains the reference in the field of powder characterisation and experimental toxicology, it is not today suitable for the case of nanoaerosols in the work place atmosphere, since a relatively large amount of matter has to be collected, and the sampling time would be much too long for this kind of measurement. However, further work could be done along these lines. In nanoaerosol metrology, surface area measurements are mainly made by ion diffusion charging–electrometer (DCE), where the area is obtained by measuring the level of collisions between atoms (or molecules) and particles (current measurement). In particular, an active area is defined as the area

Table 7.1. Available instruments and methods for characterising exposure to nanoaerosols

Metric	Instrument (or method)	Measurement		Coupling	Remarks
		Real time	Granulometry		
Direct mass	Samplers	No	Yes (some)	Yes	The only instruments with cutoff diameter $< 100\,\text{nm}$ are cascade impactors (low pressure, microslot) or coupled impactor–diffusion battery systems. Samples collected by granulometric class for analysis. Equivalent aerodynamic or diffusion diameter
	TEOM	Yes	No	No	This is a tapered element oscillating microbalance. Good sensitivity ($\sim \mu g/m^3$). Continuous measurement and short response time (∼s). Size selector placed upstream. Personal version under development
Computed mass	ELPI	Yes	Yes (10–10 000 nm)	Yes	From current distribution if density and shape of particles are known. Continuous measurement and short response time (∼s). Equivalent aerodynamic diameter. Samples collected by granulometric class for analysis
	SMPS, FMPS	Yes	Yes (1–1 000 nm, 6–600 nm)	No	From number distribution if density and shape are known. Equivalent electrical mobility diameter. Many versions of SMPS are commercially available. Continuous measurement and short response time: FMPS ∼s, longer for SMPS ∼60 s. Radioactive source for SMPS
Direct number	CNC	Yes	No	No	Size selector placed upstream. Simple to use but delicate
	SMPS	Yes	Yes (1–1 000 nm)	No	Direct number because an SMPS includes a CNC. Radioactive source
	Electron microscope	No	Yes	–	Requires specific collection system (see Sect. 7.2.5)

Computed number	ELPI	Yes	Yes (1–10 000 nm)	Yes	From current distribution if particle shape and density known. As above
	FMPS	Yes	Yes (6–600 nm)	No	From current distribution if particle shape and density known. New system, so little feedback as yet
Direct area	Diffusion charging	Yes	No	No	Various commercial systems (active area, deposited area). Size selector upstream. Little feedback as yet
	ELPI	Yes	Yes (1–10 000 nm)	Yes	From current distribution
	Electron microscope	No	Yes	–	Various approaches. Overlap must be taken into account for aggregates and possible relationship with specific surface area [51]
Computed area	SMPS, FMPS	Yes	Yes (1–1 000 nm), (6–600 nm)	No	From the number distribution. SMPS: for open agglomerates, the equivalent mobility diameter is well correlated with the projected area diameter
	ELPI	Yes	Yes (1–10 000 nm)	Yes	From the number distribution
	Joint SMPS/ELPI	No	No	Yes ELPI	From difference between equivalent aerodynamic and electrical mobility diameters

subjected to interactions with the ions of the carrier gas. As a consequence, it is important to note that the geometric area is not equal to the active area. Recently, a study has been carried out on different nanoparticles (C, Al, Ag, and Cu) to discover the response functions of three commercially available instruments [52].

7.2.4 Sampling and Deposition of Particles

In all these methods, measurements are made by continuous air sampling, i.e., each instrument requires an internal or external pump. Once captured by the sampling orifice, the nanoaerosol particles are transported by the flow of air and processed in various ways inside the instrument, e.g., electrical charging, drying, condensation of a vapour, or subjected to various forces, e.g., impaction, electrophoresis, thermophoresis, before passing into a detector, e.g., laser, electrometer, or else collected on a substrate of some kind, e.g., a filter or plate. Instruments are usually some distance from the measurement site for reasons of accessibility (see Fig. 7.8), so sampling tubes of various lengths must be used to guide the aerosol to them.

In general, sampling and transport tend to modify the initial characteristics of the aerosol in such a way that the concentration and granulometry may have significantly changed before it finally enters the measurement instrument. For nanoparticle aerosols, the effects are mainly confined to transport through the various tubes and connectors by diffusion or electrostatic effects. There are several ways to estimate and calculate the biases introduced by the sampling itself [12–14]. To illustrate this, Fig. 7.6 shows the effect of sampling tube length on deposition for a typical setup (tube diameter 6 mm and volume sampling rate 5 l/min). For the calculation, only deposition due to

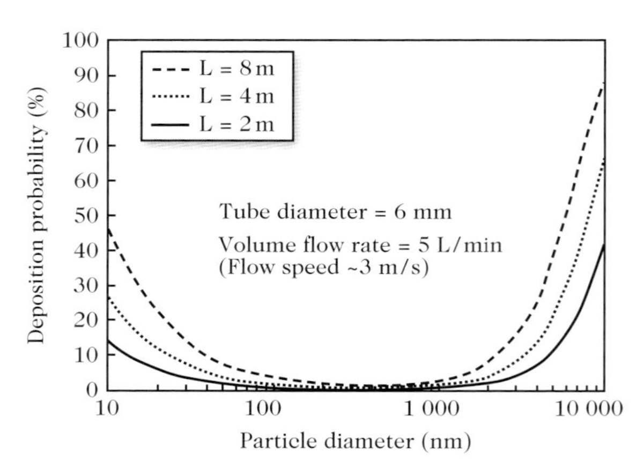

Fig. 7.6. Effect of sampling tube length on deposition for different particle diameters. The calculation takes into account sedimentation and diffusion of particles to the walls

sedimentation and diffusion to the walls have been considered, i.e., no electrostatic effect. Note that there are losses even for the shortest length, whereas the particle transit time is very short ($\sim 0.7\,\mathrm{s}$). For length $L = 2\,\mathrm{m}$, the deposition remains less than 10% over the interval from roughly 20 nm to 4 μm. This interval reduces to 50 nm–2 μm for length $L = 8\,\mathrm{m}$. With this length, the sampling efficiency for a 10 nm particle is only 50%. It is essential to take deposition into account when characterising exposure.

7.2.5 Sampling and Physicochemical Analysis

A common feature of all the instruments mentioned in Sect. 7.2.3 is that they give no indication of the chemical nature or shape of the particles, their crystal structure, or their degree of aggregation or agglomeration. Only physicochemical analysis can provide this kind of information, and this must be carried out a posteriori using laboratory techniques. Such techniques are transmission electron microscopy (TEM), scanning electron microscopy (SEM), scanning probe microscopy (SPM), mass spectrometry, X-ray diffraction, atomic absorption spectrometry, etc. Samples must be collected and prepared in certain ways before they can be implemented. This stage is not always straightforward, because each analysis technique has its own constraints, e.g., uniform particle deposition, minimal overlap, presence of elements in the collection medium that can perturb the analysis or increase the quantification limit of the given element, etc. This means that specific instruments and collection media must be used. Since it is often impossible a priori to know the concentration levels at a given work station, it is a delicate matter to optimise the sampling period. The problem is to obtain an amount of sample that is sufficient to guarantee efficient analysis while avoiding an excess of sample that might compromise it.

For transmission electron microscopy, the sample can be collected by several methods, ranging from simple filter sampling (followed by specific processing or simply placing a TEM support grid on it [53]), to the use of systems specially designed for direct collection on TEM grids. Some instruments exploit electrostatic precipitation, in which case the nanoparticles must first be charged with the right sign [54, 55], while others use thermophoretic precipitation, where the nanoparticles are taken from a hot stream then precipitated on a cold plate where the TEM grid is located [56]. Very recently, an instrument has been developed to collect nanoparticles smaller than a certain size (typically less than about 200 nm) on a TEM grid. This system exploits the joint effects of diffusion and thermophoresis [57]. Each of these collection techniques is characterised by its own intrinsic efficiency which is often difficult to identify precisely. More research is thus needed here.

Other new developments are coming up which can analyse the elemental composition of the particles in the aerosol phase. One technique known as laser induced breakdown spectroscopy (LIBS) uses a microplasma to vaporise and dissociate the particles. The elemental composition of the aerosol particles is then obtained by analysing specific atomic emissions from the bulk of the plasma. The performance of this device has recently been investigated, and the limiting size of the detected particles is around 60 nm [58].

7.2.6 Measurement Strategies and Interpretation of Results

Measurement strategy is a key point in the characterisation of nanoaerosols in the work place. This strategy must be able to identify and characterise the likely emission source(s) of nanoparticles, picking them out from the background, i.e., those nanometric particles present in the work atmosphere that are not related to the studied activity, e.g., ultrafine particles [59].

At the present time, there is no single strategy for carrying out exposure measurements. However, there is a common feature of all the strategies described in published studies, namely, the adoption of a multifaceted approach, incorporating different techniques and complementary methods among those mentioned in Sects. 7.2.3 and 7.2.5. Given the complexity of some of these, experimental studies are the prerogative of specialists in the field of nanoaerosol measurement. For this reason, it seems useful to propose a two-level strategy:

- An initial assessment determining number and mass concentrations and obtaining some indication of particle size and shape. This first assessment would use portable instruments, sampling devices to obtain samples for gravimetric analysis, and identification and characterisation by electron microscopy.
- A main assessment to characterise the many parameters insofar as possible (see Sects. 7.2.3 and 7.2.5), for the specific emissions observed in the first stage. This evaluation uses specific means and people specialised in nanoaerosol metrology to carry out the measurement campaign, analyses, and processing and interpretation of data.

The initial assessment was first described by the National Institute of Occupational Safety and Health (NIOSH) [60], then taken up by the Organisation for Economic Cooperation and Development (OECD) [61]. The process is presented in Fig. 7.7.

This initial assessment begins with an observation of the work station and its immediate and more distant environment (both inside and outside the building). The main aim here is to understand the different stages of the processes under assessment in order to locate the different possible sources of nanoparticles. At the same time, it is useful to seek sources of ultrafine particles, either inside or outside (combustion, lifting equipment, etc.), to evaluate the general air flow, and in particular any air arriving from neighbouring rooms or outside the building, so as to select suitable measurement points. Measurements are made using a condensation nucleus counter (CNC) or an optical particle counter (OPC) operating in parallel, to begin with when there is no activity, i.e., the process is stopped, if possible, then with activity,

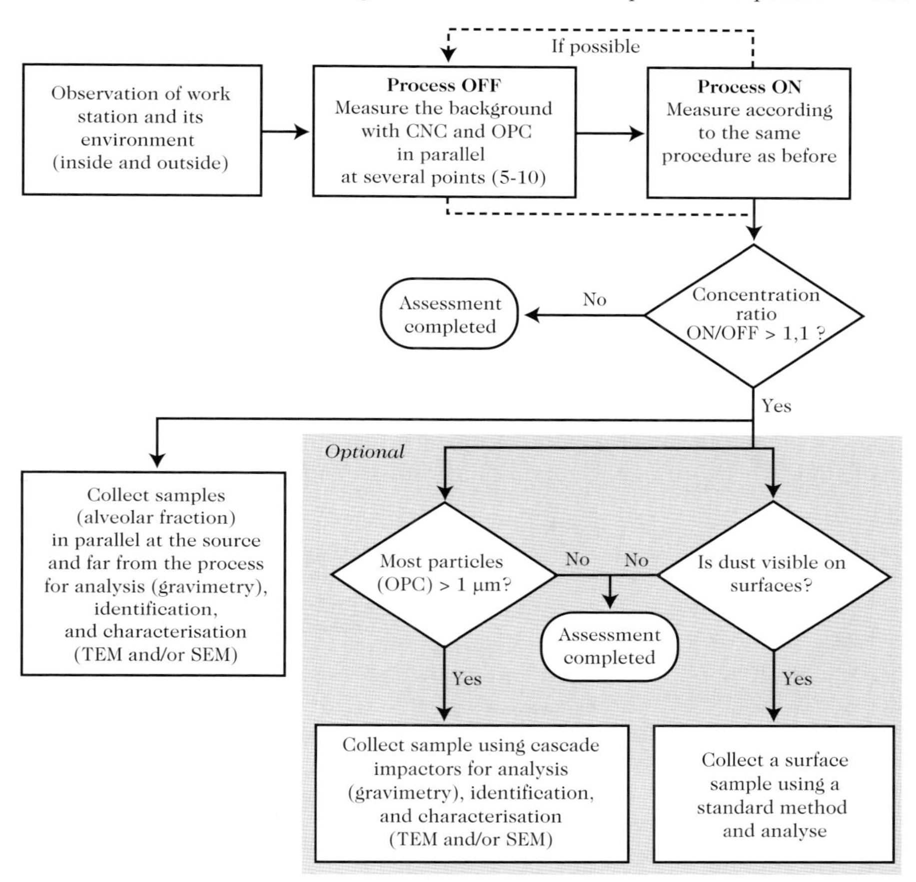

Fig. 7.7. Flow chart describing the initial assessment around a process or operation fabricating or involving nanoparticles. CNC condensation nucleus counter, OPC optical particle counter. Background refers to the ambient air background at the work station. Adapted from [60, 61]

i.e., when the process is running. If a significant rise is observed in the average concentration relative to the background (greater than 10%), samples are taken at a fixed station (possibly personal samples) with devices for measuring the alveolar fraction at the same time, for gravimetric analysis, identification, and characterisation by electron microscope. If more detailed granulometric information is required, cascade impactors can be used. Likewise, samples of surface deposits can be taken and analysed. During this initial assessment, other portable instruments can be brought in, such as a diffusion charger, providing real time measurements of surface concentrations ($\mu m^2/m^3$), or a photometer, whose response is correlated with the mass concentration. The use of relatively simple techniques means that this first stage could if necessary be conducted by people who are not aerosol measurement specialists, but who have nevertheless already handled aerosol sampling equipment, e.g., in the context of conventional exposure measurements. However, it should be stressed that the strategy based on measurements made with and without activity is only valid when the ambient air background is low and does not fluctuate.

For its part, the main assessment is a more sophisticated matter, given the instruments and methods used here to acquire data on a wide range of parameters, e.g., diameter by various techniques, real time granulometry from a few nanometers to several micrometers, number, mass, and area concentrations, specific samples for electron microscope analysis and chemical analysis techniques, etc. No typical general strategy can be laid down here, since one of the aims is precisely to adapt the experimental setup as a whole to the relevant nanoparticles, the work station, and its environment. As an example, Fig. 7.8 shows an experimental setup incorporating specific sampling lines, including where necessary conditioning stages such as drying, cooling, and

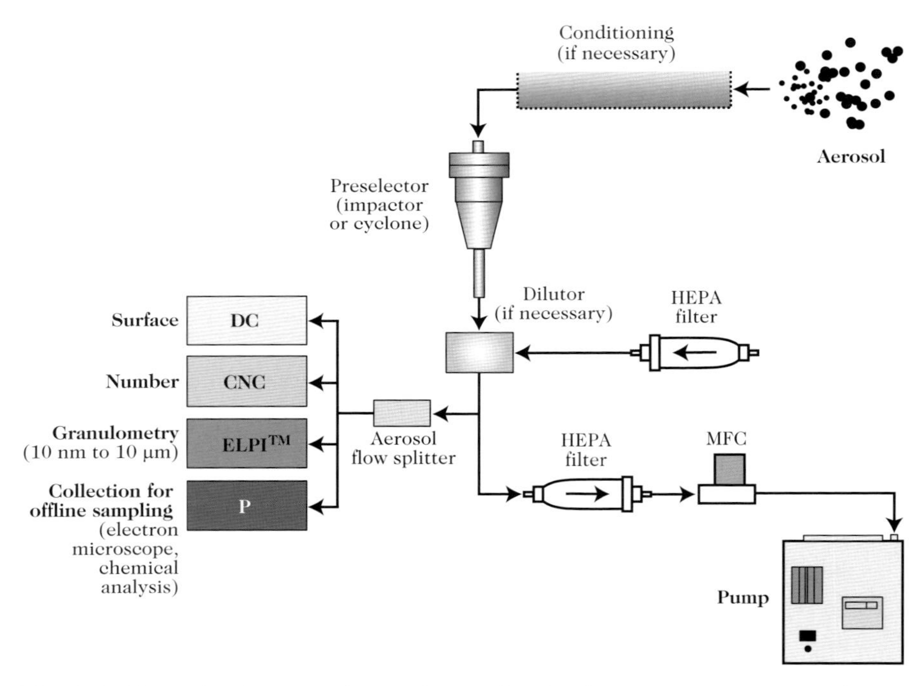

Fig. 7.8. Example of an experimental setup for simultaneous measurement of number and area concentrations and granulometry, and also to collect samples for various analyses, such as electron microscopy, X-ray diffraction, mass spectrometry, and so on. DC: diffusion charging. CNC: condensation nucleus counter. ELPI: electrical low pressure impactor. MFC: Mass flow controller. P: precipitator – simple filter or instrument using thermophoresis, diffusion, electrophoresis, etc. Adapted from [62]

so on, preselection stages removing particles with diameters above a certain cutoff diameter, dilution, and particle flow distribution.

In most work situations, the events leading to nanoparticle emission in the air may be short-lived or unstable and, in some cases, multiple emission conditions coupled with natural air movements, or air flow due to ventilation, may be encountered. These elements can increase the spatiotemporal variability of the aerosol with regard to concentration and granulometry. In addition, operators are often mobile, which makes it difficult to determine exactly when they are working at given fixed points. This is why the nanoaerosol data obtained at fixed points – and this represents today almost all studies, given the available instrumentation – cannot be directly transformed into exposure data.

Moreover, care must be taken in interpreting data obtained using the techniques described in Sect. 7.2.3, especially with regard to detection limits on particle sizes and concentrations, measured equivalent diameters, etc. For example, the result of a number concentration measurement is particularly sensitive to the efficiency of the counting instrument.

On top of this, all the techniques or methods described use computational tools of varying degrees of complexity to process the data, such as signal processing or data inversion algorithms, or data coupling when different physical principles are used.

Finally, the current concern with parameters that were until recently considered to be of secondary importance means that instrumental performances are being reassessed. This is the case for example with regard to nanoparticle morphology for the two important instruments known as the scanning mobility particle sizer (SMPS) [63] and the electrical low pressure impactor (ELPI) [64]. While the results have been published in the scientific literature, these corrections are not yet available from the instrument makers, so users must themselves integrate such adjustments into the data interpretation scheme.

7.3 Occupational Exposure

Exposure due to processes that are not designed deliberately to fabricate or manipulate nanoparticles is discussed in the next section. However, thermal processes, such as soldering, laser cutting, smelting, etc., together with certain mechanical processes involving conventional materials, e.g., milling, grinding, drilling, etc., produce more or less significant amounts of nanometric particles, aggregates, and agglomerates (ultrafine particles) to which many workers are exposed [16, 49].

7.3.1 Exposure Factors and Scenarios: Qualitative Aspects

Exposure factors are generally related to the work station (material, physical, human, and organisational), the specific task to be carried out, and the materials and tools used to do so, but also the operators and supervisors involved in the work [2].

For an operator to be exposed, there must be nanoparticle emission into the air, i.e., formation of a nanoaerosol at the source, followed by dispersion into the neighbouring environment, and transfer to the respiratory region. Furthermore, the question of exposure must be considered throughout the lifetime of a product [65, 66]:

- During fabrication (opening the reactor, collecting the product, conditioning, etc.) and handling (sampling, decanting, emptying hoppers, mixing, etc.) of nanoparticles.
- During synthesis of nanocomposites (incorporation in matrices) and their transformation (milling, grinding, etc.).
- During cleaning and maintenance of equipment (reactors, fume cupboards, glove box, filtering equipment, etc.).
- During waste disposal or recycling.

Exposure must be investigated during normal operations, but also during slowdowns, or incidents [3, 10, 18]. There are thus different situations that should be listed, depending on the immediate environment of the nanoparticles, as depicted in Fig. 7.9.

A first case concerns nanoparticles when they are dispersed in a gas (aerosol). This situation is encountered specifically in processes involving gas phase synthesis. Since such processes are carried out in closed environment such as a spray tower or reaction chamber in order to maintain the required experimental conditions, nanoparticle emission is only possible when a problem of some kind occurs with the equipment or synthesis protocol, e.g., a leak or accidental opening. After synthesis, if the reactor has not been completely emptied, the nanoparticle recovery stage is another potential source, and so is the final stage when the installation is cleaned out.

Many processes for synthesising nanoparticles produce nanopowders [26, 67]. Naturally, these also constitute a potential source of exposure whenever they are handled, or simply exposed to an air flow that may lead to their being suspended in and transported by the air. This means that any operation like decanting, sampling, recovery, weighing, mixing, drying, packaging

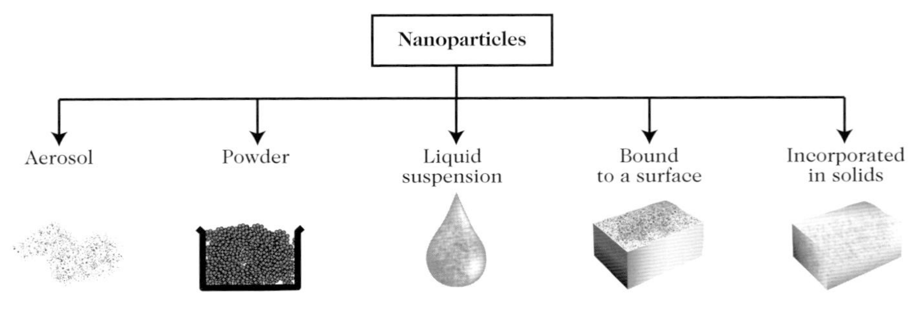

Fig. 7.9. Situations that may give rise to occupational exposure to nanoparticles. Adapted from [33]

(e.g., Big Bag packaging) may give rise to exposure. However, regarding nanopowders, a distinction should be made between those that have been fabricated, i.e., handled and incorporated, for several decades, such as carbon black, titanium dioxide, and silica, and those that have only recently been synthesised, such as carbon nanotubes and fullerenes [68]. For nanopowders that can be qualified as long-established, there already exists a significant body of knowledge concerning exposure scenarios and conditions, but it is based solely on the conventional ways of assessing exposure. It would thus be a mistake to assume that, just because titanium dioxide and silica, to name only two substances, have been produced and used for years now, exposure risks are fully understood. As discussed in Sect. 7.2, the characterisation of nanoaerosols appeals to specific and recent instrumentation, which has only been implemented in the field for at most a few years. Furthermore, the sudden interest in nanoparticles in general will result in new industrial applications, or increase some existing applications of long-established nanoparticles which may give rise to new types of exposure. In fact, exposure data is currently lacking for established nanopowders. As far as the more recent ones are concerned, a great deal of work is under way in research laboratories, and industrialisation is only just beginning. As a consequence, knowledge of real scenarios is still rather poor [18].

When nanoparticles are suspended in liquid phase, exposure is in principle reduced, as compared with nanopowders. However, decanting, gas bubbling, or sparging, along with spraying or atomisation, are operations leading to the formation of droplets which may include nanoparticles. Depending on how the droplets are generated and the environmental conditions of the operation, e.g., the humidity, the properties of the liquid and the nanoparticles, e.g., the concentration in the suspension, size, charge state, etc., the resulting aerosol may comprise a more or less significant fraction of nanoparticles (single, aggregated, or agglomerated) which could be inhaled and end up in the alveolar region. This has been demonstrated experimentally in laboratory studies [69, 70].

When nanoparticles are incorporated in a nanocomposite matrix, or deposited on a surface in the form of composite metal–polymer thin films, for example, the question of emission is relevant during fabrication and handling as a result of physical interventions like drilling, cutting, grinding, etc. This question is also relevant right through the lifetime of the final product, since damage may occur to it, e.g., abrasion [71]. For example, it has been shown recently that TiO_2 nanoparticles can be emitted by wall paints in single or aggregated form [72]. Bearing in mind the many current and future applications of nanocomposites and thin films incorporating nanoparticles, studies have already been launched for the development of experimental setups [73, 74].

As can be seen from Table 7.2, there are in principle many situations where nanoparticle exposure is possible. If the nanoparticles emitted in the

Table 7.2. Typology of potential exposure situations. Adapted from [65]

Stage	Laboratory (research and pilot)	Industry (development and production)	User industry
Quantity (mass)	mg to kg	kg to > 100 kg	kg to > 100 kg
Tasks	Manufacture, characterisation, optimisation, and development of processes, applications research	Dimensioning and optimisation of processes, applications research, production in large amounts, packaging, and transport	Deconditioning, direct use, or incorporation in various matrices, development of processes, production in large amounts, packaging, and transport
Potentially exposed staff	Researchers, laboratory technicians, students Maintenance, cleaning, and waste disposal staff	Engineers and production staff, storage and transport staff	

form of nanoaerosols have any properties suggesting that they may constitute a hazard, then the risks must be assessed as a matter of priority.

On the question of estimating the size of the working population potentially exposed, several studies have been made recently [75–80]. Globally, the figures vary from a few thousand for research and production and several hundred thousand when the activities of nanoparticle users are included. The chemical industry seems to be the sector in which the highest percentage of companies have recourse to nanoparticles, e.g., about 20% in Switzerland. While these studies attest to the existence of nanoparticles in industry, they rarely give any quantitative information about the nature and extent of exposure (concentration levels, duration, size of nanoparticles, etc.), since these studies are not generally accompanied by measurement campaigns in the field. However, they are still invaluable, since they do provide information about amounts used and stored, uses, safety strategies implemented at the work station to protect operators and the environment, etc.

So far only a few processes have reached a high level of industrial maturity, and a few relatively easy to make products can be found in industry or are commercially available (the cosmetics, textile, sports, building, and transport industries) [81]. However, given the huge potential of nanomaterials and ever increasing research effort, a large part of current industrial activity is situated well upstream. This means that those populations that have been identified and recorded as of today are quite certain to grow. The identification of populations, exposure scenarios, and safety methods to be implemented throughout the life cycle of nanomaterials must therefore remain an important line of research.

7.3.2 Practical Approach to Identify the Nanoparticulate Character of a Work Context

In a recently published document [44], the Swiss Federal Office for Public Health proposes a way to determine whether a situation should be considered as a specific risk with regard to nanoparticles. This approach takes into account several factors related to the sizes of the primary particles (PP), the aggregates, and the agglomerates, but also the stability of the latter under the given conditions (at emission, during transfer in the air, and once deposited in the respiratory tract).

Regarding the upper size limit for primary particles, this approach suggests that 500 nm would be preferable to 100 nm, which is the reference value in most publications [20–25]. There are several reasons for taking this value of 500 nm, as indicated in the discussion of the nanoparticle size range on p. 171.

Figure 7.10 is a flow chart for determining whether a given scenario involves any risk of nanoparticle exposure. The very minimum knowledge required with regard to the nanoparticle characteristics is the size of the primary particles, the presence and size of agglomerates and aggregates, and the stability of

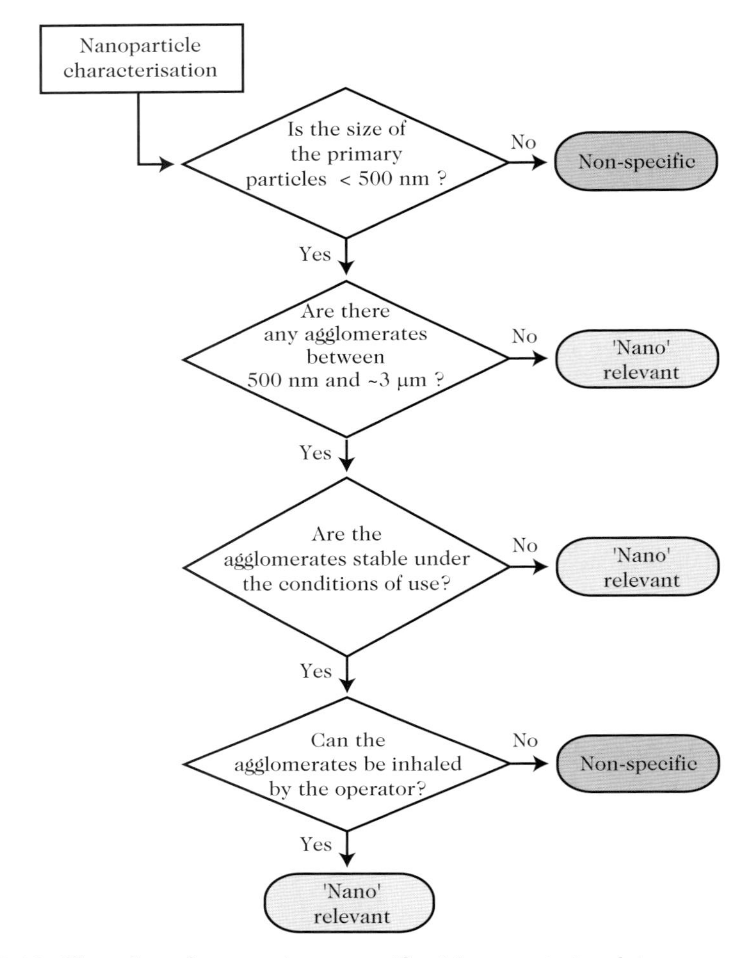

Fig. 7.10. Flow chart for assessing a specific risk scenario involving nanoparticles. Adapted from [44]

the latter, especially if they are able to break up, but also the existence of a realistic particle inhalation scenario.

This method can be implemented in the following kinds of situation:

- Research and development in state run research centers and industry.
- Production, including storage, packaging, and transport.
- All forms of use.
- Waste disposal or recycling.

If the situation is classified as a specific risk with regard to nanoparticles, a subsequent measurement campaign can be organised (see Sect. 7.2.6).

7.3.3 Emission of Nanoparticles by Powdered Materials. Nanopowders

The emission into the air of powdered materials in the form of nanopowders is a scenario suspected of giving rise to nanoparticles in the work place (see Sect. 7.3.1). There are various forms of emission: falling powders, air flows picking them up from a deposit or pile, vibration of a contaminated surface, etc. They may be found generically in a great many industrial sectors, from the chemical industry to the electronics, food, pharmaceutical, and nuclear power industries, among others.

Any given type of emission will result from competition between adhesive forces and aerodynamic or shear forces acting on the whole sample of powder particles. Many parameters come into play, such as the size, shape, and charge state of the particles, the transmitted energy, and so on. The physics involved is extremely complicated, and it is thus very difficult to make theoretical predictions about aerosol suspensions [49].

For conventional powder materials, there is a notion of the propensity of powders stored in bulk to form an aerosol, referred to as dustiness. This is in fact an index obtained from measurements on the aerosol generated by shaking the powder up in an experimental laboratory device. The test procedure is the subject of a European norm [82]. It describes two dispersion methods: shaking in a rotating drum, or in free fall in a vertical duct. The concentrations of the aerosols generated are characterised by the inhalable, thoracic, and alveolar fractions, and an index is calculated relative to the initial mass of powder used for the test (in units of mg/kg). This index provides a way of classifying powders with regard to their ability to emit dusts. It thus has a special interest for those responsible for health and safety at work, since protective measures can then be designed. But it also has advantages for industry, since it serves to optimise properties of the powder in such a way as to limit dust emissions, for example. In addition, some studies show that this index is a major determinant in occupational exposure, making it a useful parameter for a priori characterisation of exposure potential [83]. Indeed, it is for this reason that this index is included in risk assessment methods [84], even if the classification of powders may evolve depending on the method selected.

Recently, several studies have been made to adapt this idea to nanopowders, i.e., to develop a concept of nanodustiness, since the methods proposed in the European standard are unsuitable as they stand. In these studies, various approaches have been examined:

- Integrating a sampling line and specific measurements into the European method (rotating drum) [85].
- Designing a miniaturised version of the European method (rotating drum) and a sampling/measurement train [86].
- Designing a new device which integrates a specific sampling train. This new device is called the vortex shaker method (VSM) [87].

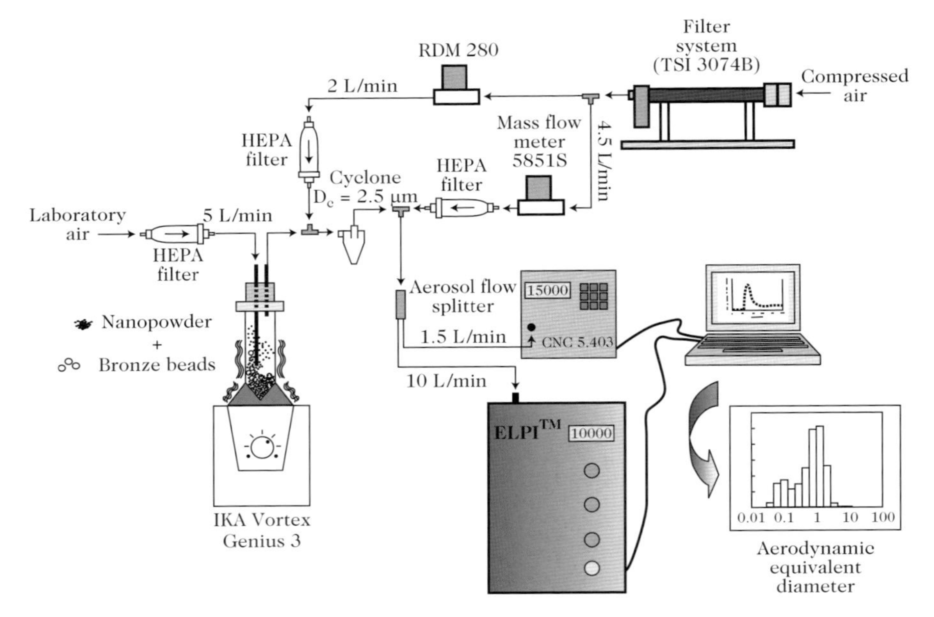

Fig. 7.11. Experimental arrangement of the vortex shaker method for studying the aptitude of a nanopowder to create an aerosol. Adapted from [62] (source INRS)

To illustrate this, Fig. 7.11 shows the experimental setup for the vortex shaker method. In a recent study, this method was used to confirm the presence of nanoparticles when handling metal oxide nanopowders produced by mechanical synthesis [62].

These different approaches have already been applied to several substances in the form of nanopowders, such as single wall and multiwall carbon, fullerenes, TiO_2, ZnO, SiO_2, Al_2O_3, and clays, to name but a few [85–89].

These devices and conditions of use have not yet been perfected, and studies must be carried out to check their performance, but also on sample trains and the measurements that should be made. In addition, it seems important to be able to integrate the determination of certain key parameters for nanopowders, such as the charge on emitted particles [89], and to set up suitable tools for systematic collection of samples for analysis by electron microscope, among other things. Finally, there is still no consensus on how to present and interpret the results. These different approaches are currently under discussion on the international level, e.g., by the Partnership for European Research (PEROSH), which includes health and safety research institutes such as the INRS in France, the IFA in Germany, the NFA in Denmark, the HSL in the United Kingdom, and the TNO in Holland, to name but a few. Furthermore, the technical committee ISO/TC 229 devoted to nanotechnologies is currently producing a document on this priority area [90].

7.4 Setting Up Reference Concentrations

At the present time there is no official occupational exposure limit (OEL) for nanoparticle exposure in French or European regulations. However, it seems important to set reference concentration values (RC) right away, in order to situate results obtained in measurement campaigns. The OEL indicates the average concentration in the air of a given pollutant which, in the current state of understanding, poses no threat to the health of the vast majority of healthy workers who are exposed to it, and this for a daily duration of 8 hours, over long periods of time. In the present case, the pollutant in question is in the form of an aerosol.

In France, for dusts with no specific effects (a category which includes TiO_2), a restrictive daily average OEL of $10\,mg/m^3$ has been laid down for the inhalable fraction and $5\,mg/m^3$ for the alveolar fraction. In addition, other OELs have been specified for various substances or types of aerosol, e.g., solder fumes [35]. It is interesting to note that the values for non-specific dusts (still qualified as inert) are the highest in Europe. For comparison, the concentrations for the inhalable and alveolar fractions are 4 and $1.5\,mg/m^3$ in Germany.

In a detailed analysis, it has been suggested that France should move toward new, lower values, viz., 5 and $2\,mg/m^3$, but these have not yet been adopted [91]. In this same analysis, it was also indicated that more work should be done to set up a reference value for nanoparticles. Note that, regarding the value proposed for the alveolar fraction, i.e., $2\,g/m^3$, it was stipulated that this should contain a very low proportion of nanoparticles.

Since the end of 2005, the NIOSH has proposed a specific limit for so-called ultrafine or nanostructured titanium dioxide, corresponding to the fraction less than 100 nm in diameter [92]. This concentration is $RV(nano) = 0.1\,mg/m^3$, fifteen times lower than the value corresponding to the alveolar fraction $OEL(Alv) = 1.5\,mg/m^3$. This recommendation resulted from an exhaustive and critical analysis of toxicological and epidemiological data for nanoscale and pigmentary TiO_2. It is interesting to note in this document that the best metric would have been area concentration, but that, in the absence of any validated instrument, and considering the urgent need to recommend a reference value, a mass concentration was specified instead.

More recently, the British Standards Institution (BSI) has produced a document in which reference concentrations (RC) are specified for nanoparticles of all types [93]. The values here would not appear to result from as rigorous and careful an investigation as the NIOSH TiO_2 limits. They should thus be treated with great caution. For all products in insoluble form, the BSI report suggests using the ratio RC(nano)/OEL(Alv) obtained by the NIOSH for TiO_2, viz.,

$$RC(nano) = \frac{OEL(Alv)}{15} \quad (mg/m^3) \,. \tag{7.1}$$

In the same document, the BSI also suggests using a reference number concentration equal to 20 000 /cm^3, which must be distinguished from the background. As before, this value does not seem to be based on any toxicological argument. Other values have also been proposed for soluble products, fibre products, etc.

Recently, a rigorous investigation has been carried out to adapt current OELs for products in their standard forms to the same products in nanometric form [94]. The authors indicate that, even in the absence of precise data regarding the toxicity of the products they consider, it is possible to adjust OELs by considering that the micro and nano forms of the same product differ with regard to several parameters that can be determined as a result of measurement or computation, such as the specific surface area (SSA), the deposited fraction (DF), either total or localised in the respiratory system, and the surface reactivity (SR), e.g., radical activity. A simple quantitative model then leads to the following formula:

$$\mathrm{RC(nano)} = \mathrm{OEL(micro)} \times \frac{\mathrm{SSA_{micro}} \times \mathrm{DF_{micro}} \times \mathrm{SR_{micro}}}{\mathrm{SSA_{nano}} \times \mathrm{DF_{nano}} \times \mathrm{SR_{nano}}} \times \frac{1}{\mathrm{PF}} \quad (\mathrm{mg/m^3})\,. \tag{7.2}$$

The protection factor (PF) is not actually known, but it expresses the degree of certainty that one attributes to using the micro OEL value for nanoparticles.

Finally, another pragmatic approach can be envisaged which leads to specific mass concentration reference values. It is based on the following assumptions:

- The toxicity does not change.
- The exposure remains the same.
- The effect of nanoparticles can be related to their surface area or their number.

When the effect is related to their area, the following formula is obtained:

$$\mathrm{RC(nano)} = \mathrm{RC(micro)} \times \frac{d_{\mathrm{nano}} \times \rho_{\mathrm{nano}}}{d_{\mathrm{micro}} \times \rho_{\mathrm{micro}}} \quad (\mathrm{mg/m^3})\,. \tag{7.3}$$

When it is related to their number, we obtain:

$$\mathrm{RC(nano)} = \mathrm{RC(micro)} \times \frac{d^3_{\mathrm{nano}} \times \rho_{\mathrm{nano}}}{d^3_{\mathrm{micro}} \times \rho_{\mathrm{micro}}} \quad (\mathrm{mg/m^3})\,. \tag{7.4}$$

The advantage is that this approach provides two alternatives regarding the potential health effects, i.e., they may be related to surface area or to number. For example, this means that, if the diameter of the particles is divided by 10, then OEL(micro) must be divided by 10 if it is the surface area that drives toxicity, but by 1 000 if it is the number.

In recent risk assessment guidelines [44], the Swiss Federal Office for Public Health used a simplified version of (7.3) as a tool for determining the daily amount to which an operator can be safely exposed.

It seems important that a committee of experts in aerosol toxicology and metrology should be set up to evaluate these different approaches, because it has become urgent to be able to situate the results of measurement campaigns in the field. More work also needs to be done to associate well designed measurement methods with these reference concentrations.

7.5 Conclusion and Prospects

It is crucial to be able to characterise exposure when assessing occupational health risks due to nanoparticles and nanomaterials. Only studies carried out in the field in companies and research centers will be able to identify the true scenarios in which people are exposed during their work activities, and characterise exactly what these people are really exposed to in terms of nanoaerosol composition, quantity, size, structure, and so on, and with what frequency. Without such exposure data from the field, it is difficult to make a fair characterisation of the work environment, emission sources, and exposure. Furthermore, such data is also needed to set up proper risk management, optimise processes, choose and assess collective or personal protection, train staff, and raise awareness of good practice with a view to reducing exposure.

Today, instruments and methods are available to characterise nanoaerosols, most of which have been developed for research applications rather than for measurements in the field. This is why one of the most important challenges in nanoaerosol monitoring over the next few years will be the development of simple and robust techniques for measuring exposure that procure a whole range of parameters, e.g., number, area, size, etc., in real time, together with samples for analysis by electron microscope, or to determine elemental composition, and so on.

The general context of scientific uncertainty regarding nanoparticle toxicity and its effects on human health associated with many developments involving nanomaterials in research and industry and with the high level of concern expressed on all sides have ensured that a great many work situations are at present considered as potentially at risk, including situations that have been known for decades and where measures taken to protect health and safety at work have demonstrated their effectiveness for chemical products already in use. It thus seems an opportune moment to develop our understanding of the nanoaerosols present in our companies and research centers, and this over their entire life cycles.

An essential line of investigation remains the identification of populations, exposure scenarios, and preventive measures implemented in the world of research and industry over the whole life cycle of nanomaterials, because more detailed information about these various elements will make it possible to target the most relevant lines of investigation in terms of research, assistance, information, and training. In the same way, it is essential to adapt methods for assessing and managing chemical risks associated with nanomaterials, integrating in particular proposals for reference concentrations and related methods. Moreover, all this research effort should be carried out on a national and international level in a coordinated, mutualised, and transparent way.

References

1. W. Yang, J.I. Peters, R.O. Williams: Inhaled nanoparticles: A current review. Int. J. Pharm. **356**, 239–247 (2008)
2. G. Gautret de la Moriciere: *Guide du risque chimique. Identification, évaluation, maîtrise*, 4th edn., Dunod, Paris (2006)
3. M. Ricaud, O. Witschger: Les nanomatériaux. Définition, risques toxicologiques, caractérisation de l'exposition professionnelle et mesures de prévention. ED 6050, June 2009, 27 p. Downloadable from sur www.inrs.fr
4. B. Hervé-Bazin (Ed.): *Les nanoparticules. Un enjeu majeur pour la santé au travail?* Series Avis d'experts, EDP Sciences, Paris (2007)
5. B. Zou, J.G. Wilson, F.B. Zhan, Y. Zeng: Air pollution exposure assessment methods utilized for epidemiological studies. J. Environ. Monit. **11**, 475–490 (2009)
6. P.A.J. Borm, D. Robbins, S. Haubold, T. Kuhlbusch, H. Fissan, K. Donaldson, R. Schins, V. Stone, W.G. Kreyling, J. Lademann, J. Krutmann, D. Warheit, E. Oberdorster: The potential risks of nanomaterials: A review carried out for ECETOC. Particle and Fibre Toxicology 3, 35 p. (2006)
7. A.D. Maynard, R.J. Aitken, T. Butz, V. Colvin, K. Donaldson, G. Oberdörster, M.A. Philbert, J. Ryan, A. Seaton, V. Stone, S.S. Tinkle, L. Tran, N.J. Walker, D. Warheit: Safe handling of nanotechnology. Nature **444**, 267–269 (2006)
8. A. Nel, T. Xia, L. Madler, N. Li: Toxic potential of materials at the nanolevel. Science **311**, 622–627 (2006)
9. G. Oberdörster, V. Stone, K. Donaldson: Toxicology of nanoparticles: A historical perspective. Nanotoxicology **1**, 2–25 (2007)
10. C. Ostiguy, B. Soucy, G. Lapointe, C. Woods, L. Mena: *Les effets sur la santé reliés aux nanoparticules*, 2nd edn. Institut de recherche Robert-Sauvé en santé et en sécurité du travail (IRRST), Québec, Rapport R-558, 112 p. (2008)
11. P. Andujar, S. Lanone, P. Brochard, J. Boczkowski: Effets respiratoires des nanoparticules manufacturées. Rev. Mal. Respir. **26**, 625–637 (2009)
12. A. Renoux, D. Boulaud: *Les aérosols. Physique et métrologie*. Lavoisier Tec&Doc, Paris, 301 p. (1998)
13. W.C. Hinds: *Aerosol Technology. Properties, Behavior and Measurement of Airborne Particles*, 2nd edn. John Wiley, New York, 483 p. (1999)
14. P.A. Baron, K. Willeke: *Aerosol Measurement, Principles, Techniques and Applications*, 2nd edn. Wiley Interscience, New York (2001) pp. 143–195
15. J.H. Vincent: *Aerosol Science for Industrial Hygienists*. Pergamon, New York, 411 p. (1995)
16. F. Gensdarmes, O. Witschger: Caractérisation et sources des aérosols ultra-fins. In: *Les Nanoparticules. Un enjeu majeur pour la santé au travail?* EDP Sciences, Paris (2007) pp. 105–142

17. D. Mark: Occupational exposure to nanoparticles and nanotubes. In: *Nanotechnology: Consequences for Human Health and the Environment.* Cambridge, RSC Publishing (2007) pp. 50–80
18. O. Witschger: Les nanoparticules: quelles possibilités métrologiques pour caractériser l'exposition des personnes? Spectra Analyse **264**, 17–30 (2008)
19. D. Vinck: *Les nanotechnologies. Idées reçues.* Editions Le Cavalier Bleu, Paris, 127 p. (2009)
20. ASTM: *Terminology for Nanotechnology*, E 2456-06, 4 p. (2006)
21. British Standards Institution (BSI): *Vocabulary: Nanoparticles.* British Standard, PAS 71, 32 p. (2005)
22. ISO: *Workplace Atmospheres: Ultrafine, Nanoparticle and Nanostructured Aerosols – Inhalation Exposure Characterisation and Assessment.* ISO/TR 27628, 34 p. (2007)
23. BSI: *Terminology for Nanomaterials.* British Standard, PAS 136, 16 p. (2007)
24. R. Sepahvand, M. Adeli, B. Astinchap, R. Kabiri: New nanocomposites containing metal nanoparticles, carbon nanotube and polymer. J. Nanoparticle Res. **10**, 1309–1318 (2008)
25. ISO: *Nanotechnologies: Terminology and Definitions for Nano-Objects – Nanoparticle, Nanofibre and Nanoplate.* ISO/TS 27687, 7 p. (2008)
26. C. Schulze Isfort, M. Rochnia: Production and physicochemical characterisation of nanoparticles. Toxicol. Lett. **186**, 148–151 (2009)
27. Y. Ju-Nam, J.R. Lead: Manufactured nanoparticles: An overview of their chemistry, interactions and potential environmental implications. Sci. Total Environ. **400**, 396–414 (2008)
28. A. Teleki, R. Wengeler, L. Wengeler, H. Nirschl, S.E. Pratsinis: Distinguishing between aggregates and agglomerates of flame-made TiO_2 by high-pressure dispersion. Powder Technol. **181**, 292–300 (2008)
29. J. Jiang, G. Oberdörster, P. Biswas: Characterization of size, surface charge, and agglomeration state of nanoparticle dispersions for toxicological studies. J. Nanoparticle Res. **11**, 77–89 (2009)
30. A.D. Maynard, R.J. Aitken: Assessing exposure to airborne nanomaterials: Current abilities and future requirements. Nanotoxicology **1**, 26–41 (2007)
31. A.D. Maynard, E.D. Kuempel: Airborne nanostructured nanoparticles and occupational health. J. Nanoparticle Res. **7**, 587–614 (2005)
32. M. Lahmani, C. Brechignac, P. Houdy: *Nanomaterials and Nanochemistry.* Springer, Berlin, Heidelberg, 747 p. (2007)
33. S.F. Hansen, B.H. Larsen, S.I. Olsen, A. Baun: Categorization framework to aid hazard identification of nanomaterials. Nanotoxicology **1**, 243–250 (2007)
34. NF EN 481 (X43-276): *Atmosphères des lieux de travail. Définitions des fractions de taille pour le mesurage des particules en suspension dans l'air.* Paris, AFNOR, 16 p. (1993)
35. Valeurs limites d'exposition professionnelle aux agents chimiques en France, 2nd edn. INRS, ND 2098 (2005). Downloadable from www.inrs.fr
36. J.H. Vincent: *Aerosol Sampling. Science, Standards, Instrumentation and Applications.* John Wiley, Chichester, 616 p. (2007)
37. O. Witschger: Sampling of airborne dusts in workplace atmospheres. Kerntechnik **65**, 28–33 (2000)
38. J.M. Park, J.C. Rock, L. Wang, Y.C. Seo, A. Bhatnagar, S. Kim: Performance evaluation of six different aerosol samplers in a particulate matter generation chamber. Atmospheric Environ. **43**, 280–289 (2009)

39. J.F. Fabriès, P. Görner, E. Kauffer, R. Wrobel, J.-C. Vigneron: Personal thoracic CIP10-T sampler and its static version Cathia-T. Ann. Occup. Hyg. **42**, 453–465 (1998)
40. O. Witschger, S.A. Grinshpun, S. Fauvel, G. Basso: Performance of personal inhalable aerosol samplers in very slowly moving air when facing the aerosol source. Ann. Occup. Hyg. **48**, 351–368 (2004)
41. P. Görner, O. Witschger, F. Roger, R. Wrobel, J.F. Fabriès: Aerosol sampling by annular aspiration slots. J. Environ. Monit. **10**, 1437–1447 (2008)
42. O. Witschger, K. Willeke, S.A. Grinshpun, V. Aizenberg, J. Smith, P.A. Baron: Simplified method for testing personal inhalable aerosol samplers. J. Aerosol. Sci. **29**, 855–874 (1998)
43. O. Witschger, J.F. Fabries: Particules ultra-fines et santé au travail. 1. Caractéristiques et effets potentiels sur la santé. Hygiène et Sécurité du Travail, Cahiers de notes documentaires, ND 2227, *199*, 21–35 (2005) Downloadable from www.inrs.fr
44. J. Höck, H. Hofmann, H. Krug, C. Lorenz, L. Limbach, B. Nowack, M. Riediker, K. Schirmer, C. Som, W. Stark, C. Studer, N. Von Götz, S. Wengert, P. Wick: Guidelines on the precautionary matrix for synthetic nanomaterials. Federal Office for Public Health and Federal Office for the Environment, Berne (2008)
45. M. Seipenbusch, A. Binder, G. Kasper: Temporal evolution of nanoparticle aerosols in workplace exposure. Ann. Occup. Hyg. **52**, 707–716 (2008)
46. International Commission on Radiological Protection (ICRP): Publication 66: *Human Respiratory Tract Model for Radiological Protection.* Oxford, Pergamon, 24, No. 1–3, 482 p. (1994)
47. O. Witschger: Inhalation et dépôts dans les voies respiratoires. In: *Les Nanoparticules. Un enjeu majeur pour la santé au travail?* EDP Sciences, Paris (2007) pp. 191–217
48. NANODEVICE: Novel concepts, methods, and technologies for the production of portable, easy-to-use devices for the measurement and analysis of airborne engineered nanoparticles in workplace air. Funded under Seventh Framework Programme, www.ttl.fi/nanodevice
49. O. Witschger, J.F. Fabries: Particules ultra-fines et santé au travail. 2. Sources et caractérisation de l'exposition. Hygiène et Sécurité au Travail, ND 2228, 199, 37–54 (2005). Downloadable from www.inrs.fr
50. ISO: *Nanotechnologies: Health and Safety Practices in Occupational Settings Relevant to Nanotechnologies.* ISO/TR 12885, 79 p. (2008)
51. S. Bau, O. Witschger, F. Gendarmes, O. Rastoix, D. Thomas: A TEM-based method as an alternative to the BET method for measuring off-line the specific surface area of nanoaerosols. Powder Technol. **200**, 190–201 (2010)
52. S. Bau, O. Witschger, F. Gensdarmes, D. Thomas: Experimental study of the response functions of direct-reading instruments measuring surface-area concentration of airborne nanostructured particles. J. Phys. Conf. Ser. *170*, 012006 (2009)
53. S.J. Tsai, E. Earl Ada, J.A. Isaacs, M.J. Ellenbecker: Airborne nanoparticle exposures associated with the manual handling of nanoalumina and nanosilver in fume hoods. J. Nanoparticle Res. **11**, 147 (2009)
54. Y. Thomassen, W. Koch, W. Dunkhorst, D.G. Ellingsen, N.P. Skaugset, L. Jordbekken, P.A. Drabløsd, S. Weinbruche: Ultrafine particles at workplaces of a primary aluminium smelter. J. Environ. Monit. **8**, 127–133 (2006)

55. J. Dixkens, H. Fissan: Development of an electrostatic precipitator for off-line particle analysis. Aerosol Sci. Technol. **30**, 438–453 (1999)
56. M. Fierz, R. Kaegi, H. Burtscher: Theoretical and experimental evaluation of a portable electrostatic TEM sampler. Aerosol Sci. Technol. **41**, 520–528 (2007)
57. J. Lyyränen, U. Backman, U. Tapper, A. Auvinen, J. Jokiniemi: A selective nanoparticle collection device based on diffusion and thermophoresis. J. Phys. Conf. Ser. *170*, 012011, 11 p. (2009)
58. K. Park, G. Cho, J.H. Kwak: Development of an aerosol focusing laser induced breakdown spectroscopy (aerosol focusing LIBS) for determination of fine and ultrafine aerosols. Aerosol Sci. Technol. **43**, 375–386 (2009)
59. D.H. Brouver, J.H. Gijsbers, W.M. Lurvink: Personal exposure to ultrafine particles in the workplace: Exploring sampling techniques and strategies. Ann. Occup. Hyg. **48**, 439–453 (2004)
60. M. Methner, L. Hodson, C. Geraci: Nanoparticle emission assessment technique (NEAT) for the identification and measurement of potential inhalation exposure to engineered nanomaterials. Part A. J. Occup. Environ. Hyg. **7**, 127–132 (2010)
61. OECD Environment, Health and Safety Publications: Series on the safety of manufactured nanomaterials, emission assessment for the identification of sources and release of airborne manufactured nanomaterials in the workplace: Compilation of existing guidance, No. 11, 18 June 2009, ENV/JM/MONO(2009)16
62. O. Witschger, R. Wrobel, E. Gaffet, S. Costes: Evaluation of aerosol release during high-energy ball milling of metallic oxide powders. In: *Proceedings of the 3rd International Symposium on Nanotechnology, Occupational and Environmental Health*, Taipei, Taiwan (2007)
63. A.A. Lall, S.K. Friedlander: On-line measurement of ultrafine aggregate surface area and volume distributions by electrical mobility analysis: I. Theoretical analysis. J. Aerosol Sci. **37**, 260–271 (2006)
64. F.X. Ouf, P. Sillon: Charging efficiency of the electrical low pressure impactor's corona charger: Influence of the fractal morphology of nanoparticle aggregates and uncertainty analysis of experimental results. Aerosol Sci. Technol. **43**, 685–698 (2009)
65. P. Schulte, C. Geraci, R. Zumwalde, M. Hoover, E. Kuempel: Occupational risk management of engineeered nanoparticles. J. Occup. Environ. Hyg. **5**, 239–249 (2008)
66. K. Ostertag, B. Hüsing: Identification of starting points for exposure assessment in the post-sue phase of nanomaterial-containing products. J. Cleaner Product **16**, 938–948 (2008)
67. W. Luther: Industrial application of nanomaterials. Chances and risks. Future Technologies No. 54, Düsseldorf (2004)
68. L.E. Murr: Nanoparticulate materials in antiquity: The good, the bad and the ugly. Mater. Charact. **60**, 261–270 (2009)
69. S.M. Mahurin, M.D. Cheng: Generating nanoscale aggregates from colloidal nanoparticles by various aerosol spray techniques. Nanotoxicology **1**, 1–9 (2007)
70. D.A. Johnson, M.M. Methner, A.J. Kennedy, J.A. Steevens: Potential for occupational exposure to engineered carbon-based nanomaterials in environmental laboratory studies. Environ. Health Perspect. **118**, 49–54 (2010)

71. L. Reijinders: The release of TiO_2 and SiO_2 nanoparticles from nanocomposites. Polymer Degradation and Stability **94**, 873–876 (2009)
72. R. Kaegi, A. Ulrich, B. Sinnet, R. Vonbank, A. Wichser, S. Zuleeg, H. Simmler, S. Brunner, H. Vonmont, M. Burkhardt, M. Boller: Synthetic TiO_2 nanoparticle emission from exterior facades into the aquatic environment. Environ. Pollution **156**, 233–239 (2008)
73. L.Y. Hsu, H.M. Chein: Evaluation of nanoparticle emission for TiO_2 nanopowder coating materials. J. Nanoparticle Res. **9**, 157–163 (2007)
74. M. Vorbau, L. Hilleman, M. Stintz: Method for the characterization of the abrasion induced nanoparticle release into air surface coatings. J. Aerosol. Sci. **40**, 209–217 (2009)
75. Agence française de sécurité sanitaire de l'environnement et du travail (AFSSET): *Les nanomatériaux, effets sur la santé de l'homme et de l'environnement*, 248 p. (2006)
76. R.J. Aitken, K.S. Creely, C.L. Tran: *Nanoparticles: An Occupational Hygiene Review. Research Report.* HSE Books, Edinburgh, 102 p. (2004)
77. Bundesanstalt für Arbeitsschutz und Arbeitsmedezin (BAUA): Exposure to nanomaterials in Germany, 24/04/2008. Downloadable from www.baua.de
78. F. Boccuni, B. Rondinone, C. Petyx, S. Iavicoli: Potential occupational exposure to manufactured nanoparticles in Italy. J. Cleaner Production **16**, 949–956 (2007)
79. K. Schmid, M. Riediker: Use of nanoparticles in Swiss industry: A targeted survey. Environ. Sci. Technol. **42**, 2253–2260 (2008)
80. K. Schmid ,B. Danuser, M. Riediker: Swiss nano-inventory. An assessment of the usage of nanoparticles in Swiss industry. Institut universitaire romand de Santé au Travail, 27.10.2008, Final report
81. C. Weill: Nanosciences, nanotechnologies et principes de précaution. In: *Dossier droit et nanotechnologies.* Cahiers Droits, Sciences & Nanotechnologies, No. 1, CNRS Editions (2008)
82. EN 15051: *Workplace Atmospheres: Measurement of the Dustiness of Bulk Materials – Requirements and Reference Test Methods.* European Standard, 24 p. (2006)
83. D.H. Brouwer, I.H.M. Links, S.A.F. De Vreede, Y. Christopher: Size selective dustiness and exposure: Simulated workplace comparisons. Ann. Occup. Hyg. **50**, 445–452 (2006)
84. H. Marquart, H. Heussen, M. Le Feber, D. Noy, E. Tielemans, J. Schinkel, J. West, D. Van Der Schaaf: 'Stoffenmanager', a web-based control banding tool using an exposure process model. Ann. Occup. Hyg. **52**, 429–441 (2008)
85. C.J. Tsai, C.H. Wu, M.L. Leu, S.C. Chen, C.Y. Husang, P.J. Tsai, F.H. Ko: Dustiness test of nanopowders using a standard rotating drum with a modified sampling train. J. Nanoparticle Res. **11**, 121–131 (2008)
86. T. Schneider, K.A. Jensen: Combined single-drop and rotating drum dustiness test of fine to nanosize powders using a small drum. Ann. Occup. Hyg. **52**, 23–34 (2008)
87. P.A. Baron, A.D. Maynard, M. Foley: Evaluation of aerosol release during the handling of unrefined single walled carbon nanotube material. NIOSH DART-02-191, December 2002
88. I. Ogura, H. Sakurai, M. Gamo: Dustiness testing of engineered nanomaterials. J. Phys. Conf. Ser. *170*, 012003, 4 p. (2009)

89. K.A. Jensen, I.K. Koponen, P.A. Clausen, T. Schneider: Dustiness behavior of loose and compacted bentonite and organoclay powders: What is the difference in exposure risk? J. Nanoparticle. Res. **11**, 133–146 (2008)
90. ISO: *Nanomaterials. General Framework for Determining Nanoparticle Content in Nanomaterials by Generation of Aerosols.* ISO/CD 12025 (2009)
91. B. Herve-Bazin: Valeurs limites poussières totales et alvéolaires: nécessité d'une réévaluation. Hygiène et Sécurité au Travail, PR16, 55–64 (2005). Downloadable from www.inrs.fr
92. NIOSH: Evaluation of health hazard and recommandations for occupational exposure to titanium dioxide. Current Intelligence Bulletin **22**, 115 p. (2005)
93. British Standards Institution (BSI): *Nanotechnologies. Part 2: Guide to Safe Handling and Disposal of Manufactured Nanomaterials.* British Standard, PD 6699-2, 28 p. (2007)
94. E.D. Kuempel, C.L. Geraci, P.A. Schulte: Risk assessment approaches and research needs for nanomaterials: An examination of data and information from current studies. In: P.P. Simeonova, N. Opopol, M.I. Luster (eds). *Nanotechnology: Toxicological Issues and Environmental Safety*, Springer (2007) pp. 119–145

8

Monitoring Nanoaerosols and Environmental Exposure

Corinne Mandin, Olivier Le Bihan, and Olivier Aguerre-Chariol

Environmental exposure refers to exposure of the population outside the occupational context (see Chap. 7) and excluding also medical exposure.[1] The kind of exposure discussed in this chapter is due to the presence of nanoparticles in the various environmental compartments, such as the air (indoors or outdoors), water (water for drinking, bathing, etc.), soils, foodstuffs, and so on. These nanoparticles may come from the nanomaterials that contain them and upon which they bestow specific novel properties, or they may be formed unintentionally by human activities such as industry, traffic, domestic fuel combustion, etc., or natural phenomena such as forest fires, for example, or again by physicochemical reactions, e.g., the reaction between gases and particles in the air, spray formation, vapour condensation, and so on. This book is concerned with the former, namely manufactured nanoparticles, but the related questions and acquired knowledge must often be viewed from the perspective of what is already known about the latter, commonly referred to as ultrafine particles.

Different types of environmental exposure are usually classified in terms of the exposure route they involve, i.e., inhalation, ingestion (or oral route), and percutaneous route. The specific requirements of environmental measurement with regard to manufactured nanoparticles[2] are thus the possibility of measuring them in these different exposure contexts, the possibility of carrying out accurate physicochemical characterisation, e.g., size range, oxidised or

[1] Given the expected therapeutic applications of nanomaterials, intravenous and intramuscular exposure cannot be ignored when surveying possible sources of human exposure. Naturally, since these treatments are expected to be beneficial to the health, they can be considered from a different perspective.

[2] We refer here to specific needs with regard to nanoparticles. To these must be added the usual criteria intrinsic to environmental measurement, such as sensitivity, repeatability, the possibility of reaching low enough detection thresholds to distinguish a concentration that exceeds the background usually encountered in the given environment, etc.

P. Houdy et al. (eds.), *Nanoethics and Nanotoxicology*,
DOI 10.1007/978-3-642-20177-6_8,

not, soluble or not, because the resulting implications for health are directly associated with their specific properties, and finally the possibility of discriminating them from the environmental background.

The different kinds of potential or actual environmental exposure are described in Sect. 8.1. The currently available measurement techniques for characterising nanoparticles in the air, water, and soils are then discussed in Sect. 8.2.

8.1 Origin and Nature of Environmental Exposure

8.1.1 Life Cycle and Environmental Exposure

Environmental exposure to nanoparticles can occur throughout the life cycle of the nanomaterials containing or emitting them. Upstream, at the fabrication, transformation, and packaging stage, industrial sites can emit nanoparticles into the air or water under normal operating conditions: rejection into the atmosphere through air vents, rejection of aqueous effluents, etc. At the present time, this feature of the overall problem is receiving little attention, the priority being to identify and characterise the exposure of workers inside the factories.

Then, during the use of the products, nanoparticles may once again be emitted. For example, tyres incorporating nanoparticles to improve their strength without increasing the weight of the vehicle can release particles into the atmosphere by abrasion while rolling on the road. Likewise, sports equipment containing nanoparticles for the same purpose will gradually wear down. Food packaging incorporating nanoparticles to improve conservation of its contents (protection against light, bacteria, etc.) may also be able to release them into the environment [1]. Finally, when the materials are recycled or destroyed, further emissions into the environment become possible. As an example, if nanomaterials are integrated into buildings, e.g., better insulating concrete, self-cleaning glass, air purifying paints, etc., their nanocomponents may be released into the air, the ground, or indirectly into underground water systems whenever the buildings are demolished.

Nanoparticles emitted in these different ways can then diffuse into various environmental compartments, perhaps accumulating in some of them, and thereby leading to exposure of the general population. Note that physicochemical phenomena such as agglomeration and aggregation may also take place, modifying the conditions of exposure.

8.1.2 Exposure Routes

Inhalation is often the first exposure route to be mentioned. Indeed, the first scientific investigations regarding nanoparticle toxicity are a natural continuation of experimental toxicology, e.g., of diesel particles, and epidemiological studies of urban particulate pollution, and so are precisely concerned with this exposure route. The latter studies, after successively investigating black smoke, then particles with median aerodynamic diameter 10 μm, have evolved in synchrony with changes in regulations and technology to the investigation of fine particles (2.5 μm), or particle number rather than mass indicators, thus highlighting the possible role, apparently non-negligible, of ultrafine particles on respiratory health [2]. With the sudden interest in nanomaterials and the potential emission of nanoparticles into the environment, it was not long before the question of inhalation exposure was raised.

In another domain, while nanomaterials are already present on the world cosmetics market, there is very little published data about exposure by the percutaneous route. Such exposure may be direct when cosmetic or personal hygiene products are used, but also through direct contact with textiles incorporating nanoparticles to improve anti-soil, anti-crease, antibacterial, or insulation properties. Exposure may also be qualified as indirect, when nanoparticles are emitted into bathing water, e.g., from sunscreen creams. Generally speaking, for the substances to which humans are exposed through the environment or consumer products, this exposure route is usually of minor importance in comparison to the inhalation or oral routes. However, in the context of nanomaterials, it requires particular attention because of the very small diameters of these particles, since this will considerably facilitate cutaneous penetration, especially in the case of fragile or damaged skin [3–6].

Finally, exposure by the oral route through food and drinking water must also be considered, all the more so in that the small size of nanoparticles means that they can cross the intestinal barrier. Since there are now plans to use nanoparticles to decontaminate soils (their high chemical reactivity allows them to degrade pollutants accumulated in the ground), there is a possibility of their transfer by lixiviation or runoff to underground or surface water systems. And from there, they may find their way into the domestic water supply. In the same way, there is a possibility that nanomaterials used as food additives, e.g., preservatives, taste enhancers, texturing agents, etc., or in food packaging might be ingested by consumers [1].

8.2 Characterising Environmental Exposure

In order to get an accurate, quantitative appreciation of the exposure of the general population, it is useful first to quantify concentration levels in a given compartment and describe the physicochemical characteristics of the relevant nanoparticles. The air compartment is the best understood at the present time, while the water and soil compartments are less well understood, but also deserve attention.

8.2.1 Nanoparticle Measurement in Air

The air compartment is the one for which instrumentation is the most highly developed. Given the importance of ultrafine particles in the air and occupational exposure of industrial workers for considerations of health and safety [7, 8], aerosol measurement has made considerable progress over the past few years.

Up to now, for larger particles, the standard metric for measuring concentrations in air was mass. For example, the mass concentrations of the fractions PM_{10} and $PM_{2.5}$ (particles of median aerodynamic diameter less than 10 and 2.5 μm, respectively) are subject to regulatory limiting values in the ambient air. Only PM_{10} measurements were required by regulation until the new European directive 2008/50/CE came into effect on 21 May 2008, regarding the quality of the ambient air in the programme *Clean Air for Europe*. This directive now requires measurement of $PM_{2.5}$. In addition, these granulometric fractions were the ones used as particulate atmospheric pollution indicators for large scale epidemiological studies which have identified some of the short and long term health effects of particles [9]. However, where nanoparticles are concerned, it turns out that the mass is no longer the most relevant criterion for monitoring concentrations and related exposure levels, so mass measurement techniques will not be further discussed in this chapter.

The first experimental toxicology studies seem to point to the important role played by specific surface area. Indeed, owing to their smaller diameters, nanoparticles have a much higher specific surface area, i.e., a much higher surface area per unit mass of particles, and this increases their potential for interacting with human tissues and biological fluids. This in turn confers upon them special toxicological properties compared with much larger particles having the same chemical composition. This surface reactivity also allows nanoparticles to adsorb substances that may themselves be toxic. For these reasons, equipment has been designed to measure particle specific surface area concentrations in the air, these being expressed in units of $\mu m^2/cm^3$.

Given a sample of a material, the specific surface area is usually measured by specific adsorption of a gas, e.g., the Brunauer–Emmett–Teller (BET) method. But for particles suspended in the air, this method is not suitable. Three other techniques are then available. The diffusion charging technique exploits the idea of fixing ions on the particles by Brownian diffusion. The particles are then collected on a filter and the electrical charge is measured by an electrometer. The specific surface area concentration is then a function of the current, proportional to the specific surface area, and the sampling rate. Photoelectric chargers use a similar idea, being based on detection of the electrical charge acquired by photoelectric effect, i.e., due to the emission of an electron by a particle previously excited by ultraviolet radiation. Finally, the third technique uses an epiphaniometer. The idea is to fix radioactive lead atoms (^{211}Pb), from an actinium source (^{227}Ac), onto the particles. The particles are collected on a filter and an alpha particle detector measures the decay of the lead atoms. Given that radioactive sources are needed here, this device would appear to have limited use in the field. An alternative to these three techniques that does not require further equipment to support it (provided that a granulometer is already present) is simply to deduce the specific surface area from the granulometric distributions and assumptions about the particle geometry. These assumptions necessarily make the results somewhat unreliable.

So far, there have been few measurements of specific surface area concentrations, either in the ambient air outdoors or in closed environments. Experiments carried out in an experimental house at the *Centre scientifique et technique du bâtiment* in France over the period 2007–2008 have provided first estimates of the concentrations to be expected in indoor air in the presence of domestic particle sources (measurements made using an Aerotrak 9 000) [10, 11]. The concentration in the living room was of the order of $15\,\mu m^2/cm^3$ in the absence of any operating source, whereas it exceeded $5\,000\,\mu m^2/cm^3$ for short periods while meat was being cooked on a gas burner. Given the scarcity of such measurements today, these values cannot be related to those that might commonly be encountered in the air of people's living areas.

Another important parameter for describing airborne nanoparticles is the number concentration, expressed by the number of particles per cm^3, in a given diameter range intrinsic to the measuring device. Condensation particle counters (CPC) are the most widely used instruments. They are generally rather compact and quiet, sometimes energy sufficient, and operate continuously, counting particles with diameters from a few nanometers (between 3 and 20 nm) to a few micrometers over short time intervals. They work by increasing the size of the particles and then using an optical system to detect them. Concentrations from a few particles to $100\,000/cm^3$ can be measured. Above this concentration range, coincidence phenomena are observed. A group of particles that are all very close to one another in the detection volume can be considered as constituting a single particle, and this results in an underestimate of the actual concentration (although this can be corrected by calculation). Since they are easy to carry, CPCs could be used to measure personal concentrations.[3]

First estimates of number concentrations are available for ultrafine particles. In the context of the multicenter European study entitled *Relationship between ultrafine and fine particulate matter in indoor and outdoor air*

[3] We speak of personal measurements when the measuring device is carried by the person whose exposure is to be assessed. This avoids having to measure the air concentrations in all the places frequented by the subject, both outdoors and indoors, along with a full analysis of the time spent by the person in each of these places, from which the exposure level would eventually be reconstructed by calculation. Measurement of personal exposure is an ideal that can rarely be put into practice in epidemiological studies, owing to the weight, bulk, noise level, and/or cost of the necessary equipment.

and respiratory health, abbreviated to RUPIOH, the concentrations inside 152 residences (located in Helsinki, Athens, Amsterdam, and Birmingham) and also outdoors (in the immediate vicinity of the residence and at a fixed station between 3 and 8 km away) were measured using CPCs (range 7 nm–3 μm) over the period 2002–2004 [12, 13]. The lowest concentrations were measured in Helsinki, with an average of 3 000 particles/cm^3 indoors and 4 500 particles/cm^3 outdoors. The highest were measured in Amsterdam, with indoor and outdoor averages of 12 000 and 26 300 particles/cm^3, respectively. As soon as indoor sources become active, e.g., cooking, tobacco smoking, incense, candles, etc., the concentrations can very quickly exceed 100 000 particles/cm^3 during the active phase, and they can even reach a million particles per cm^3 in the kitchen when cooking is in progress, for certain ventilation configurations [10]. Finally, in 36 Munich schools, the number concentrations measured over the period 2004–5 using CPCs (range 10–500 nm) varied between 2 600 and 12 200 particles/cm^3 (median concentration equal to 5 660 particles/cm^3) [14].

It is also possible to measure the size distribution of the nanoparticles, i.e., obtain number concentrations per diameter interval (or diameter channel), where the diameter can be the equivalent electrical mobility diameter, aerodynamic diameter, diffusion diameter, etc. The necessary equipment is usually rather bulky, and this complicates use in the field, in contrast to the CPC.

The scanning mobility particle sizer (SMPS) is one of the most widely used devices. It operates in real time, with short time intervals. After capture, the particles are charged by ions created artificially in the gas. Successive cycles of variable voltage between two electrodes then modify the trajectories of the charged particles. For a given voltage, only those particles with a certain electrical mobility leave this part of the device, the differential mobility analyser (DMA), to be counted by a condensation particle counter placed downstream. The different voltages applied successively are thus used to sort the particles according to their equivalent electrical mobility diameter. The diameter range is 10–500 nm.

The electrical low pressure impactor (ELPI) is a spectrometer measuring the real-time size distribution of particles with aerodynamic diameters lying between a few nanometers and 10 μm. The idea is to charge the particles electrically, then do an inertial classification in a cascade impactor. During impaction of the charged particles, at each level, a current is created and recorded by an electrometer, whence the number concentration can be deduced after conversion.

Finally, diffusion batteries can measure the diffusion or thermodynamic diameter distribution. This uses the fact that particles of lowest diameter will have a higher diffusion coefficient. The particles are conducted through parallel tubes of different lengths. The fraction passing through each tube is measured using particle counters, and this provides a way of obtaining the distribution of the diffusion coefficients, and from there the distribution of the diffusion diameters. The geometry of these setups may vary. One arrangement consists of grids, increasing in number at each level of the battery. Granulometry is measured for particles with diameters in the range 3–150 nm.

To end this review, one should mention microscopy techniques, such as scanning electron microscopy and transmission electron microscopy, providing offline visualisation of sampled nanoparticles to determine their morphology. These techniques are not specific to nanoparticles and sometimes need to be adapted, especially with regard to sampling (suitable substrate, and suitable collection period to obtain an observable amount, i.e., neither too much nor too little). Note that impactors can be used in combination with microscopy, obtaining samples for microscopic analysis in each of the diameter ranges corresponding to the levels of the impactor.

Using microscopy, the chemical composition of the nanoparticles can be determined by laser induced breakdown spectroscopy (LIBS). This real time technique, also used to analyse solid or liquid matrices, works by focusing short laser pulses in the aerosol in order to transform the matter into plasma. The spectrum emitted by the atoms in this plasma is analysed to deduce the original elemental composition of the particles. The size distribution and number of particles are obtained by coupling to an SMPS. For the moment, this device has only been used in the work place to detect possible leaks from the fabrication process [15].

In conclusion, it is crucial to identify the right metric to characterise airborne nanoparticles and associated human exposure. Given the specificity of nanoparticles, some authors suggest using a set of relevant parameters, rather than just one [16]. Another challenge in this field is to discriminate between manufactured nanoparticles and ultrafine particles already present in the ambient air [17], in order to qualify or assess their specific impacts. It would then be possible to deal with them on an ad hoc basis when the need arose. LIBS and tracking techniques look promising here.

For complementary details regarding airborne particle measurement, the reader may refer to Chap. 7 on occupational exposure to nanoaerosols. Reviews can also be found in [18, 19].

8.2.2 Nanoparticle Measurement in Water

In comparison with airborne nanoparticle measurement, the case for detection in water is still less advanced [20, 21]. It consists in characterising morphological, chemical, and physicochemical properties of nanoparticles via the following analytical techniques:

- Observation by atomic force microscopy, electron microscopy, etc.
- Analytical centrifugation, measuring the sedimentation rate in a given fluid.
- Radiation scattering, measuring single particle and aggregate sizes.
- X-ray diffraction, determining crystal structure.

- Adsorption measurements, determining specific surface area.
- Inductively coupled plasma (ICP), X-ray fluorescence, and atomic absorption techniques to determine chemical composition.
- Zetametry to determine surface properties.

These relatively sophisticated and costly techniques require offline analysis in the laboratory and cannot be used to make in situ measurements. They are still in the development stage as regards nanoparticle applications. The same question remains about the relevance of waterborne nanoparticle measurement, since agglomeration phenomena may in the end favour the disappearance of the nanoparticle fraction.

8.2.3 Nanoparticle Measurement in Soil

As for waterborne nanoparticles, the measurement of nanoparticles in soils consists in isolating the nanoparticle fraction before undertaking any analysis of size or chemical composition. Sample preparation techniques result in an aqueous solution, whereupon the techniques listed in the last section can be applied [20].

However, while the analytical methods seem to be adequate, it remains very difficult to identify and hence quantify the contribution of manufactured nanoparticles, owing to the intrinsic non-uniformity and complexity of soils. This major difficulty with identification suggests that observation techniques such as transmission electron microscopy, scanning electron microscopy, atomic force microscopy, etc., are the best suited, even the only suitable methods, even more so than for the air and water compartments.

8.3 Conclusion

As far as environmental measurement is concerned, the main challenge for the near future, whatever the compartment, air, water, or soil, is the specific characterisation of nanoparticles from nanomaterials and nanotechnologies. It seems crucial to obtain an exhaustive knowledge before they are put to use, i.e., before they can be emitted into the environment and the general population exposed.

References

1. H. Bouwmeester, S. Dekkers, M.Y. Noordam, et al.: Review of health safety aspects of nanotechnologies in food production. Regul. Toxicol. Pharmacol. **53**, 52–62 (2009)
2. A. Lefranc, S. Larrieu: Particules ultrafines et santé: Apport des études épidémiologiques. Environnement, Risques & Santé **7**, 349–355 (2008)

3. C. Bennat, C.C. Müller-Goymann: Skin penetration and stabilization of formulations containing microfine titanium dioxide as physical UV filter. Int. J. Cosmet. Sci. **22**, 271–283 (2000)
4. F. Pflücker: The human stratum corneum layer: An effective barrier against dermal uptake of different forms of topically applied micronised titanium dioxide. Skin Pharmacol. Appl. Skin Physiol. **14** (Suppl. 1), 92–97 (2001)
5. S. Schulz, P. Gehr, V. Hof, et al.: Distribution of sunscreens on skin. Advanced Drug Delivery Reviews **54** (Suppl. 1), S157–S163 (2002)
6. F. Menzel, T. Reinert, J. Vogt, et al.: Investigations of percutaneous uptake of ultrafine TiO_2 particles at the high energy ion nanoprobe LIPSION. Nuclear Instruments and Methods in Physics Research B **219–220**, 82–86 (2004)
7. R.J. Delfino, C. Sioutas, S. Malik: Potential role of ultrafine particles in associations between airborne particle mass and cardiovascular health. Environ. Health Perspect. **113**, 934–946 (2006)
8. A. Ibald-Mulli, H.E. Wichmann, W. Kreyling, A. Peters: Epidemiological evidence on health effects of ultrafine particles. J. Aerosol Med. **15**, 189–201 (2002)
9. C.A. Pope, D.W. Dockery: Health effects of fine particulate air pollution: Lines that connect. Journal of the Air & Waste Management Association **56**, 709–742 (2006)
10. X. Ji, O. Le Bihan, O. Ramalho, et al.: Particules ultrafines dans l'environnement domestique: Niveaux, déterminants et variabilités – Projet NANOP. Actes du 23ième Congrès français des aérosols, 16–17 January 2008
11. X. Ji, M. Nicolas, O. Le Bihan, et al.: Caractérisation des particules générées par la combustion d'encens. Actes du 24ième Congrès français des aérosols, 14–15 January 2009
12. A. Puustinen, K. Hameri, J. Pekkanen, et al.: Spatial variation of particle number and mass over four European cities. Atmospheric Environment **41**, 6622–6636 (2007)
13. G. Hoek, G. Kos, R. Harrison, et al.: Indoor–outdoor relationships of particle number and mass in four European cities. Atmospheric Environment **42**, 156–169 (2008)
14. H. Fromme, D. Twardella, S. Dietrich, et al.: Particulate matter in the indoor air of classrooms: Exploratory results from Munich and surrounding area. Atmospheric Environment **41**, 854–866 (2007)
15. T. Amodeo, C. Dutouquet, F. Tenegal, et al.: On-line monitoring of composite nanoparticles synthesized in a pre-industrial laser pyrolysis reactor using laser-induced breakdown spectroscopy. Spectrochimica Acta Part B **63**, 1183–1190 (2008)
16. A.D. Maynard, R.J. Aitken: Assessing exposure to airborne nanomaterials: Current abilities and future requirements. Nanotoxicology **1**, 26–41 (2007)
17. T.M. Peters, S. Elzey, R. Johnson, et al.: Airborne monitoring to distinguish engineered nanomaterials from incidental particles for environmental health and safety. J. Occup. Environ. Hyg. **6**, 73–81 (2009)
18. IRSST: *Guide de bonnes pratiques favorisant la gestion des risques reliés aux nanoparticules de synthèse*. Institut de recherche Robert-Sauvé en santé et en sécurité du travail, rapport référencé R-586 (2008)
19. B. Hervé-Bazin et al.: *Les nanoparticules, un enjeu majeur pour la santé au travail*. EDP Sciences, Paris (2007)

20. AFSSET: *Les nanomatériaux, effets sur la santé de l'homme et sur l'environnement.* Agence française de sécurité sanitaire de l'environnement et du travail (2006)
21. A.B. Boxall, K. Tiede, Q. Chaudhry: Engineered nanomaterials in soils and water: How do they behave and could they pose a risk to human health? Nanomedicine **2**, 919–927 (2007)

9

Nanoparticles and Nanomaterials: Assessing the Risk to Human Health

Denis Bard

The particular physical and chemical properties of nanoscale materials are becoming better and better understood all the time. Scientific, industrial, and medical applications are on the increase. The *Woodrow Wilson International Centre for Scholars* [1] has already listed more than 800 commercial products appealing to this type of material, from cosmetics to tennis rackets and tyres. Since this list is based on voluntary declarations by the industrial sector, the figure is likely to be a serious underestimate, and it is clear that it is also likely to increase exponentially.

With regard to the possibility of a generalised potential exposure, the specific properties of nanomaterials force us to ask what risks might be associated with their completely new toxicological characteristics. The aim of this chapter is to present the state of the art in nanoparticle and nanomaterial risk assessment. However, therapeutic and diagnostic health products will not be considered, since they appeal to nanotechnologies for which the terms in the risk–benefit ratio differ from those of consumer products.

The conventional approach proceeds in four steps [2]:

- Identify the hazard potential by epidemiological observation, or failing that, by experimental studies in the laboratory, including in vivo studies on animals, be they at the organ level or at the infracellular scale. A positive conclusion regarding the possible existence of a causal relationship between exposure to the given agent and an effect considered as harmful is a necessary condition for the risk assessment process to continue.
- Estimate the dose–response relationship, ideally through an epidemiological study, otherwise using laboratory data.
- Assess exposure, i.e., who is exposed, under what circumstances, and to what levels.
- Characterise the risk, i.e., estimate the impacts on health, either by estimating the probability of the health event occurring for each level of exposure or by estimating the absolute number of cases related to exposure.

P. Houdy et al. (eds.), *Nanoethics and Nanotoxicology*,
DOI 10.1007/978-3-642-20177-6_9,

However, prior to any risk assessment, the physical and chemical characteristics of the given agent must be specified. It should then be possible to predict its behaviour in the various environmental compartments and select the relevant exposure routes. For example, a compound with very low vapour pressure will be not be significantly inhaled, and a molecule that is insoluble in water will reduce the importance of drinking water as exposure medium.

In the case of nano-objects, whether they be man made or unintentionally produced, it goes without saying that the size parameter must be checked. This is no trivial matter. The literature indicates that nano-objects may tend to agglomerate or aggregate when released at the moment of production. However, according to the *Scientific Committee on Emerging and Newly Identified Health Risks* (SCENIHR), the resulting increase in size does not remove their surface reactivity properties, characteristic of the nanometric length scale [3]. Apart from the size parameter, it is crucial to determine other chemical characteristics, such as the composition, including the presence and proportion of impurities, notably metal impurities, but also the surface chemistry, crystal structure, and so on. Likewise for physical characteristics, such as specific surface area, morphology, stability, and so on.

9.1 Identifying the Hazards

As mentioned earlier, manufactured nano-objects are already used in many applications, exploiting the size factor, but also sometimes chemical composition. This is the case, for example, with cosmetic products incorporating nanometric TiO_2 particles, only released on the market relatively recently. It is conceivable that epidemiological studies considering the short term effects of manufactured nano-objects on the respiratory or cardiovascular systems, for example, or on reproduction, could have been carried out, but there is no trace of them in the literature. On the other hand, most of the promising applications of nanotechnology remain to be invented and developed, e.g., grafting functional groups, synthesis of certain specific surface coatings. These specific functionalities are likely to be associated with particular kinds of toxicological property, although that is only guesswork at this stage.

On the other hand, several epidemiological studies have been done on the short term effects of unintentionally produced nanoparticles in the atmosphere (called ultrafine particles in the related literature) with average aerodynamic diameters less than or equal to 100 nm (see Table 9.1).

Ultrafine particles may be produced by combustion phenomena, in particular internal combustion engines and especially diesel motors, but also combustion of the biomass. They may also come from chemical or photochemical reactions between gases, often of natural origin, which produce a gas-to-solid formation of nanoparticle dimensions. This is the case, for example, with radon and its decay products. But it should be noted that, while there are many epidemiological studies about the risks of radon, they are only concerned with

Table 9.1. Nanoparticle sources

	Anthropogenic sources	
Natural sources	Unintentional	Intentional
Incomplete combustion (fires, volcanos) Conversion of gas into particles (radon) Vapour condensation Mechanical sources such as erosion and sea spray Biological sources such as viruses and macromolecules	Internal combustion engines Power stations (non-nuclear) Incinerators Aircraft Metallurgical processes, including welding smokes Heated surfaces Cooking processes, including domestic ones Electric motors	Size, shape, and functionalities determined by intended application

risks due to the associated ionising radiation and the problem of dosimetry, and do not take into account possible effects of the resulting nanoparticles (the size effect). For this reason, they will not be further discussed here. Another possible origin of ultrafine particles is the condensation of a vapour, by the spraying of a liquid salt solution, followed by evaporation of the liquid. They can also be produced by erosion.

The twenty years of epidemiological studies so far published on the effects of airborne ultrafine particles have been reviewed in [4], and the subject is discussed in more detail in Chap. 6. The overall conclusion tends to be that a rise in the number concentration of ultrafine particles on a given day is associated in certain studies with an increase in mortality, taking into account all causes except accidents, with more marked effects for cardiovascular or respiratory causes. However, other studies have not discovered such correlations.

The deduction of a causality relation in epidemiology, an observational science, is a delicate undertaking, which requires one to evaluate the conclusions drawn from a variety of different arguments, using the approach advocated by Hill [5]. One of the points to be considered is consistency between epidemiological results and data obtained from other disciplines, in particular experimental investigations, and in the case of nano-objects, toxicology. Toxicological results available on the effects of nano-objects are discussed elsewhere in this book. What we may say at this stage, following A. Lefranc and S. Larrieu [4], is that there is consistency between cardiovascular and respiratory risks related to the inhalation of ultrafine particles and the understanding obtained through exposure science (documented uptake of ultrafine particles by the airways, with penetration as far as the pulmonary alveoli, alveolar-capillary barrier crossing, and systemic diffusion into the blood circulation) [3,6] and through toxicology: the ensuing oxidative stress and inflammation are the most commonly invoked action mechanisms, both for inhaled particles [7] and dermally administered single wall carbon nanotubes [8]. However, it is still not clear precisely what role is played by the nanometric dimension or the shape of the given nano-object, e.g., almost spherical entities or fibres of different lengths or stiffnesses [9], or the toxicity of any impurities, and in particular metals, that may be present [8].

Concluding that there is a causal relationship in epidemiology involves interpreting the available data. Such a conclusion can only reasonably be reached by the combined efforts of a group of specialists, as happens for example at the *International Agency for Research on Cancer* (IARC) when they classify chemical or physical agents as possibly, probably, or certainly human carcinogenic. Indeed, such an assessment could not be left to a single individual, especially in the case of agents like nano-objects which already involve large populations.

On the basis of the arguments presented above, we shall take it that there is a justified presumption of causality, so that we may discuss the next part of the risk assessment process.

We must now consider the predictive value of the many experimental results obtained in vitro and discussed in previous chapters. As emphasised by the SCENIHR [3], these experiments are extremely useful for carrying out an initial screening of effects, and they sometimes bring significant understanding of the relevant action mechanisms. The available data raises two problems. The first is that the doses or concentrations used do not necessarily reflect what could reasonably be anticipated for human exposure, and they are often much higher. It is a delicate matter to transpose this type of observation to human exposure without sound arguments, especially in the fields of toxicokinetics and toxicodynamics. The second point is that the data available at the present time refer to the short or very short term, and it is problematic to draw conclusions about long term effects of exposure to nano-objects, e.g., for carcinogen risk assessment. On the other hand, it is important to note the warning from the *Haut Conseil de la Santé publique* in France [9]: "The observation of an early and persistent inflammatory response lasting several days, analogous in certain ways to the one induced by asbestos (oxidative stress, signalling pathways, granuloma formation attesting to frustrated phagocytosis by macrophages, and so on) should be considered as a mechanism likely to contribute to carcinogenesis." These effects would not appear to depend on solubilisation products, in particular, metal impurities. Since they seem to vary with the length of the carbon nanotubes, the *Haut Conseil* suggests the possibility of a fibre effect, well described for asbestos.

However, it is important to individualise genotoxicity studies. Indeed, the characterisation of a carcinogen as genotoxic or otherwise determines to a large extent the kind of model chosen for the dose–response relationship discussed below. Two recent reviews [10,11] and the SCENIHR [3] indicate that the different genotoxicity or mutagenicity tests carried out on various nano-objects, such as fullerenes C_{60}, single wall carbon nanotubes, cobalt–chromium alloy nanoparticles, TiO_2 nanoparticles, metal oxide V_2O_3 nanoparticles, carbon black, and diesel engine exhaust, lead to positive results. However, the results obtained are not constant from one experiment to the next, and the SCENIHR experts do not draw any conclusion [3].

9.2 Dose–Response Relationship

In the field of risk assessment, a distinction is made between deterministic and stochastic, i.e., random, events [12].

The seriousness of a deterministic effect is proportional to the dose. Such effects are considered to have a dose threshold below which there is no effect. In this case the problem is to determine – or strictly speaking, to approximate – the threshold. Various methods are available, each with its own logic, e.g., choice of a dose with no effect from a set of data, or determination of a benchmark dose from a model of the experimental data, or regression on qualitative variables. Whatever method is selected, the resulting value, generally obtained from experimental data, will be weighted by one or more safety or uncertainty factors in order to introduce a safety margin, depending largely on uncertainty over the quality of the data used and especially on the need to transpose animal data to humans. This value, known as the toxicological reference value, represents the part of risk management that is based on scientific fact, notably with regard to fixing standards. The deterministic approach applies in principle to any type of toxic effect, including carcinogenic effects, provided that they are not genotoxic.

In the stochastic approach, it is postulated that there is no threshold dose, i.e., that any dose involves some level of risk. This risk, or rather this extra risk, may be very small as far as a single individual is concerned, e.g., 10^{-5} for a whole lifetime, but it may have important consequences for public health on a nationwide level. The same level of individual risk, if it were applicable to the whole French population of 63 million people, would lead to 630 cancers over a lifetime fixed by convention at 70 years. It would be impossible to demonstrate this postulate a priori in the current state of science, and it is thus the subject of endless controversy [13].

The result of a stochastic approach is a determination of the slope of the dose–risk relationship, which conventionally constitutes the toxicological reference value in this case.

Genotoxic carcinogens fall under the stochastic approach. In this case, an increasing dose raises the frequency of cancers, and their seriousness is not taken into account. This is therefore a population approach, inaccessible in principle to sensorial experience, whereas the deterministic approach is sometimes accessible to it, notably in the case of physical agents. (Everyone can check that the damage caused by a burn depends on the product of the temperature and the contact time, and that there is indeed a threshold for this product to produce an effect.) The relevance of this population or frequency approach is clearly demonstrated by the example of tobacco consumption in relation to the risk of lung cancer: the more a population smokes, the more frequent lung cancer becomes, provided of course that one takes into account the latency time for cancer development.

The problem in the case of nano-objects is that there is no evidence upon which to base the choice of model, deterministic or stochastic, for the carcinogenic risk. The available data would not justify a formal conclusion of genotoxicity for manufactured nano-objects or ultrafine particles, despite the growing understanding of the mechanisms whereby nano-objects can form potentially genotoxic reactive oxygen species.

With regard to other toxic effects, the data gathered so far is experimental, sometimes obtained for acellular systems, or better cellular systems, but only rarely in vivo [7]. In the latter case, the administered doses are generally very high, which tells us little about the possible dose–response relationship in humans.

A dose–response relationship has indeed been constructed for mortality and hospital admissions due to respiratory and cardiovascular problems [14], but it was based on measurements of particles with average aerodynamic diameters less than 10 μm (PM_{10}). While it is clear that ultrafine particles contribute to this risk, their intrinsic importance cannot be quantified.

To sum up, we do not have the information that would be needed to construct a dose–response relationship for nano-objects, whatever hazard we have in mind, but it seems wise to consider that an approach without threshold is justified on the basis of what we know about the action mechanisms, at least as far as carcinogenic risk is concerned.

9.3 Exposure

We now have an increasing amount of data regarding exposure to ultrafine particles, whether these concern outdoor or indoor concentrations. Questions of measurement and sampling remain crucial (see Chaps. 7 and 8), making it a delicate matter to compare measurements of airborne concentrations both in indoor air and ambient air. To give an order of magnitude, values in the literature for number concentrations of ultrafine particles in indoor air vary from 5 000 cm^{-3} on average [15, 16] to 80 000 cm^{-3} at peak if some ultrafine particle source is set in operation, such as the cooking of food in a restaurant [17]. In outdoor air, number concentrations can vary considerably from one town to another, e.g., from 15 000 cm^{-3} in Dutch cities [18] to 51 000 cm^{-3} in Rome [19].

As far as other media are concerned, e.g., foodstuffs, drinking water, and any matrix relevant to exposure via the percutaneous route, it is in principle manufactured nano-objects that become increasingly relevant with the progress made in nanotechnologies. There is a lack of data on exposure by the oral and percutaneous routes.

9.4 Conclusion

One result that should feature in any risk assessment is the qualification and then, as far as possible, the quantification of uncertainties, essential for identifying research priorities, for example. It follows from what has been said that the uncertainties involved in identifying hazards and estimating the dose–response relationship or the exposure–response relationship are so great and the public health stakes so high, that research must be carried out simultaneously on all these aspects.

Expert committees set up in various countries agree that the principle of precaution is in order. For example, the *Haut Comité de la Santé publique* in France demands in the name of the precautionary principle that the production of carbon nanotubes and their use to make intermediate products or consumer and health products should be carried out under conditions of strict confinement to protect workers from exposure whenever these activities involve a risk of aerosolization and/or dispersion. This recommendation applies to research centers using carbon nanotubes. The Royal Society in the United Kingdom and the *Comité de la prévention et de la précaution* in France advocate similar guidelines [20, 21].

Naturally, it is also important to check that this cautious approach is effective by setting up systems to monitor occupational exposure. If it is indeed the case, the residual risks relating to occupational exposure should be kept to a low level. As a result, they would be very difficult to demonstrate epidemiologically, which implies that epidemiology will not be able to play any role in establishing a causal link between manufactured nano-objects and any harmful effects, and nor will it be able to construct a dose–response relationship. In the final analysis, the assessment of risks due to exposure to manufactured nano-objects will rest mainly on experimental data. New methods of predictive toxicology must therefore be developed.

References

1. www.wilsoncenter.org
2. National Research Council: Committee on the Institutional Means for the Assessment of Risks to Public Health. Risk Assessment in the Federal Government: Managing the Process. National Academy Press, Washington (DC) (1983)
3. Scientific Committee on Emerging and Newly Identified Health Risks (SCENIHR): Risk assessment of products of nanotechnologies. In: European Commission HCD, Brussels (2009)
4. A. Lefranc, S. Larrieu: Particules ultrafines et santé: Apport des études épidémiologiques. Environnement, Risques & Santé **7**, 349–355 (2008)
5. A.B. Hill: The environment and disease: Association or causation? Proceedings of the Royal Society of Medicine **58**, 295–300 (1965)

6. G. Oberdorster, Z. Sharp, V. Atudorei, et al.: Extrapulmonary translocation of ultrafine carbon particles following whole-body inhalation exposure of rats. J. Toxicol. Environ. Health **65**, 1531–1543 (2002)
7. J.G. Ayres, P. Borm, F.R. Cassee, et al.: Evaluating the toxicity of airborne particulate matter and nanoparticles by measuring oxidative stress potential: A workshop report and consensus statement. Inhal. Toxicol. **20**, 75–99 (2008)
8. A.R. Murray, E. Kisin, S.S. Leonard, et al.: Oxidative stress and inflammatory response in dermal toxicity of single-walled carbon nanotubes. Toxicology **257**, 161–171 (2009)
9. Haut Conseil de la Santé publique: Avis relatif à la sécurité des travailleurs lors de l'exposition aux nanotubes de carbone. Paris (2009)
10. L. Gonzalez, D. Lison, M. Kirsch-Volders: Genotoxicity of engineered nanomaterials: A critical review. Nanotoxicology **2**, 252–273 (2008)
11. R. Landsiedel, M. Kapp, M. Schulz, et al.: Genotoxicity investigations on nanomaterials: Methods, preparation and characterization of test material, potential artifacts and limitations – many questions, some answers. Mutation Res. **681**, 241–258 (2009)
12. D. Bard: Extrapoler des hautes doses aux faibles doses. In: D. Bard, A. Cicolella, M. Jouan, et al. *Sciences et décision en santé environnementale*. Société française de santé publique, Vandœuvre-lès-Nancy (1997) pp. 139–151
13. D. Bard: Les effets des faibles doses: Un débat épistémologique et ses conséquences décisionnelles. Environnement, Risques & Santé **5**, 65–68 (2006)
14. N. Künzli, R. Kaiser, S. Medina, et al.: Public health impact of outdoor and traffic-related air pollution: A European assessment. Lancet **356**, 795–801 (2000)
15. S. Weichenthal, A. Dufresne, C. Infante-Rivard: Indoor ultrafine particles and childhood asthma: Exploring a potential public health concern. Indoor Air **17**, 81–91 (2007)
16. S. Weichenthal, A. Dufresne, C. Infante-Rivard, L. Joseph: Characterizing and predicting ultrafine particle counts in Canadian classrooms during the winter months: Model development and evaluation. Environ. Res. **106**, 349–360 (2008)
17. W. Ott, L. Wallace: Ultrafine particle exposures in homes, automobiles, and restaurants. Epidemiology **19**, S129–S130 (2008)
18. H. Boogaard, G. Borgman, J. Kamminga, H. Hoek: Exposure to ultrafine particles and noise during cycling and driving in 11 Dutch cities. Epidemiology **19**, S110 (2008)
19. F. Forastiere, M. Stafoggia, S. Picciotto, et al.: A case-crossover analysis of out-of-hospital coronary deaths and air pollution in Rome, Italy. Am. J. Respir. Crit. Care Med. **172**, 1549–1555 (2005)
20. Comité de la prévention et de la précaution: *Nanotechnologies, nanoparticules: Quels dangers, quels risques?* Ministère de l'Ecologie et du Développement Durable, Paris (2006)
21. The Royal Society & The Royal Academy of Engineering: *Nanoscience and Nanotechnologies*. The Royal Society, London (2004)

10

Technical Risk Prevention in the Workplace

Myriam Ricaud

Nanotechnology has become a major economic and technological issue today. Indeed, nanometric dimensions give matter novel physical, chemical, and biological properties with a host of applications. Nanotechnology is thus having an increasing impact on new and emerging industries, such as computing, electronics, aerospace, and alternative energy supplies, but also on traditional forms of industry such as the automobile, aeronautics, food, pharmaceutical, and cosmetics sectors. In this way, nanotechnology has led to both gradual and radical innovation in many areas of industry: biochips, drug delivery, self-cleaning and antipollution concretes, antibacterial clothing, antiscratch paints, and the list continues [1–3].

The world nanotechnology market, still in its infancy as recently as 2001, was then reckoned at a little more than 40 billion euros by the European Commission. By 2008, the world market for nanotechnological products had come close to 700 billion euros. Today, around 1 500 companies hold the world nanotechnology market and more than 1 000 everyday consumer products incorporating nanoparticles or using nanotechnologies in some way are already commercialised around the world [4]. The boom in this sector may thus represent around 10% of factory jobs by 2015 [5].

These observations no doubt explain why such colossal budgets have been devoted to research and development around the world. Carried along by this momentum and by the pressure of competition, and encouraged by a general lack of interest in safety procedures in many areas of research, the available funds have been oriented mainly toward applications, with little left over for questions of hygiene and safety [6]. But the wide range of industrial achievements in this field suggest that occupational exposure to nanoparticles has already become a reality, and not only in research centers, but also in production units and places where such products are used. At the present time, however, there is no well established or consensual method for making measurements or characterising occupational exposure in and around processes involving nanoparticles [7–9]. Our knowledge of the health effects of manufactured nanoparticles also remains inadequate [10–13], even though the risks

P. Houdy et al. (eds.), *Nanoethics and Nanotoxicology*,

DOI 10.1007/978-3-642-20177-6_10,

of toxicity to humans from ultrafine particles in the ambient air are already well documented. The introduction of strict safety measures, especially with regard to the reduction of exposure, remains the only way to limit the risks of occupational pathologies arising in the relevant industrial sectors [14].

10.1 Prevention

When matter is broken down into nanometric pieces, unexpected properties appear, often totally different from those of the very same material on the microscopic or macroscopic length scales. Nanotechnology thus leads to the synthesis of materials whose fundamental properties may be drastically modified, and they must then be treated in the same way as new chemical products [14]. Nanoparticles thus comprise a new family of chemical agents, with many differences of composition, dimensional characteristics, and physicochemical properties.

Risk assessment is the basis for defining, selecting, and implementing technical and organisational safety measures. The decision to carry out a risk assessment presupposes a clear understanding of the hazards for health and safety, and the levels of occupational exposure. On the other hand, toxicological data relating to nanoparticles remains fragmentary. Most come from studies, usually of limited scope, carried out on cells or animals, i.e., difficult to extrapolate to humans [10–12]. Situations in which workers are exposed to nanoparticles do exist, in companies and research centers, but at the present time very little data have been published [15, 16]. It is highly likely that we must wait many years before we find out exactly which types of nanoparticle, and precisely what associated doses, pose a real threat to humans and their environment. Indeed, assessment of the potential health effects after exposure to a chemical agent must take into account the magnitude and duration of exposure, biopersistence, intrinsic toxicity depending on the target(s), and parameters expressing interindividual variability.

Owing to the many uncertainties relating to nanoparticle toxicity and the lack of data regarding occupational exposure, it is a delicate matter to set up a quantitative assessment of the risks in the majority of work situations. The key here then is to develop and implement good practices and well adapted safety strategies when these new chemical products have to be handled, taking into account the following points:

- The chemical nature and specific properties of the product, e.g., size, size distribution, specific surface area, levels of aggregation and agglomeration, surface reactivity, morphology, porosity, crystal structure, solubility, surface treatments, etc.
- The amount of product to be handled. (Amounts of product and the kind of staff concerned vary significantly with the type of activity, e.g., research and development, industrial production, and so on.)

- Particularities of the processes involved.
- The way the work is carried out.

These preventive strategies and safety practices in the workplace should aim to reduce worker exposure to as low a level as possible. Indeed, given that our understanding of nanoparticle toxicity is still so limited, risk prevention must be based largely on limiting occupation exposure, not only levels, but also durations, and numbers of workers exposed, etc. Preventive strategies target all professional activities in which workers are exposed or likely to be exposed to nanoparticles, including production, reception and storage of raw materials, packaging and dispatch of finished products, transfer of intermediate products where relevant, sampling, use, incorporation into matrices, cleaning and maintenance of premises and installations, waste handling, tooling nanocomposites, and so on.

For the huge number of existing or future nanoparticles, the correct attitude to adopt must be based essentially on the specification and implementation of work safety practices, and these in turn must evolve as more reliable information is published regarding the adverse biological effects of nanoparticles. Such safety practices are not so very different from those recommended for any activity exposing people to hazardous chemical products, but they take on a particular importance owing to the outstanding capacity of nanoparticles for persistence and diffusion (aerosolisation and dispersion) in the ambient air of the workplace. Particular attention must be paid to nanoparticles for which there is as yet little toxicological data, or for which research has already demonstrated toxic effects, notably on animals.

It should be borne in mind that there is no generic nanoparticle. Nanoparticles are complex structures, each with their own potential for toxic effects. It is therefore important to recommend and apply a case-by-case risk management policy [17, 18]. When data are available for particles of micrometric or higher dimensions of the same chemical nature, the minimal assumption when setting up preventive measures is that the corresponding nanoparticles will exhibit at least the same level of toxicity, if they are not actually more hazardous.

At the present time, there are no specific regulations applicable to nanoparticles. However, since nanoparticles are chemical products, the general rules for chemical risk prevention specified in a country's labour regulations will apply, e.g., in France, articles R. 4412-1 to R. 4412-58 of the *Code du travail*. Specific safety regulations must also be adopted for activites involving class 1 or 2 nanoparticles known to be carcinogenic, mutagenic, or toxic for reproduction (R. 4412-59 to R. 4412-93 of the *Code du travail*).

The general scheme laid down by the *Code de travail* in France for safety at work operates in six steps:

1. Identify the hazards due to the chemical agent.
2. Avoid risks, if possible by eliminating them.

3. Assess health and work safety risks which cannot be avoided, taking into consideration the processes used and way of working (evaluate the nature and importance of the risks).
4. Set up measures to prevent or limit risks, using personal safety equipment only to complement collective safety systems, or when the latter are inadequate.
5. Check the efficiency of the safety measures adopted.
6. Train and inform workers.

The main steps that can be taken to implement this scheme are as follows (see Fig. 10.1) [9, 14, 19–23]:

- Modify the process or activity in such as way as to no longer produce or use the hazardous substance.
- Replace the hazardous substance by another, less toxic one.
- Optimise the process to obtain as low a dust level as possible in order to limit exposure. Favour closed systems and automated techniques.
- Capture pollutants at source, e.g., using a local ventilation system.
- Filter the air before rejecting it from the factory premises.
- Use personal safety gear if collective safety measures prove inadequate.
- Collect and process waste.
- Train and inform exposed employees about risks and their prevention, i.e., provide workers with the necessary information to carry out the various tasks under optimal safety conditions.
- Regularly monitor worker exposure, recording and keeping all relevant information, e.g., type of nanoparticle, characteristics, quantities, operations and tasks, means of prevention, etc.

Fig. 10.1. Plan of action for implementing safety measures. Adapted from V. Causse/INRS

In the workplace, it is always possible to evaluate various factors that may contribute to risks:

- Amounts of product handled.
- The physical state of the products, e.g., liquid suspension, powder, gel, etc.
- The processes implemented.
- The tendencies of the products to end up in the air or on work surfaces, i.e., to form aerosols, droplets, etc. (dissemination of products).
- Potential exposure of workers via different routes, viz., inhalation, cutaneous contact, ingestion, but also exposure frequency and duration, etc.
- Chemical properties, including chemical composition, solubility, reactivity, etc., and physical factors, such as size distribution, shape, degree of aggregation and agglomeration, etc., of the product. This information can come from various sources, such as safety data sheets provided by the manufacturer, reviews in the scientific literature, etc.
- Toxicological data, including identification of situations where coexposure to several products may heighten the risk.
- Exposure data, including emission sources, dust levels, etc.
- True performance of safety systems, e.g., ventilation, filtration, personal safety equipment, etc., set up for handled products.
- Risks of fire and explosion.

The whole process of risk prevention should be initiated as early as possible, when the equipment, operating mode, and place of work are being designed. The idea is to integrate the technique, work organisation, working conditions, labour relations, and environment into one coherent system. This effort should be reiterated and refined on a regular basis in order to take into account further data from risk assessment, as well as any modifications in working conditions. A permanent watchdog should be set up to keep safety measures in phase with technical and organisational developments.

10.2 Changing the Process

When a risk assessment reveals exposure to some chemical product that is considered hazardous, the first step to take, whenever it is technically possible, is to try to eliminate the risk. If that cannot be done, substitution is the next option, either by a different product, or by another, less hazardous process. Elimination or substitution are generally complicated to implement. They take time, and they have a cost both financially and in terms of human resources. Indeed, they can lead to important changes in the work station, such as the installation of new equipment or processes, reassessment of risks, adaptation of safety measures to the new working conditions, etc.

In the case of nanoparticles, which are generally used precisely because of the unusual properties they bestow upon the products incorporating them, the substitution approach can be implemented in the following ways:

- Manipulate the nanoparticles in the form of a liquid suspension, a gel, in an aggregated or agglomerated state, in tablets, or incorporated in inorganic or organic matrices, rather than in powder form.
- Favour liquid phase synthesis, rather than vapour phase or mechanical methods.
- Modify installations so as to allow continuous rather than intermittent production.
- Eliminate or limit certain critical operations such as decanting, weighing, sampling, etc.
- Optimise processes to use the smallest possible amounts of nanoparticles.
- Replace out-of-date installations to reduce the frequency of dysfunction, leakage, or ignition sources.

10.3 Collective Safety Measures

The aim of collective safety measures is to limit dispersion and accumulation of nanoparticles in the workplace atmosphere. When designing the systems for implementing this, it is important to take into account the specific properties of nanoaerosols, which behave in some ways similarly to a gas, while still of course being made up of particles.

10.3.1 Closed Systems

The production of nanoparticles, and in particular nanopowders, requires total isolation of the process, especially in cases where the material is fibrous, carcinogenic, mutagenic, or toxic for reproduction. It is important to associate work in a closed system with mechanisation or automation whenever the situation allows this, with a view to limiting the intervention and hence exposure of operators.

A closed system allows total confinement of nanoparticles when they are being synthesised or used. This avoids all contact between operators and nanoparticles. A system is defined as closed when all the operations of the process, including production, analysis, recovery of the product, cleaning, storage, etc., are carried out in a totally airtight housing. These systems generally involve mechanisation of the process, and even automation of certain tasks: transfer of products by conveyor belt, mechanised sampling, keeping reactors closed during cleaning, etc. Mechanisation provides a way of eliminating handling between the different stages of the process and maintaining confinement. For its part, automation avoids operator exposure during certain critical tasks which may generate aerosols or droplets. Examples here are:

- Grinding, transfer, sampling, suspension of nanopowders, or incorporation of nanopowders in an organic or inorganic matrix.
- Decanting, shaking, mixing, or drying of a liquid suspension containing nanoparticles.

- Filling or emptying a reactor.
- Milling, cutting, polishing, drilling, etc., of nanocomposites.
- Packaging (bagging), storage, and transport of products.
- Collection, packaging, storage, and transport of waste.

It is important to remain vigilant in the maintenance of closed systems. Indeed, as soon as it becomes necessary to open the work space, the workers in charge of maintenance operations may be exposed to nanoparticles.

In the case of particularly polluting processes, operators can also work from isolated control posts with controlled atmosphere. The process is then piloted from a distance (remote control).

10.3.2 Ventilating and Purifying the Workplace Atmosphere

When it is technically impossible to set up a fully closed system, e.g., owing to unsuitable premises, bulky equipment, etc., enclosure must be envisaged. Enclosure consists in setting up physical barriers such as partition walls or hooding, and it is systematically combined with a nanoparticle capture system. Enclosure can be total (glove box, fume cupboard, etc.), with the possibility of occasional opening for intervention inside the enclosure, or it can be partial (a simple wall, etc.). Enclosure provides a way of handling nanoparticles in isolated or ventilated rooms or installations, and hence avoiding their dissemination throughout the whole workplace atmosphere.

In the laboratory, it is recommended to use total enclosure in the form of a glove box or fume cupboard (see Fig. 10.2) [24]. Some laminar flow systems

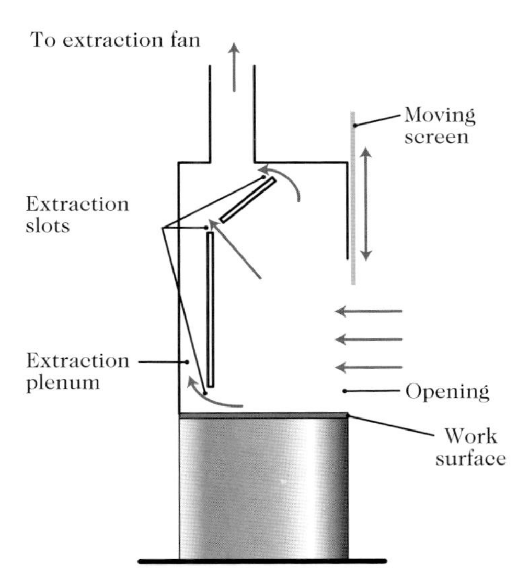

Fig. 10.2. Fume cupboard. Source INRS

can also be used. It is then easier to handle the products and the air flow is less pronounced. Laminar flow systems are cabinets, sometimes at reduced pressure, in which ventilation of the work space is achieved by a one-way flow, usually vertical and top to bottom, of filtered air [25]. This kind of equipment, sometimes specifically designed for nanoparticles [26], provides four-fold protection: for the operator, for the work space, for the environment, and for the product itself.

In workshops where manual sampling, weighing, conditioning, and tooling operations cannot be carried out in a fume cupboard or a glove box, and when they are neither mechanised nor automated, it is advisable to restrict them to a room or cubicle in which the pressure is slightly lower than elsewhere on the premises and equipped with a local extraction fan. These local ventilation systems aim to pick up any product that is released as soon as it is produced, as close to the emission source as possible, and as efficiently as possible, taking into account the nature, characteristics, and flow of the nanoparticles, as well as air flows.

Devices for capture at source can be mobile or otherwise: suction nozzles (see Fig. 10.3), suction hoppers, suction rings, ventilated tables, tables with suction, etc. Capture at source is already widely used for metal cutting and soldering processes, which are known to generate particles of nanometric sizes.

Source capture systems of proven efficiency for gases and vapours should be equally effective for capturing nanoaerosols, provided that the mouth of the device is well placed and an adequate capture rate is continually maintained. They must be chosen to suit the size and the type of operation to be carried out. The performance of such local ventilation systems thus depends intimately on their design and dimensions, efficient replacement of extracted air, good maintenance, and good work procedures. Such systems must be checked periodically, particularly for correct air flow.

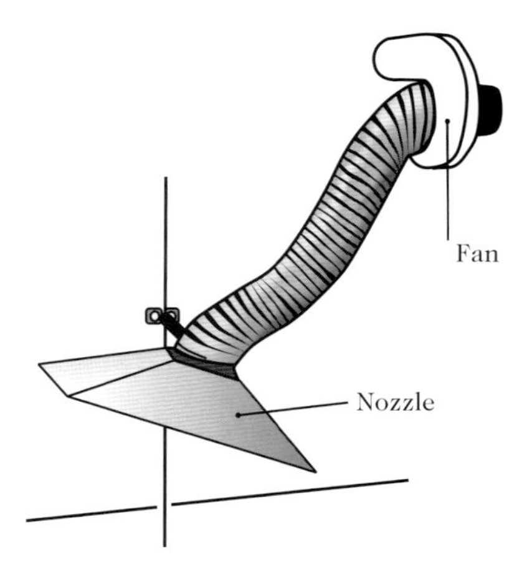

Fig. 10.3. Mobile suction nozzle. Source INRS

Local ventilation by suction at source must abide by 9 simple principles [27]:

- Enclose the area where pollutants are produced as far as possible.
- Capture as close as possible to the emission area.
- Position the device in such a way that the operator is never located between it and the source of pollution.
- Exploit the natural motion of the pollutants.
- Induce a sufficient air flow rate.
- Distribute the air flow rates uniformly in the capture area.
- Balance air outflows by corresponding air inflows.
- Avoid draughts and feelings of thermal discomfort.
- Reject polluted air well away from inflows of new air.

There is nevertheless very little scientific literature assessing the efficiency of commercially available enclosure and local ventilation systems when it comes to dealing with nanoparticles, either specific to these pollutants or otherwise. One of the rare studies published recently on the subject focuses on fume hoods [28]. It shows that nanoparticles can be transferred from inside (emission source) to outside several fume cupboards, via their openings. Whenever there is such transfer, it is always possible that an operator situated in the immediate vicinity may be exposed, and likewise for others operating in the same neighbourhood or on the same premises. When the fume hood is cleaned, e.g., with a damp cloth following contamination with nanoparticles, this may also result in the dissemination of particles into the laboratory atmosphere. Quite generally, particle transfer would appear to be a complex function of a great many parameters related to air flow (direction, rate, etc.), the fume cupboard (type, configuration, air flow compensation system, aperture), the activity (type, position of operator, hand and arm movements, etc.), and also the powder (granulometry, ease of dispersion, etc.). The authors note that there is an optimal frontal flow rate for fume cupboards, in the range 0.4–0.6 m/s, where transfer is minimal. This air speed is already recommended by various organisations [29]. Above and below this speed, nanoparticles may be transferred out of the fume hood. This article also raises the question of whether the current method for assessing fume hood performance with respect to nanoparticle transfer is adequate, and suggests that more systematic studies need to be carried out on fume cupboard performance with respect to nanoparticle handling.

As far as general ventilation is concerned, it should not be considered as the main air renewal system unless local ventilation methods are technically impossible. Indeed, general ventilation works by bringing in new air and thereby diluting pollutants, reducing their concentrations to as low a level as possible. However, it does not reduce the total amount of pollutants emitted on the premises. Exclusive use of general ventilation is generally unsatisfactory and results in the presence of residual pollution.

Finally, the use of portable mechanical tools like saws, drills, and so on, equipped with integrated capture systems for pollutants and highly efficient filters remains another possibility, mainly for tooling nanocomposites.

10.3.3 Air Filtering in the Workplace

The air in areas where nanoparticles are synthesised or used must be filtered before rejection into the atmosphere. The most widespread filtering technique uses fibrous media, thanks to their efficiency, low cost, and high level of adaptability. A fibre filter comprises metal, natural, or synthetic fibres (usually glass, polyester, cellulose, etc.). Filtration results from complex interactions between the aerosol and the fibres of the filter. The physics of these interactions depends on several parameters, such as the kind of aerosol (size of particles, concentration, electrical charges, etc.), the kind of medium (size distribution of the fibres, adherence properties, electrostatic charges, etc.), and the thermodynamic characteristics of the air. This complexity is increased by the fact that the performance of the filtering medium may evolve during the filtering process (clogging of the filter). When clean, a fibre filter is characterised by the pressure drop between upstream and downstream of the filter, and also its initial efficiency E, defined as the ratio of the number of particles retained by the filter to the number of particles counted upstream. For highly efficient filters, the penetration P is the preferred parameter, defined as the ratio of the number of particles counted downstream of the filter to the number counted upstream. The two parameters are related by $E = 1 - P$.

In aerosol filtration, it is a widespread error to assume that particle capture in fibre filters is due solely to a sieving effect, i.e., that collected particles are all bigger than the pores of the filter. However, it turns out that particle capture by fibre filters depends on several physical mechanisms. When there is no external force field apart from gravity, the most important particle capture mechanisms are [9, 30, 31]:

- Inertial impaction (see Fig. 10.4). This capture mechanism dominates for large particles, i.e., those with diameters greater than 1 μm. Owing to its

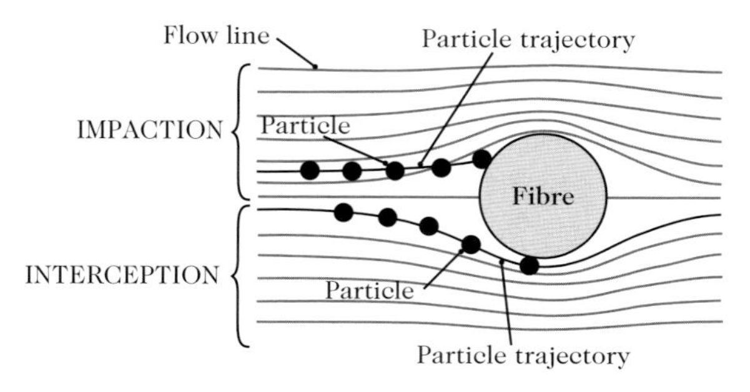

Fig. 10.4. Particle capture by impaction and interception [31]. Source INRS

inertia, the particle no longer follows the flow line around the fibre, and ends up impacting the fibre surface.

- Direct interception (see Fig. 10.4). This capture mechanism involves particles of diameter greater than 0.1 μm. The particle follows the flow line and is intercepted by a fibre whenever it comes closer than a distance equal to its radius.
- Brownian diffusion (see Fig. 10.5). This mechanism is important for small particles, i.e., with diameters less than 0.1 μm. At these sizes, particles undergo Brownian motion. If their trajectories pass close enough to a fibre, they may enter into contact with it under the influence of the Brownian motion, and subsequently adhere. The random Brownian displacements increase the probability of collisions between particles and filter fibres.
- Electrostatic effects (see Fig. 10.6). There will be an attractive force called the Coulomb force between a particle and a fibre provided that both of them are charged, a polarisation force between an electrically neutral particle and a charged fibre, and an image force between a charged particle and an electrically neutral fibre.

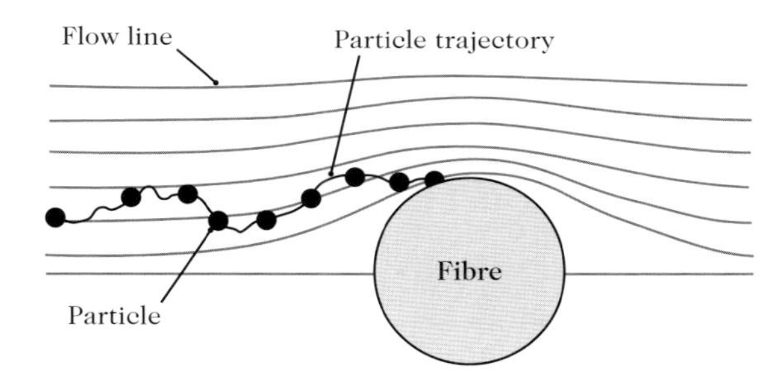

Fig. 10.5. Particle capture by diffusion [31]. Source INRS

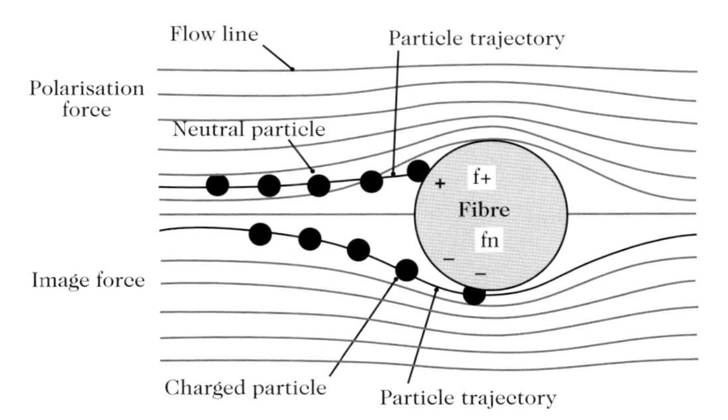

Fig. 10.6. Particle capture by polarisation [31]. Source INRS

The total collection efficiency of a fibre filter is the resultant of the three main particle capture mechanisms listed above, viz., inertia, interception, and diffusion. Figure 10.7 illustrates the influence of particle size on these three mechanisms and on the initial efficiency of the filter.

For particles with sizes in the range 100–500 nm, the efficiency is minimal. This range corresponds to particles that are too large for the diffusion effect to be effective and too small for the interception and impaction mechanisms to play an important role. This size of particle is said to be the most penetrating particle size (MPPS). These are therefore the most difficult particles to capture. It is in this particulate size range that the efficiencies of high efficiency particulate air (HEPA) filters and ultra low penetration air (ULPA) filters are determined, as measured using the standardised method EN 1822-5 [32] (see Table 10.1).

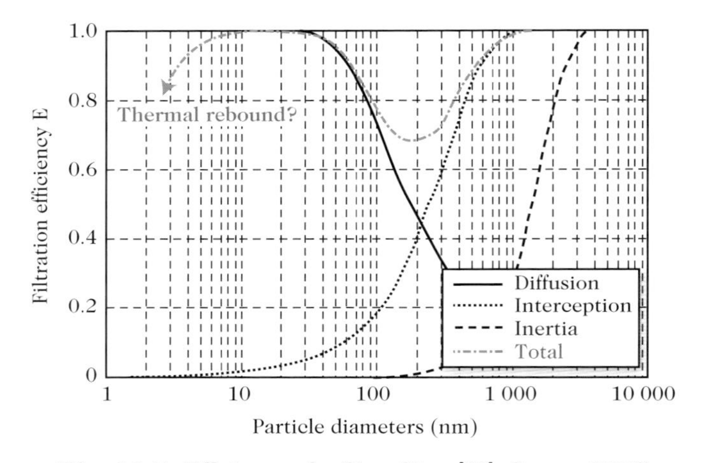

Fig. 10.7. Efficiency of a fibre filter [30]. Source INRS

Table 10.1. Classification of HEPA filters (H10–H14) and ULPA filters (U15–U17) according to the norm EN 1822-1 [33]. Local value refers to the minimal local efficiency tolerated with regard to leakages

	Overall value		Local value	
Filter class	Efficiency (%)	Penetration (%)	Efficiency (%)	Penetration (%)
H10	85	15	–	–
H11	95	5	–	–
H12	99.5	0.5	97.5	2.5
H13	99.95	0.05	99.75	0.25
H14	99.995	0.005	99.975	0.025
U15	99.9995	0.0005	99.9975	0.0025
U16	99.99995	0.00005	99.99975	0.00025
U17	99.999995	0.000005	99.9999	0.0001

In the nanometric particle range, the collection mechanism is therefore diffusion. There are many formulas, both empirical and theoretical, to estimate diffusion efficiency. They all converge and agree with experiment to the effect that fibre filter efficiency increases as particle size decreases. However, a theoretical study dating to 1991 [34] throws doubt on this conclusion for particles smaller than 10 nm, introducing the notion of a thermal rebound at the surface of the medium, suggesting that particles smaller than 10 nm might be less likely to adhere to the filter fibres due to too high an impact velocity. This drop in efficiency for particles smaller than 10 nm (see Fig. 10.7) has only been reported in a limited number of publications. Experimental studies published recently on the subject indicate an absence of thermal rebound for particles larger than 3 nm and thus confirm that fibre filters constitute an effective nanoparticle barrier [30, 35–37].

Whenever the size of particles, aggregates, or agglomerates is greater than 3 nm, they can be captured by fibre filters (see Fig. 10.8). In the fields of personal, occupational, and environmental safety, it is recommended to use high efficiency particulate air filters in the category known as 'absolute', above H13 according to the norm EN 1822-1 [33]. On the other hand, given that there have been so few studies, and those with contradictory conclusions, some questions still remain over the capture efficiency of fibre filters for particles smaller than 3 nm.

10.4 Organisational Measures

10.4.1 Work Area

The work area must be clearly signed and marked out, and restricted solely to those employees directly concerned with the synthesis or use of nanoparticles and nanomaterials, the idea being to limit the number of employees who might

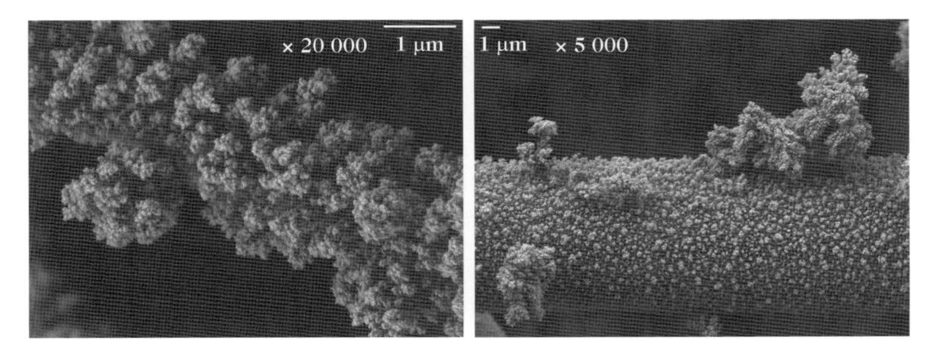

Fig. 10.8. Copper nanoparticles (3–40 nm) collected on high efficiency particulate air (HEPA) filters. Supplied by the Filtration and Absorption Unit of the CNRS/INRS

be exposed. Working areas where nanoparticle exposure is possible should be clearly identified and separated from so-called clean areas. Access from one to the other should included whatever installations are required to change personal safety gear should the need arise. Double changing rooms should be envisaged next to the activity area so that everyday garments and work clothing can be kept apart, and also so that there is no risk of contamination outside working areas.

The working area and its equipment must be kept clear of any accumulated deposit of nanoparticles which might be returned into suspension in the air. To this end, all installations, floors, and work surfaces (preferably non-porous) must be regularly and carefully dusted and cleaned using damp cloths and a vacuum cleaner equipped with a high efficiency particulate air filter, above H13 as required by the norm EN 1822-1 [33]. Air jets, brushes, and brooms should be avoided, whether for cleaning equipment and the workplace, or for clearing up after accidental spillages.

Periodic cleaning and maintenance of installations minimises the risk of unplanned interruptions, dysfunction, and accidents such as leakages.

10.4.2 Personal Hygiene

It is important to have sinks and showers on the work premises for decontaminating regions of the skin that may have been exposed to nanoparticles. It is also recommended to take a shower at the end of the shift to avoid nanoparticles encrusting themselves in the skin. Dirty clothing, in particular work clothing, should not be taken home.

To avoid ingestion of nanoparticles, it should be forbidden to eat or drink on the work premises, except in areas strictly reserved for this purpose, which must be kept scrupulously clean.

10.4.3 Product Storage

The storage of nanoparticles involves particular problems owing to their granulometric characteristics and surface reactivity. The small diameter of the particles increases the sedimentation time and facilitates suspension.

Nanoparticles must be stored in totally airtight double-walled tanks or packaging, e.g., made from plastics, these being carefully closed and labelled. The labels used must mention the presence of nanoparticles and the associated potential hazards. These tanks and packages must be stored in cool, well ventilated premises, protected from sunlight and well away from any source of heat or ignition and inflammable materials.

It may be necessary to implement a storage process under controlled atmospheric conditions, e.g., in the presence of an inert gas, particularly in the case of certain metal nanoparticles (aluminium, magnesium, lithium, zirconium, etc.), in order to reduce the risks of self-inflammation and explosion.

10.4.4 Waste Processing

Waste, and in particular products that do not meet the required production criteria, packaging, ventilation filters, vacuum cleaner bags, disposable respiratory and skin protection equipment (protective suits, half-masks, etc.) and contaminated cleaning cloths must be treated as hazardous waste. They should be sorted and packed into closed, airtight, and labelled bags (double-walled plastic packaging, for example), then removed from the work area as they are produced. Labelling can be the same as on new packaging. The waste must then be processed in appropriate installations, either by burying in a storage center, incinerating, or recycling where possible.

10.4.5 Accident and Incident Management

The procedures for dealing with accidental emissions or spillages must be printed out and distributed to all employees. Accident and incident scenarios must be specified and exercise drills organised on a periodic basis. The aims of these procedures should be as follows [20]:

- To alert the emergency services.
- To identify the areas affected by incidents or accidents of varying degree (all or part of the site).
- To set up controlled access to contaminated areas.
- To make available suitable personal safety gear for anyone who needs to enter the affected area.
- To specify the cleaning of contaminated installations and surfaces (floors, walls, etc.) by systems suited to the nature and amount of product dispersed.

10.5 Personal Safety

Personal safety equipment is reserved for situations where good work practice is not applicable and where collective safety measures are inadequate. The choice of this equipment should be based on the best possible compromise between the highest safety level that can be achieved and the need to carry out the given task under conditions of maximal comfort. All personal safety equipment must be kept in good condition, and cleaned after each use when non-disposable.

10.5.1 Respiratory Protection

Respiratory protection is required each time an employee is faced with the risk of inhaling air polluted by nanoaerosols. For example, whenever the work atmosphere is not sufficiently well ventilated in workshops or laboratories producing or using nanoparticles, it is recommended to wear a respiratory protection device, bearing in mind that nanosized objects may be able to escape in the slightest leak, e.g., if there is a problem of airtightness where the face piece is in contact with the face, or a perforation, etc.

There are two families of respiratory protection device [38]:

- Filtering respiratory protection systems, which purify the ambient air by filtration. These generally consist of a face piece which encompasses the airways (nose and mouth) to varying degrees, equipped with a suitable filter. They may employ free ventilation, i.e., the air only goes through the filter due to the respiratory exchanges of the user, or assisted ventilation, where the ambient air is sucked in through a filter by means of a pump. Filters are designed to protect against specific pollutants. In the case of potential exposure to particles or droplets dispersed in the air (solid and liquid aerosols), so-called aerosol filters are used. There are three efficiency categories for aerosol filters, specified in the norm NF EN 143 [39] and depending on their filtration performances with regard to a sodium chloride aerosol made up of particles with median mass diameter 0.66 μm and with regard to a paraffin oil aerosol made up of droplets with median diameter 0.4 μm. Class 1 filters, marked P1 or FFP1, stop at least 80% of these aerosols, while class 2 filters, marked P2 or FFP2, stop at least 94%, and class 3 filters, marked P3 or FFP3, stop at least 99.95%. Assisted ventilation filtering respiratory protection devices are classified in terms of the airtightness of the whole system, i.e., face piece + fan motor + filter. They are denoted by the letters TH (turbo hood) when the face piece used is a hood, or TM (turbo mask) when it is a half-mask or full face mask.
- Isolated respiratory protection systems, where breathable air is supplied from a non-contaminated source. The user is independent of the ambient atmosphere. These comprise a face piece and an air supply system. The user can be connected by means of a tube to a compressed air supply (compressed air adduction device) or to a nearby area where the air is not contaminated (open air device).

The face piece is the part of the respiratory protection device that is directly in contact with the operator's face. It must guarantee an airtight separation between the ambient atmosphere and the inside of the device by its face seal. There are different types of face piece [38]: filtering face pieces marked FF (throwaway half-masks comprising the filtering material itself), half-masks, full face masks, and hoods (see Fig. 10.9).

An ultrafine aerosol may enter the respiratory protection device in two ways: penetration via the filtering medium or leakages, in particular through the face seal.

The performance of the filtering medium in a respiratory protection system depends on the nature of that medium, but also the aerosol and filtering conditions. In agreement with conventional filtration theory, aerosol filters have similar performance to the filters used to protect workplace and environment,

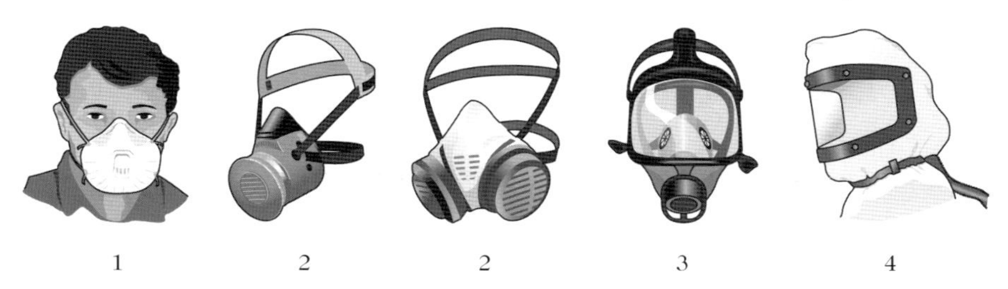

Fig. 10.9. Different types of face piece: (1) filtering face piece, (2) half-masks, (3) full face mask, and (4) hood. Source INRS

Table 10.2. Penetration of sodium chloride nanoparticles (sizes between 14 and 100 nm) through different filtering media [42]. Penetrations are given as percentages

Filtering medium	Number penetration	Mass penetration	Maximal mass penetration according to norm NF EN 143 [39]
Class 2/glass fibres	0.654	1.354	6
Class 3/glass fibres	0.007	0.018	0.05
Class 1/electrostatic	1.447	2.109	20
Class 2/electrostatic	0.290	0.543	6

described in Sect. 10.3.3. The efficiency of aerosol filters thus tends to increase as the particle size decreases [9, 30, 40], with the most penetrating fraction lying in the size range above 100 nm. In the specific case of media used in filtering face pieces (electrostatic media), studies have shown that the most penetrating fraction lies in the nanoparticle range around 30–40 nm [40, 41]. Some tests have shown that the penetration of nanoparticles through electrostatic filtering media falls off with time under laboratory conditions, whereas it increases under actual conditions of use (humidity due to respiration), but without exceeding, at least in these experiments, the threshold value specified for disposable FFP3 half-masks in up to 2 hours of use [40]. This typical effect with electrostatic filtering media, already known for micrometric particles, is also found for nanometric particles.

Finally, a study has recently been done to compare nanoparticle penetration through electrostatic filtering media and glass fibre filtering media [42]. It has shown that class 3 glass fibre filters exhibit very high efficiency for nanoparticles (see Table 10.2).

Nanoaerosol penetration through leaks has received very little attention. Theory predicts that the penetration of an ultrafine aerosol through leakages in the face piece should be less than for a gas, owing to diffusion deposition. However, no experimental confirmation has been reported [9].

For low exposure occupations, such as maintenance and cleaning of machines that have already been decontaminated, and when the ambient air contains enough oxygen (minimum 19% volume), it is thus recommended to wear an aerosol filtering respiratory protection device. When these operations are shortlasting, a half-mask or full face mask with free air supply equipped with a class 3 filter can be used (face piece equipped with a P3 filter, following the norm NF EN 143 [39], or a disposable filtering FFP3 face piece, following the norm NF EN 149 [43]). If the work is likely to last longer than one hour, it is recommended to wear a filtering respiratory protection device with assisted ventilation, and more precisely, a half-mask (TM2 P), a full face mask (TM3 P), or a hood (TH3 P) with assisted ventilation, following the norm NF EN 12942 [44]. Standard assisted ventilation respiratory protection systems operate with an air flow rate of 120 L/min. It is recommended to use assisted ventilation systems supplying air at the rate of 160 L/min to ensure that a positive pressure is maintained within the device.

For high exposure work, such as nanoparticle fabrication, manipulation, or transfer, an isolating respiratory protection system is recommended, and more exactly, a mask, hood, or protective suit, with compressed air supply.

It is important to check the efficiency of protection and good conditions of use in real working situations, and over time (checking for saturation, wear, etc.).

10.5.2 Skin Protection

There are many types of clothing, made from many different types of material, to protect against chemical products. However, the current literature remains rather limited regarding the efficiency of such clothing as a protection against nanoparticles. On the other hand, several studies presented recently (using graphite particles [40] and sodium chloride particles [45]), investigating different textile materials used for protective clothing, show that such materials function like the fibrous media used in particular for air filtering in the workplace. The most penetrating particle size (MPPS) fraction lies in the range 100–500 nm, and the penetration of the ultrafine aerosol through different textile materials among those tested tends to fall as the particle size decreases (down to at least 40 nm). These tests also show that textile materials made up of Tyvek high density polyethylene fibres (non-woven) [40] exhibit higher nanoparticle exclusion efficiencies than textile materials made from cotton or polypropylene fibres.

It thus turns out that the most suitable clothing for protection against type 5 chemical risks as specified by the norm EN ISO 13982 [46] (clothing for protection against chemical products in the form of particles) should be made from Tyvek. This kind of airtight clothing provides protection against airborne solid products. It is thus recommended to wear a single-use garment, in particular, a disposable suit with a hood, with tighteners at the neck, wrists, and ankles, without folds, lapels, or turnups, and with flap pockets.

Likewise, airtight plastic gloves are recommended when handling nanoparticles. Some results seem to indicate that gloves made from different plastic materials (nitriles, vinyls, latex, and neoprene) form an efficient physical barrier between the skin and airborne nanoparticles [40].

Finally, it may also be necessary to wear shoe covers and goggles equipped with lateral protection.

10.6 Informing Staff

10.6.1 Labelling

Labelling is the first source of essential and concise information available to the user regarding the hazards of the nanoparticles and precautions to be taken when using them. It takes into account toxicological risks and also the risk of fire and explosion. Nanoparticles do not feature as such in the tables of Appendix 6 of the EC regulation no. 1272/2008 of 16 December 2008 [classification, labelling, and packaging (CLP) regulation], which covers all hazardous materials that have been classified on the European level and are subject to harmonised labelling, i.e., those for which a vote by member states has made classification and specific labelling compulsory. The tables of Appendix 6 do not provide an exhaustive list, however. The classification and labelling of most commercially available substances have not been examined at the European level. For substances not appearing in the tables of Appendix 6, and this includes nanoparticles and nanomaterials, it is the responsibility of the manufacturer, importer, or resaler to classify and label them in accordance with their intrinsic properties.

10.6.2 Safety Data Sheets

The safety data sheet (SDS) is issued by the nanoparticle supplier and comes in addition to labelling. It informs much more fully than the label about every kind of risk the nanoparticles may involve, but also about safety measures to be respected when using them. Given that the SDS is such an important tool for risk assessment and essential input when drafting task sheets, the manufacturer should be strongly encouraged to supply one, even if it is not compulsory, as in the case of nanoparticles. The registration, evaluation and authorisation of chemical substances (REACH) regulation specifies the way the SDS should be drawn up and transmitted, and in its Appendix II, provides a guide for drafting these information sheets.

10.7 Staff Training

It is crucial to train employees and raise their awareness of risks and ways of preventing them. This concerns all employees called upon to work in the presence of nanoparticles within the company. Good practice in the field of safety is in constant evolution, especially in such a recent sector as nanomaterials, and it should never be treated as permanently acquired. Training programmes must therefore be regularly renewed.

Staff training should deal with the following issues:

- Health risks.
- Risks of fire and explosion.
- Safety measures to be respected during:
  1. manufacture, handling, transfer, sampling, recovery, packaging, and storage of products,
  2. cleaning and maintenance of equipment and premises,
  3. waste processing,
  4. all operations carried out on nanocomposites.
- Use and maintenance of collective and personal safety systems.
- Rules of hygiene laid down within the company, all of which must be scrupulously observed by the staff.
- Understanding product labels and safety data sheets.
- Steps to be taken in the event of an incident, accident, accidental spillage, etc.

Staff training is the responsibility of the company manager. It can be designed and run by the management with the participation of the medical department, employee representatives, or an organisation like the *Comité d'hygiène, de sécurité et des conditions de travail* (CHSCT) in France.

10.8 Conclusion

The huge investment in the manufacture and use of nanoparticles and nanomaterials, and the high expectations in many sectors of activity, have already led to a plethora of industrial achievements. This means that public exposure to nanoparticles has already become a reality. Although much fewer in number, the relevant part of the working population is currently potentially more exposed than the general public, carrying out operations that are more likely to emit nanoparticles, and handling much greater amounts of these materials.

Given the many unknowns with regard to these new chemical products, their potential effects on health, and the difficulties encountered in characterising exposure, a quantitative assessment of the risks turns out to be difficult to implement in most work situations. It is thus important to set up specific safety procedures in all professional environments involving nanoparticles in some way (companies, pilot units, research centers, universities, etc.) and throughout the life cycle of the resulting products. The aim of this kind of work safety practice, which must necessarily evolve as further information is published on the adverse effects of nanoparticles, is to avoid occupational exposure, or at least reduce it to a strict minimum. Collective protection and protection integrated into work procedures must always be given priority: working in closed systems, automating processes, enclosing equipment, capturing pollutants at source, filtering the air in the workplace, etc. Working conditions and worker exposure must also be regularly monitored.

References

1. M. Lahmani, C. Dupas, P. Houdy (Eds.): *Nanoscience. Nanotechnologies and Nanophysics*. Springer, Berlin, Heidelberg, New York, 823 p. (2007)
2. M. Lahmani, C. Brechignac, P. Houdy (Eds.): *Nanomaterials and Nanochemistry*. Springer, Berlin, Heidelberg, New York, 747 p. (2007)
3. W. Luther: Industrial application of nanomaterials: Chances and risks. Technological Analysis. Future Technologies Division of VDI Technologiezentrum GmbH, Düsseldorf, 112 p. (2004)
4. Woodrow Wilson International Center for Scholars, Project on Emerging Nanotechnologies. A consumer product inventory. www.nanotechproject.org/inventories (2009)
5. Lux Research: *The Nanotech Report*, 5th edn., New York (2008)
6. A.D. Maynard: Nanotechnology: A research strategy for addressing risk. Woodrow Wilson International Center for Scholars, Project on Emerging Nanotechnologies, 45 p. (2006)
7. A.D. Maynard, R.J. Aitken: Assessing exposure to airbone nanomaterials: Current abilities and future requirements. Nanotoxicology **1**, 26–41 (2007)
8. O. Witschger, J.F. Fabriès: Particules ultra-fines et santé au travail. 2. Sources et caractérisation de l'exposition. Hygiène et Sécurité au Travail, Institut national de recherche et de sécurité (INRS) **199**, 37–54 (2005)
9. B. Herve-Bazin: *Les nanoparticules: un enjeu majeur pour la santé au travail?* Les Ulis, EDP Sciences, 701 p. (2007)
10. G. Oberdoster, V. Stone, K. Donaldson: Toxicology of nanoparticles: A historical perspective. Nanotoxicology **1**, 2–25 (2007)
11. P.A.J. Borm, D. Robbins, S. Hauubold, T. Kuhlbuscht, H. Fissan, K. Donaldson, R. Schins, V. Stone, W.G. Kreyling, J. Lademann, J. Krutmann, D. Wahreit, E. Oberdorster: The potential risks of nanomaterials: A review carried out for ECETOC. Particle and Fibre Toxicology **3**, 11 (2006)
12. C. Ostiguy, B. Soucy, G. Lapointe, C. Woods, L. Mena: *Les effets sur la santé liés aux nanoparticules*, 2nd edn., Institut de recherche Robert-Sauvé en santé et sécurité du travail, Report R-558, 112 p. (2008)
13. P. Hoet, J. Boczkowski: What's new in nanotoxicology? Brief review of the 2007 literature. Nanotoxicology **2**, 171–182 (2008)
14. *Les nanomatériaux*: www.inrs.fr/dossiers/nanomateriaux.html, Institut national de recherche et de sécurité (INRS) (2007)
15. M.M. Methner, M.E. Birch, D.E. Evans, B.K. Ku, K. Crouch, M.D. Hoover: Identification and characterization of potential sources of worker exposure to carbon nanofibers during polymer composite laboratory operations. J. Occup. Environ. Hygiene **4**, 125–130 (2007)
16. J.H. Han, E.J. Lee, J.H. Lee, P.K. So, Y.H. Lee, G.N. Bae, S.B. Lee, J.H. Ji, M.H. Cho, I.J. Yu: Monitoring multiwalled carbon nanotube exposure in carbone nanotube research facility. Inhal. Toxicol. **20**, 741–749 (2008)
17. M. Ricaud, F. Roos, D. Lafon: Les nanotubes de carbone: Quels risques, quelle prévention? Hygiène et Sécurité au Travail, Institut national de recherche et de sécurité **210**, 43–57 (2008)
18. M. Ricaud: Le point des connaissances sur les silices amorphes, ED 5033. Institut national de recherche et de sécurité, 5 p. (2007)

19. C. Ostiguy, B. Roberge, L. Menard, C.A. Endo: *Guide de bonnes pratiques favorisant la gestion des risques liés aux nanoparticules de synthèse*. Institut de recherche Robert-Sauvé en santé et sécurité du travail, rapport R-586, 63 p. (2008)
20. Nanotechnologies: Guide to safe handling and disposal of manufactured nanomaterials. British Standards (BSI), 26 p. (2007)
21. Responsible production and use of nanomaterials, Verband der Chemischen Industrie (VCI)/Bundesanstalt für Arbeitsschutz und Arbeitsmedizin (BAuA), 54 p. (2008)
22. P. Schulte, C. Geraci, R. Zumwalde, M. Hoover, E. Kuempel: Occupational risk management of engineered nanoparticles. J. Occup. Environ. Hygiene **5**, 239–249 (2008)
23. ISO TR/12885: Nanotechnologies: Health and safety practices in occupational settings relevant to nanotechnologies (2008)
24. J. Triolet, J. Capois, G. Gautret De La Moriciere, X. Le Quang, J.M. Petit, J.C. Protois, M. Rocher: La conception des laboratoires de chimie. Hygiène et Sécurité au Travail, Institut national de recherche et de sécurité **188**, 7–26 (2002)
25. I. Balty, B. Belhanini, H. Clermont, J. C. Cornu, M.A. Jacquet, J.C. Texte: Postes de sécurité microbiologique, postes de sécurité cytotoxique. Hygiène et Sécurité au Travail, Institut national de recherche et de sécurité **193**, 37–52 (2003)
26. P. Bombardier: The first solution for nanoparticles handling designed by Faure Ingénierie: Description of the PSPN (poste de sécurité pour particules nanostructurées). Nanosafe, Grenoble (2008)
27. Principes généraux de ventilation, guide pratique de ventilation No. 0, ED 695, Institut national de recherche et de sécurité, 36 p. (1989)
28. S.J. Tsai, E. Ada, J.A.Isaacs, M. J. Ellenbecker: Airborne nanoparticle exposures associated with the manual handling of nanoalumina and nanosilver in fume hoods. J. Nanopart. Res. **11**, 147–161 (2009)
29. Sorbonnes de laboratoire, guide pratique de ventilation No. 18, ED 795, Institut national de recherche et de sécurité, 25 p. (2009)
30. D. Thomas, G. Mouret, S. Calle-Chazelet, D. Bemer: Filtration des nanoparticules: Un problème de taille. Hygiène et Sécurité au Travail, Institut national de recherche et de sécurité **211**, 13–19 (2008)
31. D. Thomas: Etude de la filtration des aérosols par des filtres à fibres. Habilitation à diriger des recherches, spécialité Génie des procédés, université Henri-Poincaré (2001)
32. NF EN 1822-5: Filtres à air à très haute efficacité et filtres à air à très faible pénétration (HEPA et ULPA). Part 5: Mesure de l'efficacité de l'élément filtrant (2000)
33. NF EN 1822-1: Filtres à air à très haute efficacité et filtres à air à très faible pénétration (HEPA et ULPA). Part 1: Classification, essais de performance et marquage (1998)
34. H.C. Wang, G. Kasper: Filtration efficiency of nanometer-size aerosol particles. J. Aerosol. Sci. **22**, 31–41 (1991)
35. M. Heim, B.J. Mullins, M. Wild, J. Meyer, G. Kasper: Filtration efficiency of aerosol particles below 20 nanometers. Aerosol Sci. Technol. **39**, 782–789 (2005)

36. S.H. Huang, C.W. Chen, C.P. Chang, C.Y. Lai, C.C. Chen: Penetration of 4.5 nm to 10 μm aerosol particles through fibrous filters. J. Aerosol Sci. **38**, 719–727 (2007)
37. S.C. Kim, M.S. Harrington, D.Y.H. Pui: Experimental study of nanoparticles penetration through commercial filter media. J. Nanopart. Res. **9**, 117–125 (2007)
38. P. Hure, M. Guimon: Les appareils de protection respiratoire, choix et utilisation, ED 780. Institut national de recherche et de sécurité, 54 p. (2003)
39. NF EN 143: Appareils de protection respiratoire. Filtres à particules. Exigences, essais, marquage (2000)
40. L. Golanski, A. Guillot, F. Tardif: Are conventional protective devices such as fibrous filter media, cartridge for respirators, protective clothing and gloves also efficient for nanoaerosols? Nanosafe 2 (2008)
41. S. Rengasamy, R. Verbofsky, W.P. King, R.E. Shaffer: Nanoparticle penetration through NIOSH-approved N95 filtering-facepiece respirators. J. Internat. Soc. Respir. Protect. **24**, 49–59 (2007)
42. C. Mohlmann, J. Pelzer, M. Berges: Efficiency of respiratory filters against ultrafine particles. Third International Symposium on Nanotechnology. Occup. Environ. Health, Taipei (2007)
43. NF EN 149: Appareils de protection respiratoire. Demi-masques filtrants contre les particules. Exigences, essais, marquage (2001)
44. NF EN 12942: Appareils de protection respiratoire. Appareils filtrants à ventilation assistée avec masques complets, demi-masques ou quarts de masques. Exigences, essais, marquage (1998)
45. S.H. Huang, Y.H. Huang, C.W. Chen, C.P. Chang. Nanoparticle penetration through protective clothing materials, Third International Symposium on Nanotechnology. Occup. Environ. Health, Taipei (2007)
46. DIN EN ISO 13982: Protective clothing for use against solid particulates. Part 1: Performance requirements for chemical protective clothing providing protection to the full body against airborne solid particulates. Part 2: Test method of determination of inward leakage of aerosols of fine particles into suits (2005)

11

Occupational Exposure to Nanoparticles and Medical Safety

Patrick Brochard, Daniel Bloch, and Jean-Claude Pairon

The problem of occupational exposure to nanoparticles (NP) has raised many questions which remain unanswered today:

- When airborne NPs, either dissociated or more commonly in the form of aggregates, are inhaled by humans, will they produce a biological and/or tissular response where they are deposited, i.e., in the respiratory tract, or at some distance from the deposition area, i.e., an indirect effect secondary to the inflammatory response of the respiratory tract or a direct effect due to translocation of nanoparticles through the biological membranes?
- Do these responses predict a harmful effect on health in the short or long term?
- Are the biological responses large enough to be detected by simple, robust, and non-invasive tests?
- Can the results be interpreted on a personal level in the context of a consultation with the factory doctor, or only on a collective level in the context of an epidemiological investigation?

It is not easy to answer these questions at the present time, but it is important for factory doctors to understand what is at stake, and the limitations on what they can do about it. And the stakes are high, since the presence of a hazard established on the basis of experimental models implies that action must be taken in the work environment, and that the relevant workers must be kept fully informed on a regular basis. But the limitations are also great, since in the absence of a clearly specified risk for humans, the principle of precaution must be applied, despite the difficulties that entails, notably due to the social and economic context.

Under such conditions, the only option is to appeal to our present state of knowledge and take into account all the questions arising from the available data. We shall not discuss the analysis of experimental data, which can be found in Chaps. 13, 15, and 16. The data must be interpreted on a case-by-case basis for each type of nanoparticle tested, taking into account the precise physicochemical characteristics of each sample.

P. Houdy et al. (eds.), *Nanoethics and Nanotoxicology*,
DOI 10.1007/978-3-642-20177-6_11,

The aim here will nevertheless be to outline the points that should alert the factory doctor and guide him or her to the best decisions in the current state of uncertainty.

11.1 Should We Organise a Specific Occupational Safety Programme for Workers Coming into Contact with Nanoparticles?

The development of nanotechnology raises questions about occupational safety programmes, in particular for workers potentially exposed to NPs. While primary safety measures remain the backbone of these programmes, it is important to investigate the need for secondary measures based on medical surveillance of exposed workers involved in NP production, but also direct NP applications (incorporation in intermediate products such as composite materials) and indirect uses of NPs (implementation, tooling, or destruction of products containing materials with an NP content, e.g., in the automobile industry).

The idea of specifically increasing medical surveillance is based on the notions of hazard and risk. In 2006, a safety and precaution committee set up by the French Ministry of the Environment and Sustainable Development concluded that data available at the beginning of 2006 were sufficient to consider that NPs represent a hazard owing to the biological reactivity occasioned by their small size, independently of their chemical nature. However, it was impossible to assess the risk for humans due to the lack of published data. On the other hand, it seemed necessary at the time to apply the precautionary principle, in particular with regard to primary safety measures [1]. These conclusions have not been called into question by any more recent data published in the field of NPs or nanomaterials.

Application of the precautionary principle begins with primary safety measures, i.e., the implementation of collective and personal safety measures to control worker exposure. At the most, it may be necessary to apply the principle of substitution for the most strongly suspected NPs. Even in the absence of occupational exposure limits, the means must be developed to monitor potential exposure levels and the impact of collective safety measures. Monitoring personal exposure involves specifying in medical files whether the person has been working in potentially contaminated areas, but also the physicochemical characteristics of the NPs, the nature of the process, and, whenever possible, quantitative data (ambient measurements and personal measurements). This kind of exposure monitoring does not raise major difficulties in NP production units. On the other hand, it requires the tracking of nanoparticles in materials and in articles containing those materials, since the NPs may be emitted during tooling activities or when processing waste. This in turn requires mention of the NPs in documents such as safety data sheets drawn up by manufacturers.

The principles of these technical safety strategies and their practical application in the workplace are discussed in Chap. 10.

11.2 What Should Be the Basis for Organising Specific Medical Monitoring?

11.2.1 What Experimental Toxicological Data Should Be Considered?

The main biological response is the induction of intracellular oxidative stress which, when antioxidant defences are overcome, results in a cytotoxic response (apoptosis and necrosis), and in certain cases, DNA modifications similar to those observed for known genotoxic particles. These intracellular events are also associated with a local or systemic proinflammatory response. Finally, it seems likely that the small size of NPs facilitates their translocation through biological membranes such as the alveolar-capillary barrier, the intestinal mucous membrane, and the epidermis, and that it can contribute to inducing systemic effects [2]. All these points are further developed in later chapters, and in particular in Chaps. 12 and 13.

Manufactured NPs are already very diverse and the prospects for future developments are huge. However, the available toxicological data concerns only a small selection of NPs (see Chaps. 15 and 16). For this reason factory doctors must be able to handle these uncertain situations and adapt their practices by exploiting the predictive power of certain effects deduced from the available studies [3–6]. In vitro experimental models and in vivo studies on animals help to identify the characteristics of NPs that are related to the biological response:

- *Size.* This conditions the way the NP is internalised by cells and its translocation through biological membranes, but it also explains the high surface reactivity of NPs.
- *Shape.* This can cause a fibre effect, as in Stanton's model [7], which showed that long thin particles (diameter $< 0.25\,\mu m$ and length $> 8\,\mu m$) were the most reactive, regardless of their chemical composition.
- *Surface reactivity.* This depends on the chemical composition of the underlying NP and functionalisation where appropriate.
- *Biopersistence.* This determines the lifespan of the NP in tissues, and hence the possibility of a long-lasting inflammatory reaction at sites where NPs are retained.

Tissue responses described in in vivo studies concern above all the respiratory systems (inflammatory reaction and pulmonary fibrosis), but also the cardiovascular system (activation of coagulation factors, inflammatory response of vascular walls, changes in heart rate) and the central nervous system (inflammatory response). Other systemic effects remain poorly documented, in particular, the renal response and effects on reproduction.

11.2.2 What Human Data Should Be Considered?

Clinical Research

There is still very little data for humans. Some studies have investigated the short term response observed after controlled exposure of healthy or sick subjects to NPs (mainly carbon black NPs). Table 11.1 collects the main results of published studies.

Clinical and Epidemiological Data

There have not yet been any published epidemiologial studies of occupational exposure to manufactured NPs. Two examples of industrial sectors concerned with manufactured NPs can nevertheless be discussed here: the manufacture and use of carbon black and titanium dioxide. The results must be interpreted with caution as far as NPs are concerned. The exposure situations are complex owing to the fact that the NPs are often associated with larger particles and other airborne contaminants.

Carbon Black

Several epidemiological studies have been done in sectors producing or using carbon black, in particular to assess carcinogenic effects. In 1996, carbon blacks were classified by the International Agency for Research on Cancer (IARC) in category 2B (possibly human carcinogenic on the basis of inadequate epidemiological data for humans and adequate experimental data for animals). The assessment was reviewed in 2006 in the light of more recent epidemiological studies, and the classification was confirmed [23].

Morbidity Studies [24–26]. Several cross-sectional studies in the UK have carried out longitudinal analyses of the respiratory function, chest radiographs, and respiratory symptoms on cohorts of workers in the European carbon black industry.

Generally speaking, these studies suggest that exposure to carbon black can affect respiratory tests of the obstructive syndrome type [reduction of the maximal expiratory volume second (MEVS) or the MEF 25–75], and also symptoms associating coughing and bronchial hypersecretion which are related rather with recent or current exposure than with cumulative exposure levels. On the other hand, cumulative exposure would appear to be responsible in a few cases for the appearance of small opaque regions on the chest radiograph. Note that, in these studies, exposure assessments did not specify the level of exposure to NPs, since the measurements only generally took into account the inhalable fraction, or at best the alveolar fraction.

Table 11.1. Studies on healthy or sick volunteers exposed to nanoparticles. COPD chronic obstructive pulmonary disease, CB carbon black, MEF 25–75 mean expiratory flow rate between 25 and 75% of vital capacity, CODC diffusion capacity for CO, ECG electrocardiogram, FRE functional respiratory exploration, DF deposited fraction, BCF blood count formula, UF ultrafine

Nanoparticle	Subjects	Parameter studied	Results	Reference
UF CB (25 nm) 10 μg/m^3	Healthy, resting	DF	DF = 0.66	Frampton 2001 [8]
UF CB tagged with Tc99 (100 nm)	Healthy	Biodistribution	Fast translocation into circulation and urine (Tc dissociates?)	Nemmar et al. 2002 [9]
UF CB (25 nm) 10 μg/m^3	Healthy, rest/exercise	DF	DF (rest) = 0.66 DF (exercise) = 0.80	Daigle et al. 2003 [10]
UF CB 10–50 μg/m^3	Healthy and asthmatic, rest/exercise	FRE, exhaled NO, induced sputum	Change in CODC and MEF 25–75	Pietropaoli et al. 2004 [11]
UF CB (25 nm)	Asthmatic, rest/exercise	DF	DF (rest) = 0.76 for healthy subject, higher with exercise or asthma	Chalupa et al. 2004 [12]
UF CB 10 and 25 μg/m^3	Healthy and asthmatic, rest/exercise	FRE, ECG, BCF, exhaled NO, induced sputum	Discrete changes in BCF, ICAM-1, ECG. No change in inflammatory markers	Frampton et al. 2004 [13]
UF vs. fine ZnO 500 μg/m^3	Healthy, resting	Peripheral leukocytes, hemostasis, ECG	No difference between UF/F and pretest values	Beckett et al. 2005 [14]
UF CB 10–50 μg/m^3 (11×10^6 p/cm^3)	Healthy and asthmatic	BCF	Altered expression of adhesion molecules (higher retention of lung leukocytes)	Frampton 2005 [15]

(*Continued*)

Table 11.1. (Continued)

Nanoparticle	Subjects	Parameter studied	Results	Reference
UF CB 50 $\mu g/m^3$ and SO_2 200 ppb	Healthy and stable angina	ECG, inflammatory and coagulation markers	No effects due to particles alone	Routledge et al. 2006 [16]
UF CB (100 nm) tagged with Tc99	Healthy	Biodistribution	No extrapulmonary translocation at 70 hr	Wiebert et al. 2006 [17]
UF CB (4–20 nm), aggregates 100 nm, tagged with Tc99	Healthy	Biodistribution	No extrathoracic translocation at 6 hr	Mills et al. 2006 [18]
Hydrophobic particles (diethylexylsebacate) and hygroscopic particles (NaCl)	Healthy, rest/exercise	DF	Higher DF for hydrophobic particles and during exercise	Londahl et al. 2007 [19]
UF CB (100 nm) tagged with Tc99	Healthy non-smokers, smokers, and COPD	Biodistribution	No extrathoracic translocation at 48 hr	Moller et al. 2008 [20]
UF CB (25 nm) 50 $\mu g/m^3$	Healthy, rest/exercise	Vein plethysmography	Change in peripheral blood flow	Shah et al. 2008 [21]
UF CB (25 nm) 10 and 25 $\mu g/m^3$	Healthy, rest/exercise	ECG	Increased vagal tone (ST elevation, QT shortening)	Zareba et al. 2009 [22]

Study by Hodgson and Jones [27], and by Sorahan et al. [28]. This mortality study was carried out on about 1 500 male subjects working in carbon black production in the United Kingdom from 1947, with monitoring initially up until 1968. It revealed an excess of broncho-pulmonary cancers that was not statistically significant. Continued later [28] with medical monitoring up until 1974, it then revealed a statistically significant excess of lung cancers (SMR = 1.7, 95% confidence interval 1.3–2.2), but without finding a relation between the cancer excess and cumulative exposure. Furthermore, studies of mortality by non-malignant respiratory pathologies did not show significant differences relative to national reference levels. On this point, the authors conclude that the results have limited validity owing to the small sample and lack of exposure data, but that it is nevertheless justified to consider that non-cancerous respiratory pathologies, if there are any, are infrequent and not serious.

Other Carcinogenicity Studies [23]. This new assessment of epidemiological data is based essentially on seven studies:

- Two studies in the UK and German carbon black production sectors revealed a statistically significant excess of broncho-pulmonary cancers in a cohort of exposed workers, without determining the dose–response relationship.
- A large scale US cohort study on the work forces in fifteen production units did not find significant evidence for an increased risk of lung cancer. However, this study did not take into account exposure levels.
- A German study in the rubber industry showed a significant increase in pulmonary and gastric cancers. Carbon black exposure was rather poorly characterised and the increased risk disappeared after adjusting for exposure to other associated chemical agents, viz., nitrosamines, asbestos, and talc.
- In a cohort study in the US to investigate the effects of exposure to formaldehyde in ten factories, exposure to other substances, including carbon black, was characterised in order to identify the specific effects of the associated chemical substances. No significant excess of cancers could be put down to the carbon black exposure.
- An Italian study was done on a cohort of dockers who had handled bags of carbon black. Apart from cases of mesotheliomas and melanomas probably unrelated to carbon black exposure, a statistically significant excess of bladder cancers was identified, while other cancer locations, such as the lungs and stomach, exhibited no excess.
- A case-control study carried out in Montreal (Canada) identified a statistically significant excess of lung, kidney, and oesophagus cancers that was related to high exposure to carbon black, but no excess of cancers at other locations, e.g., stomach, colon, bladder, etc.

In conclusion, the IARC confirmed their classification of carbon black in category 2B in 2006.

Titanium Dioxide (TiO_2)

Several large scale epidemiological studies have been carried out in the sectors of TiO_2 production and use. This has been produced on an industrial scale for more than 60 years. Most of the production is pigmentary TiO_2, which is made in the form of aggregates of diameter 200–300 nm, or larger agglomerates, and is generally classified in the category of fine particles, rather than ultrafine particles. Nanometric TiO_2 comprises single particles of diameter around 20 nm, which may also occur in the ambient work atmosphere in the form of much larger agglomerates. A review of these studies was produced by the National Institute for Occupation, Safety and Health (NIOSH) in 2005 [29], and another by the IARC in 2006, but care must be taken in extrapolating them to the potential effects of ultrafine TiO_2.

Study by Chen and Fayerweather [30]. This study, carried out in the US on 1 576 people exposed to TiO_2 between 1956 and 1985, investigated the incidence of lung cancers and chronic respiratory diseases such as chronic bronchitis, pleural plaque, and pulmonary fibrosis. Morbidity data was collected over the period 1956–1985, and death records over the period 1935–1983. No significant increase in the risk of benign or cancerous respiratory pathologies was detected that could be attributed to exposure to TiO_2. Chest radiographs taken for a subpopulation of 398 workers did not identify pulmonary fibrosis, but a few cases of poorly defined nodules and pleural thickening, although not in significantly increased proportion compared with a control population. Given the disparate quality of the data, notably regarding exposure and morbidity/mortality records, the NIOSH considered that it was difficult to draw conclusions from this study.

Study by Fryzek et al. [31]. Fryzek et al. did a retrospective mortality study on a population of 4 241 workers in four TiO_2 factories in the US, employed for at least 6 months since 1 January 1960. Mortality was observed up until 31 December 1980. Exposures were estimated by atmospheric samples and uniform exposure groups were established by reconstituting careers according to the different posts occupied. Average TiO_2 concentrations decreased from 4.6 mg/m^3 between 1976 and 1980 to 1.1 mg/m^3 between 1996 and 2000. Certain posts such as packaging involved exposure levels 3 to 6 times higher than other posts.

Globally speaking, the mortality rate was lower than expected (general population) and paradoxically, even lower in the most exposed groups. No increase in mortality due to respiratory cancers was identified, including among the most exposed groups. The results of this study have been questioned, however, with criticism in particular of the statistical analyses [32].

Study by Boffetta et al. [33]. This case-control investigation was carried out in Montreal on 857 cases of lung cancer diagnosed between 1976 and 1985, comparing them with a control group comprising 533 healthy men and 533 men suffering from a non-lung cancer. Exposures to TiO_2 were estimated from answers to a questionnaire which inquired particularly about the subjects' occupations, and established exposure levels from parameters like the probability, duration, and frequency of exposure for different occupations. According to the authors, no correlation could be established between exposure to TiO_2 and an increased risk of lung cancer. However, an erroneous assessment of exposure levels inherent in the questionnaire method, combined with the small statistical sample, may limit the validity of these conclusions.

Study by Boffetta et al. [34]. This retrospective mortality study involved 15 017 workers (14 331 men) employed for at least one month in 11 production factories in 6 European countries between 1927 and 2001. Exposure was estimated from the professional careers and the results of ambient air measurements made since the 1990s. The observed mortality rates were compared with national average values. While in some countries an excess of deaths due to lung cancer was observed, this excess could be explained by a higher national average for this type of cancer. According to the authors, the data do not suggest that a carcinogenic effect can be attributed to TiO_2 exposure.

To conclude, in view of the inadequate epidemiological data and on the basis of experimental data on animals, the IARC classified TiO_2 in category 2B in 2006.

Case Studies

Recent monitoring of workers in a small Chinese printing house, exposed to nanometric particles (30 nm) from a paste containing polyacrylic esters, described 7 cases of diffuse pneumopathies associated with pleural and pericardial effusion [35]. Very little information is available about exposure, but the report suggests very high levels of contamination (spray process, no collective or personal safety system, poorly aired premises) over a period of 5–13 months. While the causal relationship between the ultrafine particles and respiratory damage remains hypothetical due to the presence of other airborne contaminants in the workplace, a transmission electron microscope analysis of pleural fluids, bronchoalveolar lavage fluids, and pleuropulmonary biopsies revealed particles of the same size as those observed in the incriminated powder (no chemical characterisation). These results attest to the importance of intracellular internalisation (pulmonary epithelial and mesothelial cells) and translocation into the pleural space.

Effects of Environmental Pollution

Epidemiological studies on populations exposed to unintentionally produced ultrafine particles (atmospheric pollution and general pollution, specific occupational environments involving welding and oxyacetylene cutting, for example) are more difficult to use owing to the highly complex nature of the aerosols emitted and the particular types of population studied. In addition, size distribution measurements are almost nonexistent and exposure is measured by mass concentration, at best on the basis of $PM_{2.5}$ in studies of atmospheric pollution [36] and, in the occupational environment, on the basis of the alveolar fraction [37]. And neither do studies done on humans under controlled experimental exposure conditions on volunteers (concentrated atmospheric particles, diesel particles) make it possible to account for the specific role of ultrafine particles. This is because of the highly complex composition of the aerosols, where the organic and metal compounds adsorbed at the surface of ultrafine particles probably play a key role [38–40].

11.3 Implementing Medical Surveillance

11.3.1 Available Biomarkers

A very important area of respiratory research concerns the use of biomarkers to monitor chronic respiratory disease. This is particularly relevant in the case of chronic inflammatory conditions such as asthma or chronic obstructive pulmonary disease (COPD). It can be useful to monitor certain markers reflecting the degree of inflammation of the airways when these markers are easy to observe by non-invasive techniques (exhaled air, exhaled air condensate, induced sputum), and when they provide a way of following the patient's evolution, either spontaneous or under therapeutic treatment [41]. The subjects are their own control and it is temporal variations that are interesting. Among these markers, the most widely used for these conditions are the exhaled fraction of nitric oxide (F_ENO) and in particular its alveolar fraction (C_{ALV}), leukotriene B4 (LTB4), the pH and temperature of the exhaled air, 8-isoprostane measurement, and cytological analysis of the induced sputum. It is essential to standardise collection techniques and recommendations have been drawn up [42–48].

It is even more interesting to use biomarkers to detect infraclinical alteration due to airborne contaminants of recognised pathogenic potential. Tobacco provides a good example. For example, Malerba et al. [49] have analysed non-invasive methods that study biomarkers associated with infraclinical stages of chronic bronchitis in smokers: F_ENO, LTB4, interleukine(IL)-10, myeloperoxidase, certain matrix metalloproteinases (MMP-9, MMP-12), vascular endothelial growth factor (VEGF), exhaled hydrogen peroxide, malondialdehyde (MDA), and neutrophils in sputum [50].

At the present time there are no formal recommendations concerning biomarkers that can be used for populations exposed to airborne contaminants in the workplace. The most interesting discussion of biomarkers and inhaled particles in humans concerns the risk of pneumoconiosis. A recent review of the literature [51] dealing with the well established risks for humans of inorganic particle inhalation (crystalline silica, coal mine dust, asbestos) assesses the usefulness of certain biomarkers with regard to the determinism of these pneumoconioses (mainly silicosis) due to their involvement in the mechanisms of the disease, their validation in humans or animal models, and their predictive value with regard to the development of the disease. Table 11.2 summarises the main markers appearing in publications on pneumoconiosis.

These biomarkers of respiratory effects nevertheless remain important tools in the field of research. The lack of standardisation and of normal values in the non-exposed population and in the population exempt from respiratory disease means that these tests cannot yet be used for the medical surveillance of people exposed to airborne contaminants even where the risk is established, and a fortiori when the risk is not established.

The systemic effects of nanoparticles are also worrying due to the recognition of the cardiovascular consequences of pulmonary inflammatory disease, related to the diffusion of inflammatory mediators or the systemic translocation of airborne contaminants. It is now established that subjects suffering from chronic obstructive pulmonary disease are more likely to develop cardiovascular disease [52] in relation with the level of inflammation [53]. Cardiovascular effects due to ultrafine particles coming from atmospheric pollution are now well established, both in epidemiological studies [20, 57] and in studies on humans exposed to controlled aerosols of atmospheric particles or diesel particles [39, 40, 55–60].

There is still very little data concerning measurement of the systemic effects of NPs in humans. The only available data concern the studies on volunteers already cited (see Table 11.1). Only a very short term response has been recorded. The first studies by Rochester and coworkers [13, 61], with healthy and asthmatic volunteers exposed at rest or exercising to an aerosol of ultrafine carbon particles (25 nm, 10–50 $\mu g/m^3$), did not identify any changes in a range of blood markers for inflammation (IL-6, soluble fractions of selectin E, L, or P, of the adhesion molcule sICAM-1, of the vascular cell adhesion molecule sVCAM-1, of the CD40 ligand sCD40L) and coagulation (factor VII, fibrinogen, Willebrand factor), but they observed modifications in the cell markers of the peripheral blood cells (monocytes, lymphocytes, and eosinophiles). The absence of any response by peripheral inflammatory and coagulation markers was also mentioned in an analogous study on healthy volunteers exposed to ultrafine carbon particles (50 $\mu g/m^3$) with or without coexposure to SO_2 [16].

Only in vivo studies have observed inflammatory and coagulation biomarkers after airway exposure [62]. The explanatory model put forward, in particular from the analysis of the pulmonary and systemic response to carbon nanotubes on the TaqMan array genes [63], invokes the relationship between the respiratory response (pulmonary expression of many genes coding for mediators of inflammation, oxidative stress, tissue remodelling, and thrombosis) and the systemic response measured by the expression of peripheral blood genes (soluble inflammatory and coagulation factors, activation of inflammatory blood cells). The authors suggest that these combined studies of peripheral blood genes and circulating soluble proteins could be used in epidemiological studies and for the medical surveillance of exposed populations [62].

Table 11.2. Pneumoconiosis biomarkers [51]

Type of biomarker	Description
Exposure	Measurement of crystalline silica in lung tissue and BAL fluid Measurement of asbestos fibres and/or asbestos bodies in sputum, lung tissue, and BAL fluid
Effect: early response	Oxidative stress markers: • generation of reactive oxygen species • NF-κB activation • serum antioxidant capacity • reduced glutathione and isoprostane in the serum • erythrocyte SOD and glutathione peroxidase activity • lymphocyte DNA alteration Plasma neopterin Protein 16 of Clara cells (CC 16)
Effect: late response	Lysosomal and cytoplasmic enzymes (β-N-acetyl glucosaminidase, β-glucuronidase, lactate dehydrogenase, alkaline phosphatase) Angiotensin conversion enzyme (ACE) Inflammatory and fibrosis markers: • cytokine tumour necrosis factor-α (TNF-α), IL-1 • chemokine IL-8, cytokine IL-6 • other macrophage chemoattractants (MIP) • other neutrophil chemoattractants (CNC) • other monocyte chemoattractants (MCP) • other circulating leukocyte chemoattractants (ICAM) • anti-inflammatory cytokine (IL-10) • lymphokine interferon-γ (IFN-γ), IL-4 • growth factors (PDGF, IGF-1, bFGF, EGF, TGF) • fibronectin • carbohydrate antigen 19-9 (CA 19-9) • elastin • collagen synthesis enzymes (PIIIP) • collagen degradation enzymes (MMP) • acute phase C reactive protein (CRP) • apoptosis markers (Fas)
Susceptibility	Polymorphism of IL-1 and IL-1 receptor antagonist Polymorphism of TNF-α Polymorphism of lymphotoxin-α (LTA) Polymorphism of MnSOD and glutathione S-transferase (GST) Polymorphism of the DNA repair enzyme OGG1

Regarding NP exposure biomarkers, there are still very few human data available. Recall the Chinese clinical study which demonstrates the usefulness of transmission electron microscopy observations of the particles present in the respiratory cells and pleural fluids. This information, although limited here by the absence of chemical characterisation of the particles, opens up interesting prospects, but rather for clinical studies on unhealthy subjects (use of bronchoalveolar lavage fluids for patients suspected of having a disease related to inhalation of these particles) than for the surveillance of exposed subjects (possible use of induced sputum).

Finally, the human data so far obtained suggests the possibility of monitoring exposure through the translocation of particles (or the soluble fractions making them up) into the urine. Such data are still controversial for humans [9, 63–71] and at the present time there is no urinary exposure biomarker.

11.4 Publications on Medical Surveillance

Despite the many questions still open, Seaton [72] has provided a clear statement of the problems in a discussion of the experience acquired in two areas that received wide media coverage, the asbestos tragedy and the lessons learned from research on atmospheric pollution. In particular, he describes the need to respond to certain situations which are already observed in the workplace (periods of accidential inhalation during dysfunction of the manufacturing process or when working on materials that may emit NPs, regular inhalation of small doses in poorly controlled processes). Finally, more recently, this author has pointed out the difficulties for the powers that be to redirect regulatory requirements in a situation characterised by such a high level of uncertainty [73].

The NP production industry [74] has also pointed out the problems involved in setting up a specific medical surveillance programme owing to the fact that there is almost no available information about the effects on humans, the non-specificity of effects observed on experimental models and their relevance for humans, and the total absence of information regarding the prevalence of the expected health effects. Some possible courses of action have been suggested, such as measuring heart rate variations, monitoring coagulation factors, and analysing certain extracellular inflammatory markers, although these are all tests whose sensitivity and predictive value remains to be correctly established. In addition, the authors view these tests as non-specific and difficult to implement on a personal level. For this reason, they insist on the need to optimise primary safety programmes, while they do not suggest any particular protocol for medical surveillance, but instead stress the need to develop exposure records which could prove extremely useful for setting up epidemiological monitoring.

More recently, the NIOSH has made recommendations (February 2009) on the basis of data in the literature and considering the main lines of occupational health surveillance programmes [75]. The report reviews the main features of medical surveillance programmes: medical check-up before beginning the activity, including a detailed occupational history, regular medical check-ups, medical examinations after accidental exposure, full information for the worker regarding risks and the kinds of symptoms to expect, and the maintenance of a permanent medical file. The difficulty still lies in determining the kind of screening to be carried out during this surveillance. It must satisfy certain criteria regarding such tests [76].

One specific point has been raised about NPs: when the chemical composition of the NPs involves a chemical element that is already subject to some specific regulations, these should still be applied. There again, the NIOSH stresses the preeminence of primary safety programmes, which must also take into account the risks specific to NPs (explosion, fire). Finally, the US agency concludes as to the inadequacy of scientific arguments to justify a specific medical programme. It also points out that published toxicological data only concerns a small proportion of currently manufactured NPs, and that an understanding of the relevant toxicity mechanisms is not sufficient to make reliable extrapolations on the sole basis of predictive factors such as size, shape, biopersistence, intrinsic surface reactivity, and surface reactivity after functionalisation, etc. Finally, the three point recommendations of the NIOSH are based above all on controlling exposure in the workplace, pursuing toxicological research to test effect biomarkers and clinical screening, and setting up prospective epidemiological studies associated with exposure records [77]. This proposal has also been put forward by the NP working group set up by the *Institut de recherche en santé publique* (IReSP) in France.

11.5 What Is on Offer in France?

There are three situations in which medical surveillance of workers exposed to manufactured nanoparticles needs to be discussed.

11.5.1 Surveillance Protocols for Cross-Sectional and Cohort Epidemiological Studies

What kind of surveillance should be introduced in a cross-sectional epidemiological study (exposure/no exposure) or in a cohort study (longitudinal monitoring)? The advantage here is to have large enough groups of subjects to study the distribution of biological, clinical, or functional indicators between exposure groups, assuming that the confounding variables of the occupational environment (e.g., other harmful effects associated with the work station) or outside work (e.g., smoking, past medical record, etc.) can be correctly accounted for.

The chosen effect indicators will depend on the hypothesis under investigation and the distribution of values expected in the non-exposed population.

The methods for collecting these indicators, e.g., questionnaires, biological samples, functional examinations, imaging, must respect the usual criteria for studies carried out on a general population (non-invasive examinations, easy to implement in the workplace).

The IReSP is currently running a working group on these issues, and a protocol for epidemiological surveillance should be drawn up by the *Institut de veille sanitaire* (InVS) in France. The aim is to be able to monitor any medium or long term effects of occupational exposure to NPs on general health. Furthermore, it aims to encourage more detailed studies to investigate specific research hypotheses. The relevant nanomaterials initially identified are carbon nanotubes, owing to the fact that they are similar in shape to asbestos fibres, but also carbon black, titanium dioxide, and amorphous silicas, produced in large amounts in France.

11.5.2 Surveillance Protocols for Clinical Research Studies Screening the Effects of Controlled Exposure in the Laboratory

The advantage here is to use the subject as his/her own control (comparison before and after exposure), for small samples of the population and with perfectly characterised exposure. Once again, the choice of parameters studied depends on the hypotheses to be tested and the usual clinical research criteria.

11.5.3 Surveillance Protocols in the Workplace

A distinction should be made between the different circumstances for setting up surveillance.

Medical Check-up When Starting a New Job Involving Nanoparticles

The aims are twofold:

1. To identify medical contraindications for taking up the given job or task. In the case of NPs, the medical examination must examine any possible contraindications with regard to wearing a personal respiratory protection system. There is still no consensus over medical aptitude restrictions related to the risk of aggravating a preexisting pathology (respiratory failure, coronary heart disease). It is the job of the factory doctor to determine this aptitude depending on the safety measures effective at the work station. A specific problem concerns pregnant women. There are no human risk assessment data, but some experimental results establish transplacental transfer of NPs [78]. Under these conditions, it would make sense to strengthen safety measures at the work station and avoid the person taking up this post if there should be any problems controlling exposure.

2. Inform the employee about hazards and risks involved in the work station and explain the available means of protection (collective and personal safety measures). This information should take into account current uncertainties regarding the hazards due to the biological reactivity of NPs in experimental models and the almost total absence of human risk assessment data. However, the safety rules are clearly established and discussed further in Chap. 10.

Medical Check-up Following an Incident Involving a Single Accidental Exposure

These medical check-ups are not compulsory, but they can be arranged with the agreement of the employee to examine any consequences of the incident on simple clinical parameters (functional symptoms, clinical signs, respiratory function parameters). At the present time no diagnostic protocol has been drawn up for this type of incident.

A variant on this medical check-up can be proposed within the framework of a before and after study. These are traditional clinical studies at the beginning and end of a shift, recording parameters before the work begins (Monday morning) and at the end of the shift (the same day or at the end of the week). These studies satisfy the same principles as clinical studies in the laboratory, except that exposures are not controlled and the choice of parameters to be studied must take into account the problems of measurement in the workplace.

Regular Medical Check-ups

This kind of periodic check-up is compulsory, with stipulated frequency according to regulations. It should record all events occurring since the previous check-up. In the particular case of employees exposed to NPs, no special protocol is laid down. Once again, the idea is to identify any modification in the state of health that may be related to the work station and check for the appearance of any contraindication for pursuing the given activity as a result of a modification in the state of health, whatever the cause.

Medical Check-up upon Return to Work after Sick Leave

This compulsory check-up is also laid down by regulations and must be implemented after sick leave resulting from a work accident or occupational illness, and after any sick leave lasting more than 21 days. The idea is on the one hand to determine any link between work and the health event, and on the other to see whether there are any consequences that can be considered as a new contraindication (or medical aptitude restriction) to resuming work. Once again, the problem of respiratory and cardiovascular diseases arises. The factory doctor must judge each case on its own merits, referring above all to the exposure assessment at the given work station and to the possible evolution of the health problem that resulted in sick leave.

Medical Check-up Requested by Employee or Employee's Doctors

Finally, the employee can at any time request a consultation with a factory doctor. This kind of check-up focuses on the specific questions raised directly by the employee or via his/her GP. The content of the medical should therefore be centered on the questions raised. Finally, the factory doctor may also request to see the employee if he/she considers that medical surveillance should be increased due to some specific event, e.g., short but repeated periods of sick leave, information and if necessary adaptation of the work station at the beginning of a pregnancy.

11.6 Conclusion

At the present time there are no consensual recommendations regarding the type of medical surveillance that must be carried out by the factory doctor as a result of some occupational activity that may expose an employee to NPs. However, the factory doctor must adapt his/her approach to suit each situation involving a medical check-up. This theme needs to be considered on a national level in order to put forward recommendations. Whatever the conclusions, it is important to remember that primary safety measures must remain a priority of the safety programme regarding the use of NPs in the workplace.

For NPs as for many other potential hazards, the workplace is an important environment for studying the effects of exposure, simply because exposure levels are generally higher than for the general population. The identification of any effects observed in the workplace and conclusions drawn from the surveillance programmes discussed here could also usefully be extended to other, non-occupational situations such as domestic exposure, DIY activities, and so on.

Appendix: Table of Acronyms

BAL	Bronchoalveolar lavage
bFGF	Basic fibroblast growth factor
CNC	Cytokine induced neutrophil chemoattractant
EGF	Epidermal growth factor
IARC	International Agency for Research on Cancer
ICAM	Intracellular adhesion molecule
IGF-1	Insulin-like growth factor
InVs	Institut de veille sanitaire, France
IReSP	Institut de recherche en santé publique, France
MCP	Monocyte chemoattractant protein
MIP	Macrophage inflammatory protein

MMP	Matrix metalloproteinase
NF-κB	Nuclear factor κB
NIOSH	National Institute for Occupational Safety and Health
PDGF	Platelet-derived growth factor
PIIIP	Procollagen III peptide
SMR	Standardized mortality ratio
SOD	Superoxide dismutase
TGF	Transforming growth factor

References

1. Comité de la prévention et de la précaution (CPP): Nanotechnologies, nanoparticules: quels dangers, quels risques? Ministère de l'Ecologie et du Développement durable, 1–64, www.ecologie.gouv.fr (2006)
2. W.G. Kreyling, M. Semmler-Behnke, J. Seitz, W. Scymczak, A. Wenk, P. Mayer, S. Takenaka, G. Oberdorster: Size dependence of the translocation of inhaled iridium and carbon nanoparticle aggregates from the lung of rats to the blood and secondary target organs. Inhal. Toxicol. **21** (S1), 55–60 (2009)
3. T. Xia, N. Li, A.E. Nel: Potential health impact of nanoparticles. Annu. Rev. Public Health **30**, 137–150 (2009)
4. A. Nel, T. Xia, L. Mädler, N. Li: Toxic potential of materials at the nano level. Science **311**, 622–627 (2006)
5. K. Donaldson, P.J. Borm, G. Oberdorster, K.E. Pinkerton, V. Stone, C.L. Tran: Concordance between in vitro and in vivo dosimetry in the proinflammatory effects of low-toxicity, low-solubility particles: The key role of the proximal alveolar region. Inhal. Toxicol. **20**, 53–62 (2008)
6. G. Oberdorster, E. Oberdorster, J. Oberdorster: Nanotoxicology: An emerging discipline evolving from studies of ultrafine particles. Environ. Health Perspect. **113**, 823–839 (2005)
7. M.F. Stanton, C. Wrench: Mechanisms of mesothelioma induction with asbestos and fibrous glass. J. Natl. Cancer Inst. **48**, 797–821 (1972)
8. M.W. Frampton: Systemic and cardiovascular effects of airway injury and inflammation: Ultrafine particle exposure in humans. Environ. Health Perspect. **109** (Suppl. 4), 529–532 (2001)
9. A. Nemmar, P.H. Hoet, B. Vanquickenborne, D. Dinsdale, M. Thomeer, M.F. Hoylaerts, H. Vanbilloen, L. Mortelmans, B. Nemery: Passage of inhaled particles into the blood circulation in humans. Circulation **105**, 411–414 (2002)
10. C.C. Daigle, D.C. Chalupa, F.R. Gibb, P.E. Morrow, G. Oberdorster, M.J. Utell, M.W. Frampton: Ultrafine particle deposition in humans during rest and exercise. Inhal. Toxicol. **15**, 539–552 (2003)
11. A.P. Pietropaoli, M.W. Frampton, R.W. Hyde, P.E. Morrow, G. Oberdorster, C. Cox, D.M. Speers, L.M. Frasier, D.C. Chalupa, L.S. Huang, M.J. Utell: Pulmonary function, diffusing capacity, and inflammation in healthy and asthmatic subjects exposed to ultrafine particles. Inhal. Toxicol. **16** (Suppl. 1), 59–72 (2004)

12. D.C. Chalupa, P.E. Morrow, G. Oberdorster, M.J. Utell, M.W. Frampton: Ultrafine particle deposition in subjects with asthma. Environ. Health Perspect. **112**, 879–882 (2004)
13. M.W. Frampton, M.J. Utell, W. Zareba, G. Oberdorster, C. Cox, L.S. Huang, P.E. Morrow, F.E. Lee, D. Chalupa, L.M. Frasier, D.M. Speers, J. Stewart: Effects of exposure to ultrafine carbon particles in healthy subjects and subjects with asthma. Res. Respir. Health Eff. Inst. **126**, 1–47 (2004)
14. W.S. Beckett, D.F. Chalupa, A. Pauly-Brown, D.M. Speers, J.C. Stewart, M.W. Frampton, M.J. Utell, L.S. Huang, C. Cox, W. Zareba, G. Oberdorster: Comparing inhaled ultrafine versus fine zinc oxide particles in healthy adults: A human inhalation study. Am. J. Respir. Crit. Care Med. **171**, 1129–1135 (2005)
15. M.W. Frampton: Inflammation and airborne particles. Clin. Occup. Environ. Med. **5**, 797–815 (2006)
16. H.C. Routledge, S. Manney, R.M. Harrison, J.G. Ayres, J.N. Townend: Effect of inhaled sulphur dioxide and carbon particles on heart rate variability and markers of inflammation and coagulation in human subjects. Heart **92**, 220–227 (2006)
17. P. Wiebert, A. Sanchez-Crespo, R. Falk, K. Philipson, A. Lundin, S. Larsson, W. Moller, W.G. Kreyling, M. Svartengren: No significant translocation of inhaled 35-nm carbon particles to the circulation in humans. Inhal. Toxicol. **18**, 741–747 (2006)
18. N.L. Mills, N. Amin, S.D. Robinson, A. Anand, J. Davies, D. Patel, J.M. de la Fuente, F.R. Cassee, N.A. Boon, W. Macnee, A.M. Millar, K. Donaldson, D.E. Newby: Do inhaled carbon nanoparticles translocate directly into the circulation in humans? Am. J. Respir. Crit. Care Med. **173**, 426–431 (2006)
19. J. Londahl, A. Massling, J. Pagels, E. Swietlicki, E. Vaclavik, S. Loft: Size-resolved respiratory-tract deposition of fine and ultrafine hydrophobic and hygroscopic aerosol particles during rest and exercise. Inhal. Toxicol. **19**, 109–116 (2007)
20. W. Moller, K. Felten, K. Sommerer, G. Scheuch, G. Meyer, P. Meyer, K. Haussinger, W.G. Kreyling: Deposition, retention, and translocation of ultrafine particles from the central airways and lung periphery. Am. J. Respir. Crit. Care Med. **177**, 426–432 (2008)
21. A.P. Shah, A.P. Pietropaoli, L.M. Frasier, D.M. Speers, D.C. Chalupa, J.M. Delehanty, L.S. Huang, M.J. Utell, M.W. Frampton: Effect of inhaled carbon ultrafine particles on reactive hyperemia in healthy human subjects. Environ. Health Perspect. **116**, 375–380 (2008)
22. W. Zareba, J.P. Couderc, G. Oberdorster, D. Chalupa, C. Cox, L.S. Huang, A. Peters, M.J. Utell, M.W. Frampton: ECG parameters and exposure to carbon ultrafine particles in young healthy subjects. Inhal. Toxicol. **21**, 223–233 (2009)
23. IARC 2006: http://monographs.iarc.fr/ENG/Meetings/93-carbonblack.pdf, http://monographs.iarc.fr/ENG/Meetings/93-titaniumdioxide.pdf
24. K. Gardiner, M. van Tongeren, M. Harrington: Respiratory health effects from exposure to carbon black: Results of the phase 2 and 3 cross-sectional studies in the European carbon black manufacturing industry. Occup. Environ. Med. **58**, 496–503 (2001)
25. M.J. Van Tongeren, K. Gardiner, C.E. Rossiter, J. Beach, P. Harber, M.J. Harrington: Longitudinal analyses of chest radiographs from the European Carbon Black Respiratory Morbidity Study. Eur. Respir. J. **20**, 417–425 (2002)

26. K. Gardiner, N.W. Trethowan, J.M. Harrington, C.E. Rossiter, I.A. Calvert: Respiratory health effects of carbon black: A survey of European carbon black workers. Br. J. Ind. Med. **50**, 1082–1096 (1993)
27. J.T. Hodgson, R.D. Jones: A mortality study of carbon black workers employed at five United Kingdom factories between 1947 and 1980. Arch. Environ. Health **40**, 261–268 (1985)
28. T. Sorahan, L. Hamilton, M. van Tongeren, K. Gardiner, J.M. Harrington: A cohort mortality study of UK carbon black workers, 1951–1996. Am. J. Ind. Med. **39**, 158–170 (2001)
29. NIOSH Evaluation of Health Hazard and Recommendations for Occupational Exposure to Titanium Dioxide 2005: www.cdc.gov/niosh/review/public/TIo2/pdfs/TIO2Draft.pdf
30. J.L. Chen, W.E. Fayerweather: Epidemiologic study of workers exposed to titanium dioxide. J. Occup. Med. **30**, 937–942 (1988)
31. J.P. Fryzek, B. Chadda, D. Marano, K. White, S. Schweitzer, J.K. McLaughlin, W.J. Blot: A cohort mortality study among titanium dioxide manufacturing workers in the United States. J. Occup. Environ. Med./Am. Coll. Occup. Environ. Med. **45**, 400–409 (2003)
32. J.J. Beaumont, M.S. Sandy, C.D. Sherman: Titanium dioxide and lung cancer. J. Occup. Environ. Med. **46**, 759 (2004); erratum: 1189
33. P. Boffetta, V. Gaborieau, L. Nadon, M.F. Parent, E. Weiderpass, J. Siemiatycki: Exposure to titanium dioxide and risk of lung cancer in a population-based study from Montreal. Scand. J. Work. Environ. Health. **27**, 227–232 (2001)
34. P. Boffetta, A. Soutar, J.W. Cherrie, F. Granath, A. Andersen, A. Anttila, M. Blettner, V. Gaborieau, S.J. Klug, S. Langard, D. Luce, F. Merletti, B. Miller, D. Mirabelli, E. Pukkala, H.O. Adami, E. Weiderpass: Mortality among workers employed in the titanium dioxide production industry in Europe. Cancer Causes Control **15**, 697–706 (2004)
35. Y. Song, X. Li, X. Du: Exposure to nanoparticles is related to pleural effusion, pulmonary fibrosis and granuloma. Eur. Respir. J. **34**, 559–567 (2009)
36. C.A. Pope, R.T. Burnett, G.D. Thurston, M.J. Thun, E.E. Calle, D. Krewski, J.J. Godleski: Cardiovascular mortality and long-term exposure to particulate air pollution: Epidemiological evidence of general pathophysiological pathways of disease. Circulation **109**, 71–77 (2004)
37. J.M. Antonini, A.B. Lewis, J.R. Roberts, D.A. Whaley: Pulmonary effects of welding fumes: Review of worker and experimental animal studies. Am. J. Ind. Med. **43**, 350–360 (2003)
38. A. Peretz, J.H. Sullivan, D.F. Leotta, C.A. Trenga, F.N. Sands, J. Allen, C. Carlsten, C.W. Wilkinson, E.A. Gill, J.D. Kaufman: Diesel exhaust inhalation elicits acute vasoconstriction in vivo. Environ. Health Perspect. **116**, 937–942 (2008)
39. E.V. Brauner, L. Forchhammer, P. Moller, J. Simonsen, M. Glasius, P. Wahlin, O. Raaschou-Nielsen, S. Loft: Exposure to ultrafine particles from ambient air and oxidative stress-induced DNA damage. Environ. Health Perspect. **115**, 1177–1182 (2007)
40. J.M. Samet, A. Rappold, D. Graff, W.E. Cascio, J.H. Berntsen, Y.C. Huang, M. Herbst, M. Bassett, T. Montilla, M.J. Hazucha, P.A. Bromberg, R.B. Devlin: Concentrated ambient ultrafine particle exposure induces cardiac changes in young healthy volunteers. Am. J. Respir. Crit. Care Med. **179**, 1034–1042 (2009)

41. S.A. Kharitonov, P.J. Barnes: Exhaled biomarkers. Chest **130**, 1541–1546 (2006)
42. B. Balbi, P. Pignatti, M. Corradi, P. Baiardi, L. Bianchi, G. Brunetti, A. Radaeli, G. Moscato, A. Mutti, A. Spanevello, M. Malerba: Bronchoalveolar lavage, sputum and exhaled clinically relevant inflammatory markers: Values in healthy adults. Eur. Respir. J. **30**, 769–781 (2007)
43. M.C. Levesque, D.W. Hauswirth, S. Mervin-Blake, C.A. Fernandez, K.B. Patch, K.M. Alexander, S. Allgood, P.D. McNair, A.S. Allen, J.S. Sundy: Determinants of exhaled nitric oxide levels in healthy, nonsmoking African American adults. J. Allergy Clin. Immunol. **121**, 396–402, e3 (2008)
44. I. Horvath, J. Hunt, P.J. Barnes, K. Alving, A. Antczak, E. Baraldi, G. Becher, W.J. van Beurden, M. Corradi, R. Dekhuijzen, R.A. Dweik, T. Dwyer, R. Effros, S. Erzurum, B. Gaston, C. Gessner, A. Greening, L.P. Ho, J. Hohlfeld, Q. Jobsis, D. Laskowski, S. Loukides, D. Marlin, P. Montuschi, A.C. Olin, A.E. Redington, P. Reinhold, E.L. van Rensen, I. Rubinstein, P. Silkoff, K. Toren, G. Vass, C. Vogelberg, H. Wirtz: Exhaled breath condensate: Methodological recommendations and unresolved questions. Eur. Respir. J. **26**, 523–548 (2005)
45. P.P. Rosias, C.M. Robroeks, A. Kester, G. Jden Hartog, W.K. Wodzig, G.T. Rijkers, L.J.V. Zimmermann, C.P. van Schayck, Q. Jobsis, E. Dompeling: Biomarker reproducibility in exhaled breath condensate collected with different condensers. Eur. Respir. J. **31**, 934–942 (2008)
46. H. Knobloch, G. Becher, M. Decker, P. Reinhold: Evaluation of H_2O_2 and pH in exhaled breath condensate samples: Methodical and physiological aspects. Biomarkers **13**, 319–341 (2008)
47. D.W. Hauswirth, J.S. Sundy, S. Mervin-Blake, C.A. Fernandez, K.B. Patch, K.M. Alexander, S. Allgood, P.D. McNair, M.C. Levesque: Normative values for exhaled breath condensate pH and its relationship to exhaled nitric oxide in healthy African Americans. J. Allergy Clin. Immunol. **122**, 101–106 (2008)
48. M.J. Cruz, S. Sanchez-Vidaurre, P.V. Romero, F. Morell, X. Munoz: Impact of age on pH, 8-isoprostane, and nitrogen oxides in exhaled breath condensate. Chest **135**, 462–467 (2009)
49. M. Malerba, B. Ragnoli, M. Corradi: Non-invasive methods to assess biomarkers of exposure and early stage of pulmonary disease in smoking subjects. Monaldi Arch. Chest Dis. **69**, 128–133 (2008)
50. K. Hildebrandt, R. Ruckerl, W. Koenig, A. Schneider, M. Pitz, J. Heinrich, V. Marder, M. Frampton, G. Oberdorster, H.E. Wichmann, A. Peters: Short-term effects of air pollution: A panel study of blood markers in patients with chronic pulmonary disease. Part. Fibre Toxicol. **6**, 25 (2009)
51. M. Gulumian, P.J. Borm, V. Vallyathan, V. Castranova, K. Donaldson, G. Nelson, J. Murray: Mechanistically identified suitable biomarkers of exposure, effect, and susceptibility for silicosis and coal-worker's pneumoconiosis: A comprehensive review. J. Toxicol. Environ. Health B Crit. Rev. **9**, 357–395 (2006)
52. D.M. Mannino, D. Thorn, A. Swensen, F. Holguin: Prevalence and outcomes of diabetes, hypertension and cardiovascular disease in COPD. Eur. Respir. J. **32**, 962–969 (2008)
53. F.L. Fimognari, S. Scarlata, M.E. Conte, R.A. Incalzi: Mechanisms of atherothrombosis in chronic obstructive pulmonary disease. Int. J. Chron. Obstruct. Pulmon. Dis. **3**, 89–96 (2008)
54. R.D. Brook, B. Franklin, W. Cascio, Y. Hong, G. Howard, M. Lipsett, R. Luepker, M. Mittleman, J. Samet, S.C. Smith, I. Tager: Air pollution and

cardiovascular disease: A statement for healthcare professionals from the Expert Panel on Population and Prevention Science of the American Heart Association. Circulation **109**, 2655–2671 (2004)
55. A. Peretz, E.C. Peck, T.K. Bammler, R.P. Beyer, J.H. Sullivan, C.A. Trenga, S. Srinouanprachnah, F.M. Farin, J.D. Kaufman: Diesel exhaust inhalation and assessment of peripheral blood mononuclear cell gene transcription effects: An exploratory study of healthy human volunteers. Inhal. Toxicol. **19**, 1107–1119 (2007)
56. A.J. Lucking, M. Lundback, N.L. Mills, D. Faratian, S.L. Barath, J. Pourazar, F.R. Cassee, K. Donaldson, N.A. Boon, J.J. Badimon, T. Sandstrom, A. Blomberg, D.E. Newby: Diesel exhaust inhalation increases thrombus formation in man. Eur. Heart J. **29**, 3043–3051 (2008)
57. H. Tornqvist, N.L. Mills, M. Gonzalez, M.R. Miller, S.D. Robinson, I.L. Megson, W. Macnee, K. Donaldson, S. Soderberg, D.E. Newby, T. Sandstrom, A. Blomberg: Persistent endothelial dysfunction in humans after diesel exhaust inhalation. Am. J. Respir. Crit. Care Med. **176**, 395–400 (2007)
58. N. Mills, N. Amin, S. Robinson, et al.: Do inhaled carbon nanoparticles translocate directly into the circulation in humans? Am. J. Respir. Crit. Care Med. **173**, 426–431 (2006)
59. N.L. Mills, S.D. Robinson, P.H. Fokkens, D.L. Leseman, M.R. Miller, D. Anderson, E.J. Freney, M.R. Heal, R.J. Donovan, A. Blomberg, T. Sandstrom, W. MacNee, N.A. Boon, K. Donaldson, D.E. Newby, F.R. Cassee: Exposure to concentrated ambient particles does not affect vascular function in patients with coronary heart disease. Environ. Health Perspect. **116**, 709–715 (2008)
60. A.J. Ghio, Y.C. Huang: Exposure to concentrated ambient particles (CAPs): A review. Inhal. Toxicol. **16**, 53–59 (2004)
61. M.W. Frampton, J.C. Stewart, G. Oberdorster, P.E. Morrow, D. Chalupa, A.P. Pietropaoli, L.M. Frasier, D.M. Speers, C. Cox, L.S. Huang, M.J. Utell: Inhalation of ultrafine particles alters blood leukocyte expression of adhesion molecules in humans. Environ. Health Perspect. **114**, 51–58 (2006)
62. P.P. Simeonova, A. Erdely: Engineered nanoparticle respiratory exposure and potential risks for cardiovascular toxicity: Predictive tests and biomarkers. Inhal. Toxicol. **21** (S1), 68–73 (2009)
63. A. Erdely, T. Hulderman, R. Salmen, A. Liston, P.C. Zeidler-Erdely, D. Schwegler-Berry, V. Castranova, S. Koyama, Y.A. Kim, M. Endo, P.P. Simeonova: Cross-talk between lung and systemic circulation during carbon nanotube respiratory exposure. Potential biomarkers. Nano Lett. **9**, 36–43 (2009)
64. M. Semmler, J. Seitz, F. Erbe, P. Mayer, J. Heyder, G. Oberdorster, W.G. Kreyling: Long-term clearance kinetics of inhaled ultrafine insoluble iridium particles from the rat lung, including transient translocation into secondary organs. Inhal. Toxicol. **16**, 453–459 (2004)
65. P. Wiebert, A. Sanchez-Crespo, J. Seitz, R. Falk, K. Philipson, W.G. Kreyling, W. Moller, K. Sommerer, S. Larsson, M. Svartengren: Negligible clearance of ultrafine particles retained in healthy and affected human lungs. Eur. Respir. J. **28**, 286–290 (2006)
66. Z. Chen, H. Chen, H. Meng, G. Xing, X. Gao, B. Sun, X. Shi, H. Yuan, C. Zhang, R. Liu, F. Zhao, Y. Zhao, X. Fang: Bio-distribution and metabolic paths of silica coated CdSeS quantum dots. Toxicol. Appl. Pharmacol. **230**, 364–371 (2008)

67. X. He, H. Nie, K. Wang, W. Tan, X. Wu, P. Zhang: In vivo study of biodistribution and urinary excretion of surface-modified silica nanoparticles. Anal. Chem. **80**, 9597–9603 (2008)
68. M.T. Zhu, W.Y. Feng, Y. Wang, B. Wang, M. Wang, H. Ouyang, Y.L. Zhao, Z.F. Chai: Particokinetics and extrapulmonary translocation of intratracheally instilled ferric oxide nanoparticles in rats and the potential health risk assessment. Toxicol. Sci. **107**, 342–351 (2009)
69. A.A. Burns, J. Vider, H. Ow, E. Herz, O. Penate-Medina, M. Baumgart, S.M. Larson, U. Wiesner, M. Bradbury: Fluorescent silica nanoparticles with efficient urinary excretion for nanomedicine. Nano Lett. **9**, 442–448 (2009)
70. W.S. Cho, M. Cho, S.R. Kim, M. Choi, J.Y. Lee, B.S. Han, S.N. Park, M.K. Yu, S. Jon, J. Jeong: Pulmonary toxicity and kinetic study of Cy5.5-conjugated superparamagnetic iron oxide nanoparticles by optical imaging. Toxicol. Appl. Pharmacol **239**, 106–115 (2009)
71. K. Sarlo, K.L. Blackburn, E.D. Clark, J. Grothaus, J. Chaney, S. Neu, J. Flood, D. Abbott, C. Bohne, K. Casey, C. Fryer, M. Kuhn: Tissue distribution of 20 nm, 100 nm and 1000 nm fluorescent polystyrene latex nanospheres following acute systemic or acute and repeat airway exposure in the rat. Toxicology **263**, 117–126 (2009)
72. A. Seaton: Nanotechnology and the occupational physician. Occup. Med. **56**, 312–316 (2006)
73. A. Seaton, L. Tran, R. Aitken, K. Donaldson: Nanoparticles, human health hazard and regulation. J. Roy. Soc. Interface **7** (Suppl. 1), S119–S129 (2010)
74. M. Nasterlack, A. Zober, C. Oberlinner: Considerations on occupational medical surveillance in employees handling nanoparticles. Int. Arch. Occup. Environ. Health **81**, 721–726 (2008)
75. NIOSH: Interim guidance for medical screening and hazard surveillance for workers potentially exposed to engineered nanoparticles. Current Intelligence Bulletin 60. National Institute for Occupational Safety and Health. Department of Health and Human Services, February 2009. Publication No. 2009-116 (2009)
76. W.E. Halperin, J. Ratcliffe, T.M. Frazier, L. Wilson, S.P. Becker, P.A. Schulte: Medical screening in the workplace: Proposed principles. J. Occup. Med. **28**, 547–552 (1986)
77. P.A. Schulte, M.K. Schubauer-Berigan, C. Mayweather, C.L. Geraci, R. Zumwalde, J.L. McKernan: Issues in the development of epidemiologic studies of workers exposed to engineered nanoparticles. J. Occup. Environ. Med. **51**, 323–335 (2009)
78. M. Saunders: Transplacental transport of nanomaterials. Wiley Interdiscip. Rev. Nanomed. Nanobiotechnol. **1**, 671–684 (2009)

Part II

Nanotoxicity: Experimental Toxicology of Nanoparticles and Their Impact on the Environment

Research on the environmental impacts of nanomaterials is still in its infancy.[1] Indeed, work on humans is more advanced, because there is an urgent need to understand the consequences for those in direct contact with nanoparticles.

Transfer into ecosystems, complex interactions with solutes and organic molecules present in high concentrations in soil and water, and toxicity studies on micro-organisms and multicelled organisms serving as bioindicators for pollution are examples of cross-disciplinary subjects requiring us to understand what happens to nanoparticles in aqueous media from the atomic scale (redox changes, dissolution, transport of pollutants to the surface) to that of the porous medium, before investigating bioavailability, which involves a tiny fraction of these nanoparticles.

Toxicity (cytotoxicity and genotoxicity) studies are then possible. One of the difficulties is that, in order to be credible, one must work on the trophic chains transferring over long periods very small quantities of nanoparticles resulting from the degradation of nanoengineered materials.

[1] Introduction by Jean-Yves Bottero, Director of Research at the CNRS and the CEREGE.

12

Surface Reactivity of Manufactured Nanoparticles

Mélanie Auffan, Jérôme Rose, Corinne Chanéac, Jean-Pierre Jolivet, Armand Masion, and Mark R. Wiesner, and Jean-Yves Bottero

Manufactured nanoparticles are usually defined to be any intentionally produced particles with:

1. at least one space dimension in the range 1–100 nm,
2. novel or enhanced properties compared with larger particles of the same chemical composition.

This definition is quoted and accepted by many institutions [1–4]. However, given the surge of interest in nanotechnology today, the prefix 'nano' is often used inappropriately and in situations where the two parts of this definition are not fulfilled. Very often, only (1) referring to the size is actually respected. However, we shall shown in this chapter that it is part (2) of the definition that catches the essence of what constitutes a nanoparticle.

A clear example is provided by gold particles. Gold is well known for its constancy, and in particular, its resistance to oxidation [5]. And yet gold nanoparticles smaller than 5 nm in diameter prove to be excellent catalysts [6,7]. There are many nanoscale effects of this kind, such as the fluorescence of quantum dots [8,9], the reduced melting temperatures of gold or tin nanoparticles [10–12], or the extraordinary hardness of carbon nanotubes [13]. This shows that our physical, chemical, and thermodynamic understanding of these materials on the macroscopic scale cannot simply be transferred to the nanometric scale.

However, not all nanoparticles in the range 1–100 nm will exhibit different properties to their larger counterparts. There is a critical size, considerably smaller than 100 nm, below which properties begin to change. We shall show in this chapter that nanoscale effects arise when the number of atoms at the surface of the nanoparticle is large enough to generate an excess surface energy [14] and hence modify their crystal structure [15–18].

The two parts of the above definition are essential when assessing the risks for humans and the environment that may arise from our nanotechnological activities. Owing to their small size (point 1), nanoparticles may be able to cross protective biological barriers, e.g., the spleen, kidneys, and

P. Houdy et al. (eds.), *Nanoethics and Nanotoxicology*,

DOI 10.1007/978-3-642-20177-6_12,

cell membranes, and hence reach parts of the organism that would be inaccessible to larger particles [19]. On the other hand, the unique properties of nanoparticles (point 2) may induce their own form of toxicity. In the interaction between nanoparticles and living organisms, an increased surface reactivity and modified physicochemical properties, e.g., adsorption/desorption, generation of reactive oxygen species, electronic exchange, may well modify biological responses. In this case, it is quite impossible simply to transfer our knowledge of the toxicity of macroscopic particles to the case of nanoparticles.

The aim in this chapter is to discuss the relationships between size-dependent structural changes and the surface reactivity of particles. We shall then be able to pick out those nanoparticles that will exhibit a different interface reactivity that might influence their ecotoxicity. We shall focus on metal nanoparticles and metal oxide nanoparticles which are the ones most widely used in research and development [20–24] and for which there is increasing concern over their ecotoxic impact [1, 19, 25–36].

12.1 Nanoscale Description of Surfaces

One particularity of nanoparticles is that they have a very large total surface area for a given mass, precisely because of their small size. Indeed, if we consider 1 cm^3 of particles with diameter 5 nm, the total surface area is 1 200 m^2, as compared with 10 cm^2 for particles with diameter 5 μm. But the surface is the contact zone between the particles and their environment, a zone commonly called an interface. This is a place where chemical exchanges (anions, cations) and electrochemical exchanges (electrons, protons) can take place, not to mention adsorption/desorption reactions, all of which involve species in solution, whether they be polluting cations or biological molecules. The surface thus plays a key role in the toxicity of nanoparticles.

12.1.1 Surface Atoms

When describing the surface of a nanoparticle, we must take into account the large number of atoms located at the surface as compared with the number of atoms in the core, provided that we only consider length scales smaller than around 20 nm [37]. These surface atoms occupy a special position, between solid and solution, and this underlies their reactivity. For example, consider metal oxide nanoparticles, so important today in the field of manufactured products (sunscreen creams, catalytic converters, photovoltaic cells, tyres, etc.) and particularly interesting for their high chemical stability and low solubility in aqueous media at biological pH values. The surface atoms are oxygen atoms belonging to the crystal planes bounding the particle. They have reduced coordination number owing to the fact that the periodicity has been broken. This mismatched atomic coordination perturbs the balance of charge at the surface in such a way that the surface oxygen atoms carry an unbalanced partial charge. This in turn gives the system a more or less basic character, depending on the equilibrium and protonation constants in solution, determined by

$$\mathrm{M}_n\text{–}\mathrm{O}^{(nv-2)} + \mathrm{H}^+ \longleftrightarrow \mathrm{M}_n\text{–}\mathrm{OH}^{(nv-1)} \qquad (K_{n,1}) ,$$
$$\mathrm{M}_n\text{–}\mathrm{OH}^{(nv-1)} + \mathrm{H}^+ \longleftrightarrow \mathrm{M}_n\text{–}\mathrm{OH}_2^{(nv)} \qquad (K_{n,2}) ,$$

where $K_{n,1}$ and $K_{n,2}$ are the protonation constants, n is the number of metal cations bound to the oxygen atom, and $v = z/N$ is the formal bond valence defined by Pauling [38], with z the formal charge of the cation and N the cation coordination.

These equilibria explain the origin of the surface charge, but they also reveal its full complexity, since the protonation constants of the surface oxygen atoms depend on the characteristics (i.e., size, formal charge) of the cations to which they are bound and also their type of coordination (i.e., terminal atoms, or μ_2 or μ_3 bridging atoms). There is no experimental method for measuring these constants, but they can be evaluated more or less accurately using various models, e.g., the multisite complexation model (MUSIC) [39]. In its most elaborate version MUSIC2, this model evaluates the electrostatic interactions involved in protonation of surface sites (metal–proton and oxygen–proton interactions), taking into account structural details (lengths of metal–oxygen bonds) and also the solvation of surface groups by hydrogen bonds [40]. Hence, given the crystal structure of a particle and the nature of its exposed crystal faces, a rather detailed description of the surface can be obtained on the atomic scale, from which the acidity can be estimated.

Figure 12.2 gives a description of the faces (001), (101), and (010) of boehmite (γ-AlOOH), a γ-alumina precursor that is very widely used in the petrol industry as catalyst substrate. The particle morphology is shown in Fig. 12.1. Boehmite has orthorhombic structure, and space group *Cmcm* with lattice parameters $a = 0.2868$ nm, $b = 1.2214$ nm, and $c = 0.3694$ nm. The structure is lamellar with sheets stacked along the b-axis.

The face (001) contains three types of surface group, mono-, di-, and tricoordinated, of the same density (5.71 groups/nm^2). The monocoordinated sites μ_1 are in aquo form $\mathrm{Al_2\text{-}OH_2}$, with a residual charge of $^+0.53$ below pH 9.28 ($\mathrm{pK}_{1.2}$) and in hydroxo form $\mathrm{Al_1\text{-}OH}$ with charge $^-0.47$ above pH 9.28. The value $\mathrm{pK}_{1.1}$ is very high, indicating that the oxo form of the μ_1 groups is not observed on the usual pH scale. The dicoordinated sites μ_2 are in the hydroxo form and are not charged, whatever the pH. They are chemically inert and

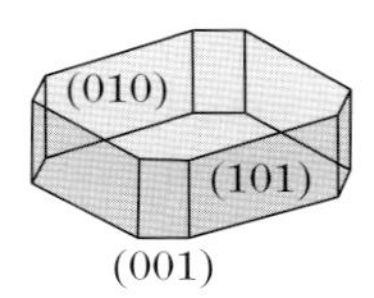

Fig. 12.1. Particle morphology of boehmite (γ-AlOOH)

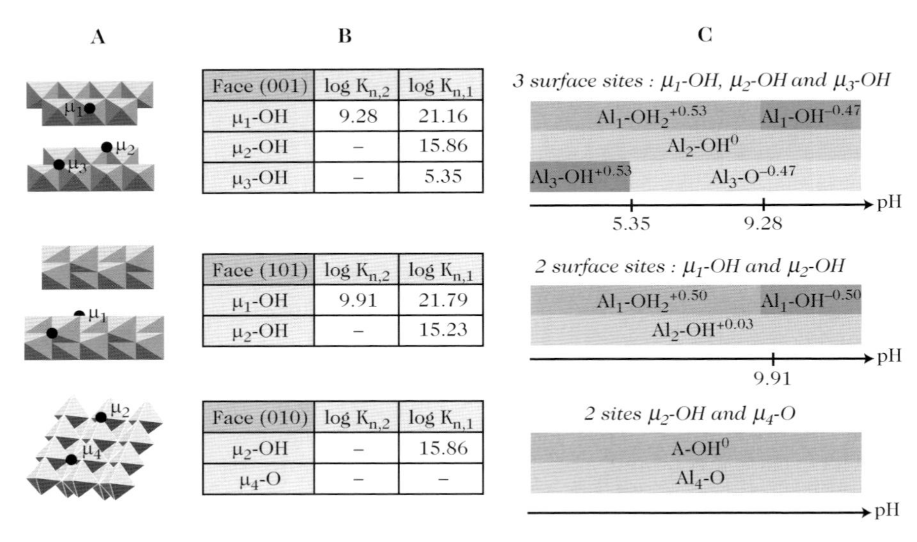

Face (001)	log $K_{n,2}$	log $K_{n,1}$
μ_1-OH	9.28	21.16
μ_2-OH	–	15.86
μ_3-OH	–	5.35

Face (101)	log $K_{n,2}$	log $K_{n,1}$
μ_1-OH	9.91	21.79
μ_2-OH	–	15.23

Face (010)	log $K_{n,2}$	log $K_{n,1}$
μ_2-OH	–	15.86
μ_4-O	–	–

Fig. 12.2. Structural characteristics of boehmite. (**A**) Description of surface sites on faces (001), (101), and (010). (**B**) Acidity constants for each site, as calculated by the MUSIC2 model. (**C**) Degree of protonation of sites at different pH values

do not therefore contribute to the charge on this face. The μ_3 groups have charge $^+0.5$ for their hydroxo form, below pH 5.35, and $^-0.5$ for their oxo form, above pH 5.35. The isoelectric point (pH at which the charges balance exactly) for this face is 8.1. Note that, for a given site, at most one of the acidity constants is active, which means that it is impossible to associate two protons in succession with a single surface oxygen atom. Furthermore, it can be shown that the acidity of a site increases with the number of cations with which it is coordinated. The face (101) only has μ_1-OH and μ_2-OH sites, with density 7 groups/nm^2, and only the singly coordinated groups contribute to surface acidity. Its isoelectric point is 10. The face (010) is charged over the whole pH scale. This description shows the fundamental differences in acidity, and hence electrostatic charge, between the faces of a single particle. Each face thus has its own surface reactivity.

12.1.2 Excess Surface Energy

Given that surface energy is the energy required to create surface area by cleaving a crystal, its contribution will be highly dependent on the size of the domains that are formed. Calculations made for a crystal of NaCl [41] show that this energy is negligible for micrometric crystals (< 1 J/g), and that it becomes quite considerable for nanometric particles (560 J/g). The surface energy contribution is thus highly unfavourable to the thermodynamic stabilisation of very small particles. They must thus be considered as metastable systems.

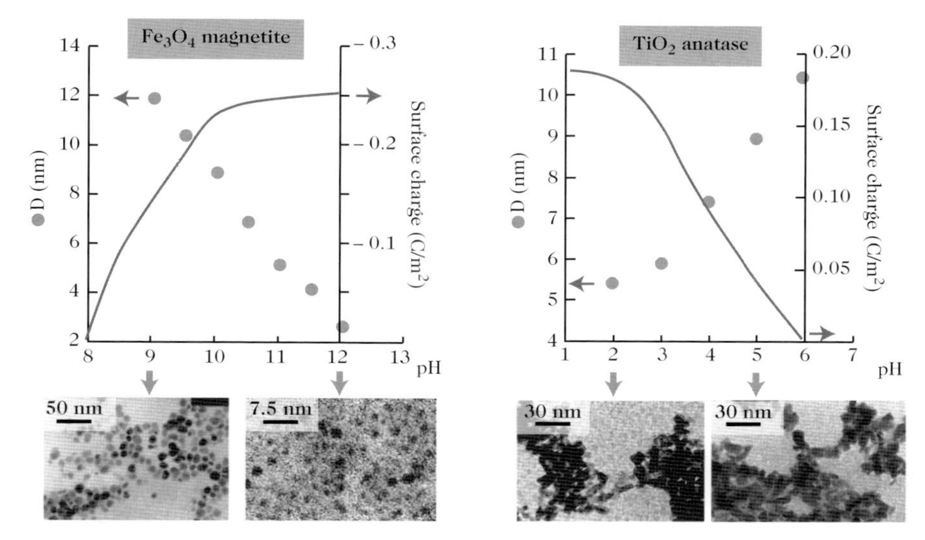

Fig. 12.3. pH dependence of nanoparticle size and surface charge for magnetite (*left*) and anatase (*right*). Particles were obtained by precipitation in aqueous solution under controlled pH

From the expression for the surface enthalpy, $\Delta G^0_{\text{surface}} = \gamma A$, where γ is the surface tension and A the interfacial area, there are two ways of minimising this energy contribution. The first is for the particles to grow in size by Ostwald[1] ripening or by aggregation to reduce the solid–solution interface. The second possibility consists in reducing the interfacial tension by specific adsorption reactions. As an example, gold colloidal suspensions obtained using the method developed by Turkevitch et al. [42] are stabilised by adsorption of acetate groups. In the case of metal oxides, Jolivet et al. [14] have proposed an expression for calculating the pH dependence of the interfacial tension in a way that takes into account the protonation–deprotonation equilibria of the surface sites described above. This semi-quantitative model shows that the drop in interfacial tension will be bigger for higher surface charges, in agreement with the experimental size variations of magnetite (Fe_3O_4) nanoparticles [14] and anatase (TiO_2) nanoparticles [43] (see Fig. 12.3).

The surface energy is thus the driving force for nanoparticle evolution in solution. It is thus essential to treat nanoparticles as metastable entities, which may undergo transformations depending on the nature of the dispersing solution or any other modification of their environment.

[1] Ostwald ripening is a spontaneous phenomenon in which the smallest particles dissolve and then recrystallise on the bigger ones, in such a way that the average particle size increases, without significant modification of the shape of the size distribution.

12.1.3 Surface Properties

The first point to mention is the tendency of nanoparticles to form colloidal suspensions, which can remain stable over very long periods. Precisely because of their small mass, they can compensate for it by specific interactions with the solution. There are two types of stabilisation depending on whether they involve electrostatic or steric interactions [40, 44].

The electrostatic stabilisation of nanoparticles, described by the DLVO theory (due to Derjaguin and Landau, Verwey and Overbeek), is specific to charged species in aqueous solution. This is a kinetic stabilisation method which results from the balancing of attractive forces intrinsic to the chemical nature of the solid (Van der Waals forces) and repulsive forces of electrostatic origin between the charged surfaces. Under such equilibrium conditions, the particles disperse spontaneously under the action of Brownian motion. Electrostatic stabilisation requires strict control of the pH and the nature and concentration of the ions in solution, but it has the great advantage that the particle surface is bare, in the sense that the surface atoms are in direct contact with the solution.

Steric stabilisation involves a modification of the chemical nature of the surface by adsorption or grafting of a coupling agent, usually a polymer, which can form a shell around the particles. This shell guarantees solvation of the particles and prevents them from aggregating by steric repulsion. By virtue of the modified surface state, collisions between particles (induced by Van der Waals forces) are quasi-elastic and redisperse the particles. The quality of the dispersion will depend on the compatibility between the polymer chains at the surface in relation to the nature of the solvent, since this guarantees good shell swelling and surface coverage. Steric stabilisation has many advantages: it is a thermodynamic stabilisation method that applies to many systems, including uncharged systems, and to the dispersion of particles in non-polar solvents.

Generally speaking, nanoparticle surface properties are closely linked to the chemical properties of the surface atoms. The acid–base properties (Lewis or Bronsted) and redox properties of surface atoms lead to the following possibilities for the surface atoms:

- To form surface complexes that can modify the hydrophilic–hydrophobic balance.
- To form hydrogen bonds, typically in the case of OH or NH surface groups, which facilitate particle solvation.
- To be the site of electron transfer reactions (inner or outer sphere) and ion transfers which may be able to propagate into the particle core.
- To adsorb cations from the solution, which may subsequently serve as heterogeneous nucleation sites for the formation of core–shell systems.

Considering more specifically interactions with biological media, one should mention the ability of many metal surface cations to produce free radicals[2] in much greater quantities than for larger particles of the same chemical composition.

12.2 Relation Between Surface Energy and Control of Size and Shape

Adsorption of protons or complexing molecules, which can stabilise nanoparticle dispersions, also plays an important role in controlling the size and shape of the particles if these reactions are initiated during their synthesis. This is exemplified by the synthesis of boehmite γ-AlOOH, precursor of γ-alumina,[3] in aqueous solution.

12.2.1 Proton Adsorption–Desorption

The protonation–deprotonation equilibria of surface oxygen sites contribute to the electrostatic surface charge of the forming particles, and hence to the energy of the exposed crystal faces. In the case of isotropic particles, where all faces have similar energies, lowering the interfacial tension by increasing the surface charge provides a way of controlling the size of the nanoparticles, as shown in Fig. 12.3 for magnetite and anatase. For geometrically anisotropic particles, the energy behaviour of each face must be taken into account in order to specify the equilibrium morphology of the particle, using Wulff's theorem [37]. The exposed faces will be such that the total surface energy is as small as possible. The pH dependence of the energy of the usual faces of boehmite (γ-AlOOH), plotted in Fig. 12.4, shows that each face has its own energy behaviour.

Note the striking behaviour of the face (010), with low energy which remains constant over the whole pH range. The development of this face is thus favoured. Indeed, this explains the platelike morphology often obtained for this aluminium hydroxide. For each pH, the relative values of each face must therefore be considered in order to specify the equilibrium morphology. Syntheses carried out at pH 4.5, 6.5, and 11.5 lead to different particle morphologies (see Fig. 12.4). At pH 4.5, the low energies of the faces (001) and

[2] A free radical is an entity with one or more unpaired electrons in its outer shell, giving it a high reactivity.

[3] The transition aluminas, and in particular γ-alumina, are widely used as catalysts or active phase supports in refining and petrochemistry. The catalytic properties depend on the texture of the material (specific surface area, porosity), but also on the physicochemical surface properties, which depend on the crystal faces. γ-alumina is obtained from boehmite γ-AlOOH by heat treatment via a topotactic transformation. The nature of the exposed crystal faces is thus related to those of the initial compound. This means that the surface properties of γ-alumina are already determined when the boehmite is synthesised.

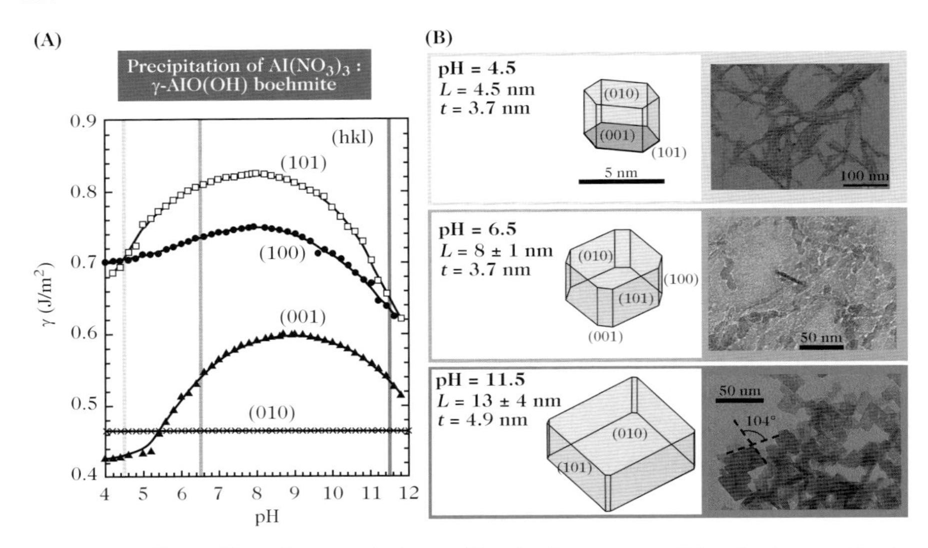

Fig. 12.4. Controlling the morphology of boehmite nanoparticles during synthesis. (**A**) pH dependence of the surface energies of the usual faces of boehmite. From [45]. (**B**) Equilibrium morphologies of the particles for different pH values

(010) result in pseudo-isotropic particles developing preferentially these two types of face. At pH 6.5, the extent of face (001) falls off considerably owing to its increased energy. The pseudo-hexagonal morphology of the particles is then governed by the energy of the base face (010). At pH 11.5, the significant drop in the energy of the lateral faces (101) results in an increase in the size of the diamond-shaped platelets.

These results show that the development of certain crystal faces can be favoured by controlling the acidity, which in turn affects the interfacial tension, thereby determining the size and shape of the resulting particles.

12.2.2 Polyol Adsorption

Nanoparticle morphologies can also be guided by adsorption of organic or inorganic complexing agents. By selecting the amount and chemical nature of the ligands, adsorption can be encouraged at certain surface sites, thereby lowering the interfacial tension there and favouring growth of the corresponding crystal faces. Boehmite has been precipitated out in the presence of various polyols (C2 to C6) comprising a saturated alkyl chain with one alcohol function per carbon atom [46]. Syntheses at pH 11.5 with molar ratio $Cn/Al = 0.07$ lead to size and shape variations depending on the nature of the polyol (see Fig. 12.5). The longer the alkyl chain, the greater the size reduction of the boehmite particles. Dulcitol (C6) doubles the specific surface area, which reaches $360\,m^2/g$ with particle dimensions of $6 \times 6 \times 4\,nm^3$ instead of $15 \times 15 \times 7\,nm^3$ for the reference material, synthesised without polyol.

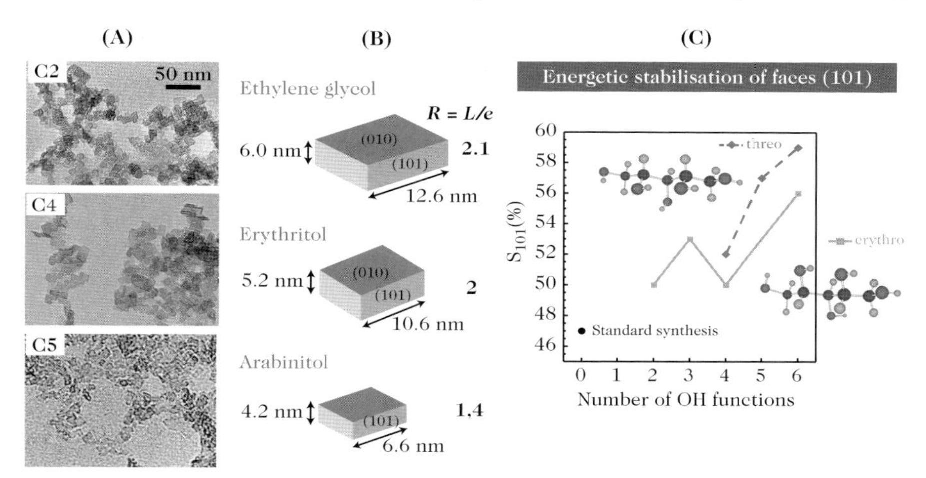

Fig. 12.5. Boehmite nanoparticles synthesised in the presence of polyols. (**A**) Variation of size and (**B**) variation of morphology and anisotropy ratio ($R = L/t$) as a function of the polyol chain length. (**C**) Percentage of face (101) for varying polyol stereochemistry

Particle size reduction does not occur equally in all directions, i.e., it is not homothetic. So for example, the faces (101) vary between 47 and 59%. This leads to changes in the anisotropy ratio $R = L/t$ of the platelets, where L is their length and t their thickness.

The effect of the polyol also depends on stereochemical details. It tends to be reinforced when all the alcohol functions lie on the same side of the chain (threo isomer). The adsorption isotherms for preshaped boehmite particles indicate the existence at the maximum of one monolayer with a low adsorption energy. The adsorption mechanism involves the formation of hydrogen bonds between the hydroxylated groups carried by surface aluminium atoms and the alcohol functions. The polyol seems to flatten itself against a given face for which the distances between surface sites match the positions of the alcohol groups along the chain, thereby explaining the pronounced effect of long-chain alcohols and threo isomers.

Physisorption of molecules therefore suffices to modify the surface energy and favour the selective growth of certain faces of a given particle.

12.3 Surface Reactivity and Photocatalysis

Nanoparticle surface reactivities can assume different forms, as already discussed, and as will be discussed further in Sect. 12.4. One such form is the photocatalytic effect. Indeed, this can be considered as a reactivity induced by an influence from outside the nanoparticle, whereas chemical reactivity with respect to adsorbing molecules (see Sect. 12.4) occurs spontaneously.

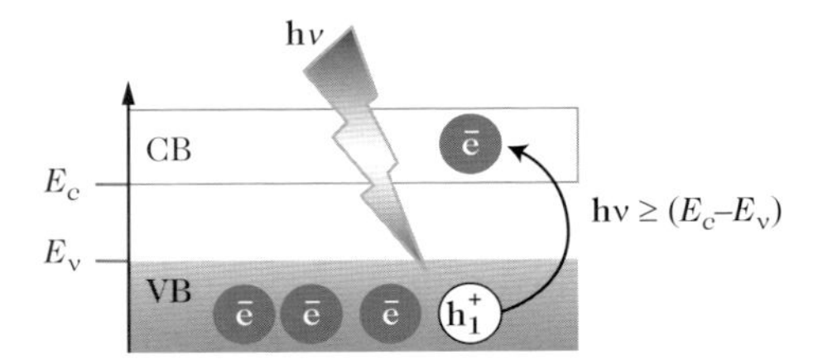

Fig. 12.6. Creation of an electron $\overline{e}$/hole h_f^+ pair following irradiation by a light ray of energy $h\nu$ greater than or equal to the difference between the energy E_c of the conduction band and the energy E_v of the valence band

The photocatalytic effect is not unique to nanoparticles, but there is a clear modification of these effects depending on the size of the particles, as we shall see below.

12.3.1 Photocatalysis: Definition

Photocatalysis is usually described as a photochemical and catalytic reaction occurring at the surface of a solid, generally a semiconductor [47–49]. The word 'photochemical' implies that this reaction only occurs in the presence of light, while the word 'catalytic' implies that the catalyst (the solid here) is regenerated. This reaction is initiated by the transition of an electron from the valence band (VB) to the conduction band[4] (CB) under the effect of a light ray carrying energy $h\nu$ greater than or equal to the energy difference between the two bands (see Fig. 12.6). In the case of ZnO and TiO_2 (anatase), this energy is on average 3.2 eV (= 388 nm), whereas for CdS, it is 2.6 eV (= 517 nm). It results in the creation of a free hole h_f^+ in the valence band. This electron–hole pair $\{\overline{e}/h_f^+\}$ can occur in many materials, e.g., TiO_2, ZnO, CdS, CeO_2, etc.

This pair will nevertheless attract one another mutually, with a stronger force in a medium of lower dielectric constant. Bound together by this attractive electrostatic interaction, the pair is called an exciton. Recombination of the electron and hole is a very rapid process, so the exciton is characterised by a short lifetime. Some materials like TiO_2 have a high enough dielectric constant to guarantee a good separation of electron and hole, whence some excitons may succeed in reaching the surface of the material. At the surface, the available electrons and holes can react with acceptor and donor species to form radicals. Two reactions will then take place there: an oxidation reaction due to the photogenerated hole and a reduction reaction due to the photogenerated electron. Different species may thus appear, as summarised in Fig. 12.7 in an aqueous medium. The electron–hole pair can react with (1) terminal oxygen atoms belonging to the solid or (2) adsorbed external compounds (water

[4] The valence band is the band of highest energy that is fully occupied. The conduction band is the band of lowest energy that is empty or only partially filled.

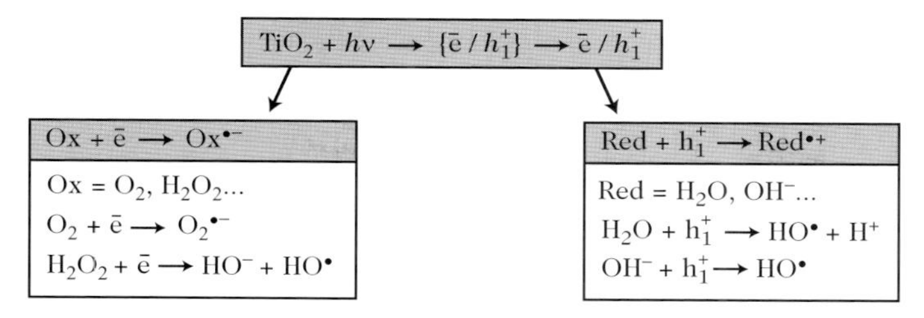

Fig. 12.7. Oxidation and reduction reactions that can occur at the surface of TiO_2

molecules or organic compounds). This can be illustrated by the direct or indirect oxidation of methanol. Oxidation can take place through a photogenerated $OH^\bullet$ as in (12.1) or through a hole which acts directly on the methanol as in (12.2):

$$\begin{aligned} CH_3OH + OH^\bullet &\longrightarrow CH_2OH^\bullet + H_2O \, , \\ CH_2OH^\bullet &\longrightarrow HCHO + H^+ + \bar{e} \, , \end{aligned} \tag{12.1}$$

$$\begin{aligned} CH_3OH + h_f^+ \longrightarrow CH_3OH^+ &\longrightarrow CH_2OH^\bullet + H^+ \, , \\ CH_2OH^\bullet &\longrightarrow HCHO + H^+ + \bar{e} \, . \end{aligned} \tag{12.2}$$

However, even though these photocatalytic surface (redox) reactions were discovered over 80 years ago [50–52], and the decomposition of water by photocatalysis at the surface of a TiO_2 electrode was demonstrated in 1972 [50], we still do not fully understand the reaction mechanisms and the debate continues [53]. It goes beyond the scope of this chapter to detail the complex mechanisms underlying photocatalysis, but in the next section, we shall discuss some aspects of the problem that are related to particle size.

12.3.2 Effect of Particle Size on Photocatalysis

We shall focus here on the case of titanium dioxide, because it is certainly the material with the most photocatalytic applications at the present time. Titanium dioxide exists in three main (polymorphic) forms: rutile, anatase, and brookite. The first studies in the 1960s showed that the photocatalytic activity depends on the mineralogical form of the solid. Kato et al. [54] found that anatase had greater photocatalytic activity for the oxidation of hydrocarbons and alcohols than rutile. Now the stability of the three forms of TiO_2 depends on particle size [55]. It has recently been shown that rutile is the most thermodynamically stable phase for particle sizes larger than 35 nm, and anatase for sizes smaller than 11 nm, while brookite turns out to be the most stable in the intermediate size range. The exact reason for these differences has not yet been fully elucidated, but it is certainly related to the nature of the exposed crystallographic faces.

Apart from this indirect size effect, several authors [56, 57] were able to show that there is an optimal size for photo-oxidation of organic substrates for a given polymorph. Almquist and Biswas [56] found the optimum to lie in the range 25–40 nm, whereas Wang et al. [57] located it around 11 nm. The authors considered that there were opposing effects between the large specific surface area when the size decreases, and hence a larger amount of adsorbed molecules, and the greater proximity of the electron–hole pairs, leading to a higher recombination rate before surface reactions could occur.

The band transition energy or band gap $E_g = E_c - E_v$ defined above can also be affected by changing the size of the photocatalyst particles [56,58–60]. The absorption spectra of many semiconductor nanoparticles, generally called quantum dots, exhibit modifications. Size reduction results in a spectral shift toward the blue (blueshift). This is explained by examining the expression for E_c as a solution of the Schrödinger equation [61] using an appropriate Hamiltonian, viz.,

$$E_c \approx \frac{\pi^2 \hbar^2}{2R^2} \frac{1}{\mu} - \frac{1.8e^2}{\varepsilon R} ,$$

where R is the particle radius, ε the dielectric constant of the solid, and μ the reduced mass of the exciton given by

$$\frac{1}{\mu} = \frac{1}{m^*_{\mathrm{e}^-}} + \frac{1}{m^*_{\mathrm{h}^+}} ,$$

with $m^*_{\mathrm{e}^-}$ the effective mass of the electron and $m^*_{\mathrm{h}^+}$ the effective mass of the hole. According to this, when the radius decreases, E_c increases. This has been shown experimentally on several occasions for various materials, including anatase, where E_c varies between 3.2 and 3.5 eV for particles smaller than 16 nm (see Fig. 12.8). When E_c increases, this raises the photocatalytic activity by increasing the redox potential, the key parameter for electron

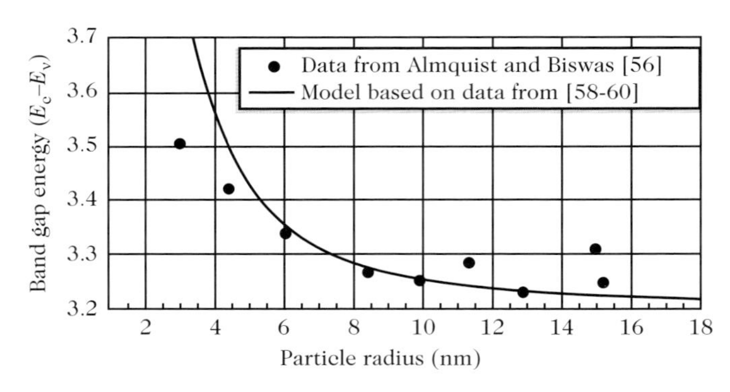

Fig. 12.8. Size dependence of the band gap for anatase particles. Experimental data extracted from [56] and modelled according to [58–60]

transfer. As mentioned in the introduction, it is interesting to note that these changes only concern very small particles, and it is not valid in the size range generally taken to define nanoparticles (1–100 nm).

12.3.3 Environmental Applications of Photocatalysis

The investigations of Frank and Bard [62, 63] seem to be the first to suggest using TiO_2 under UV illumination to purify water by decomposing pollutants. Since then, there has been a surge of literature on air and water purification. One advantage of photocatalysis with TiO_2 is that it uses only TiO_2 and UV illumination, either artificial or natural. This reduces the cost as compared with other oxidation processes, e.g., using ozone. However, it is generally accepted that photocatalysis processes based on TiO_2 can only be used to treat effluents containing very low contaminant concentrations, owing to the rather average efficiency of this process [53].

As mentioned in Sect. 12.3.1, pollutants can be oxidised or reduced directly or indirectly. Applications generally involve oxidation of organic pollutants, the ultimate aim being CO_2 formation by the following reactions:

$$\begin{aligned} \mathrm{R{-}COOH} &\longleftrightarrow \mathrm{R{-}COO^-} + \mathrm{H^+} , \\ \mathrm{R{-}COO^-} + \mathrm{OH^\bullet} &\longrightarrow \mathrm{R{-}COO^\bullet} + \mathrm{OH^-} , \\ \mathrm{R{-}COO^\bullet} &\longrightarrow \mathrm{R^\bullet} + \mathrm{CO_2} \longrightarrow \mathrm{R{-}COOH} \longrightarrow \cdots \longrightarrow \mathrm{CO_2} + \mathrm{H_2O} . \end{aligned}$$

However, the reactions occurring at the surface of a photocatalyst can be much more complex, producing many reaction intermediates. In some cases, in particular the treatment of pesticides, the intermediates may be more toxic than the contaminant itself. There are many processes using UV lamps, but the most promising developments aim to use solar radiation. For example, it has been shown that a reactor with parabolic collector [64–66] can treat municipal or agricultural effluents up to 50 mg/l, without using an artificial light source. At these concentrations, mineralisation is total. Similar efficiency was obtained by a cascade process based on solar irradiation [67]. Apart from the mineralisation of organic compounds, photocatalytic treatment seems to be a promising way of disinfecting water. For example, Rincon and Pulgarin [68] have shown that the parabolic collector process could disinfect lake water contaminated by *Escherichia coli* K12 (10^6 CFU/mL) in 3 hours.

Environmental applications are not restricted to the treatment of liquid effluents, but can also be used to purify gases, e.g., air, and to eliminate odours. There are already applications to treat volatile organic compounds such as formaldehydes, toluene, and so on. Photocatalysts can also be associated with cloths to deodorize certain industrial processes, e.g., composting plants.

12.4 Surface Reactivity and Adsorption–Desorption

12.4.1 Pollutant Adsorption. Arsenic

Nanoparticles are recognised for their ability to hold ions at their surface. This strong affinity for ions is in part due to their high specific surface area. However, the adsorption capacity cannot be explained simply by an increase in specific surface area, i.e., the adsorption capacity per gram. Another important parameter is the surface reactivity of nanoparticles, i.e., the adsorption capacity per nm^2 of surface. Recent studies have shown that iron oxide nanoparticles of diameter 10 nm can hold up to 10 As^{III}/nm^2 [69], 13As^{V}/nm^2 [69], or 22–34Co^{II}/nm^2 [70], whereas microscopic iron oxides can hold on average 1–4 atoms/nm^2 [71, 72]. The only way to explain such a high adsorption capacity per nm^2 is that adsorption mechanisms at the surface of nanoparticles differ from conventional mechanisms occuring at the surface of microparticles. However, the exact nature of the mechanisms underlying this nanoscale effect are still poorly understood.

One way to probe adsorption sites on particles is to use specific chemical probes [1, 73–75]. For example, As^{III} has been used to probe the surface of nanomaghemites of diameter 6 nm. Although the adsorption of arsenic on oxides has been widely covered [76–81], novel sites have been discovered on iron oxides with diameters smaller than 10 nm. These sites arise through a structural modification of the surface of maghemite particles when their size is decreased [17]. Whereas micromaghemite particle surfaces are composed of iron octahedra and tetrahedra, $[Fe]_{oct}$ and $[Fe]_{tetra}$, respectively, nanomaghemites have a predominantly octahedral surface, rich in vacancies at tetrahedral positions [17]. These vacancies are highly reactive potential adsorption sites for arsenic. By X-ray diffraction and absorption spectroscopy (K threshold of As), it has been shown that arsenic fills the tetrahedral vacancies. Two sites have been discovered: (1) a site with tetrahedral positioning at the surface of a ring of 5–6$[Fe]_{oct}$ and (2) a site with tetrahedral positioning at the surface of an $[Fe]_{oct}$ trimer (see Fig. 12.9).

In the same way as crystal growth (see Sect. 12.1.2), the adsorption of arsenic on these reactive sites has the effect of stabilising the nanoparticles thermodynamically. Whereas the main driving forces for adsorption on macroscopic particles are electrostatic attraction and chemical affinity, when it comes to nanoparticles, another force must be taken into account, due to the reduction of surface energy. If the surface of 6 nm maghemite nanoparticles is saturated by a layer of As^{III} polyhedra in crystallographic positions, their diameter increases by 0.5 nm. This significant growth reduces the surface pressure of the nanoparticles by 8%. In comparison, the adsorption of an As^{III} monolayer at the surface of 20 nm maghemite particles only decreases the surface pressure by 2%. In conclusion, the nanoscale effects observed during pollutant adsorption have two origins:

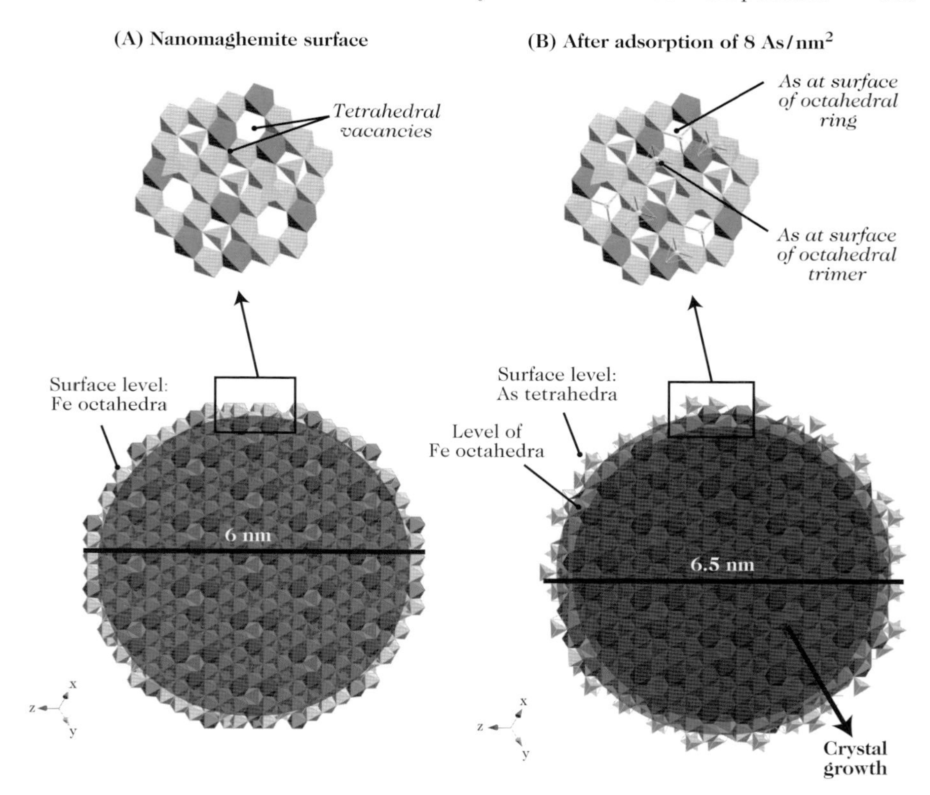

Fig. 12.9. Adsorption of As^{III} at the surface of nanomaghemites (see *colour plate*). Surface structure (**A**) before and (**B**) after adsorption. This interpretation is based on structural information obtained by X-ray diffraction and absorption spectroscopy. Adapted from [73]

1. an atomic rearrangement of the the nanoparticle surface creating new adsorption sites,
2. a significant reduction of the excess surface energy of the nanoparticles.

12.4.2 Catalysis. MoS_2 Particles

As with metal oxide nanoparticles, the presence of highly reactive sites at the surface of metal sulfide nanoparticles leads to novel properties. This is the case notably for MoS_2 particles, which are chemically inert on the macroscopic scale, but which become highly efficient catalysts when in the form of nanoparticles dispersed on a substrate. These nanoparticles are used in petrol or biomass refining processes to extract heteroatoms like sulfur and oxygen. These reactions are carried out under hydrogen pressure.

The changes in the properties of MoS_2 are directly due to a modification in the crystal structure when the particle size is decreased. The catalytic activity is attributed to the presence of vacancy sites, highly reactive to sulfur, on the nanoparticle edges. These edges have different atomic structure and stoichiometry from the catalytically inert base faces (0001). Lauritsen et al. [82] have shown that vacancy formation is size dependent. The $(10\bar{1}0)$ edges of clusters of 21 of Mo atoms contain Mo and S atoms, whereas the $(\bar{1}010)$ edges of smaller clusters contain S atoms and vacancies. These vacancies form spontaneously at the surface, which does indeed show that their formation energy does not depend solely on the nature of the edges $(10\bar{1}0)$ or $(\bar{1}010)$, or on the ratio of Mo to S, but predominantly on the size of the particles. In addition, two other key parameters for the catalytic activity of MoS_2, viz., the binding energy ΔE_s of sulfur and the dissociation capacity of hydrogen, are also size dependent.

12.4.3 Dissolution and Salting out of Toxic Ions

Recent studies have shown that the chemical instability of nanoparticles is one of the main factors leading to their toxicity [27]. Chemically unstable inorganic nanoparticles can be oxidised or reduced in biological media, then salt out toxic ions in the vicinity of living organisms. The toxicity of quantum dots in particular is associated with their oxidation/dissolution in biological media and the salting out of toxic Cd^{2+} or Se^{2-} ions [83]. Salting out of Fe^{2+} ions also plays a role in the toxicity of iron-containing nanoparticles [25].

The driving force behind dissolution depends mainly on the solubility of the crystal in a given environment and the concentration gradient between particles and solution [84]. There is thus no doubt that, for a given mass, the kinetics of dissolution will be proportional to the specific surface area and hence faster for nanoparticles than for micrometric particles. But the main question concerns the relation between the solubility of the crystals (K_b) and the size of the particles. From the thermodynamic point of view, K_b is often taken as the solubility product (K_{sp}). The relation between K_b and K_{sp} is given by

$$\ln K_b = \ln K_{sp} + c\frac{\gamma}{l} ,$$

where γ is the surface tension, l the characteristic length of the crystal, and c a constant. However, several studies [85–87] have shown that this approximation is only valid for crystals with diameters greater than 25 nm. When the size of the crystal falls below 25 nm, the change in crystal structure as a function of the size and the appearance of crystal defects can no longer be neglected, and this approximation is no longer justified. Furthermore, the surface tension [88] and activation energy relevant to the dissolution process [87] are also size dependent. Taken together, these features imply that, for nanoparticles smaller than 25 nm, in addition to the specific surface area effect, thermodynamic factors will also favour their dissolution [84], and hence their potential toxicity.

12.5 Conclusion

The examples discussed here show that nanoparticles in the size range 20–100 nm have properties that do not differ drastically from those of larger particles. However, there is a critical size below 20–30 nm for which nanoscale effects begin to arise. These changes in the properties come about for two reasons:

1. A change in the crystal structure of the particles when their size is reduced, e.g., an atomic rearrangement at the surface, presence of crystal defects, appearance of vacancies, changed morphology (see Sect. 12.2).
2. Thermodynamic stabilisation of the nanoparticles (see Sect. 12.1).

These novel properties raise new questions about their ecotoxicity:

- Can the crystal modifications responsible for the catalytic properties of nanoparticles (see Sect. 12.4.2) facilitate electron and ion transfer, and hence perturb electron transport in the bacterial respiratory chain?
- Can the strong adsorption of ions at the the nanoparticle surface (see Sect. 12.4.1) modify the flow of ions between the intra- and extracellular media?
- Can the enhanced dissolution of nanoparticles (see Sect. 12.4.3) cause increased salting out of toxic ions within an organism?
- Can the photocatalytic properties of nanoparticles (see Sect. 12.3) increase the generation of reactive oxygen species by nanoparticles and cause them to induce greater oxidative stress than macroscopic particles?

Acknowledgements

The authors would like to thank the *Centre national de la recherche scientifique* (CNRS) and the *Commissariat à l'énergie atomique* (CEA) in France for financing the GDRi, International Consortium for the Environmental Implications of NanoTechnology (iCEINT) directed by Jean-Yves Bottero, and also the National Science Foundation (NSF) and the US Environmental Protection Agency (EPA) for financing the Center for the Environmental Implications of NanoTechnology (CEINT) (EF-0830093) directed by Mark R.Wiesner.

References

1. Académies des Sciences et Technologies de Paris: Nanoscience, Nanotechnologies. www.academie-sciences.fr, **18** (2004)
2. S.F. Hansen , B.H. Larsen , S.I. Olsen, A. Baun: Categorization framework to aid hazard identification of nanomaterials. Nanotoxicology **1**, 243–250 (2007)

3. Nanoscale Science Engineering and Technology Subcommittee T.C.o.T., National Science and Technology Committee: The National Nanotechnology Initiative Strategy Plan. National Nanotechnology Coordination Office, Arlington, VA (2004)
4. Royal Society of London: Nanoscience, and nanotechnology: Opportunities and uncertainties. www.nanotec.org.uk, **19** (2004)
5. M. Daniel, D. Astruc: Gold nanoparticles: Assembly, supramolecular chemistry, quantum-size-related properties, and applications toward biology, catalysis, and nanotechnology. Chem. Inform. **104**, 293–346 (2004)
6. M. Haruta: Size- and support-dependency in the catalysis of gold. Catalysis Today **36**, 153–166 (1997)
7. T.K. Sau, A. Pal, T. Pal: Size regime dependent catalysis by gold nanoparticles for the reduction of eosin. J. Phys. Chem. B **105**, 9266–9272 (2001)
8. A.P. Alivisatos: Semiconductor clusters, nanocrystals, and quantum dots. Science **271**, 933–937 (1996)
9. K.L. Kelly, E. Coronado, L.L. Zhao, G.C. Schatz: The optical properties of metal nanoparticles: The influence of size, shape, and dielectric environment. J. Phys. Chem. B **107**, 668–677 (2003)
10. K. Dick, T. Dhanasekaran, Z. Zhang, D. Meisel: Size-dependent melting of silica-encapsulated gold nanoparticles. J. Am. Chem. Soc. **124**, 2312–2317 (2002)
11. S.L. Lai, J.Y. Guo, V. Petrova, G. Ramanath, L.H. Allen: Size-dependent melting properties of small tin particles: Nanocalorimetric measurements. Phys. Rev. Lett. **77**, 99–102 (1996)
12. M. Zhang, M.Y. Efremov, F. Schiettekatte, E.A. Olson, A.T. Kwan, S.L. Lai, T. Wisleder, J.E. Greene, L.H. Allen: Size-dependent melting point depression of nanostructures: Nanocalorimetric measurements. Phys. Rev. B **62**, 10548–10557 (2000)
13. Z. Wang, Y. Zhao, K. Tait, X. Liao, D. Schiferl, C. Zha, R.T. Downs, J. Qian, Y. Zhu, T. Shen: A quenchable superhard carbon phase synthesized by cold compression of carbon nanotubes. Proc. Natl. Acad. Sci. USA **101**, 13699–13702 (2004)
14. J.P. Jolivet, C. Froidefond, A. Pottier, C. Chanéac, S. Cassaignon, E. Tronc, P. Euzen: Size tailoring of oxide nanoparticles by precipitation in aqueous medium. A semi-quantitative modelling. J. Mater. Chem. **14**, 3281–3288 (2004)
15. P. Ayyub, V.R. Palkar, S. Chattopadhyay, M. Multani: Effect of crystal size reduction on lattice symmetry and cooperative properties. Phys. Rev. B **51**, 6135–6138 (1995)
16. J.F. Banfield, A. Navrotsky: Nanoparticles and the environment. Reviews in mineralogy and geochemistry. Geochimica et Cosmochimica Acta **67**, 1753 (2001)
17. S. Brice-Profeta, M.A. Arrio, E. Tronc, N. Menguy, I. Letard, C. Cartier dit Moulin, M. Nogues, C. Chaneac, J.P. Jolivet, P. Sainctavit: Magnetic order in γ-Fe_2O_3 nanoparticles: A XMCD study. J. Magn. Magn. Mater. **288**, 354–365 (2005)
18. R. Lamber, S. Wetjen, N.I. Jaeger: Size dependence of the lattice parameter of small palladium particles. Phys. Rev. B **51**, 10968–10971 (1995)
19. G. Oberdörster, E. Oberdörster, J. Oberdörster: Nanotoxicology: An emerging discipline evolving from studies of ultrafine particles. Environ. Health Perspect. **113**, 823–839 (2005)

20. J.Y. Bottero, J. Rose, M.R. Wiesner: Nanotechnologies: Tools for sustainability in a new wave of water treatment processes. Integrated Environmental Assessment and Management **2**, 391–395 (2006)
21. D.F. Emerich, C.G. Thanos: The pinpoint promise of nanoparticle-based drug delivery and molecular diagnosis. Biomolecular Engineering **23**, 171–184 (2006)
22. A.K. Gupta, M. Gupta: Synthesis and surface engineering of iron oxide nanoparticles for biomedical applications. Biomaterials **26**, 3995–4051 (2005)
23. D.A. Pereira de Abreu, P. Paseiro Losada, I. Angulo, J.M. Cruz: Development of new polyolefin films with nanoclays for application in food packaging. European Polymer Journal **43**, 2229–2243 (2007)
24. S.K. Sahoo, V. Labhasetwar: Nanotech approaches to drug delivery and imaging. Drug Discov. Today **8**, 1112–1120 (2003)
25. M. Auffan, W. Achouak, J. Rose, C. Chaneac, D.T. Waite, A. Masion, J. Woicik, M.R. Wiesner, J.Y. Bottero: Relation between the redox state of iron-based nanoparticles and their cytotoxicity towards *Escherichia coli*. Environ. Sci. Technol. **42**, 6730–6735 (2008)
26. M. Auffan, J. Rose, T. Orsiere, M. De Meo, A. Thill, O. Zeyons, O. Proux, A. Masion, P. Chaurand, O. Spalla, A. Botta, M.R. Wiesner, J.Y. Bottero: CeO_2 nanoparticles induce DNA damage towards human dermal fibroblasts in vitro. Nanotoxicology **3**, 161–171 (2009)
27. M. Auffan, J. Rose, M.R. Wiesner, J.Y. Bottero: Chemical stability of metallic nanoparticles: A parameter controlling their potential toxicity in vitro. Environmental Pollution **157**, 1127–1133 (2009)
28. J.D. Fortner, D.Y. Lyon, C.M. Sayes, A.M. Boyd, J.C. Falkner, E.M. Hotze, L.B. Alemany, Y.J. Tao, W. Guo, K.D. Ausman, V.L. Colvin, J.B. Hughes: C_{60} in water: Nanocrystal formation and microbial response. Environ. Sci. Technol. **39**, 4307–4316 (2005)
29. S. Lanone, J. Boczkowski: Biomedical applications and potential health risks of nanomaterials: Molecular mechanisms. Curr. Mol. Med. **6**, 651–663 (2006)
30. W.R. Moore, J.M. Genet: Antibacterial activity of gutta-percha cones attributed to the zinc oxide component. Oral Surg. **53**, 508–517 (1982)
31. A. Nel, T. Xia, L. Madler, N. Li: Toxic potential of materials at the nanolevel. Science **311**, 622–627 (2006)
32. E. Oberdörster: Manufactured nanomaterials (fullerene, C_{60}) induce oxidative stress in the brain of juvenile largemouth bass. Environ. Health Perpect. **112**, 1058–1062 (2004)
33. C.M. Sayes, V. Colvin: The differential cytotoxicity of water soluble fullerenes. Nano Lett. **4**, 1881–1887 (2004)
34. A. Thill, O. Zeyons, O. Spalla, F. Chauvat, J. Rose, M. Auffan, A.M. Flank: Cytotoxicity of CeO_2 nanoparticles for *Escherichia coli*. Physicochemical insight into the cytotoxicity mechanism. Environ. Sci. Technol. **40**, 6151–6156 (2006)
35. D.B. Warheit, T.R. Webb, C.M. Sayes, V.L. Colvin, K.L. Reed: Pulmonary instillation studies with nanoscale TiO_2 rods and dots in rats: Toxicity is not dependent upon particle size and surface area. Toxicol. Sci. **91**, 227–236 (2006)
36. M.R. Wiesner, G.V. Lowry, P.J.J. Alvarez: Assessing the risks of manufactured nanomaterials. Environ. Sci. Technol. **40**, 4337–4345 (2006)
37. M. Lahmani, C. Brechignac, P. Houdy: *Nanomaterials and Nanochemistry*, Springer, Heidelberg, Berlin, New York (2006) Chap. 1

38. L. Pauling: The principles determining the structure of complex ionic crystals. J. Am. Chem. Soc. **51**, 1010–1026 (1929)
39. T. Hiemstra, P. Venema, W.H.V. Riemsdijk: Intrinsic proton affinity of reactive surface groups of metal (hydr)oxides: The bond valence principle. J. Coll. Interf. Sci. **184**, 680–692 (1996)
40. J.P. Jolivet, C. Chaneac, E. Tronc: Iron oxide chemistry. From molecular clusters to extended solid networks. Chem. Commun. 481–487 (2004)
41. A.W. Adamson, A.P. Gast: *Physical Chemistry of Surfaces*. John Wiley, New York (1997)
42. J. Turkevitch, P.C. Stevenson, J. Hillier: A study of the nucleation and growth processes in the synthesis of colloidal gold. J. Discuss. Faraday Soc. **11**, 55–75 (1951)
43. A.S. Pottier, S. Cassaignon, C. Chaneac, F. Villain, E. Tronc, J.P. Jolivet: Size tailoring of TiO_2 anatase nanoparticles in aqueous medium and synthesis of nanocomposites. Characterization by Raman spectroscopy. J. Mater. Chem. **13**, 877–882 (2003)
44. G. Cao: *Nanostructures and Nanomaterials: Synthesis, Properties and Applications*. Imperial College Press, London (2004)
45. P. Euzen, P. Raybaud, X. Krokidis, H. Toulhoat, J.L. Le Loaer, J.P. Jolivet, C. Froidefond: *Handbook of Porous Materials*. Wiley-VCH, Chichester (2002) pp. 1592–1677
46. D. Chiche: *Heterogeneous Catalysts*. Elsevier (2006)
47. E. Pelizzetti, N. Serpone: *Homogeneous and Heterogeneous Photocatalysis*. Reidel Publishing Company, Dordrecht (1986)
48. N. Serpone, E. Pelizzetti: *Photocatalysis: Fundamentals and Applications*. John Wiley, New York (1989)
49. A. Fujishima: *TiO_2 Photocatalysis: Fundamentals and Applications*. BKC, Tokyo (1999)
50. A. Fujishima, K. Honda: Electrochemical photolysis of water at a semiconductor electrode. Nature **238**, 37–38 (1972)
51. C. Renz: Photoreactions of oxides of titanium, cerium and earth acids. Helv. Chim. Acta **4**, 961–968 (1921)
52. E. Baur, A. Perret: On the action of light on dissolved silver salts in the presence of zinc oxide. Helv. Chim. Acta **7**, 910–915 (1924)
53. A. Fujishima: TiO_2 photocatalysis and related surface phenomena. Surface Science Reports **63**, 515–582 (2008)
54. S. Kato, F. Mashio: TiO_2 photocatalyzed oxidation of tetraline in liquid phase. J. Chem. Soc. Japan, Indust. Chem. Sect. **67**, 1136–1140 (1964)
55. M.R. Ranade, A. Navrotsky, H.Z. Zhang, J.F. Banfield, S.H. Elder, A. Zaban, P.H. Borse, S.K. Kulkarni, G.S. Doran, H.J. Whitfield: Energetics of nanocrystalline TiO_2. Proc. Natl. Acad. Sci. USA **99** (Suppl. 2), 6476–6481 (2002)
56. C.B. Almquist, P. Biswas: Role of synthesis method and particle size of nanostructured TiO_2 on its photoactivity. Journal of Catalysis **212**, 145–156 (2002)
57. C.C. Wang, Z. Zhang, J.Y. Ying: Photocatalytic decomposition of halogenated organics over nanocrystalline titania. Nanostructured Materials **9**, 583–586 (1997)
58. N. Serpone, D. Lawless, R. Khairutdinov: Size effects of the photophysical properties of colloidal anatase TiO_2 particles: Size quantization versus transitions in this indirect semiconductor? J. Phys. Chem. **99**, 16646–16654 (1995)

59. A.J. Nozik: *Photocatalytic Purification and Treatment of Water and Air*. Elsevier, Amsterdam (1993)
60. A.L. Linsebigler, G. Lu, J.T. Yates: Photocatalysis on TiO_2 surfaces: Principles, mechanisms, and selected results. Chem. Rev. **95**, 735–758 (1995)
61. M. Hoffmann et al.: *Reactive Oxygen Species Generation on Nanoparticulate Material. Environmental Nanotechnology: Applications and Impacts of Nanomaterials*. McGraw Hill, New York (2007)
62. S.N. Frank, A.J. Bard: Heterogeneous photocatalytic oxidation of cyanide ion in acqueous solutions at titanium dioxide powder. J. Am. Chem. Soc. **99**, 303–304 (1977)
63. S.N. Frank, A.J. Bard: Heterogeneous photocatalytic oxidation of cyanide and sulfite in aqueous solutions of semiconductor powders. J. Phys. Chem. **81**, 1484–1488 (1977)
64. M.H. Pérez, G. Peñuela, M.I. Maldonado, O. Malato, P. Fernández-Ibáñez, I. Oller, W. Gernjak, S. Malato: Degradation of pesticides in water using solar advanced oxidation processes. Appl. Catal. B Environ. **64**, 272–281 (2006)
65. I. Oller, W. Gernjak, M.I. Maldonado, L.A. Perez-Estrada, J.A. Sanchez-Perez, S. Malato: Solar photocatalytic degradation of some hazardous water-soluble pesticides at pilot-plant scale. J. Hazard. Mater. **138**, 507–517 (2006)
66. M. Kositzi, I. Poulios, S. Malato, J. Caceres, A. Campos: Solar photocatalytic treatment of synthetic municipal wastewater. Water Res. **38**, 1147–1154 (2004)
67. C. Guillarda, J. Disdiera, C. Monnet, J. Dussaud, S. Malato, J. Blancoc, M.I. Maldonadoa, J.M. Herrmann: Solar efficiency of a new deposited titania photocatalyst: Chlorophenol, pesticide and dye removal applications. Appl. Catal. B Environ. **46**, 319–332 (2003)
68. A.G. Rincón, C. Pulgarin: Field *E. coli* inactivation in absence and presence of TiO_2. Is solar UV dose appropriate parameter to standardization of solar water disinfection? Solar Energy **77**, 635–648 (2004)
69. S. Yean, L. Cong, C.T. Yavuz, J.T. Mayo, W.W. Yu, A.T. Kan, V.L. Colvin, M.B. Tomson: Effect of magnetic particle size on adsorption and desorption of arsenite and arsenate. J. Mater. Res. **20**, 3255–3264 (2005)
70. A. Uheida, G. Salazar-Alvarez, E. Bjorkman, Z. Yu, M. Muhammed: Fe_3O_4 and γ-Fe_2O_3 nanoparticles for the adsorption of Co^{2+} from aqueous solution. Journal of Colloid and Interface Science **298**, 501–507 (2006)
71. H.A. Al-Abadleh, V.H. Grassian: Oxide surfaces as environmental interfaces. Surface Science Reports **52**, 63–161 (2003)
72. L. Sigg, P. Behra, G.N. Stumm: *Chimie des milieux aquatiques, chimie des eaux naturelles et des interfaces dans l'environnement*. Dunod, Paris (2000) pp. 350–390
73. M. Auffan, J. Rose, O. Proux, D. Borschneck, A. Masion, P. Chaurand, J.L. Hazemann, C. Chaneac, J.P. Jolivet, M.R. Wiesner, A. VanGeen, J.Y. Bottero: Enhanced adsorption of arsenic onto nano-maghemites: As(III) as a probe of the surface structure and heterogeneity. Langmuir **24**, 3215–3222 (2008)
74. A.S. Madden, M.F. Hochella, T.P. Luxton: Insights for size-dependent reactivity of hematite nanomineral surfaces through Cu^{2+} sorption. Geochim. Cosmochim. Acta **70**, 4095–4104 (2006)
75. F. Villieras, L.J. Michot, F. Bardot, M. Chamerois, C. Eypert-Blaison, M. Francois, G. Gerard, J.M. Cases: Surface heterogeneity of minerals. Comptes Rendus Geosciences **334**, 597–609 (2002)

76. A. Manceau: The mechanism of anion adsorption on iron oxides: Evidence for the bonding of arsenate tetrahedra on free $Fe(O,OH)_6$ edges. Geochim. Cosmochim. Acta **59**, 3647–3653 (1995)
77. B.A. Manning, S. Goldberg: Modeling arsenate competitive adsorption on kaolinite, montmorillonite and illite. Clays Clay Miner. **44**, 609–623 (1996)
78. D.M. Sherman, S.R. Randall: Surface complexation of arsenic(V) to iron(III) (hydr)oxides: Structural mechanism from ab initio molecular geometries and EXAFS spectroscopy. Geochim. Cosmochim. Acta **67**, 4223–4230 (2003)
79. S. Thoral, J. Rose, J.M. Garnier, A. vanGeen, P. Refait, A. Traverse, E. Fonda, D. Nahon, J.Y. Bottero: XAS Study of iron and arsenic speciation during Fe(II) oxidation in the presence of As(III). Environ. Sci. Technol. **39**, 9478–9485 (2005)
80. G.A. Waychunas, J.A. Davis, C.C. Fuller: Geometry of sorbed arsenate on ferrihydrite and crystalline FeOOH: Re-evaluation of EXAFS results and topological factors in predicting sorbate geometry, and evidence for monodentate complexes. Geochim. Cosmochim. Acta **59**, 3655–3661 (1995)
81. G.A. Waychunas, B.A. Rea, C.C. Fuller, J.A. Davis: Surface chemistry of ferrihydrite: Part 1. EXAFS studies of the geometry of coprecipitated and adsorbed arsenate. Geochim. Cosmochim. Acta **57**, 2251–2259 (1993)
82. J.V. Lauritsen, J. Kibsgaard, S. Helveg, H. Topsoe, B.S. Clausen, E. Laegsgaard, F. Besenbacher: Size-dependent structure of MoS_2 nanocrystals. Nat. Nano. **2**, 53–58 (2007)
83. A.M. Derfus, W.C.W. Chan, S.N. Bhatia: Probing the cytotoxicity of semiconductor quantum dots. Nano Lett. **4**, 11–18 (2004)
84. P. Borm, F.C. Klaessig, T.D. Landry, B. Moudgil, J. Pauluhn, K. Thomas, R. Trottier, S. Wood: Research strategies for safety evaluation of nanomaterials. Part V. Role of dissolution in biological fate and effects of nanoscale particles. Toxicol. Sci. **90**, 23–32 (2006)
85. C. Fan, J. Chen, Y. Chen, J. Ji, H.H. Teng: Relationship between solubility and solubility product: The roles of crystal sizes and crystallographic directions. Geochim. Cosmochim. Acta **70**, 3820–3829 (2006)
86. A.L. Rogach, D.V. Talapin, E.V. Shevchenko, A. Kornowski, M. Haase, H. Weller: Organization of matter on different size scales: Monodisperse nanocrystals and their superstructures. Advanced Functional Materials **12**, 653–664 (2002)
87. D.V. Talapin, A.L. Rogach, M. Haase, H. Weller: Evolution of an ensemble of nanoparticles in a colloidal solution: Theoretical study. J. Phys. Rev. B **105**, 12278–12285 (2001)
88. A.N. Goldstein, C.M. Echer, A.P. Alivisatos: Melting in semiconductor nanocrystals. Science **256**, 1425–1427 (1992)

13

Fate of Nanoparticles in Aqueous Media

Jérôme Labille and Jean-Yves Bottero

The European Commission estimated the world nanotechnology market at slightly over 40 billion euros in 2001. In 2010–2015, according to the estimates of the US National Science Foundation (NSF), the worldwide economic stakes due to the advent of nanotechnology could run as high as 1 000 billion dollars per year across all sectors [1]. This growth is founded on the multitude of potential applications of nanotechnologies. Considering the exponential increase in mass production of nanomaterials that this implies, it becomes important to ask what would become of such materials should they be released into the environment. Indeed, nanoparticles are very small, comparable in size to a virus, suggesting a high level of mobility in the environment and living organisms, down to the smallest length scale, that is, internalisation by living cells. For this reason, investigations into this specific problem, underway since the beginning of the 2000s when nanotechnology came into its own, must be organised upstream if possible, but at worst in parallel with research and development, and as far as possible in collaboration with it.

The aim of this chapter is to identify all the factors believed to control the transport and fate of nanoparticles in the environment. This is a key consideration if we are to understand the parameters determining their bioavailability and potential toxicity. Indeed, depending on whether the nanoparticles released into the environment are perfectly dispersed in the aqueous phase or end up trapped in some kind of collecting medium, the effects on living organisms will be completely different and access to the food chain will be modified, whence the overall environmental impact of their life cycle can vary from one situation to another.

Many factors influence the various scenarios, both environmental and intrinsic to the nanoparticles themselves. However, we may nevertheless classify them into two ever opposing tendencies, namely, dispersion of the nanoparticles favouring their mobility, or aggregation, attachment, sorption, or trapping on a surface (see Fig. 13.1).

P. Houdy et al. (eds.), *Nanoethics and Nanotoxicology*,
DOI 10.1007/978-3-642-20177-6_13,

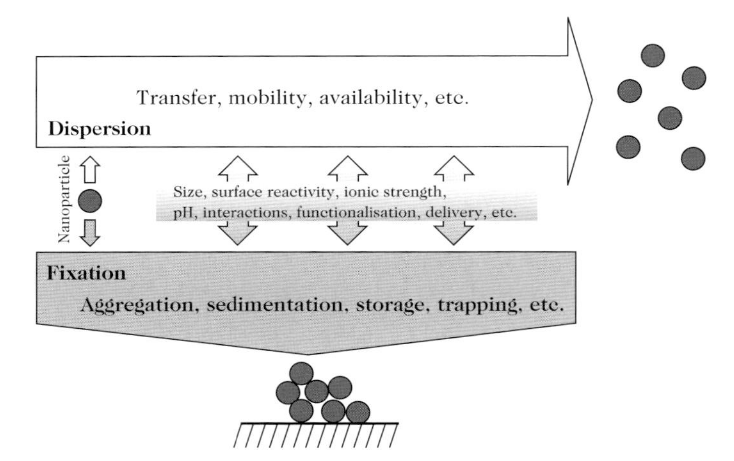

Fig. 13.1. Parameter controlling the transport, fate, and bioavailability of nanoparticles in an aqueous environment

Hence, in order of increasing complexity:

1. The intrinsic properties of the nanoparticles, and notably their surface properties, determine their affinity for the dispersing medium, and hence their tendency to agglomerate or disperse, depending for example on their hydrophilic or hydrophobic nature.
2. The ionic strength and pH of the dispersing medium largely control the stability of the nanoparticles in suspension via interparticle electrostatic repulsion.
3. The interaction of the nanoparticles with surrounding dissolved or particulate matter is likely to modify their surface properties and hence their behaviour, but it may also integrate them into other, larger scale matter transfer processes, such as the cycle of natural mineral elements or the food chain, for example.
4. Finally, the aquatic environment is far from being a homogeneous and continuous liquid phase, and there are many systems that could constitute a barrier to nanoparticle migration. Disordered organic or inorganic porous media, such as soils or biofilms, constitute compartments within which the opposing effects of nanoparticle mobility and nanoparticle attachment or trapping will determine their life cycle in the environment.

The behaviour of particles in water, especially particles in the colloidal size category, i.e., in the micrometer range, has been widely studied in the literature. The many physicochemical approaches for understanding and modelling their behaviour in suspension provide a good knowledge base for the new problem of understanding and predicting the behaviour of nanoparticles in aquatic media. However, the very definition of nanoparticles and nanomaterials, namely, that at this length scale, these materials exhibit quite different, even unheard-of properties compared with those at larger length scales, and in particular a much higher surface reactivity, implies that their behaviour in the environment may differ significantly from the predictions of colloid physicochemistry.

This chapter is essentially built up around the incomplete analogy between nanoparticles and colloids, benefiting from the conventional physicochemical understanding of colloids when it is indeed applicable to nanoparticles, and focusing on its limitations or proposing further considerations when problems arise with this approach.

13.1 Specific Features of Nanoparticles

In moving from the microscale to the nanoscale, various changes occur in the physicochemical properties characterising particles. These novel properties, unique to nanoparticles, are likely to have a considerable bearing on their transport within the environment.

13.1.1 Increased Specific Surface Area

The smaller the size, the greater the ratio of surface area to volume, and the more the surface properties will be enhanced. For example, for spherical particles, we have

$$S_{\text{spe}} = \frac{3}{\rho r}\,, \tag{13.1}$$

where S_{spe} is the specific surface area expressed in units of area per unit mass, ρ is the density of the spherical object, and r is its radius. This shows how the specific surface area can increase enormously as the particle radius decreases, opening the possibility of a very large number of physicochemical interactions at the interface with the surrounding medium. In addition to this high adsorption capacity, the surface reactivity also tends to increase.

13.1.2 Size-Related Surface Energy

When the particle size is decreased, the proportion of its atoms at the surface increases exponentially. So within the nanoparticle family, we may distinguish the category of ultrasmall nanoparticles, with diameters less than 20 nm, in which the proportion of atoms at the surface becomes significant, even predominant.

Nanoparticles in this category have a particularly high surface reactivity and tend to exhibit novel properties (see Chap. II). This is easily explained thermodynamically, from the point of view of minimising the surface free energy as given by the Young–Laplace equation

$$\Delta P = \frac{2\gamma}{r}, \tag{13.2}$$

where ΔP is the pressure change across the surface, γ the interfacial tension, and r the particle radius. The surface free energy thus increases as the particle size decreases. In the case of ultrasmall nanoparticles, the surface pressure exerted on the objects is such that specific atomic rearrangements may sometimes occur at the interface of the nanoparticle and the environment. This in turn leads to novel properties, and in particular, an enhanced surface reactivity which tends to minimise the interfacial tension.

13.1.3 Consequences for the Fate of Nanoparticles in the Environment

A high adsorption capacity combined with an enhanced surface reactivity are the two main properties characterising nanoparticles. Their effects may be expressed in various ways in the environment, depending on the type of nanoparticle and the composition of the system.

The tendency of dispersed nanoparticles to reduce their surface free energy by reorganising themselves into structures with lower specific surface area is certainly the property most frequently encountered in the natural environment. One way to achieve this is through the phenomenon of aggregation, i.e., the particles tend to stick together to form agglomerates and thereby lower the area of contact with the solvent. Aggregation dynamics is a physicochemical phenomenon that has been widely discussed and modelled in the literature. However, none of the present models take into account surface reactivity, a prerequisite for adapting them to nanoparticles.

At the same time, other reactions can increase the size of the nanoparticles, such as growth by attachment of dissolved species. For example, iron oxide nanoparticles such as nanomaghemite with diameter less than 20 nm exhibit an adsorption capacity for the arsenate ion which exceeds a monolayer of adsorbed ions. Indeed, a careful investigation has demonstrated that the arsenate ions adsorb preferentially at the octahedral sites of the iron oxide crystal system, in such a way as to simulate crystal growth [2], thereby reducing the radius of curvature of the nanoparticle, and hence also its surface tension.

Finally, local atomic rearrangement provides another process whereby ultrasmall nanoparticles can reduce the surface tension. For example, depending on the conditions of temperature and pressure, titanium dioxide adopts the most stable crystal phase available. This means that, at the surface of 14 nm nanoparticles, the rutile mineralogical form is less stable and the system tends to reorganise into the anatase form [3].

13.2 Nanoparticle Dispersion and Transport in Aqueous Media

13.2.1 Nanoparticle Surface Properties and Affinity for Water

The intrinsic surface properties of nanoparticles constitute the main parameters determining their affinity for a dispersing aqueous medium. The particle may be hydrophilic or hydrophobic. Hydrophilic nanoparticles exhibit a high affinity for water, with total wetting, so their immersion meets no energy resistance. They disperse very easily and uniformly through the aqueous phase. On the other hand, when they come into contact with the aqueous phase, hydrophobic nanoparticles tend to agglomerate in order to reduce the particle–water contact area. The result is the formation of clusters, which tend to float to the surface of the water or sink, depending on their density.

Metal Oxides

The main manufactured metal oxides the most widely used in nanotechnology are titanium dioxide TiO_2, cerium dioxide CeO_2, silica SiO_2, the iron oxides Fe_3O_4 (maghemite and magnetite), and zinc oxide ZnO. These minerals have the particularity of existing naturally in varying amounts. However, differences in the size of the particles (natural SiO_2 is coarser), their crystal phase (natural TiO_2 tends to be amorphous), or their magnetic properties (Fe_3O_4) can generally be used to identify man-made metal oxides.

By virtue of their atomic structure, these minerals exhibit free coordinations at their surface, and these can attach dissociated water molecules at their surface by chemisorption. This in turn gives rise to oxo or hydroxo groups, whose protonation–deprotonation reactions give these surfaces an amphoteric charge. The oxide is characterised by its pH at zero charge, or point of zero charge (PZC) (see Table 13.1):

$$\equiv\text{S–O}^- + \text{H}_3\text{O}^+ \iff \equiv\text{S–OH} + \text{H}_2\text{O} \iff \equiv\text{S–OH}_2^+ + \text{OH}^- \,. \tag{13.3}$$

This surface speciation of metal oxides gives them a highly hydrophilic tendency, and they disperse easily in aqueous media.

Table 13.1. PZC values of various metal oxides

Oxide	PZC
TiO_2	5.4–6.9
CeO_2	8.1–8.6
SiO_2	2–4
Fe_3O_4	6.5–7
ZnO	9–10

Carbon-Containing Nanoparticles

Fullerenes

Fullerenes are a very special kind of nanoparticle. Due to their aromatic structure, they are also referred to as molecules. Discovered in 1985 [4], there is still much to learn about this particular crystal form of carbon, and indeed a great deal of research is currently under way to better understand all its properties. These nanoparticles, and more precisely, those denoted by C_{60}, are indeed highly reactive chemically, by virtue of their aromatic structure and mechanically strained spatial conformation [5].

The question of the affinity of the fullerenes for aqueous media is another subject of debate. Initially, these nanoparticles were considered to be totally insoluble in water and perfectly hydrophobic, along the lines of the other crystal phases of carbon, i.e., graphite and diamond. Indeed, fullerenes tend to be organophilic, with high solubility in certain organic solvents in the benzene and naphthalene families [6, 7].

However, more and more counterexamples are coming to light in the literature. There are many ways to obtain a stable dispersion of C_{60} nanoparticles in an aqueous phase, without using surfactants or modifying the surface of the nanoparticles. These methods usually involve liquid/liquid extraction under mechanical shaking, where the nanoparticles are transferred from an organic solvent into water [8–11]. The result is an aqueous dispersion of fullerenes in the form of clusters of the order of a hundred nanometers across ($n\mathrm{C}_{60}$), raising interesting questions of how the water molecules arrange themselves at the surface of the $n\mathrm{C}_{60}$ clusters. A much less efficient method for obtaining stable dispersions of $n\mathrm{C}_{60}$, and with a much lower yield, involves simply shaking a mixture of pure fullerenes and H_2O for a sufficiently long time (20 h of ultrasound or 15 days of magnetic shaking) [12].

The effective affinity of fullerenes for water was finally brought out by producing an adsorption isotherm for water vapour on previously dehydrated C_{60} powder (see Fig. 13.2). The adsorption of water molecules on the surface of the powder, which begins even for low partial pressures of water vapour, and its multilayer arrangement at $P/P_0 > 0.7$, are characteristic of a hydrophilic material. In addition, the hysteresis observed during desorption, which lasts right through to total drying of the sample ($P/P_0 = 0$, $\mathrm{H_2O_{ads}} = 1$ monolayer), indicates an irreversible change in the hydration state of the fullerene, and hence probably in its surface chemistry. Indeed, the first monolayer of adsorbed water remains strongly bound to the surface. Fourier transform infrared spectroscopy and proton nuclear magnetic resonance spectroscopy have demonstrated hydroxylation of the fullerene during this wetting stage [13]. The hydrophilicity resulting from this reaction is what allows the $n\mathrm{C}_{60}$ clusters to disperse in water.

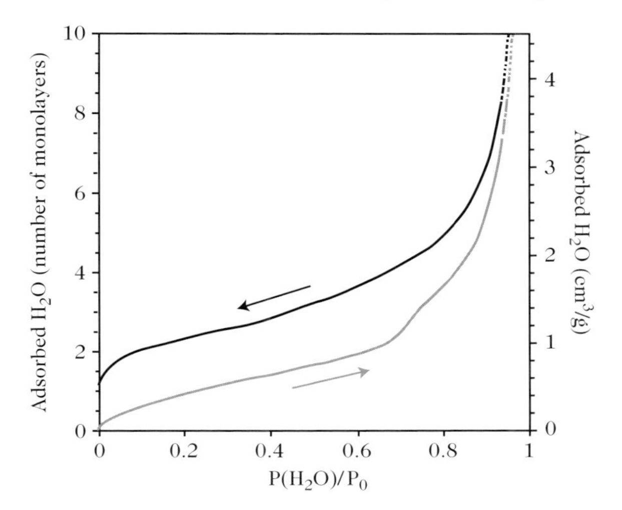

Fig. 13.2. Adsorption (*grey*) and desorption (*black*) isotherms of water vapour on a powder of pure fullerene C_{60}

Carbon Nanotubes

Carbon nanotubes are rolled up graphene sheets and as a consequence have a quite similar chemistry and atomic organisation to the fullerenes in many respects. However, it turns out that minor differences of a structural or chemical kind between nanotubes and fullerenes give rise to their totally different behaviour in the aqueous phase. Indeed, the high chemical reactivity of the fullerenes, allowing their hydroxylation, comes in part from their highly electrophilic aromatic atomic structure, with its mechanically highly strained C–C bonds. The last feature does not occur to the same extent in carbon nanotubes. For this reason, they have a very low affinity for water, characteristic of a totally hydrophobic material.

On the other hand, a second feature of carbon nanotubes distinguishing them from the fullerenes can in fact favour an increased affinity for water, namely, their generally rather low level of purity. Indeed, the metal catalysts used to synthesise carbon nanotubes, like iron or nickel, end up integrated within them in the form of defects in their atomic structure. But these defects are a potential source of locally higher affinity for the aqueous phase. For example, the oxidation of iron atom defects in carbon nanotubes can favour their dispersion in water. So the overall affinity of a nanotube for water is inversely proportional to its degree of purity.

Metal Nanoparticles

Pure metal nanoparticles like silver, gold, or iron nanoparticles, are widely used in nanotechnology for their various specific properties. The evolving

Table 13.2. Redox potentials of different metals

Oxidant/reductant	E^0 (V)
Au^{3+}/Au	+1.50
O_2/H_2O	+1.23
Hg^{2+}/Hg	+0.85
Ag^+/Ag	+0.80
Fe^{3+}/Fe^{2+}	+0.77
Cu^{2+}/Cu	+0.34
Pb^{2+}/Pb	−0.13
Sn^{2+}/Sn	−0.14
Ni^{2+}/Ni	−0.23
Cd^{2+}/Cd	−0.40
Fe^{2+}/Fe	−0.44
Zn^{2+}/Zn	−0.76
Al^{3+}/Al	−1.66

reactivity of surface species in an aqueous medium, in other words the stability of their purely metallic particulate state, depends largely on their redox potential. The lower the redox potential compared with that of the oxidising surrounding medium, e.g., the pair O_2/H_2O, the more likely the metal is to oxidise, and conversely. Hence, among the three examples of metal nanoparticles mentioned above, iron and silver tend to oxidise to Fe^{2+} then Fe^{3+} and Ag^+, respectively, whereas gold is of course perfectly stable (see Table 13.2). However, the natural surrounding medium is sometimes locally superoxidizing, in areas under biotic influence, for example, and this may result locally in the destabilisation of the metals by oxidation.

13.2.2 Stability of Nanoparticles in Suspension Dispersion and Aggregation

The dynamics of nanoparticle aggregation or deposition can be broken down into two stages:

- The nanoparticles are physically brought into collision with a surface.
- If the conditions are right, the nanoparticle attaches itself somehow to this surface.

In the case of aggregation, the surface in question may be that of an identical or totally different neighbouring particle, or the surface of a growing aggregate. In the case of deposition, the surface will be that of a fixed collector on which the nanoparticles accumulate.

The result of this two-stage dynamics depends largely on the nanoparticle transport conditions (diffusion) on the one hand, and on the balance of interactions between nanoparticle and surface (attractions and repulsions) on the other. By understanding the factors determining the colloidal stability of nanoparticle suspensions, one can:

- predict their behaviour in an aqueous environment on the basis of given environmental conditions,
- invent water decontamination processes capable of removing the nanoparticles by exploiting flocculation/sedimentation mechanisms, for example [14].

Forces at Interfaces

Van der Waals Forces

The attractive Van der Waals forces are present in all systems. They result from the short range forces between atoms due to instantaneous fluctuations of the ionic clouds around their nuclei. In energy terms, the interaction between two spheres of radius r with distance x between their centres [15] is given by

$$V_{\mathrm{vdw}}(x) = -\frac{A}{6}\left[\frac{r^2}{x^2-4r^2} + \frac{2r^2}{x^2} + \ln\left(1-\frac{4r^2}{x^2}\right)\right] , \qquad (13.4)$$

where A is the Hamaker constant, a characteristic of the material from which the particles are made and the interstitial material. Typical values are $A \sim 10^{-20}$ J. Hence, V_{vdw} varies as $-1/x^6$.

Electrostatic Repulsion

In an electrolyte solution, the charge of a particle in suspension is neutralised by a cloud of counterions concentrated at its surface. This cloud is organised into a layer of ions adsorbed on the surface of the particle, called the Stern layer, and an outer diffuse layer. This structure is referred to as the electrical double layer (see Fig. 13.3).

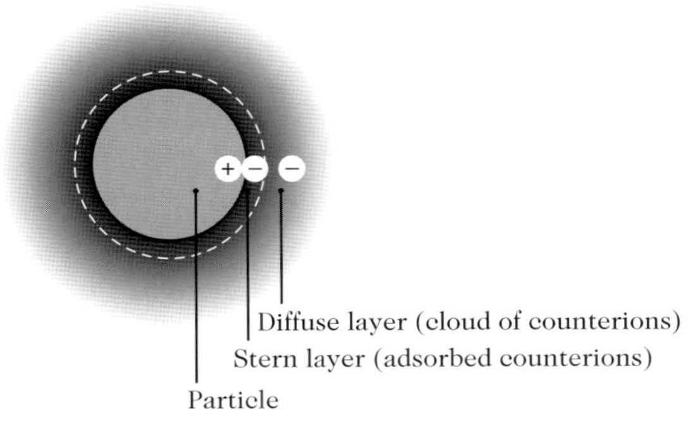

Fig. 13.3. Double layer distribution of counterions (negative charges) from the solution at the surface of a positively charged particle

The concentration of counterions in the diffuse layer decreases exponentially on moving away from the particle, until it reaches the value of the ionic background. Their distribution is modelled by the Poisson–Boltzmann equation (13.5), where the solvent is treated as an ideal continuum of dielectric constant ε, and the ions as point charges q that do not interact with one another, but only directly with the particles:

$$\nabla^2 \Psi(x) = -\frac{1}{\varepsilon} \sum_i q_i c_i^\infty \exp\left[-\frac{q_i \Psi(x)}{kT}\right] , \tag{13.5}$$

where Ψ is the charge potential, x the distance from the particles, c_i^∞ the concentration of the ions i, k the Boltzmann constant ($k = 1.38 \times 10^{-23}$ J/K), and T the temperature.

There is no exact analytic solution to the Poisson–Boltzmann differential equation. Gouy (1910) and Chapman (1913) derived it for the simple case of a plane, uniformly charged particle of infinite extent in an infinite, homogeneous, symmetrical electrolyte. Their equation thus reduces (13.5) from three dimensions to just one:

$$\frac{\mathrm{d}^2 \Psi}{\mathrm{d}^2 x} = -\frac{q C^\infty}{\varepsilon} \sinh \frac{q \Psi(x)}{kT} , \tag{13.6}$$

where ε is the dielectric constant of the medium.

The Debye–Hückel approximation brings in one more assumption, namely that the surface potential Ψ_0 does not exceed 25 mV and that the potential at infinite distance from the surface is zero. This allows one to linearise the Poisson–Boltzmann equation to obtain

$$\Psi(r) = \Psi_0 \exp(-\kappa r) , \tag{13.7}$$

where κ^{-1} is the Debye length, directly related to the concentration and charge of the ions, defined by

$$\kappa^2 = \frac{\sum_i q_i^2 c_i^\infty}{\varepsilon k T} . \tag{13.8}$$

For $T = 298$ K and in water with a uniform electrolyte, this gives

$$\kappa = 3.288 \sqrt{I} , \tag{13.9}$$

where I is the ionic strength.

Hence repulsive forces of electrostatic nature arise between neighbouring particles of the same charge sign in suspension, because their electrical double layers repel one another. It can be shown that the electrostatic repulsion energy V_r between two particles is given as a function of the distance x between them by [16]

$$V_\mathrm{r}(x) = \frac{\pi \varepsilon r_1 r_2}{r_1 + r_2} \left\{ (\Psi_1 + \Psi_2)^2 \ln\left[1 + \exp(\kappa x)\right] + (\Psi_1 - \Psi_2)^2 \ln\left[1 - \exp(-\kappa x)\right] \right\} , \tag{13.10}$$

where Ψ_i is the surface potential of particle i and r is its radius.

For identical, parallel, plane surfaces and for $\kappa x \gg 1$, an approximate expression for $V_{\mathrm{r}}(x)$ is

$$V_{\mathrm{r}}(x) = \frac{64NckT}{\kappa} \tanh^2\left(\frac{e\Psi_0}{4kT}\right) \exp(-\kappa x) , \tag{13.11}$$

where N is Avogadro's number. V_{r} thus increases exponentially as the distance between the two planes is reduced down to a distance close to κ^{-1} and/or when κ^{-1} decreases with the ionic strength.

DLVO Theory

The DLVO theory due to Derjaguin, Landau, Verwey, and Overbeek [17, 18] expresses the balance of forces in the vicinity of the particle–electrolyte interface, taking into account the electrostatic repulsions between the electrical double layers, the short range Van der Waals attractions, and the very short range Born repulsions (impossible interpenetration of atoms).

When the sum of the attractive forces (negative potential energy) and repulsive forces (positive potential energy) at the interface between a particle and a charged surface is plotted as a function of the distance between them, the typical profile of the DLVO energy balance is as shown in Fig. 13.4. There are three regions characteristic of the dominant interactions depending on the distance on the one hand and the energy balance relative to the interface on the other.

At almost zero separation (III), the surfaces are in contact, the Van der Waals forces dominate, and the interaction is strongly attractive. A large

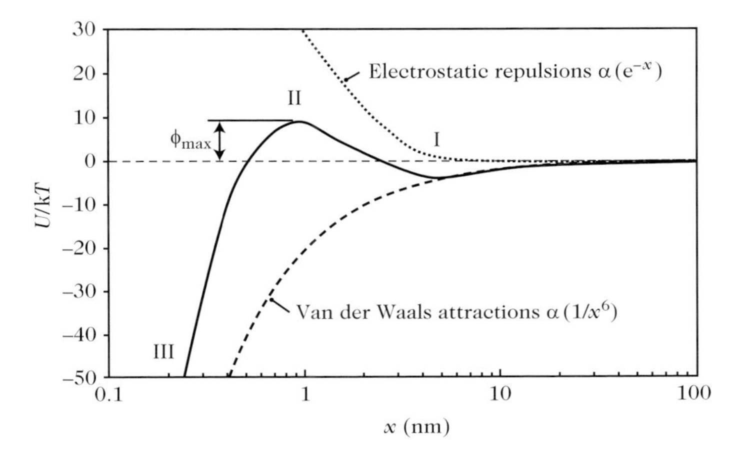

Fig. 13.4. Example of the interaction between a particle and a plane surface, illustrating the three characteristic regions: secondary attractive minimum (I), repulsive energy barrier (II), and primary attractive minimum (III). The amplitude of the energy barrier is defined by ϕ_{max}

energy would be required to separate the surfaces, of the kind rarely encountered in the environment. This is the primary energy minimum.

At a certain separation distance between the surfaces, the contribution of the electrostatic forces creates a repulsive energy barrier (II) which tends to prevent either approach toward or separation from the surface. It can be overcome when the system has enough energy, e.g., under the effects of shaking, heating, or shear, which increase the potential energy of the particles. Whether or not the repulsive energy barrier can be overcome also depends on the height $\phi_{\max}$ of the barrier, of course, itself related to the magnitude of the repulsive forces. Now the latter may actually be non-existent, in the case of opposite or zero charges between the particle and the nearby surface. In this case, the interactions are purely attractive at all distances, or else they may be screened, depending on the physicochemical properties of the medium, such as the pH or the ionic strength (a phenomenon discussed in more detail below), which therefore also influence the balance of forces at the interfaces. In most cases, when the attractive and repulsive contributions coexist with the same order of magnitude, the overall profile of the force system gives rise to a secondary attractive minimum (I) at the foot of the energy barrier, corresponding to a distance at which the interaction between the particles is weakly attractive.

The DLVO theory applies to highly simplified systems, where the particles are rigid and carry uniform surface charge, the electrolyte is symmetric, and the charges can be treated as points. In reality, these conditions are not usually satisfied, and experimental observations often deviate from the predictions of the theory. As a result, the DLVO theory has been more and more refined and improved to suit the specific features of the systems under investigation.

In the case of nanoparticles, the surface tension parameter which tends to favour the aggregation of ultrasmall particles in suspension is a determining factor which must necessarily be taken into account when computing the balance of forces at interfaces.

Effects Due to Salinity and Counterions

Since the Van der Waals attractive forces are constant, the balance of forces varies essentially due to the repulsive electrostatic forces, since they depend on conditions like the pH and the ionic strength in the medium. The Debye length, and hence the range of the electrostatic repulsions, varies inversely with the ionic strength of the medium (13.7–13.9). So for low ionic strength, interparticle repulsions are strong and this favours dispersion. On the other hand, when the ionic strength of the medium increases, the repulsions are reduced to shorter range, and the energy barrier of the system is lowered. The Van der Waals attractions then become the dominant forces in the system, so that when particles collide they will tend to stick together and form macroscopic aggregates, subjected to sedimentation under the effects of gravity. This aggregation mechanism is a form of colloidal destabilisation by coagulation.

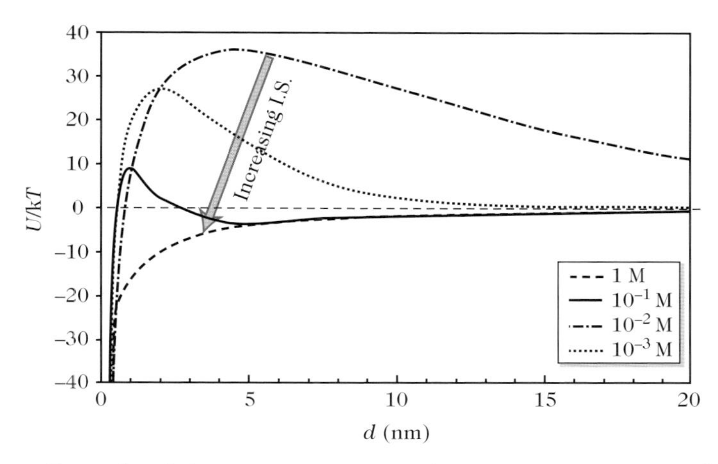

Fig. 13.5. Example of the interaction energy between two charged surfaces, calculated using the DLVO theory for different concentrations of monovalent salt. $T = 20°\mathrm{C}$, $A = 10^{-20}$, and particle radius $= 50\,\mathrm{nm}$

This effect of the salt concentration on the balance of forces at interfaces is illustrated in Fig. 13.5, where the repulsive energy barrier falls quickly as the ionic strength increases, to the benefit of the shorter range Van der Waals attractions.

The salt concentration required to achieve such a destabilisation is called the critical coagulation concentration (CCC) [19–21]. As can be seen from the calculation of the Debye length, which involves the ionic strength, compression of the double layer is more pronounced for higher counterion valence z. The efficiency of coagulation is thus increased and the CCC reduced when there are multivalent ions in solution. The empirical Schulze–Hardy rule, i.e., CCC $\propto z^{-6}$, describes this tendency for highly dilute suspensions, where the particles can be treated as infinitely far apart, the electrolyte symmetrical, the surfaces planar, and the surface potential constant [22].

The aggregation of a nanoparticle suspension is exemplified in Fig. 13.6 for $n\mathrm{C}_{60}$ nanoparticles in the presence of two different salts, NaCl and $CaCl_2$. When the salt concentration increases above the CCC, the diffusion coefficient $D_{n\mathrm{C}_{60}}$ of the entities in suspension drops suddenly, indicating that the nanoparticles will aggregate. The CCC values measured with $CaCl_2$ and NaCl are $(2\pm0.4)\times10^{-3}$ and $(1\pm0.4)\times10^{-1}$ mol/L, respectively. These values agree with the salt valence effect predicted by the Schulze–Hardy relation for ideal systems. This means that, for this precise example with nanoparticles, the above-mentioned classical theory for the dynamics of colloids and surfaces is applicable to predict their behaviour in suspension. Since $n\mathrm{C}_{60}$ nanoparticles have an average size of about 100 nm, the typical high reactivity of nanoparticles, observed in particular for sizes smaller than 20 nm, is not expected in this

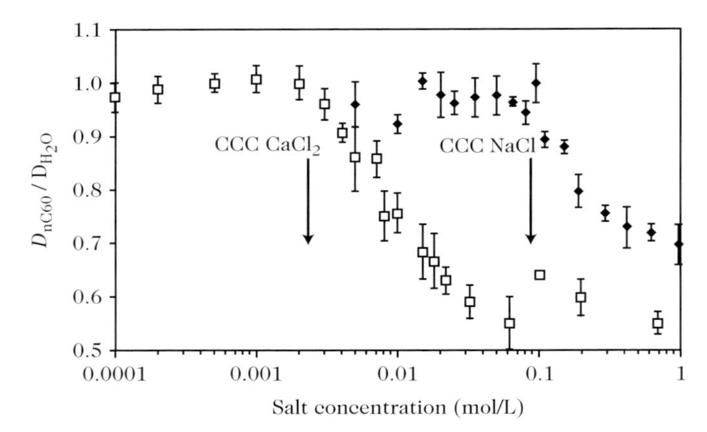

Fig. 13.6. Average diffusion coefficient of nC_{60} in aqueous suspension for different values of the salt concentration and valence: $CaCl_2$ (*squares*) and NaCl (*diamonds*). Data are normalised with respect to the reference value D_{H_2O} measured in water in the absence of the salt. *Arrows* indicate the critical coagulation concentrations of each salt

case, and the dynamics of the particles in suspension can indeed be described by the classical theory.

It should also be noted that the CCC values for nC_{60} plotted in Fig. 13.6 are of the same order of magnitude as the salinities encountered in sea water (NaCl) and fresh water ($CaCl_2$). This implies that the stability of the nanoparticle suspension is very sensitive to the salinity of the aqueous medium. For example, natural areas where waters with very different salinities mix together, e.g., estuaries, are likely to be the scene of significant nanoparticle aggregation/sedimentation phenomena.

Effects Due to pH

When the nanoparticle surface charge is amphoteric, e.g., metal oxides, nC_{60}, it can be amplified, reversed, or eliminated, depending on the pH. The pH thus plays a decisive role in the stability of the dispersion, by controlling interparticle electrostatic repulsions.

The surface electric potential of the particles, or ζ potential, controls these electrostatic interactions. It can be calculated by subjecting the particle suspension to an electric current under which the particles migrate toward the electrode of opposite charge. The rate of migration, or electrophoretic mobility, is then measured. It gives the ζ potential directly.

The ζ potential thus varies with the pH as a result of surface protonation/deprotonation, and vanishes at a pH value defined as the isoelectric point (IEP) (see Figs. 13.7 and 13.8), characteristic of the surface chemistry of the material. Near this point, the absence of electrostatic interparticle

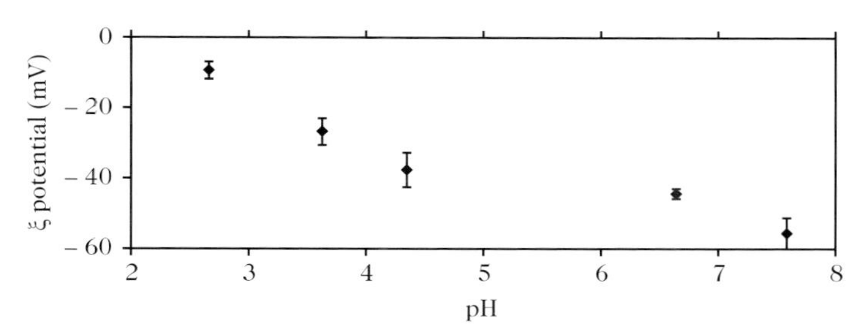

Fig. 13.7. pH dependence of the ζ potential of nC_{60} in aqueous suspension. The isoelectric point is near pH = 2

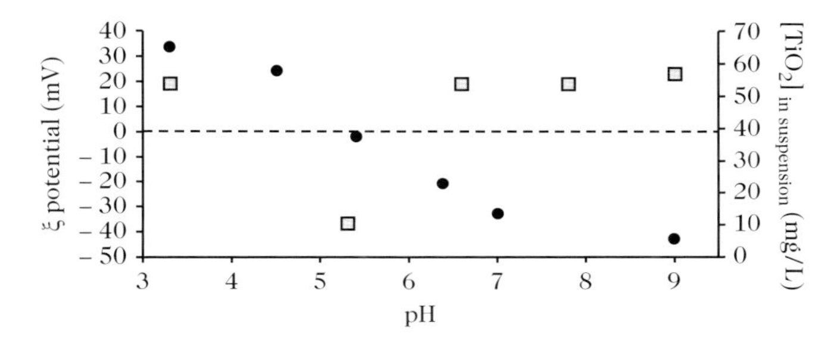

Fig. 13.8. pH dependence of the ζ potential (*disks*) of TiO_2 nanoparticles and correlation with the concentration of nanoparticles in suspension (*squares*). The isoelectric point close to pH = 5.5 corresponds to the minimal concentration of nanoparticles in suspension due to aggregation/sedimentation. From [23]

repulsions destabilises the dispersion and the nanoparticles aggregate (see Fig. 13.8).

Point of Zero Charge (PZC) and Isoelectric Point (IEP). The point of zero charge is the pH for which the surface charge σ_0, probed directly on the particle surface by proton titration, vanishes. In contrast the isoelectric point is the pH for which the electric or ζ potential, measured in the vicinity of the hydrodynamic cutoff plane of the particle by electrophoretic mobility, vanishes (see Fig. 13.9). The titration yields an intrinsic surface charge density, whereas the ζ potential takes into account electrostatic interactions between the particle and its neighbourhood.

Interactions with Dissolved and Particulate Matter

The fate of nanoparticles when released into the environment can be affected by many physicochemical processes. In rhizospheric or aquatic natural environments, their potential interaction with surrounding dissolved or particulate matter, either natural or man-made, is likely to modify their physicochemical properties, and hence their stability in suspension. The various processes that may come about through such interactions depend essentially on the affinity and the size ratio between the nanoparticles and the other elements of the

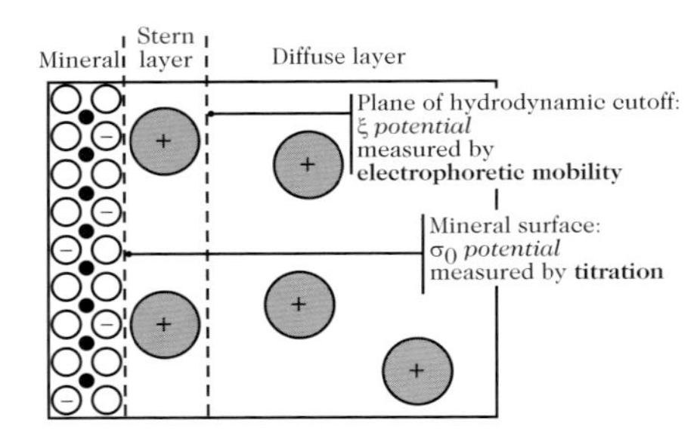

Fig. 13.9. Point of zero charge and isoelectric point

environment. For example, a distinction can be made between the particulate matter in suspension belonging to the colloidal fraction, in the micrometer range, which includes fine inorganic particles and organic macromolecules, and dissolved matter in solution, which includes small organic or inorganic molecules.

Nanoparticle–Colloid Interactions. Whether the colloids are organic or inorganic, their micrometric size and high surface reactivity means that they have a high specific surface area and that this will be a key feature for a great many physicochemical interactions at the interfaces encountered in the environment. Inorganic particles, such as clays, or organic macromolecules (polysaccharides) are particularly abundant in natural systems, and these constitute the most reactive colloids [24–26].

For nanoparticles, with their enhanced surface reactivity and even greater specific surface area [see (13.1)], these colloids in suspension constitute a considerable potential interaction area. In the case of adsorption, the transfer of the nanoparticles into the environment, and possibly into living organisms, will necessarily be determined by the carrier, and will thus be governed by new conditions of colloidal stability with regard to the ionic strength, pH, and hydrodynamic context. For example, they may be trapped into the well known mechanisms of flocculation, which often control the fate of colloids in suspension, and subsequently sediment out in situ, when the local hydrodynamic situation allows it.

The affinity of the nanoparticles for colloids depends mainly on the physicochemical properties of the respective surfaces. In general, it is the nanoparticle–colloid electrostatic interactions that dominate. For example, opposite surface charges favour rapid adsorption, whereas when the surface charges are of the same sign, efficient adsorption requires screening of the electrostatic repulsions by an increase in salinity. But these key parameters for adsorption, viz., the sign and density of the surface charge, are directly dependent on the pH of the solution in relation to the isoelectric points of the surfaces that come into contact. Hence, surface chemistry, salinity, and pH are the three interdependent parameters determining the affinity of nanoparticles for the surrounding particulate matter.

When the colloids are made up of organic macromolecules, the complexity of the nanoparticle adsorption mechanism is increased due to the incidence of the spatial conformation of the macromolecules. Indeed, flexible macromolecules will tend to tangle up around the nanoparticles, yielding a stable multipoint adsorption, whereas rigid and extended macromolecules will tend to reduce the number of points of contact with the nanoparticle and hence discourage adsorption. Moreover, the rigidity/flexibility of the macromolecules depends to a large extent on the electrostatic interactions between the charged sites on them. There are therefore many possible interactions between nanoparticles and macromolecules, depending on the size, charge density, and charge sign of each, and also on the ionic strength and pH of the medium [27–29].

When the interaction conditions are such that the adsorbed macromolecules have one end of a strand or elongated loops stretching out to a certain distance from the nanoparticle, it may happen that the free parts of these molecules interact with neighbouring nanoparticles. The macromolecule then forms interparticle bridges, which may initiate aggregation and sedimentation. This is called bridging flocculation (see Fig. 13.10).

A specific example of nanoparticle flocculation by macromolecular bridging is shown in Fig. 13.11 for nC_{60} in the presence of bacterial polysaccharide. Size distribution measurements were made by dynamic light scattering. In pure water without salt, no aggregation is observed, because the interparticle electrostatic repulsions keep them well separated. However, when the salinity

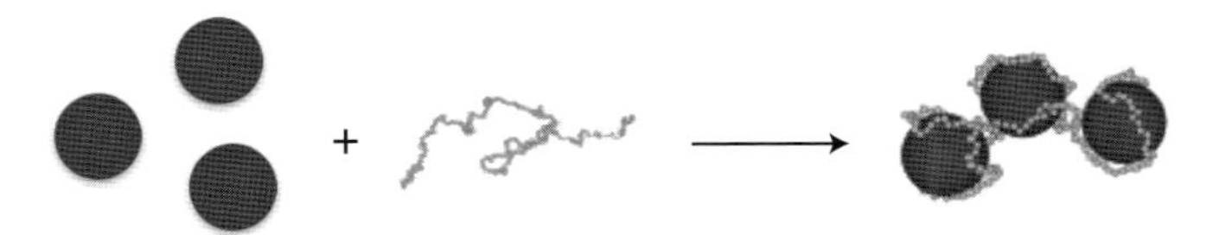

Fig. 13.10. Example of bridging flocculation resulting from the adsorption of a macromolecule onto several originally dispersed nanoparticles. Adapted from [28]

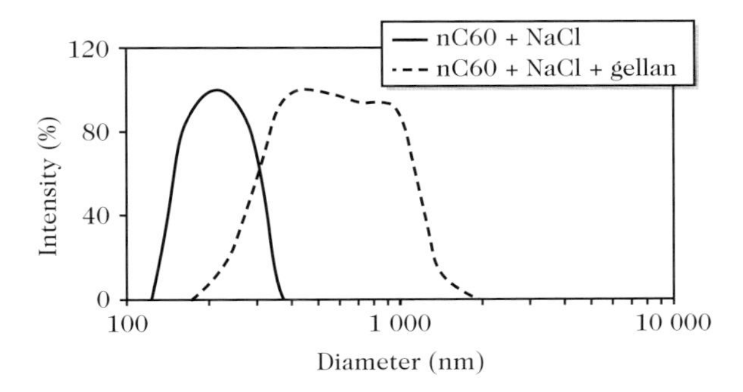

Fig. 13.11. Size distributions of nC_{60} in NaCl (3.5×10^{-3} M), with or without polysaccharide gellan. Measurements made by dynamic light scattering

of the solution is taken to 3.5×10^{-3} mol/L of NaCl, i.e., just below the CCC, the interparticle electrostatic repulsions are partially screened, allowing them to come closer together, but not to coagulate. In this case, flocculation by polysaccharide gellan is observed, with an average size increase from 200 to 400–1 000 nm. Indeed, for bridging flocculation to occur, the separation between two neighbouring particles must not exceed the free strand length of the polymer.

Interactions Between Nanoparticles and Dissolved Molecules. By virtue of their high surface reactivity, nanoparticles disseminated in suspension in an aquatic environment are also the scene of many interactions with the surrounding dissolved matter. These molecules may be organic or inorganic and their tendency to adsorb onto the nanoparticles will depend on their affinity for the nanoparticle surface. Various types of chemical, electrostatic, or physical interaction are encountered in the environment. The most common are discussed below. In every case, the adsorption of these molecules onto the nanoparticle surface has significant consequences for the fate of the ensemble in suspension in the medium, and therefore on its potential toxicity, since nanoparticle surface properties, and hence the forces at the interfaces, are completely transformed as a result of their new coating.

By definition, hydrophobic nanoparticles like carbon nanotubes have little tendency to disperse in an aqueous medium. When they are released into natural water, they thus try to reduce their area of contact with the aqueous phase, either by aggregating, or by attaching amphiphilic elements to their surface to reduce the surface energy. Small naturally occurring organic molecules such as humic substances, which have surfactant properties, are perfect for this role, making the nanotube surfaces hydrophilic and charged [30]. The result is that the nanotubes have a higher affinity for the aqueous phase and are more likely to be dispersed. In addition, owing to its biocompatibility, coating by these molecules also reduces the proven ecotoxicity of the nanoparticle as compared with its pure form [31].

In complete contrast to carbon nanotubes, metal oxide nanoparticles are perfectly hydrophilic and have a charged surface that favours their dispersion in an aqueous medium over a certain pH range far from the isoelectric point and for salinities below the critical coagulation concentration. However, their surface reactivity, which favours interaction with surrounding dissolved matter, completely changes these stabilisation/destabilisation conditions. For example, pure nanoparticles like iron oxide, which have an isoelectric point favouring aggregation around pH = 6, can remain perfectly dispersed at this pH when they are first coated with molecules that carry their own intrinsic charge (see Fig. 13.12) [32].

The layer of adsorbed molecules produces a new surface charge which induces new interparticle electrostatic repulsions, independently of the isoelectric point of the bare nanoparticles. The new isoelectric point of the coated nanoparticles depends largely on the pKa of the adsorbed molecules.

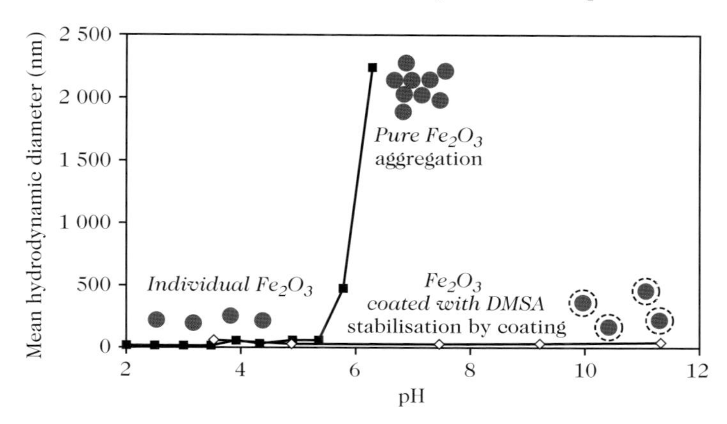

Fig. 13.12. pH dependence of the average size of maghemite iron oxide nanoparticles in the case of bare nanoparticles and nanoparticles coated with dimercaptosuccinic acid (DMSA). From [32]

Furthermore, the adsorbed molecule may also exhibit its own level of toxicity, e.g., heavy metals such as arsenic, lead, cobalt, etc., or organic pollutants such as pesticides, PCB, PAH, etc. This type of affinity with nanoparticles has been widely discussed in the literature [2]. In such cases, when a pollutant binds onto the nanoparticle, this alters the balance of species and may tend to reduce its proportion of free and labile fractions in the environment. Such an affinity can be envisaged as a potential remedy for this type of pollution. However, that would not mark the end of its life cycle, since it is then the toxicity of the nanoparticle–pollutant ensemble that must be considered in risk assessments. Indeed, nanoparticles can serve as particularly bioavailable and efficient carriers, resulting in a kind of passive toxicity.

Dynamics of Aggregation

The extent to which nanoparticles aggregate together or with other components in suspension in the medium depends largely on the balance of forces at the interfaces, as explained earlier. However, the kinetics of this aggregation is governed by two different contributions: the frequency of collisions between nanoparticles, which depends on the dominant transport mode, and the sticking efficiency during collisions, which depends on the balance of forces at the interfaces and/or the energy supplied to the system.

A collision efficiency coefficient α_{ij} can be defined to account for the ratio of these two contributions in the attachment of particles of type i to particles of type j:

$$\alpha_{ij} = \frac{\beta_{ij}^{\text{eff}}}{\beta_{ij}} , \tag{13.12}$$

where β_{ij} is the collision rate between particles of types i and j, and β_{ij}^{eff} is the true rate of efficient collisions in which the particles stick together. So α_{ij} is the ratio of the actual measured aggregation rate to the maximal aggregation rate calculated using a linear model. It varies from unity for systems with diffusion-limited aggregation (DLA), where all collisions are efficient, to zero for stable suspensions, where repulsive interparticle forces are dominant and where aggregation never occurs. An intermediate aggregation regime characterised by a value of α_{ij} strictly between 0 and 1 is referred to as reaction-limited aggregation (RLA).

The approach due to Von Smoluchowski [33] treates particle aggregation as a series of reactions, for each of which one has a system of equations. The rate of formation of aggregates of type k, with volume v_k, is calculated from

$$\frac{\mathrm{d}n_k}{\mathrm{d}t} = \frac{1}{2} \sum_{i=k-j=1}^{k-1} \alpha_{ij}\beta_{ij}n_i n_j - n_k \sum_{i=1}^{\infty} \alpha_{ik}\beta_{ik}n_i \,, \qquad (13.13)$$

where the first term corresponds to the formation of aggregates of type k resulting from collisions between individuals of types i and j with numerical concentrations n_i and n_j, respectively, and collision rate β_{ij}, while the second term describes the disappearance of aggregates of type k, with numerical concentration n_k, resulting from their collision with individuals of type i at a rate β_{ik}, leading to the formation of even bigger aggregates.

The physical motion of the particles in suspension is governed by the extent to which the system is being shaken up, and this in turn depends on the hydrodynamic flow, the temperature, the particle sizes, and the densities of the particles and the solvent. Taking these parameters into account, there are three types of particulate motion, and the collision rates β_{ij} are estimated for each case.

Brownian motion concerns particles in suspension that are not significantly affected by gravity, i.e., with nanometric to micrometric sizes. The particle motions are random. When their transport is solely due to this Brownian contribution, i.e., when no other energy is supplied by mechanical shaking, we speak of perikinetic aggregation. The particle collision frequency $\beta_{ij\,\text{peri}}$ then depends on the diffusion coefficient of the particles, hence on their radii r_i and r_j, the fluid viscosity μ, and the temperature T:

$$\beta_{ij\,\text{peri}} = \frac{2kT}{3\mu}\left(\frac{1}{r_j} + \frac{1}{r_j}\right)(r_i + r_j) \,, \qquad (13.14)$$

where k is the Boltzmann constant.

In contrast, when the particle transport is constrained by a shear gradient, aggregation is modelled under the assumption that the particles have straight trajectories in laminar flow. Here we speak of orthokinetic aggregation. Considering spherical particles, the collision range for a particle of type j with a particle of type i is a sphere concentric with i of diameter $d_i + d_j$, where

d_i and d_j are the diameters of particles i and j, respectively. The collision frequency $\beta_{ij\,\mathrm{ortho}}$ between i and j is then given by

$$\beta_{ij\,\mathrm{ortho}} = \frac{1}{6}(d_i + d_j)^3 G_\mathrm{m} \,, \tag{13.15}$$

where G_m is the mean velocity gradient of the fluid.

Particles or aggregates of different sizes are affected to different extents by gravitational attraction. The most massive thus tend to sediment out more quickly than the less massive ones, and may drag the latter along with them after collision and attachment. The sedimentation rate v satisfies Stokes' law:

$$v = \frac{g}{18\mu}\left(\rho_\mathrm{s} - \rho_\mathrm{l}\right) d_i^2 \,, \tag{13.16}$$

where ρ_s is the density of the particles, ρ_l is the fluid density, and g is the acceleration due to gravity. If the system is treated with a rectilinear collision model, the collision frequency is defined as $\beta_{ij\,\mathrm{sd}}$, given by

$$\beta_{ij\,\mathrm{sd}} = \frac{\pi g\left(\rho_\mathrm{s} - \rho_\mathrm{l}\right)}{72\mu}(d_i + d_j)^3 |d_i - d_j| \,. \tag{13.17}$$

However, Stokes' law is not always applicable to calculate the sedimentation rate of aggregates. Recent studies have shown that it tends to underestimate the rate in certain cases, because it does not take into account the porosity and permeability of the aggregates with respect to the solvent [34].

In order to assess the global nanoparticle aggregation kinetics, one must take into account the net effect of all the interparticle collision frequencies induced by these different contributions to particle motion. Figure 13.13 shows how the orthokinetic and perikinetic contributions to the collision frequency β_{ii} of particles i with one another depend on their size and the mean velocity gradient G_m. One can then estimate which type of motion will dominate as far as the overall collision frequency is concerned. When $\beta_{ii\,\mathrm{peri}} \gg \beta_{ii\,\mathrm{ortho}}$, the collision frequency is limited by Brownian motion, and conversely. Note that $\beta_{ii\,\mathrm{peri}}$ is constant because it depends solely on the temperature and viscosity of the medium, and vanishes for particle sizes above 5 μm, because Brownian motion is then negligible. Logically, mechanical shaking strongly favours the collision of the biggest particles. This is why $\beta_{ii\,\mathrm{ortho}}$ increases with the particle radius. For each particle size, there is therefore a threshold velocity gradient $G_\mathrm{m}^\mathrm{threshold}$ beyond which the orthokinetic contribution becomes significant compared with Brownian motion. For 1 μm particles, $G_\mathrm{m}^\mathrm{threshold}$ is of the order of $30\,\mathrm{s}^{-1}$, typical of the values encountered in natural aquatic environments with little turbulence. This means that the dynamics of colloids is sensitive to the hydrodynamic situation in the medium. For large nanoparticules, with diameters of the order of 100 nm, $G_\mathrm{m}^\mathrm{threshold}$ is of the order of $1\,000\,\mathrm{s}^{-1}$, typical of the values encountered in systems with infinite shear, such as waterfalls or turbines. Finally, for nanoparticles with diameters of around 30 nm, $G_\mathrm{m}^\mathrm{threshold}$

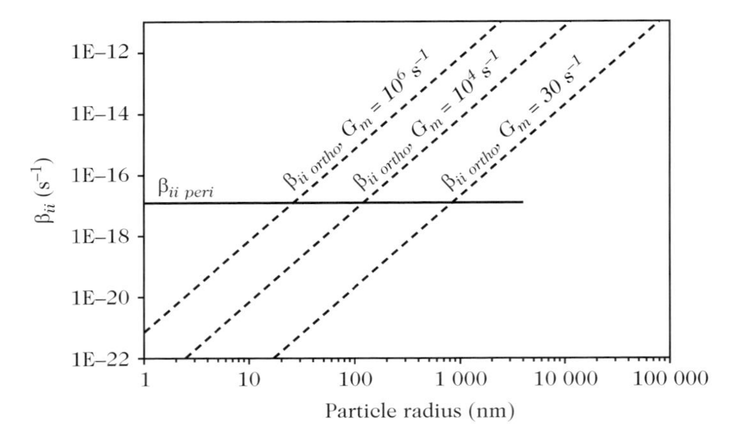

Fig. 13.13. Calculated interparticle collision frequencies as a function of particle size and mean velocity gradient, with the contributions of Brownian and orthokinetic motions ($\alpha = 1$, $T = 298\,\mathrm{K}$)

reaches an extreme theoretical value of the order of $10^6\,\mathrm{s}^{-1}$ that never occurs in the natural environment. It would thus appear that nanoparticle collision frequencies are essentially limited by Brownian motion, and that any hydrodynamic contribution to their aggregation dynamics classically encountered in the environment is negligible.

However, while the hydrodynamics of the medium has no effect on nanoparticle diffusion dynamics on the mesoscopic scale, it remains an important factor for the large scale transfer in a preferred flow direction, because Brownian motion is purely random as far as direction is concerned.

13.2.3 Nanoparticle Mobility and Attachment in Water-Saturated Porous Media

The stability of nanoparticles in suspension in an aqueous medium depends on many physicochemical considerations at the interfaces, and this in turn depends on the composition of the surrounding medium. However, since the natural environment is never homogeneous and continuous, but tends to be organised into different reservoirs with variable scale and degrees of confinement, further parameters intrinsic to each reservoir will constrain the fate of nanoparticles in suspension. We shall be concerned here with disordered, porous media, which constitute a selective natural barrier to nanoparticle migration, and which are very common in the natural environment. These media may have largely inorganic composition, as in sandy aquifers, whose permeability to nanoparticles determines their transfer to supplies of fresh water for human consumption. They may also be on a much smaller scale, but much more widespread and purely organic, like the biofilm secreted by microbial fauna on almost all non-sterile surfaces exposed to humidity.

In the next section we discuss various experimental and theoretical methods for understanding nanoparticle mobility in porous media, depending on whether the medium is granular and inorganic or more like an organic gel.

Flow in a Porous Inorganic Medium

The behaviour of nanoparticles in suspension in a saturated porous medium obeys several laws. Similarly to the dynamics of aggregation in a dilute medium, the fate of nanoparticles in an aqueous medium is governed by the efficiency of transport through the medium on the one hand, and the efficiency of attachment to the surface of stationary spherical grains (collectors) on the other. So the probability η of a nanoparticle coming into contact with a collector and attaching to it can be written as the product of the efficiency of transport η_0 in the stationary phase and the efficiency of attachment α [35], i.e.,

$$\eta = \eta_0 \alpha \,. \tag{13.18}$$

In the case of nanoparticles, it is essentially the diffusive contribution that determines the transport efficiency η_0, because the contributions from interception and gravity are negligible [36]. Now the efficiency of transport by diffusion depends on the porosity of the stationary phase, the size ratio of mobile nanoparticles and grains making up the stationary phase, and the rate of approach of the particle toward the surface.

The attachment efficiency α varies from 0 when the nanoparticle–surface interaction is totally repulsive to 1 when it is strongly attractive. α thus expresses the proportion of contacts which induce attachment compared with the total number of contacts. Similarly to the estimate of the collision efficiency in the model for aggregation in a dilute medium, the attachment efficiency also depends here on the balance of forces exerted at the nanoparticle–collector interface. The latter are of the same kind as in a dilute medium, i.e., we must consider a balance between short-range Van der Waals or chemical attractions and electrostatic repulsions. However, due to confinement by the pore system within which the nanoparticles are diffusing, the range of these forces at the interfaces, of the same order of magnitude as the pore size, is likely to significantly restrict their transport.

The attachment efficiency α of nanoparticles in a porous medium is estimated empirically, since conventional particle transport models do not provide a good description of experimental observations. So the mobility of nanoparticles passing through a column of length L and radius r_c, characterised by uniform porosity ε, is expressed via the attachment efficiency α given by [36]

$$\alpha = \frac{4r_\mathrm{c}}{3(1-\varepsilon)\eta_0 L} \ln(n_L/n_0) \,, \tag{13.19}$$

where n_L and n_0 are the numerical particle concentrations measured experimentally at the outlet and inlet, respectively, of the medium of length L.

Using this equation, it is also interesting to extract the distance L as a variable in order to estimate the filtering efficiency of the porous medium, or in other words, the penetration depth required for the nanoparticle concentration to drop by a given proportion.

There is still not much data in the literature describing nanoparticle transfer through saturated porous media [37–40]. The first results tend to indicate the importance of electrostatic interactions between nanoparticles and stationary phase. Indeed, the effects of pH and ionic strength on surface charges would appear to agree with reduced mobility when these charges are opposite in sign or screened, and increased mobility when they are of the same sign or enhanced.

A typical batch experiment is shown in Fig. 13.14. Natural sand is added to a TiO_2 nanoparticle suspension, the mixture is shaken up for one hour, then left to separate out. The nanoparticle concentration is measured before introducing the sand and after separation. The resulting concentration ratio is explained by the surface charges of the sand and the nanoparticles. The sand is negatively charged over the whole pH range studied (the IEP of pure silica is 2), whereas TiO_2 has an IEP of 5.5. When the surface charges have the same sign, i.e., when pH > 5.5 (II), there is no change in the concentration of nanoparticles in suspension after introducing the sand, hence no attractive interaction. On the other hand, when the surface charges have opposite signs, i.e., when pH < 5.5 (I), the concentration of nanoparticles in suspension falls sharply, indicating a high level of nanoparticle adsorption onto the sand.

The diffusion column experiment is conventionally used to study the parameters determining the mobility or attachment of nanoparticles in a saturated porous medium. A mobile electrolyte phase circulates at constant rate through a column filled with a stationary phase of known and controlled pore volume. At time 0, a volume of nanoparticles in suspension is injected

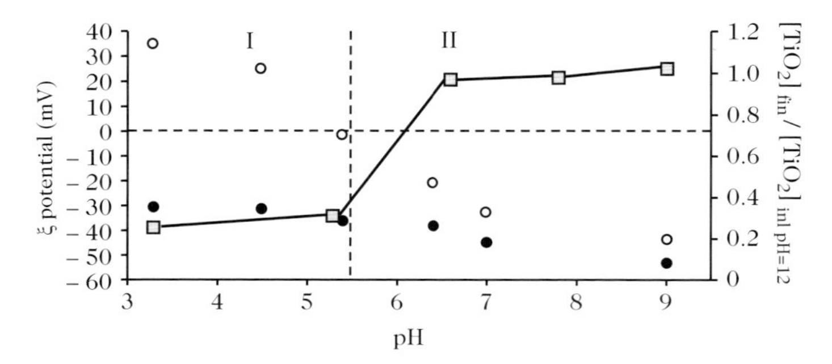

Fig. 13.14. pH dependence of the ratio of concentrations of TiO_2 nanoparticles in suspension before introducing sand and after separation (*squares*). Correlation with surface charges on the sand (*black disks*) and on the TiO_2 (*white disks*), measured by streaming potential and electrophoretic mobility, respectively. Adapted from [23]

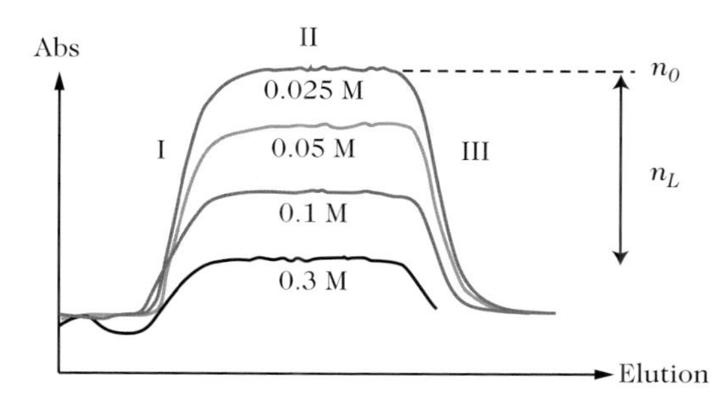

Fig. 13.15. Experimental breakthrough curves obtained for nC_{60} nanoparticles in a column filled with silica microspheres, for different NaCl concentrations. Nanoparticle number concentrations are measured by UV/visible spectroscopy at the characteristic wavelength 270 nm for absorption by nC_{60}

into the circuit upstream of the column. The nanoparticle concentration n_L is measured in the mobile phase at the column outlet as time goes by. The resulting evolution is called a breakthrough curve, typically box shaped as in Fig. 13.15. The rise phase (I) corresponds to the appearance of nanoparticles at the column outlet. Their concentration increases suddenly from zero to a maximal value corresponding to the fraction that is not attached to the stationary phase. The curve then levels out (II) and this plateau persists until the full volume of nanoparticles injected at the column inlet has had time to flow through. The end of the injection is characterised by a sudden drop (III) in the nanoparticle concentration to return to the initial value. In this way, many parameters can vary in the mobile phase (pH, ionic strength, hydrodynamic flow) and in the stationary phase (porosity, mineralogy), and their effects on nanoparticle attachment as reckoned by the ratio n_L/n_0 can be investigated.

The effect of the ionic strength on nanoparticle attachment through electrostatic forces is illustrated in Figs. 13.15 and 13.16 for the example of nC_{60} nanoparticles in a column of an ideal porous medium made up of silica beads. When the salt concentration increases, electrostatic repulsions between the negatively charged nanoparticles and the negatively charged silica beads are screened, attractive interactions between nC_{60} and silica begin to dominate, the proportion of nanoparticles leaving the column drops, and the attachment coefficient rises.

A critical salt concentration can be defined beyond which α increases. It is interesting to compare this with the critical coagulation concentration measured in a dilute medium. The critical attachment concentrations measured here for nC_{60} nanoparticles are 3×10^{-4} and 3×10^{-2} mol/L for $CaCl_2$ and NaCl, respectively, whereas the critical coagulation concentrations are 2×10^{-3} and 10^{-1} mol/L (see Fig. 13.6). The fact that nanoparticle attachment in a porous medium occurs at a lower salt concentration than the one required for self-aggregation of the nanoparticles by coagulation suggests that,

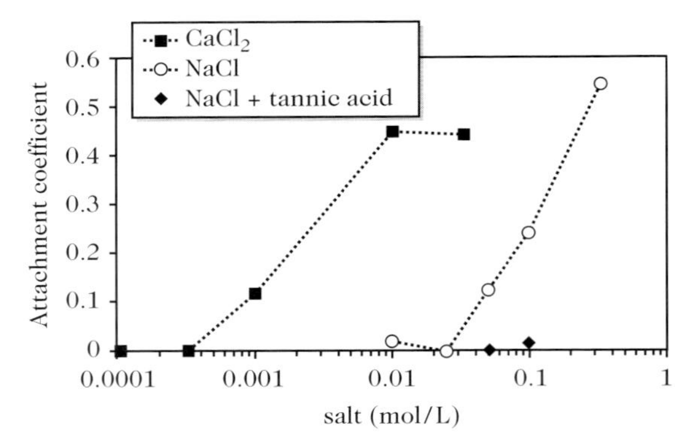

Fig. 13.16. Dependence of the attachment coefficient of $n\mathrm{C}_{60}$ nanoparticles in a column filled with silica beads on the concentrations of NaCl and $\mathrm{CaCl_2}$, with or without tannic acid

in the present case, the driving force behind attachment is the electrostatic nanoparticle–bead interaction, rather than mechanical trapping in the pore network of post-formed nanoparticle aggregates.

Furthermore, through its affinity for mobile or stationary inorganic surfaces, the presence of organic matter in the porous medium is likely to affect nanoparticle mobility. In a similar way to what is observed in dilute media, when organic molecules coat the nanoparticle or collector surface, the balance of the interactions between them will be modified by the new properties of each surface. Figure 13.16 shows how tannic acid introduced in the mobile phase tends to increase the mobility of $n\mathrm{C}_{60}$ nanoparticles, despite the fact that the critical attachment concentration has been exceeded.

Diffusion in a Porous Organic Medium

In the natural environment, most non-sterile surfaces exposed to damp conditions are colonised by microbial organisms. These produce a hydrated exopolymer which covers them and allows them to stick to the surface. This organic matrix is mainly made up of polysaccharides, proteins, and nucleic acids, and forms a thin film on the surface called a biofilm [41]. This is particularly crucial in freshwater rhizosphere environments, where the roots of plants and bacteria are coated with this biological substance and it plays the role of a selective barrier or filter. The bioavailability of nanoparticles dispersed in the environment for these organisms, and hence their access to the food chain, is thus heavily dependent on the ability of these biofilms to retain them. Attractive interactions between the nanoparticles and the fibres making up the biofilm favour retention [42–45], whereas repulsive interactions will allow diffusion. Finally, steric hindrance also plays a decisive role with respect to the maximal size of objects allowed to diffuse through the biofilm.

Most of the literature on diffusion in organic porous media like gels is devoted to compounds such as macromolecules, proteins, metals, or viruses. Special tools are needed for this type of study, such as in situ labelling to detect diffusing nanosolutes. Fluorescence correlation spectroscopy appears to be the best suited to measuring the diffusion coefficient of low concentration fluorescent nanometric entities in situ in an organic medium such as a gel, biofilm, or intracellular medium [46–50]. This method is indeed highly sensitive and proves to be efficient even at trace concentrations. When fluorescent labelling is not possible, methods like diffusive gradient in thin films (DGT) or voltametry can also be used to measure the diffusion coefficients of solutes in gels covered with microelectrodes [51].

In contrast to the recent, largely phenomenological methods for studying the mobility of nanoparticles in a flow through a granular porous medium (see the discussion of flow in an inorganic porous medium on p. 313), the random diffusion of nanosolutes in a disordered porous network is described and modelled by many laws and theories, widely discussed in the literature.

Diffusion in Disordered Media

In a uniform Euclidean system of arbitrary dimensions, the random diffusion of a nanoparticle A is described by the classic Fick law [52]

$$\overline{x^2(t)}^A = 2dD^At \; , \tag{13.20}$$

where $\overline{x^2(t)}^A$ is the mean squared displacement of A at time t, D^A is the diffusion coefficient of A, and d is the number of dimensions. However, in disordered porous media such as soils, biofilms, or bacterial flocs, many environmental parameters restrict the movement of diffusing particles, and classical transport theories are no longer applicable. The diffusion is said to be anomalous. Scientific understanding of this phenomenon is mainly based on structural models of disordered media. The theory of factals [53], based on the self-similarity of structures at different scales, and models of percolation [52, 54] are among the best suited to this task. A detailed review of these numerical approaches can be found in [55].

The diffusion properties of nanosolutes in gels have also been modelled [48, 56–60]. For example, in an agarose gel, Fatin-Rouge et al. identified three distinct parameters controlling the diffusive motion of nanoparticles [48, 60]. They model the distribution of nanoparticles A in a water-saturated gel by a global partition coefficient Φ given by

$$\Phi = \vartheta\gamma\pi = \frac{[A]_{\mathrm{g}}}{[A]_{\mathrm{w}}} \; , \tag{13.21}$$

where $[A]_{\mathrm{g}}$ and $[A]_{\mathrm{w}}$ are the nanoparticle concentrations in the gel and in the solution, respectively, and ϑ, γ, and π are the contributions from purely steric, chemical, and electrostatic effects, respectively.

Steric Hindrance

In a disordered medium, when the size R_A of the diffusing nanoparticle is not infinitely small compared with the size R_P of the interconnected pores, the presence of tortuosity and dead end pores creates steric obstruction which slows down the macroscopic diffusion of particle A. A first estimate of ϑ has been proposed [61], based on the assumption that R_A and R_P are indeed monodisperse and that A is spherical, whence ϑ is given by

$$\vartheta = (1 - \phi)\left(1 - \frac{R_A}{R_P}\right)^2 , \tag{13.22}$$

where ϕ is the volume fraction of the gel. Even though R_P is rarely monodisperse in natural gels and biofilms, this relation gives a good approximation to the steric contribution.

Another, better suited approach considers the reduction of $\overline{x^2(t)}^A$. Indeed, (13.20) is not generally valid, and the anomalous diffusion law can be written in the form [55, 62–66]

$$\overline{x^2(t)}^A = \Gamma t^{2/d_w} , \tag{13.23}$$

where d_w is the fractal dimension of diffusion and Γ the transport coefficient. For random normal diffusion, one has $d_w = 2$, as in (13.20), whereas for anomalous diffusion, $d_w > 2$ corresponds to the slowing down of diffusive particle transport caused by steric hindrance. From (13.23), d_w can be calculated by measuring the characteristic time $t_c(x)^A$ required for these nanoparticles to cover a distance x in the disordered medium (see Fig. 13.17).

Owing to the steric contribution, nanoparticles above a certain critical size cannot diffuse through the pore network. This critical size is of the order of the mean diameter of the interconnected pores in the network. In bacterial biofilms, Lacroix-Gueu et al. [46] have demonstrated the anomalous diffusion of latex and bacteriophages of radius 55 nm. In an agarose gel of mass concentration around 1.5%, steric hindrance comes into play for diffusing entities larger than 10 nm, and a critical size was identified at around 70 nm [48, 67]. However, the critical size is logically inversely related to the gel density. Now it has been shown that an organic gel may restructure locally at its interface with the environment in such a way that the fibres arrange themselves into a denser structure, thus lowering the critical size of diffusing objects [50].

Electrostatic Interactions

Most gels, flocs, or biofilms encountered in the environment are mainly composed of polysaccharides and humic substances. These highly reactive components carry anionic charge over a wide pH range, by virtue of the acid groups they contain [68, 69]. This induces an overall negative charge in the diffusing medium, which can be modelled by the Donnan potential Ψ_D, i.e., the

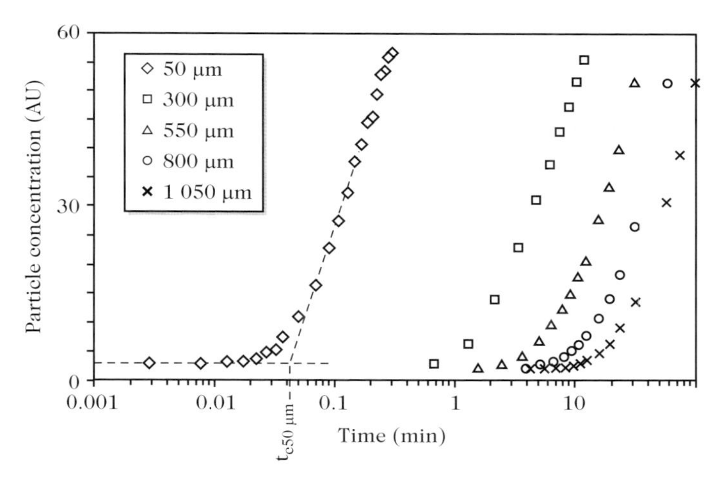

Fig. 13.17. Concentration profile as a function of time and distance in an agarose gel (1.5 wt%) for nanoparticles (protein R-phycoerythrin, $R_h \approx 4.5$ nm) incorporated at time 0 and distance 0. The value of t_c obtained from the intersection between the horizontal line representing the background and linear regression of the concentration change on a semi-logarithmic scale is the characteristic time at which the nanoparticles reach the corresponding distance. Data obtained by fluorescence correlation spectroscopy [50]

potential difference between the core of the gel and the solution at an infinite distance from it [70]:

$$\psi_D = \frac{RT}{zF} \sinh^{-1} \frac{\rho}{2zFc} , \tag{13.24}$$

where ρ is the charge density of the gel, c and z are the molar concentration and charge of the electrolyte in the solution at an infinite distance from the gel, F is the Faraday constant, R is the perfect gas constant, and T is the temperature.

The diffusive motion of charged nanoparticles in a charged gel is thus affected by an electrostatic contribution π which can be described by a Boltzmann distribution of the form

$$\pi = \exp\left(-\frac{Z_A F \psi_D}{RT}\right) , \tag{13.25}$$

where Z_A is the electric charge of particles A.

For example, the effect of a negatively charged agarose gel on the diffusive motion of positively charged nanoparticles is illustrated in Fig. 13.18. The anionicity of the gel is neutralised by protonation for $pH < 3$, and screened for $pH > 10.5$ by the addition of basic molecules required to obtain this pH. When the pH lies between 3 and 10.5, the influence of the gel charge is enhanced, the Donnan potential is maximal (see Fig. 13.18A), the partition coefficient

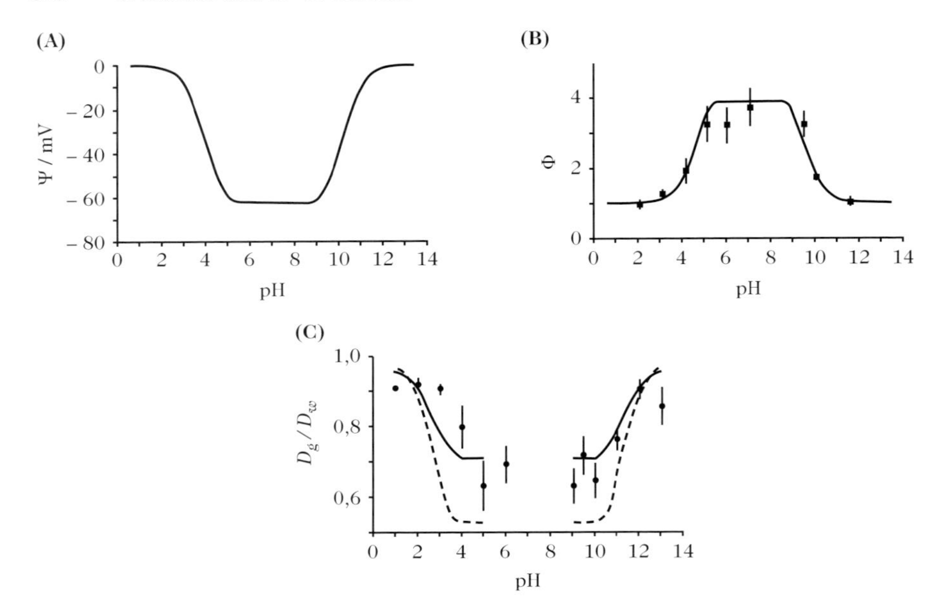

Fig. 13.18. Effects of pH on diffusive motion of a positively charged rhodamine R6G^{2+} nanosolute in a negatively charged agarose gel. The partition coefficient Φ (**B**) governed by electrostatic attractions is inversely related to the Donnan potential Ψ_{D} (**A**), and the diffusion coefficient of the solute in the gel normalised by its value in water, viz., $D_{\mathrm{g}}/D_{\mathrm{w}}$ (**C**). Data adapted from [60]

Φ is maximal (see Fig. 13.18B), and the diffusion coefficient D_{g} is minimal (see Fig. 13.18C), owing to electrostatic attractions with particles of opposite charge.

The electrostatic contribution π is thus easily destroyed by neutralising the electrostatic interactions by adjusting the pH or increasing the ionic strength. In this precise case, the authors were also able to specify the critical salt concentration required to achieve this, of the order of 10^{-3} mol/L of NaCl. This level of salinity is of the same order as the level in river water, which suggests that, in the natural environment, nanoparticle diffusion in natural biogels will be highly dependent on this parameter.

Specific Chemical Interactions

When the nanoparticles A exhibit a specific chemical affinity for certain sites S on the biogel fibres, adsorption occurs and the SA complex forms with equilibrium constant K_A^{int} given by

$$K_A^{\mathrm{int}} = \frac{[SA]}{[A]_{\mathrm{P}}[S]} , \tag{13.26}$$

where $[SA]$ and $[S]$ are the concentrations of the SA complexes and free sites S, and $[A]_{\mathrm{P}}$ is the concentration of particles A in the pores of the gel.

If we can assume that $\vartheta = 1$, the expression for Φ can be reformulated as follows by combining (13.25) and (13.26):

$$\Phi = \frac{[SA] + [A]_{\mathrm{P}}}{[A]_{\mathrm{W}}} = \exp\left(-\frac{Z_{\mathrm{A}} F \Psi_{\mathrm{D}}}{RT}\right)\left(1 + K_A^{\mathrm{int}}[S]\right) = \pi\gamma \,. \tag{13.27}$$

An expression for the chemical contribution γ can then be obtained, viz.,

$$\gamma = 1 + K_A^{\mathrm{int}}[S] \,. \tag{13.28}$$

Equations (13.22), (13.25), and (13.28) can be used to estimate the steric, electrostatic, and chemical contributions governing the distribution of nanoparticles between a porous medium and the solution surrounding it. Although the complexity of environmental conditions makes it difficult to check the validity of the approximations made in this simplified approach, it nevertheless has the advantage of taking into consideration all the parameters likely to control nanoparticle diffusion in the medium. This means that the key roles played by steric hindrance, pH, ionic strength, and the chemistry of the surrounding medium can be demonstrated and estimated.

13.3 Conclusion

For many nanoparticles with diameters greater than a few tens of nanometers, most of the models and theoretical relations established to describe the behaviour of colloids in suspension remain applicable. However, for ultrasmall nanoparticles in the range 20–40 nm, with novel surface properties of the kind described in Chap. II, the conventional physicochemical approach is no longer adequate to explain the dynamics in suspension, because new contributions begin to have an effect on their behaviour. Many differences can be put down to the increased surface tension. This makes the elementary dispersion of nanoparticles in the medium unstable, favouring aggregation, attachment, or complexation which tend to reduce this surface energy.

Nanoparticle transport is largely dominated by random diffusion due to Brownian motion. While this gives them high mobility in liquids, it also increases the probabilities of collision with and attachment to neighbouring surfaces, thereby hindering dispersion. The stability of nanoparticles in suspension, or conversely their aggregation kinetics, determines the time scale over which they belong to the nanometric scale and behave as would be expected of that scale. Beyond this point, the formation of micrometric aggregates classifies the resulting entities in the colloid family, subjected to its own dynamics, where hydrodynamic and gravitational contributions largely dominate stability in suspension. Moreover, aggregation of nanoparticles tends to reduce their dwell time in the environment, and hence also their bioavailability.

There are many natural surfaces likely to collect nanoparticles. These range from grains of sand in aquifers to biofilms on plant roots, not to mention bacterial flocs in suspension in water. These also perturb the fate of nanoparticles in the environment, even reducing their persistence. The heterogeneity of these surfaces plays an important role, and must be characterised in order to better understand and predict the fate of nanoparticles in such media.

Whether they be in a dilute medium dominated by competition between dispersion and aggregation or in a porous and saturated confined medium where the balance between mobility and attachment becomes a factor, the final fate of nanoparticles always results from the balance of forces at contacting surfaces. Attractive interactions favour aggregation or attachment, whereas globally repulsive forces stabilise dispersion and mobility of the particles. There are many factors that can cause this competitition to go one way or the other, and they may be of intrinsic, environmental, or even anthropic origins. However, it is the surface properties of the nanoparticles that determine the nature of the interactions at the interface and also the affinity for the surrounding medium and matter. This is the key parameter, while environmental factors such as the pH or ionic strength have a secondary influence on the balance of competing forces. It is therefore essential in any study of the fate of nanoparticles in the environment to characterise above all the changes in the surface properties of the nano-objects in all the relevant environmental compartments. The influence of any man-made, but possibly modified coatings, or natural coatings formed during the life cycle of the nanomaterial in the environment is therefore of prime importance, since surface properties will depend directly upon such factors.

Finally, as far as the environment is concerned, an understanding of these various processes controlling the behaviour of nanoparticles in suspension should prove very useful in developing procedures for dealing with water pollution by nanoparticles. It should be possible to devise controlled processes for extracting nanoparticles from solution, exploiting mechanisms like flocculation, filtration, or even flotation [14].

References

1. AFSSET: Les nanomatériaux. Effets sur la santé de l'homme et sur l'environnement. Saisine AFSSET 2005/010 (2006)
2. M. Auffan et al.: Langmuir **24**, 3215 (2008)
3. H.Z. Zhang, J.F. Banfield: J. Mater. Chem. **8**, 2073 (1998)
4. H.W. Kroto, J.R. Heath, S.C. O'Brien, R.F. Curl, R.E. Smalley: Nature **318**, 162 (1985)
5. R.C. Haddon: Science **261**, 1545 (1993)
6. R.G. Alargova, S. Deguchi, K. Tsujii: J. Am. Chem. Soc. **123**, 10460 (2001)
7. Y. Marcus et al.: J. Phys. Chem. B **105**, 2499 (2001)
8. G.V. Andrievsky, V.K. Klochkova, A.B. Bordyuhb, G.I. Dovbeshkoc: Chem. Phys. Lett. **364**, 8 (2002)

9. J.A. Brant, J. Labille, J.Y. Bottero, M.R. Wiesner: Langmuir **22**, 3878 (2006)
10. S. Deguchi, R.G. Alargova, K. Tsujii: Langmuir **17**, 6013 (2001)
11. W.A. Scrivens, J.M. Tour, K.E. Creek, L. Pirisi: J. Am. Chem. Soc. **116**, 4517 (1994)
12. J. Labille et al.: Fuller. Nanotub. Car. N. **14**, 307 (2006)
13. J. Labille et al.: *Environmental Pollution*, (ASAP 2010)
14. NANOSEP: In: ANR Programme PRECODD. BRGM, CEREGE, LISBP, LMSGC, INERIS, CIRSEE, IMFT, Arkema France, 2009–2011
15. H.C. Hamaker: Physica **4**, 1058 (1937)
16. R. Hogg, T.W. Healy, D.W. Fuerstenau: Translation Faraday Society **62**, 1638 (1966)
17. B. Derjaguin, L. Landau: Acta Phys. Chim. **14**, 633 (1961)
18. E.J.W. Verwey, J.T.G. Overbeek: *Theory of the Stability of Lyophobic Colloids*. (Elsevier 1948)
19. Y. Adachi: Advances in Colloid and Interface Science **56**, 1 (1995)
20. K.H. Gardner: In: *Interfacial Forces and Fields. Theory and Applications*, Surfactant Science Series. (Marcel Dekker, New York 1999) pp. 509–550
21. H. Van Olphen: *An Introduction to Clay Colloid Chemistry*, 2nd edn. (John Wiley, New York 1977)
22. P.C. Hiemenz, R. Rajagopalan: *Principles of Colloid and Surface Chemistry*, 3rd edn. (Marcel Dekker, New York, Basel, Hong Kong 1997) p. 605
23. N. Solovitch-Vella, J. Labille, J. Rose, P. Chaurand, D. Borschneck, M.R. Wiesner, J.Y. Bottero: Environmental Science and Technology (in press)
24. J. Labille, F. Thomas, I. Bihannic, C. Santaella: Clay Miner. **38**, 173 (2003)
25. J. Labille, F. Thomas, M. Milas, C. Vanhaverbaeke: J. Coll. Interf. Sci. **284**, 149 (2005)
26. C.E. Clapp, W.W. Emerson: Soil Science **114**, 210 (1972)
27. S. Ulrich, A. Laguecir, S. Stoll: J. Nanopart. Res. **6**, 595 (2004)
28. S. Ulrich, M. Seijo, A. Laguecir, S. Stoll: J. Phys. Chem. B **110**, 20954 (2006)
29. S. Ulrich, M. Seijo, S. Stoll: Curr. Opin. Colloid Int. **11**, 268 (2006)
30. M.A. Chappell et al.: Environmental Pollution **157**, 1081 (2009)
31. A.J. Kennedy et al.: Environ. Toxicol. Chem. **27**, 1932 (2008)
32. M. Auffan et al.: Environ. Sci. Technol. **40**, 4367 (2006)
33. M. Von Smoluchowski: Zeitschrift für Physikalische Chemie **92**, 129 (1917)
34. K.H. Gardner, T.L. Theis, T.C. Young: Colloid Surface A **141**, 237 (1998)
35. R.W. Newburry: In: *The Ecology of Aquatic Insects*, ed. by V.H. Resh, D.M. Rosenberg. (Praeger, New York 1984) pp. 323–357
36. N. Tufenkji, M. Elimelech: Environ. Sci. Technol. **38**, 529 (2004)
37. B. Espinasse, E.M. Hotze, M.R. Wiesner: Environ. Sci. Technol. **41**, 7396 (2007)
38. H.F. Lecoanet, J.Y. Bottero, M.R. Wiesner: Environ. Sci. Technol. **38**, 5164 (2004)
39. H.F. Lecoanet, M.R. Wiesner: Environ. Sci. Technol. **38**, 4377 (2004)
40. N. Tufenkji, M. Elimelech: Langmuir **21**, 841 (2005)
41. J.W. Costerton et al.: Annu. Rev. Microbiol. **41**, 435 (1987)
42. T.J. Battin, L.A. Kaplan, J.D. Newbold, C.M.E. Hansen: Nature **426**, 439 (2003)
43. R.N. Jordan: Biofilms **310**, 393 (1999)
44. T. Galle, B. Van Lagen, A. Kurtenbach, R. Bierl: Environ. Sci. Technol. **38**, 4496 (2004)

45. S. Ballance et al.: Canadian Journal of Fisheries and Aquatic Sciences **58**, 1708 (2001)
46. P. Lacroix-Gueu, R. Briandet, S. Leveque-Fort, M.N. Bellon-Fontaine, M.P. Fontaine-Aupart: Comptes Rendus Biologies **328**, 1065 (2005)
47. E. Guiot et al.: Photochemistry and Photobiology **75**, 570 (2002)
48. N. Fatin-Rouge, K. Starchev, J. Buffle: Biophys. J. **86**, 2710 (2004)
49. E. Haustein, P. Schwille: Methods **29**, 153 (2003)
50. J. Labille, N. Fatin-Rouge, J. Buffle: Langmuir **23**, 2083 (2007)
51. H. Zhang, W. Davison: Analytica Chimica Acta **398**, 329 (1999)
52. S. Havlin, D. Ben-Avraham: Adv. Phys. **36**, 695 (1987)
53. B. Mandelbrot: *The Fractal Geometry of Nature.* (W.H. Freeman 1982)
54. D. Stauffer: *Introduction to Percolation Theory.* (T. Francis, London 1985)
55. S. Havlin, D. Ben-Avraham: Adv. Phys. **51**, 187 (2002)
56. B. Amsden: Polym. Gels Networks **31**, 13 (1998)
57. N. Fatin-Rouge, K. Wilkinson, J. Buffle: J. Phys. Chem. B **110**, 20133 (2006)
58. E.M. Johnson, D.A. Berk, R.K. Jain, W.M. Deen: Biophys. J. **70**, 1017 (1996)
59. A. Pluen, P. A. Netti, K. J. Rakesh, D. A. Berk: Biophys. J. **77**, 542 (1999)
60. N. Fatin-Rouge, A. Milon, J. Buffle, R.R. Goulet, A. Tessier: J. Phys. Chem. B **107**, 12126 (2003)
61. C. Giddings, E. Kucera, C.P. Russel, M.N. Myers: J. Phys. Chem. **72**, 4397 (1968)
62. S. Alexander, R. Orbach: J. Phys. Lett. **43**, L625 (1982)
63. I. Webman: Physical Review Letters **47**, 1496 (1981)
64. D. Benavraham, S. Havlin: J. Phys. A Mathematical and General **15**, L691 (1982)
65. Y. Gefen, A. Aharony, S. Alexander: Phys. Rev. Lett. **50**, 77 (1983)
66. R. Rammal, G. Toulouse: J. Phys. Lett. **44**, L13 (1983)
67. A. Pluen, P.A. Netti, R.K. Jain, D.A. Berk: Biophys. J. **77**, 542 (1999)
68. M. Rinaudo, I. Roure, M. Milas, A. Malovikova: Internat. J. Poly. Anal. Charact. **4**, 57 (1997)
69. J. Labille, F. Thomas, M. Milas, C. Vanhaverbeke: J. Coll. Interf. Sci. **284**, 149 (2005)
70. R.M. Rosenberg: *Principles of Physical Chemistry.* (Oxford University Press, New York 1977)

14

Ecotoxicology: Nanoparticle Reactivity and Living Organisms

Mélanie Auffan, Emmanuel Flahaut, Antoine Thill, Florence Mouchet, Marie Carrière, Laury Gauthier, Wafa Achouak, Jérôme Rose, and Mark R. Wiesner, and Jean-Yves Bottero

Nanotechnology is a major source of innovation with important economic consequences. However, the potential risks for health and the environment have raised questions on national, European, and international levels. Past experience of sanitary, technological, and environmental risks has shown that it is not a good policy to attempt to deal with them after the fact. It is thus crucial to assess the risks as early on as possible. A particular problem is the potential dissmination of mass produced man-made nanoparticles into the environment [1, 2]. Nanomaterials represent a particular hazard for humans due to their ability to penetrate and subsequently damage living organisms [3]. Indeed, the data available at the present time shows that some nanomaterials, especially insoluble particles, can cross biological barriers and distribute themselves within living organisms.

The surge of interest in nanoparticles is a result of their unique properties, or nano-effects, often radically different to those of the same macroscopic materials (see Chap. II). The main cause underlying the change in properties is the very high surface to volume ratio. A nanoparticle of diameter 6 nm will have 35% of its atoms at the surface and hence an exceptionally high interfacial reactivity. These novel properties on the nanoscale lie at the heart of current scientific work on drug delivery, tumour targeting, the replacement of silicon in microelectronics by carbon nanoparticles, the synthesis of tougher materials, and many other projects. Considering the huge range of applications, it seems reasonable to expect their dissemination in the environment at each step in their life cycle, from design through production to use and disposal of finished products. As a consequence, it is important to study the risks for the biological components of the various repository media, and in particular concentrating media, such as the aquatic compartment.

By definition, a toxic product is a chemical compound which can harm the environment by affecting the biological organisms that occupy it, including human beings. Owing to their novel properties, the ecotoxicological impact of nanoparticles cannot be studied in the same way as other xenobiotics in the environment, e.g., pesticides, medicines, etc. Nanoparticles have mass, charge,

P. Houdy et al. (eds.), *Nanoethics and Nanotoxicology*,
DOI 10.1007/978-3-642-20177-6_14,

and above all surface area. They are subject to the phenomena of classical and quantum physics. Their reactivity means that their surface atoms are labile, easily change their redox state, and highly reactive with respect to compounds in the aqueous phase.

It was because nanoparticles were seen as conventional pollutants that the first nanotoxicological investigations often led to contradictory results [4, 5], and consequent controversy between research groups. These differences arose because the properties of nanoparticles, and the conditions of exposure of organisms, were poorly controlled. Most of the physicochemical properties of nanoparticles have a potential impact on their interaction with living beings. Among the most significant are their chemical nature, crystal structure, specific surface area, size, and morphology (e.g., spherical, acicular, fibre), surface charge, surface functionalisation (presence of chemical functions), and state of aggregation. A poor understanding of the physicochemical behaviour of nanoparticles is likely to lead to an erroneous interpretation of ecotoxicological data. For example, the differences observed over the last few years in the biological effects of carbon nanoparticles (C_{60}) can be imputed at least in part to the presence of residues from the synthesis and from the organic solvent used to disperse them [6].

It is thus difficult to understand the results of studies about the ecotoxicity of nanoparticles on different organisms such as bacteria (see Sect. 14.2), aquatic organisms (see Sect. 14.3), or plants (see Sect. 14.4), without first considering their physicochemical properties (see Sect. 14.1). To illustrate the interactions between nanoparticles and organisms, this chapter will mainly discuss metal nanoparticles (e.g., Fe, Ag), metal oxide nanoparticles (TiO_2, CeO_2, Fe_3O_4, γ-Fe_2O_3, ZnO), and carbon nanoparticles (C_{60} and carbon nanotubes) which are stimulating a great deal of interest today in terms of development and applications.

14.1 Physicochemical Properties and Ecotoxicity of Nanoparticles

In most cases, the nanoparticles studied are poorly characterised, or not characterised at all, from the physicochemical point of view. However, in order to assess the potential risks due to the presence of nanoparticles in the environment, a systematic characterisation is essential. One of the main problems in interpreting published work stems from the poor understanding and/or excessive diversity of the samples. Nanoparticles can have widely different morphologies, crystal structures, and surface properties. Several different methods must therefore be combined in the research effort, including ecotoxicology, physicochemistry, and crystal chemistry. In this section, we shall show that the ecotoxic response can be very different depending on the state of aggregation of the nanoparticles (see Sect. 14.1.1), their chemical stability (see Sect. 14.1.2), or modifications to their surface (see Sect. 14.1.3).

14.1.1 Nanoparticle Aggregation

Nanoparticles are not thermodynamically stable systems. One can define an interfacial tension which gives this dispersed state a high free energy. Without stabilisation via electrostatic repulsion (surface charge) and/or steric repulsion (adsorbed molecules), nanoparticles will agglomerate and hence be eliminated from the suspension by precipitation or flocculation. Once stabilised, nanoparticle suspensions can remain as such for long periods, but that will depend on the physicochemical conditions in the solution. For example, an increase in the ionic strength, a change of pH, or the presence of extracellular proteins [7] can perturb the stability of nanoparticle suspensions. And this type of modification is very common in ecosystems, in particular due to biological activity.

In ecotoxicity studies, nutrient solutions in equilibrium with aquatic organisms, micro-organisms, or plants contain nutrients, organic salts, sources of carbon and energy (glucose), sources of nitrogen (amino acids), and growth factors (vitamins, fatty acids). The high surface reactivities of nanoparticles for molecules and ions in solution associated with the environmental pH close to the zero charge point of most nanoparticles [8] will significantly perturb their colloidal stability. Such is the case with maghemite (γ-Fe_2O_3) nanoparticles (see Fig. 14.1) characterised by a mean hydrodynamic diameter of 20 nm in water at pH 3, but which form 50–100 μm aggregates in cell nutrient solutions.

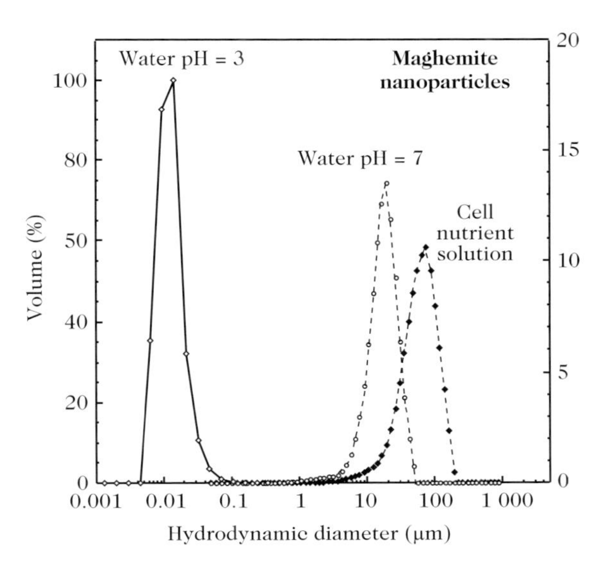

Fig. 14.1. Aggregation of nanoparticles in different aqueous media. Examples of maghemite (γ-Fe_2O_3) nanoparticles of diameter 6 nm in ultrapure water with acid pH, lower than the zero charge point (ZCP), neutral pH close to the ZCP, and in a cell nutrient solution

This aggregation of nanoparticle suspensions often contributes to the variability of the observed effects. For example, there are contradictions regarding size effects in the case of TiO_2 nanoparticles. According to Adams et al. (2006) and Verran et al. (2007), there is no effect, whereas Qi et al. (2006) find that the toxicity increases when the size of the nanoparticles is reduced [9–11]. These disagreements probably arise from the nanoparticle composition and the conditions under which the toxicity tests were carried out. In certain cases, nanoparticles may tend to aggregate, thereby reducing their contact with the given organism and hence also reducing their toxicity [10]. Sondi and Solopek-Sondi (2004) also observed that silver nanoparticles are toxic only when contact occurs on a solid medium, but not in a liquid medium where they note only slowed growth [12]. This can be explained by aggregation of the silver nanoparticles with intracellular components of dead cells. Once aggregated, their bactericidal effects are lessened and bacteria can develop normally. On the other hand, silver nanoparticle aggregation can be avoided by adding bovine serum albumin, and in this case, the bactericidal effect is maintained [13].

The destabilisation of nanoparticles in solution generally happens suddenly when physicochemical conditions are propitious. For the ionic strength, there is a critical coagulation concentration (CCC) beyond which contacts between nanoparticles cause them to stick together. The rate at which the solution is destabilised is then a question of kinetics. As a guide, one can use a simplified expression which gives the evolution of the concentration $N(t)$ of isolated particles in an initially stable suspension just after complete destabilisation, viz.,

$$\frac{1}{N(t)} = \frac{1}{N(0)} - \frac{4kT}{3\eta t} ,$$

where $N(0)$ is the initial concentration of nanoparticles in solution, η is the viscosity of the solution, k is the Boltzmann constant, and T is the temperature. For example, less than one minute is required for half of a suspension of 1 mg/L of CeO_2 nanoparticles to aggregate. This should be compared with the characteristic time for adsorption onto cells. On the other hand, it is very likely to be short compared with the modifications in the metabolism. This will affect the ecotoxicity of CeO_2 nanoparticles. Indeed, these suspensions prove to be toxic for *Escherichia coli* when their stability is maintained by working in a medium of low ionic strength. But at higher ionic strengths, CeO_2 nanoparticles aggregate and the toxic effect is no longer observed. However, we shall see later in the chapter that the colloidal stability of nanoparticles does not alone guarantee a toxic effect.

Carbon nanotubes are also prone to very strong interactions with many biological molecules, especially proteins. In fact, DNA is commonly used to stabilise carbon nanotube suspensions [14]. Moreover, it has been shown that carbon nanotubes interact with the immune system, not only in the blood complement [15], but also in the respiratory system through pulmonary surfactants [15]. Consequently, the state of aggregation of carbon nanotubes may vary in time after exposure, and in different ways depending on the target organ. Likewise, the presence in the environment of industrial surfactants such as waste water, or natural surfactants such as humic acids, is likely to significantly modify their dispersion [16].

14.1.2 Chemical Stability of Nanoparticles

Similarly to aggregation, the chemical stability of nanoparticles, e.g., with regard to dissolution, oxidation, reduction, and generation of reactive oxygen species, plays an important role with respect to ecotoxicity. For example, nanoparticles are often made from soluble materials such as ZnO or CdS, which can salt out toxic ions. This is the case with Zn^{2+} ions released when ZnO nanoparticles are dissolved, and this underlies their bactericidal effects [17, 18]. Furthermore, the solubility of materials in the form of nanoparticles can be higher than that of the bulk material due to their higher specific surface area, but also their higher surface reactivity (see Chap. II).

The effect of specific surface area on the solubility of ZnO nanoparticles, and hence on their toxicity, has been demonstrated [19, 20]. Nanoparticles of diameter 100 nm are significantly toxic at concentrations above 12 mmol/L [20], whereas nanoparticles with diameters 10–15 nm are bactericidal from 1.3 mmol/L [19]. On the other hand, identical toxic effects are observed for 30 nm ZnO nanoparticles, ZnO microparticles, and dissolved $ZnCl_2$ salts [17, 18].

Nanoparticles can also generate reactive oxygen species, e.g., TiO_2, ZnO, Fe^0, Fe_3O_4. This is due to the properties of the material, and can be enhanced by the specific properties of the nanoparticles. Reactive oxygen species can also be generated under the effects of UV radiation. This is exemplified by TiO_2 and ZnO nanoparticles which exhibit an increased bactericidal effect under irradiation [9, 21]. Reactive oxygen species are produced by reactions of the following type:

$$\begin{aligned} TiO_2 + h\upsilon &\longrightarrow TiO_2(h^+ + e^-)\,, \\ e^- + O_2 &\longrightarrow O_2^-\,, \\ O_2^- + 2H^+ + e^- &\longrightarrow H_2O_2\,, \\ H_2O_2 + O_2^- &\longrightarrow {}^\bullet OH + OH^- + O_2\,, \\ H^+ + H_2O &\longrightarrow {}^\bullet OH + H^+\,. \end{aligned}$$

Reactive oxygen species are also produced by Fenton reactions involving Fe^{2+} emitted during oxidation and dissolution–recrystallisation of iron-containing nanoparticles:

$$\begin{aligned} Fe^0 + O_2 + 2H^+ &\longrightarrow Fe^{2+} + H_2O_2\,, \\ Fe^0 + H_2O_2 &\longrightarrow Fe^{2+} + 2OH^-\,, \\ Fe^{2+} + H_2O_2 &\longrightarrow Fe^{3+} + {}^\bullet OH + OH^- \quad \text{(Fenton reaction)}\,. \end{aligned}$$

For example, nanoparticles containing only the oxidised form Fe^{3+}, e.g., maghemite, are stable and non-toxic towards *E. Coli* [22]. In contrast, those containing the reduced forms Fe^0 or Fe^{2+}, e.g., iron metal or magnetite, oxidise in solution and are highly bactericidal. On the other hand, it is silver nanoparticles containing the oxidised form Ag^+ rather than the purely metallic form Ag^0 which turn out to be toxic [13]. Moreover, for a given mass, the toxicity increases when the size of silver nanoparticles is reduced, and this is directly correlated with the increase in the fraction of Ag^+ ions at the particle surface.

Carbon nanotubes are different in this respect, because they are extremely hydrophobic and insoluble in the vast majority of solvents. However, residues of catalysts used to synthesise them (mainly transition metals like Fe, Co, and Ni) may nevertheless lead to the release of metal ions during exposure to carbon nanotubes.

14.1.3 Functionalised or Passivated Nanoparticles

Among the applications predicted for nanoparticles, some require the nanoparticle surface to be modified in order to increase their bioavailability, facilitate their dispersion in matrices, or deliver them to specific organs (as in the case of drug delivery). This happens in particular with iron oxide nanoparticles, widely used in the biomedical field. Owing to their zero charge point close to the physiological pH, these nanoparticles aggregate significantly in biological media (see Fig. 14.1). One way to limit aggregation is to create negative charges artificially, in order to generate sufficiently strong repulsive forces to keep them dispersed. A very effective molecule here is 2,3-dimercaptosuccinic acid [COOH–CH(SH)–CH(SH)–COOH] [23]. With its two thiol (–SH) functions, this molecule adsorbs strongly onto the surface of iron oxide nanoparticles via Fe–S bonds, while the $-COO^-$ groups confer a negative charge upon the nanoparticles, thereby limiting electrostatic attractions [24]. These strong chemical bonds survive prolonged suspension of iron oxide nanoparticles in biological media.

However, these surface modifications can cause drastic changes in the physicochemical properties and fate of nanoparticles in living organisms. For example, gold nanorods functionalised by specific bacterial antibodies exhibit a high level of toxicity, whereas non-functionalised gold nanorods have no toxic effect on the same bacteria [25]. In this case, bactericidal effects require direct exposure of the bacterial wall to the nanoparticles and light activation.

Carbon nanotubes are also often functionalised. There are two main types of carbon nanotube (see Fig. 14.2): single-walled carbon nanotubes (SWCNT) and multi-walled carbon nanotubes (MWCNT) made from one or more concentric tubes, respectively. Among the MWCNT, double-walled carbon nanotubes (DWCNT) are intermediary between SWCNTs and MWCNTs with regard to characteristics such as morphology and mechanical and electronic properties. DWCNTs have a major advantage over SWCNTs in that it is

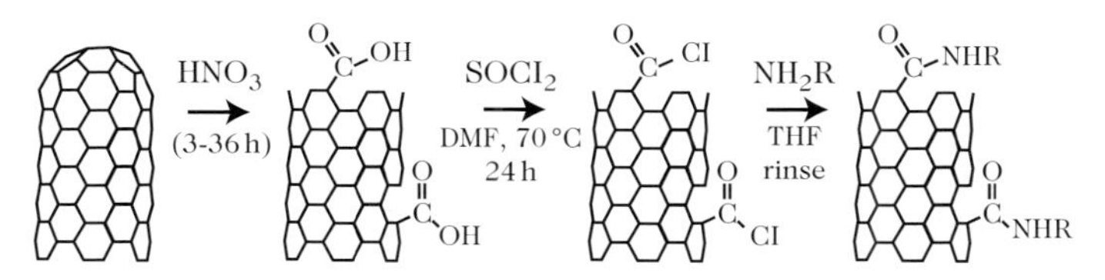

Fig. 14.2. Modification of nanoparticle surface properties. Different possibilities for functionalising the surface of carbon nanotubes following a primary oxidation stage

possible to modify their outer surface (by covalent grafting) without touching the inner tube. This means that they can be given useful surface properties, e.g., to facilitate their dispersion in a solvent, but without seriously damaging their mechanical properties (covalent functionalisation of SWCNTs partly destroys the carbon lattice) or electrical properties. Surface functionalisation of carbon nanotubes by oxygen functions can be achieved by reacting with an oxidising acid like concentrated nitric acid, for instance, or with mixtures of sulfuric acid and potassium permanganate, or other oxidising solutions. In this way, carboxylic acid and hydroxyl functions can be covalently grafted onto the surface, making the carbon nanotubes hydrophilic (see Fig. 14.2). These oxygen functions can serve as elementary building blocks for subsequent grafts of chemical functions, polymer chains, or molecules [26].

14.2 Ecotoxicity for Bacteria

There are many studies on the antibacterial properties of nanoparticles, e.g., [27]. For example, it is well established that silver and TiO_2 nanoparticles are efficient bactericides, used today to sterilise medical equipment. However, very few studies have directly investigated the harmful effects of nanoparticles on bacterial ecosystems. This is the subject of the present section.

The Cell: Basic Functional Unit of Life. One cell can function in complete autonomy in planktonic form or in a biofilm: this is the case of single-cell organisms, e.g., bacteria, archaea, micro-algae, protozoa, etc., or organisms integrated into a multicellular structure, e.g., fungal hyphae, tissues, etc.

Eukaryotic and prokaryotic cells share a highly organised structure made up essentially of four kinds of macromolecule: lipids, proteins, nucleic acids, and polysaccharides. It is the structure and organisation of these macromolecules on the cellular level that differentiates between the various organisms. A cell is always bounded by a membrane which isolates it from its surroundings and other cells. This membrane is structured in such a way as to retain chemical components and ions while at the same time allowing certain exchanges with the environment, namely the evacuation or entry of metabolites. This membrane bounds the compartment in which the essential functions of cell life take place, namely the cytoplasm. This in turn contains the nucleus or nucleoid, where the genetic information specific to the cell is stored, to be faithfully transmitted to the following generation. Most micro-organisms and plant cells have a wall, in contrast to animal cells. This outer wall beyond the cytoplasmic membrane serves mainly to maintain the cell structure, whereas animal cells have an intracellular cytoskeleton.

Unlike prokaryotic cells which do not carry organelles, the cytoplasm of eukaryotic cells contains the nucleus which houses the genome, and mitochondria and chloroplasts (in the case of photosynthesising organisms) which provide the energy the cell needs to function.

14.2.1 Bacteria

Microbial cells constitute the main part of the terrestrial biomass despite their very small size. The number of bacteria is estimated to be around 5×10^{30} cells. Bacteria lie at the base of the food chain and are one of the main components of biogeochemical cycles, e.g., nutrients, minerals. They occur in most terrestrial and aqueous environments and can survive under extreme conditions, e.g., anaerobia, extreme temperature and pH, high metal concentrations, etc. They are highly flexible in morphological and physiological terms, with a great ability to adapt to and resist changing environmental conditions and all kinds of xenobiotic. Bacteria also exhibit the highest biological specific surface area, and this is in permanent interaction and exchange with the biotic and abiotic constituents of the environment. For this reason, any investigation of nanoparticle ecotoxicology must involve detailed study of nanoparticle–bacteria interactions, and the relevant toxicity mechanisms. Furthermore, bacteria can transform and 'metabolise' nanoparticles, modifying their mobility and bioavailability in the environment, important processes that need to be monitored in the context of environmental study.

It has been well established that nanoparticles have bactericidal effects. This suggests that nanoparticles may affect the viability and diversity of micro-organisms, and as a consequence, the functioning of the whole ecosystem, if they should occur in the environment at high concentrations and in a dispersed form.

In the environment, nanoparticles will begin by interacting with bacterial exopolymers, walls, and membranes. The cytoplasmic membrane plays a decisive role in the transport of nutrients and the wall in the protection of the cell against osmotic lysis:

- The cytoplasmic membrane comprises a double phospholipid layer about 8 nm thick, which is a permeable barrier. Many proteins, called intramembrane proteins, are encased in this membrane, most being involved in the transport of nutrients, secretion of other proteins, or rejection of toxic substances. The membrane is also where respiration takes place and the scene of the electron transfer chain.
- The wall is a rigid structure made up of peptidoglycans. The structure of the wall distinguishes between Gram-negative and Gram-positive bacteria. The wall of Gram-negative bacteria is the more complex, comprising

several sheets, while that of Gram-positive bacteria has a simpler composition but is often thicker.

- Bacteria also produce a wide range of exopolysaccharides which differ by their structure and function. These exopolysaccharides serve mainly to protect bacteria from hydric stress, the defence system of the host in the case of pathogens, and toxic substances, allowing them to colonise different media and arrange themselves in biofilms.

It is essential to take into account the kinds of interactions between nanoparticles and these bacterial constituents when carrying out nanoecotoxicity studies. Bacteria also provide a useful model because they operate an extracellular electron transport system which allows them to oxidise or reduce substrates that prove too large to be internalised, such as humic acids or iron oxides. Bacteria can metabolise these substrates by shuttles which are reduced in the membrane and oxidised in the substrate and vice versa, or directly by contact with enzymes or cytochrome located in the membrane. Other bacteria produce filaments several micrometers long from proteins, called fimbriae or pili, which can reduce iron oxides. This is the case of 'nanowires' of *Geobacter sulfurreducens* [28]. It is important to consider these reactions, initiated directly by enzyme activity or indirectly by production of oxidising or reducing agents, in the transformation of nanoparticles in the environment, e.g., redox, dissolution.

14.2.2 Effects of Nanoparticles on Bacterial Viability

In most studies today, the exposure conditions of the bacteria (solid or liquid media), the toxicity tests used, e.g., colony counts, growth curves, or membrane permeability, the types of nanoparticles, e.g., size, shape, dispersant, and the bacteria chosen for study, differ widely from one research group to another. For example, the toxicity of zinc oxide nanoparticles has been studied in a gelled solid medium [20, 29], in a liquid medium [17–19, 30], and by immersion of fabrics impregnated with nanoparticles in ultrapure water with the bacteria [31]. The same goes for TiO_2 nanoparticles investigated for their bactericidal effect in a liquid suspension [10, 32], dispersed in a gelled medium [9], adsorbed on cotton fibres [21], or adsorbed onto functionalised thin films [33].

The main studies dealing with the bactericidal effects of nanoparticles are summarised in Table 14.1. It should be borne in mind that the wide range of methods used here makes it difficult to compare results. However, two paradigms arise in these ecotoxicity studies, and these will be discussed in Sects. 14.2.3 and 14.2.4.

14.2.3 Exposure of Bacteria to Nanoparticles

The first paradigm concerns the conditions under which bacteria are exposed to nanoparticles. The surface properties of cell membranes are a decisive factor

Table 14.1. Different studies investigating the bacterial ecotoxicity of nanoparticles

Nanoparticle	Biological species	Dose studied	Effects observed or parameters measured	Ref.
Ag^0	*E. coli*	10–60 mg/L	70% drop in bacterial survival above 10 mg/L. Increased membrane permeability	[12]
Ag^0	*E. coli*, *Pseudomonas aeruginosa*, *S. typhus*, and *Vibrio cholera*	25–100 mg/L	*P. aeruginosa* and *V. cholera* more resistant than *E. coli* and *S. typhus*. Above 75 mg/L, no growth observed in the four types of bacteria	[34]
Ag^0	*E. coli*	0.01–1 mg/L	Most toxic nanoparticles have triangular shape, for which 100% growth inhibition is observed for 0.6 mg/L, compared with 0.2 mg/L for spherical nanoparticles	[35]
Fe^0	*E. coli*	7–700 mg/L, 1 h	20% drop in bacterial survival above 70 mg/L and oxidative stress. Toxicity associated with oxidation of Fe^0 to Fe^{2+} and Fe^{3+}	[22]
Fe_3O_4	*E. coli*	7–700 mg/L, 1 h	20% drop in bacterial survival above 350 mg/L and oxidative stress. Toxicity associated with oxidation of Fe^{2+} to Fe^{3+}	[22]
γ-Fe_2O_3	*E. coli*	7–700 mg/L, 1 h	No significant drop in bacterial survival. Nanoparticles chemically stable	[22]
MgO	*E. coli*, *Staphylococcus aureus*	1 mg/L	Bacterial survival rate depends on particle size	[36]
MgO	*Bacillus subtilis*	0.50 g MgO, 24 h	Bactericidal effects increase with decreasing particle size. Nanoparticle surface generates high concentrations of O_2^-	[36]
ZnO	*E. coli*	8–800 mg/L	Significant perturbation of cells for concentrations above 1.3×10^{-3} mol/L. Nanoparticle internalisation is observed	[19]

ZnO	*E. coli*	100–250 mg/L	Significant bacteriostatic activity, alteration of the membrane. Coating by polyethylene glycol or polyvinylpyrolidone do not affect antibacterial activity	[18]
CeO_2	*E. coli*	0.5–500 mg/L, 3 h	50% drop in bacterial survival above 7 mg/L. Toxicity associated with reduction of Ce^{4+} to Ce^{3+}	[38]
CeO_2	*Synechocystis*	0.5–500 mg/L, 3 h	50% drop in bacterial survival without pH buffer and above 25 mg/L. Toxicity is due to changed pH	[39]
TiO_2	*E. coli*, *Bacillus megaterium*	1–400 mg/L	Growth inhibition in both bacteria in ambient lighting conditions	[40]
Pt/TiO_2	*E. coli*, *S. aureus*, *Enterococcus faecalis*	1 000 mg/L	Bactericidal effects under UV radiation: *E. coli* > *S. aureus* > *E. faecalis*. Pt(IV) increases bactericidal effects in darkness	[41]
CNT	*E. coli*	1–50 mg/L	Significant antimicrobial activity and alteration of cell membrane	[42]

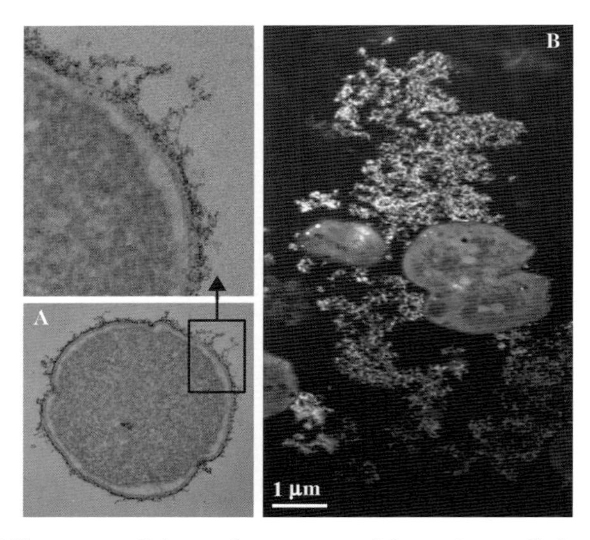

Fig. 14.3. Different conditions of exposure of bacteria to CeO_2 nanoparticles. (**A**) Adhesion onto the cell wall of *E. coli*. The nanoparticles form a monolayer covering the surface of the *E. coli*. (**B**) Adhesion in the exopolysaccharide layer of *Synechocystis*. In this case, little direct contact is observed between the nanoparticles and the bacterial wall

in nanoparticle toxicity [43]. For example, it turns out that C_{60} nanoparticles associate more strongly at the surface of Gram-negative bacteria, e.g., *E. coli*, than at the surface of Gram-positive bacteria, e.g., *B. subtilis*. It also turns out that, when nanoparticle toxicity is due to 'direct' redox effects, the proximity of the nanoparticles and the bacterial walls plays an important role (see Fig. 14.3) [38, 44]. This 'direct' redox toxicity can be inhibited or limited when the exposure of the cells to nanoparticles is modified. If the nanoparticles have aggregated and/or if their surface charge has been modified, the close contact interaction may not be able to occur and this form of toxicity is then significantly reduced. In this case, the area ratio between target cells and nanoparticles is large and one would no doubt observe effects due to size or the state of aggregation. On the other hand, when toxicity is due to an 'indirect' effect, such as the salting out of potentially toxic ions, e.g., Zn^+, Cd^{2+}, Ag^+ [17] or a change in pH [39], the exposure conditions are no longer fundamental. The important measurement for understanding toxic effects is then the nanoparticle concentration. In this case, the state of agglomeration will not be the key, even though several studies have suggested such a connection.

Some studies have also investigated bacterial communities in natural soils. Tong et al. [45] assessed the impact of adding 1 mg of C_{60} per gram of soil by carrying out DNA and fatty acid analyses, finding only a small impact on the structure of these communities. It thus turns out that C_{60} nanoparticles are less toxic under natural soil conditions [45] than under controlled laboratory conditions [43]. However, the impact of C_{60} on the physiology and functions of soil bacteria remains unknown.

14.2.4 Oxidative Stress

The second paradigm concerns nanoparticle-induced oxidative stress. Nanoparticles that are chemically unstable in biological media can produce reactive oxygen species in the vicinity of bacteria and induce significant oxidative stress. It seems that metallic nanoparticles are the most sensitive to oxidation or reduction, e.g., Fe^0, Fe_3O_4, CeO_2, and have the most marked effect on bacteria [46].

Using bacterial strains deficient in superoxide dismutase, an antioxidant, it has been shown that oxidative stress is one of the main toxicity mechanisms. For iron-containing nanoparticles, reactive oxygen species are generated through Fenton reactions, which produce hydroxyl radicals from the emitted Fe^{2+}. For example, magnetite (Fe^{2+}/Fe^{3+}) nanoparticles of radius 6 nm are highly toxic to *E. coli* from 0.7 g/L of Fe_3O_4. It has been shown using X-ray absorption spectroscopy that the surface of magnetite nanoparticles oxidises to maghemite (Fe^{3+}) after contact with the bacteria [22]. This change of phase occurs via desorption of Fe^{2+} from the structure and the creation of surface vacancies [47]. Fe^0 nanoparticles are much more sensitive to oxidation and generate toxicity at 10 times lower doses, viz., 0.07 g/L of Fe^0 [22]. They are entirely transformed into lepidocrocite (Fe^{3+}) and magnetite (Fe^{2+}/Fe^{3+}). This oxidation follows a dissolution–recrystallisation process producing a hydroxide of Fe^{2+} and Fe^{3+}, called green rust.

For CeO_2 nanoparticles, reactive oxygen species are produced in redox cycles $Ce^{4+} \longrightarrow Ce^{3+} \longrightarrow Ce^{4+}$, which occur on the nanoparticle surface [48]. These cycles underlie the catalytic properties of CeO_2 nanoparticles and are accompanied by significant electron transfer, ion transfer, and the creation of vacancies in the surface structure. In biological media, these redox cycles can induce the oxidation of certain compounds at the interface with the bacterial walls. Thill et al. (2006) showed that 50% of the *E. coli* population does not survive the presence of 0.003 g/L of CeO_2 (Ce^{4+}) nanoparticles of diameter 7 nm adsorbed on their walls [38]. This toxicity is associated with the reduction of 30% of their surface atoms into Ce^{3+}.

One consequence of the production of reactive oxygen species is that they can trigger a chain of destructive radical reactions such as lipid peroxidation, in the bacterial lipopolysaccharide layer. This happens with reactive oxygen species generated during oxidation of TiO_2 and ZnO nanoparticles [49]. In particular, Sunada et al. [33] have observed the destruction of the outer then inner membrane in *E. coli* in the presence of TiO_2 nanoparticles. Finally, an interesting example is C_{60} [50], which induces a modification in the synthesis of bacterial fatty acids. This is a mechanism for protecting the cell membrane against reactive oxygen species. *Pseudomonas putida* reduces the synthesis of conventional fatty acids in favour of cyclopropane fatty acids, while *Bacillus subtilis* synthesises more monosaturated fatty acids. Membrane fluidity is increased in both cases.

The Cell: A Chemical Factory. A cell interacts with its environment to obtain the nutrients it transforms (metabolism) in order to extract energy and to produce the macromolecules it needs to keep the cell machinery working and maintain the cell structure. It also produces metabolites that it must release into its environment.

A cell transforms chemical compounds to generate another living organism by reproducing, doubling its contents to give rise to a cell that generally has the same properties and characteristics as the mother cell. Cell division involves a stage in which the genetic material is doubled by replication of the chromosomes. The genes essential to cell division are transcribed from the DNA to make RNA, which is in turn translated into proteins with the help of ribosomes, particles composed of RNA and proteins.

All these operations are orchestrated by regulators which allow the cell to 'sense' its environment and adapt its responses and its way of life to external conditions by expressing suitable genes. The cells also communicate with one another via chemical mediators. They can move toward environments where conditions are more favourable, because most living organisms are endowed with mobility and able to move in an autonomous way, with the exception of plants, which are sessile. Living organisms evolve through genetic rearrangements which allow them to acquire new properties. This evolution takes place over several generations and can be studied in micro-organisms. Single-cell micro-organisms have the property of reproducing quickly and autonomously, reaching high population densities under laboratory conditions and producing several generations over a reasonable lapse of time, which makes them good models for studying the cell machinery and its adaptations, evolution, and limitations in the face of environmental stress.

14.3 Ecotoxicity for Aquatic Organisms

The available ecotoxicological data is rather incomplete and insufficient to draw global conclusions about the impact of nanoparticles on the aquatic environment. One particular difficulty is to evaluate the concentrations of nanoparticles which might occur in aquatic environments and which could be qualified as realistic from an environmental standpoint. This section presents the results of recent studies carried out on aquatic vertebrates and invertebrates in order to assess the ecotoxicity of carbon nanoparticles (see Sect. 14.3.1) and metal and metal oxide nanoparticles (see Sect. 14.3.2).

14.3.1 Carbon Nanoparticles

Most available studies concern the fullerenes C_{60}. These studies, summarised in Table 14.2, demonstrate the ingestion of C_{60} and its associated toxicity in several model organisms, viz., the freshwater crustacea *Daphnia magna* and *Hyalella azteca*, along with the fish *Pimephales promelas*, *Oryzias latipes*, *Danio rerio*, *Micropterus salmoides*, and *Carassius auratus* [3, 51–57].

Table 14.2. Studies on the ecotoxicology of C_{60} in crustacea, fish, and earthworms

Biological species	Concentration	Effects observed and parameters measured	Ref.
Freshwater crustacea			
Daphnia magna	0.5–1–2.5–5 mg/L for 48 h and 21 days, and 30 mg/L for 2 days	Moulting delayed after 21 days and reduced reproduction at 2.5 and 5 mg/L, respectively, in invertebrates. Reduced expression of the protein PMP70 (peroxisomal lipid transport) in *P. promelas*, but not in *O. latipes*, suggesting modifications in acyl-CoA pathways	[53]
Hyalella azteca	7 mg/L for 96 h		
Marine copepods	3.75–7.5–15–22.5 mg/L for 96 h		
Pimephales promelas	0.5 mg/L for 96 h		
Oryzias latipes	0.5 mg/L for 96 h		
Daphnia magna	0.5–1.5–10–25–50–100 mg/L for 48 h	Increased immobilisation and mortality from low doses. EC_{50} (immobilisation) = 9.3 mg/L and LC_{50} (mortality) = 10.5 mg/L	[56]
Daphnia magna and the fish *Pimephales promelas*	0.5 ppm of THF–nC_{60} and water–nC_{60} for 48 h	LC_{50} for THF–nC_{60} = 0.8 mg/L and water–nC_{60} > 35 mg/L. 100% mortality in fish exposed to THF–nC_{60}, but not in fish exposed to water–nC_{60}. Lipid peroxidation in brain and gills. Increased expression of isoenzymes CYP2 in liver of individuals exposed to water–nC_{60}	[55]
Daphnia magna	0.26 mg/L of nC_{60} and $C_{60}H_xC_{70}H_x$	Increased heart rate and predation, and reduced reproduction above 0.26 mg/L	[52]
Daphnia magna	0.04–0.18–0.26–0.35–0.44–0.51–0.7–0.88 mg/L (filtered THF–nC_{60}) and 0.2–0.45–0.9–2.25–4.5–5.4–7.2–9 mg/L (sonicated nC_{60})	Increased mortality from lowest doses. LC_{50} 48 h = 0.46 mg/L, LOEC = 0.26 mg/L and NOEC = 0.18 mg/L (THF-nC_{60}). LC_{50} 48 h = 7.9 mg/L, LOEC = 0.5 mg/L and NOEC = 0.2 mg/L (sonicated nC_{60})	[52]

(*Continued*)

Table 14.2. (Continued)

Biological species	Concentration	Effects observed and parameters measured	Ref.
Fish			
Micropterus salmoides	0.5 and 1 mg/L of THF–nC_{60} for 48 h	Lipid peroxidation in brain at 0.5 mg/L. Glutathione depletion in gills at 1 mg/L. Increased limpidity of exposure water due to bacterial activity	[3]
Danio rerio (embryos)	1.5 and 50 mg/L of THF–nC_{60} and THF–$nC_{60}(OH)_{16-18}$ for 96 h	Delayed hatching and development of larvae, reduced survival and hatch rate, and pericardial edema at 1.5 mg/L of C_{60}	[57]
Juvenile carp *Carassius auratus*	0.04–0.20–1 mg/L of nC_{60} for 32 days	Induction of the antioxidant enzymes superoxide dismutase and catalase in gills and liver. Reduced glutathione in all tissues tested. Reduced lipid peroxidation in gills and brain except at 1 mg/L in the liver. Inhibited growth at 1 mg/L	[54]
Danio rerio (embryos)	0.1–0.2–0.3–0.5 mg/L of DMSO–nC_{60} and $nC_{60}(OH)_{24}$ under illumination for 120 pfh (malformations and mortality) and 24 pfh (sublethal effects)	Mortality, altered expression of genes involved in oxidative stress. Reduced mortality, malformations and pericardial edema at 0.2 and 0.3 mg/L with light reduction. Reduced mortality and pericardial edema in embryos coexposed in the presence of glutathione precursor. Increased sensitivity of embryos coexposed to GSH inhibitors. Increased mortality in embryos coexposed in the presence of H_2O_2	[58]
Earthworms			
Eisenia veneta	1 000 mg of C_{60} per kg of food (dry weight) for 28 days	No effect on hatching or mortality at 1 000 mg of C_{60} per kg of food	[59]

DMSO dimethyl sulfoxide, EC_{50} efficient concentration for 50% of individuals exposed, pfh post-fertilization hours, LC_{50} lethal concentration for 50% of individuals exposed, LOEC lowest observed effect concentration, NOEC no observed effect concentration, TFH tetrahydrofurane

There has been little work on the ecotoxicology of carbon nanotubes in aquatic organisms (see Table 14.3). Petersen et al. (2008) demonstrated that the freshwater oligochaetes *Lumbriculus variegatus* ingest SWCNTs associated with sediment particles, identifying them in the intestine but not establishing whether they are absorbed in the tissues [60]. Roberts et al. (2007) demonstrated the ingestion of SWCNTs coated with lysophospholipids by the freshwater crustacea *Daphnia magna*, and observed mortality associated with high concentrations [61]. Templeton et al. (2006) found increased mortality and reduced fertilization rate in the estuarine copepod *Amphiascus tenuiremis*, depending on the SWCNT mixtures used [62]. Recently, Kennedy et al. (2008) identified reduced viability in the cladoceran *Ceriodaphnia dubia* exposed to raw MWCNTs, while this was not observed when these same MWCNTs were functionalised [63]. In the amphipods *Leptocheirus plumuloss* and *Hyalella azteca* exposed via sediments, they also observed that mortality increased as the size of the sediment particles decreased, although mortality here was lower for exposure to raw MWCNTs than for exposure to carbon black and active carbon. In the zebrafish *Danio rerio*, Cheng et al. found delayed hatching of eggs after exposure to SWCNTs and DWCNTs [64], and exposure to MWCNTs functionalised by bovine serum albumin [65]. In the trout *Onchorhynchus mykiss* exposed to SWCNTs, Smith et al. (2007) observed various respiratory toxicological effects and gill pathologies (hyperventilation, secretion of mucus), neuronal pathologies, and hepatic pathologies (apoptotic bodies, abnormal cell division) [66].

Two studies published recently investigate the effects of raw DWCNTs on amphibians. Amphibians and especially their larvae are excellent indicators for the health of ecosystems at the land–water interface. Studies on larvae of the axolotl *Ambystoma mexicanum* (see Fig. 14.4) revealed no sign of toxicity or genotoxicity, despite massive ingestion of DWCNTs [67]. In the xenopus *Xenopus laevis* (see Fig. 14.5), results show that, despite the mortality and growth inhibition measured at high DWCNT concentrations, associated with massive ingestion [68], no genotoxicity was observed.

In terrestrial organisms (earthworms), Scott-Fordsmand et al. (2008) showed that the exposure of *Eisenia veneta* to carbon-based nanoparticles by feeding affects neither hatch rate nor mortality at 1 000 mg C_{60}/kg dry weight of food and up to 495 mg of carbon nanotubes/kg dry weight of food [59]. In contrast, reproduction in these worms is affected from 37 mg carbon nanotubes/kg of food. Petersen et al. (2008) showed that exposure of *Eisenia foetida* to carbon nanotubes in soil induced a bioaccumulation factor twice as small as for exposure to pyrene, the chosen control molecule [69]. The authors identified carbon nanotubes in the intestine, associated with ingested soil particles. However, absorption of carbon nanotubes by tissues was not demonstrated for these organisms.

Table 14.3. Ecotoxicology of carbon nanotubes in aquatic invertebrates and vertebrates, and earthworms

Biological species	Carbon nanotube	Exposure mode	Concentration	Effects observed and parameters measured	Ref.
Invertebrates					
Freshwater oligochaete annelid	C^{14}-MWCNT	Contamination of sediment, exposure of worms for 7, 14, and 28 days	0.37 and 0.037 mg/g of dry sediment	Bioaccumulation factor lower than pyrene control. Nanotubes found in intestine, associated with ingested sediment particles. Absorption of nanotubes by tissues not demonstrated	[60]
Lumbriculus variegatus	C^{14}-SWCNT		0.03 and 0.003 mg/g of dry sediment		
Daphnia magna	SWCNT	Exposed 48 h in solution	0.1–0.5–1.5–10–50–100 mg/L	Increased immobilisation and mortality from lowest concentrations. EC_{50} (immobilisation) = 1.306 mg/L and LC_{50} (mortality) = 2.425 mg/L	[56]
	MWCNT			EC_{50} (immobilisation) = 8.723 mg/L and LC_{50} (mortality) = 22.751 mg/L	
Daphnia magna	SWCNT with lyso-phosphatidylcholine	Exposed 48 and 96 h in solution	2.5–5–10–20 mg/L with food. 0.1–0.25–0.5–1–2.5 mg/L without food	Ingestion of SWCNT coated with lysophospholipids. Daphnia can modify nanotubes by eliminating the coating lysophospholipids. 100% mortality after 48 h at 20 mg/L and 85% mortality after 96 h at 20 mg/L	[61]
Estuarine copepod *Amphiascus tenuiremis*	Purified SWCNTs and functionalised SWCNTs	Exposure in solution	0.58–0.97–1.6 and 10 mg/L	No significant effect on development and reproduction with purified SWCNTs. Increased mortality, reduced fertilization rate, reduced success in development of nauplii with functionalised SWCNTs	[62]
Estuarine copepod *Amphiascus tenuiremis* polychaete *Streblospio benedicti*	SWCNTs associated with sediment and C^{14}-SWCNTs	Coexposure to sediments and organic contaminants (COH) and aromatic hydrocarbons (PAH)	5 mg/g for 14 days	Reduced bioaccumulation of COH in *S. benedicti*. No impact of SWCNTs on accumulation of PAHs in *A. tenuiremis*. Neither model assimilates nanotubes in its tissues, although *S. benedicti* ingests C^{14}-SWCNTs (activity of C^{14} in excrement the same as in sediment)	[70]

Invertebrates (Cont.)					
Amphipods (sediment) *Hyalella azteca*, *Leptocheirus plumulosus*	Raw MWCNT, MWCNT-OH, MWCNT-COOH stabilised with organic matter	Exposure in solution	0.4–1.1–3.3–9.9 and 30% in sediment (dry weight)	Ingestion of raw, hydroxylated, and carboxylated MWCNTs, sometimes causing mortality and immobilisation. *C. dubia* EC_{50} at 48 h = 50.9 raw MWCNT	[63]
Cladoceran (water column) *Ceriodaphnia dubia*			32 and 120 mg/L of MWCNT-OH, 39.5 mg/L of raw MWCNT, and 88.9 mg/L of MWCNT-COOH		
Vertebrates					
Zebrafish *Danio rerio* (embryos)	Raw MWCNT	Exposure in solution at stage 8–16 cells for 24, 48, and 72 pfh	2.5–5–10–20–30–40–50–60–70–100–200–300 mg/L	NOEC = 40 mg/L and LOEC = 60 mg/L. Lethal effects (mortality at 72 pfh) and sublethal effects (delayed hatching, lowered blood flow rate) from 60 mg/L. 100% and no hatching above 200 mg/L. Teratogenic effects and apoptosis above 100 mg/L. Caudal and notochordal malformations at 60 and 70 mg/L. Inflammatory response by production of mucus at 60 mg/L	[71]
		Microinjection at 8 cell stage	5 ng/ml	35% mortality. Malformations from 60 mg/L	
	Raw DWCNT, raw SWCNT	Exposure in aqueous medium at stage 4 pfh for 4–96 h	SWCNT: 20–40–60–120–240–360 mg/L. DWCNT: 120–240 mg/L	Delayed hatching of eggs from 120 mg/L (SWCNT) between 52 and 72 h after fertilization and 240 mg/L (DWCNT). Delay probably induced by the Co and Ni used to synthesise the SWCNTs and which remains as a trace even after purification. Embyronic development not affected	[64]

(*Continued*)

Table 14.3. (Continued)

Biological species	Carbon nanotube	Exposure mode	Concentration	Effects observed and parameters measured	Ref.
Vertebrates (Cont.)					
Zebrafish *Danio rerio* (embryos)	Functionalised MWCNT (BSA-MWCNT) and labelled with FITC	Microinjection in egg at 1 cell stage and 72 pfh. Microangiography (venous sinus injection)	About 2 ng per embryo	No mortality or developmental defect for micro-injected eggs. In vivo biodistribution in developing eggs equivalent to control (without microinjection). Intracellular biodistribution not equivalent: accumulation of FITC-BSA-MWCNTs in nuclei of blastoderm cells; cell division and proliferation rate faster in the presence of MWCNTs. Biodistribution of FITC-BSA-MWCNTs throughout the organs and tissues of the larva after microangiography. Preferential accumulation in natatory bladder. Disappearance of signal beyond 96 h in relation with excretion of MWCNTs. Immune response induced in early stages in the presence of MWCNTs. Recruitment of lysosomal vesicles in the blastoderm. Lower survival rate of the second generation after 14 days of fertilization in the presence of MWCNTs (effect on reproductive potential)	[65]
Trout *Onchorhynchus mykiss*	SWCNT in presence of SDS	Exposure in solution at juvenile stage for 10 days	0.1–0.5 and 0.25 mg/L	Respiratory toxicological effects. Gill and neuron pathologies. Liver affected. Abnormal cell division. Raised ventilation rate and secretion of mucus. Increased activity of ATPase $Na^{+}L^{+}$ in gills and intestine. Reduced thiobarbituric acid in the brain, gills, and liver. Increase in the overall level of glutathione in the gills and liver	[66]

Vertebrates (Cont.)					
Amphibian axolotl *Ambystoma mexicanum*	Raw DWCNT	12 day exposure in solution at two-finger stage	1–10–100–125–250–500 and 1 000 mg/L	No sign of acute toxicity (mortality, growth) or genotoxicity (induction of micronuclei), despite massive ingestion of nanotubes by larvae	[67]
Amphibian xenopus *Xenopus laevis*	Raw DWCNT	12 day exposure at stage 50 of the Nieuwkoop–Faber table	10–100–500 mg/L	Mortality and delayed growth measured at high DWCNT concentrations, associated with massive ingestion of DWCNTs. No associated genotoxicity (induction of micronuclei)	[68]
Earthworms					
Eisenia foetida	C^{14}-MWCNT, C^{14}-SWCNT	Mixed with soil for 1, 7, 14, and 28 days	0.03 mg/g of dry soil and 0.3–0.03 mg/g of dry soil	Bioaccumulation factor only half the value for the pyrene control. Nanotubes present in intestine, associated with ingested soil particles. It was not shown that nanotubes were absorbed in the tissues	[69]
Eisenia veneta	DWCNT	Exposure via food for 28 days	0–50–100–300–495 mg DWCNT/kg of food (dry weight)	Reproduction affected (EC_{10}) above 37 mg DWCNT/kg of food ($EC_{50} = 176 \pm 150$ mg DWCNT/kg). No effect on hatch rate or on mortality up to 495 mg DWCNT/kg of food	[59]

BSA bovine serum albumin, FITC fluorescein isothiocyanate, pfh post-fertilization hours, NOEC no observed effect concentration, SDS sodium dodecyl sulfate

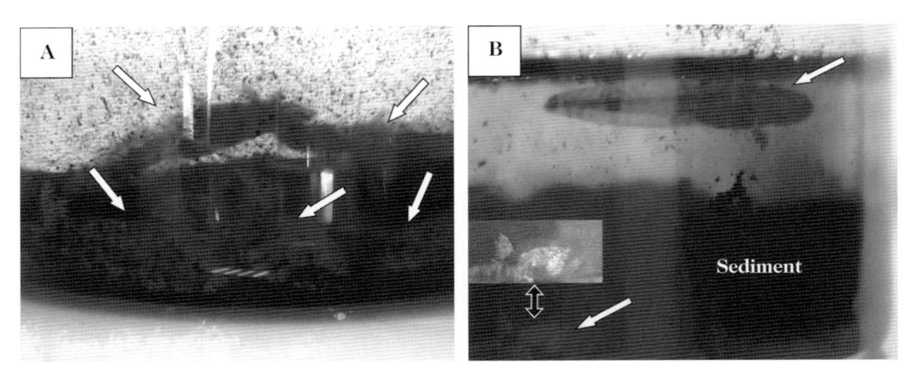

Fig. 14.4. Exposure of axolotl larvae to carbon nanotubes. Axolotl larvae (*white arrows*) in the presence of carbon nanotubes (**A**) at the beginning and (**B**) at the end of exposure (12 days). The larvae bury themselves in the 'sediment' of carbon nanotubes (**A**) and swim through the water column (**B**), probably in search of oxygen at the end of exposure. When they are at this level, they are not covered with carbon nanotubes. No toxicity is observed (mortality, growth) in larvae exposed to a broad range of concentrations of raw carbon nanotubes (1–1 000 mg/L). See colour plate

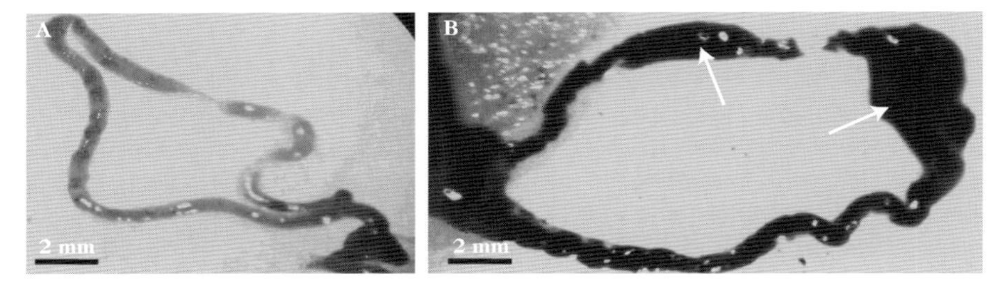

Fig. 14.5. Accumulation of carbon nanotubes in the intestine of xenopus larvae. Taken from [68], with the kind permission of Elsevier. (**A**) Control without carbon nanotubes and (**B**) after 12 days' exposure in the presence of 10 mg/L of carbon nanotubes. Dissection of the xenopus larvae revealed an accumulation of black clusters in the digestive system. The intestines of larvae exposed to carbon nanotubes in the medium have a swollen appearance (*white arrows*) (**B**) compared with the control (**A**). See colour plate

14.3.2 Metal and Metal Oxide Nanoparticles

Two interesting studies investigate exposure of the zebrafish (*Danio rerio*) to copper nanoparticles [72] and exposure of the rainbow trout (*Oncorhynchus mykiss*) to TiO_2 nanoparticles [73]. In both cases, toxicological effects were observed in the gills, with the proliferation of epithelial cells and the development of edemas in the gill filaments. However, the blood parameters of these fish were barely altered. Copper nanoparticles induced a slight vacuolisation of the hepatic cells but did not significantly modify the activity of plasma alanine aminotransferase which reflects kidney and liver damage. On the other hand, in the case of TiO_2 nanoparticles, an increase in the activity of ATPase with respect to Na^+ and K^+ and a decrease in the concentration of thiobarbituric acid reactive substances attest to possible effects on osmoregulation and oxidative stress in the fish gills. These effects are also observed in the intestine, and to a lesser extent in the brain, but not in the liver.

According to these results, both exposure routes seem to bring about accumulation of nanoparticles, although the direct route predominates over the trophic route. These studies do not inform as to the origins of the nanoparticles observed in the animals' internal organs. They may be carried there via the blood, following translocation through directly exposed gill cells, or after crossing the epithelium of the gastro-intestinal tract following trophic exposure.

14.3.3 Latex Nanoparticles

A model study involving fluorescent latex nanoparticles, which have the advantage of being easily localised within the organism, showed that they distributed themselves in the gills, blood, intestine, liver, and the kidney of the medaka (*Oryzias latipes*) [74]. Some nanoparticles were also visible in the sexual organs and the brain, despite these being protected from xenobiotics by physiological barriers (the blood–brain barrier and the blood–testicle barrier for male sexual organs). The presence of nanoparticles in the exposed fish depends on their size and aggregation state. When the salinity of the biological medium is increased, the nanoparticles aggregate, and this seems to favour their presence in organisms [74]. It seems that nanoparticles with mean hydrodynamic diameter around 470 nm enter the fish more efficiently than smaller or bulkier nanoparticles.

14.3.4 Co-contamination by Nanoparticles and Metals or Organic Pollutants

The toxicity of nanoparticles observed in organisms can be intrinsic (due to the nanoparticle alone) or indirect (nanoparticle as potential carrier) owing to their proven adsorption potential, which means that there may be pollutants at their surface or within their structure whose toxic potential may be induced, repressed, or limited. Indeed, when they come into contact with the environment, the nanoparticles will be in permanent interaction with the other components of the medium, and in particular, the contaminants. In some cases, the nanoparticles may play the role of collector, e.g., by adsorption, for certain molecules, or a masking role wherein they immobilise a non-negligible fraction of the compounds that are potentially reactive for living matter. It is thus impossible a priori to predict the potential biological effects resulting from the presence of nanoparticles in a complex environment such as a natural aquatic environment.

The literature shows that the adsorption potential of carbon-based nanoparticles like carbon nanotubes has been studied as a way of removing organic and inorganic pollutants from the air, including dioxin [75] and volatile organic compounds [76], but also from water or aqueous solutions, including fluoride [77], 1,2-dichlorobenzene [78], trihalomethanes [79], and divalent metal ions [80, 81]. However, while these studies demonstrate the efficiency of these nanoparticles for adsorbing pollutants, no study to our knowledge indicates whether this adsorption efficiency extends to biological effects, nor whether this adsorption can modify the direct effects of these pollutants in organisms.

It has been shown for carp (*Cyprinus carpio*) that coexposure to cadmium and TiO_2 nanoparticles causes a significant increase in the accumulation of cadmium in these fish [82]]. The overall accumulation of cadmium increases by 146% in the presence of 10 μg/L of nanoparticles, going from 9 to 22 μg of Cd^+ per gram of fish. The cadmium and TiO_2 nanoparticles accumulate mainly in the viscera, gills, skin, scales, and muscles. The bioconcentration factor of cadmium in the gills is 152 in the presence of TiO_2 nanoparticles, as compared with 34 in their absence. The co-accumulation of cadmium and nanoparticles in the viscera occurs either directly through the gill cell barrier, or indirectly by trophic exposure and accumulation in the gastro-intestinal tract, followed in some cases by relocation in the internal organs of the viscera.

14.4 Phytotoxicity and Translocation in Plants

Due to the fact that they do not move, plants have developed particularly effective transport systems enabling them to obtain nutrients (1) from the soil via the root system, and (2) from atmospheric gases via the stalk and leaves. These two routes for supplying nutrients can both be involved in the uptake of pollutants, and hence nanoparticles. Within the plant, nutrients are transported by xylem vessels (raw sap, rich in water and minerals) and the phloem vessels (phloem sap, rich in glucides). The main driving force for the flow of water from the root to the aerial parts of the plant is transpiration. Evaporation of water from the leaves of the plant creates a suction effect which causes a massive uptake of water, nutritive elements, and also potentially nanoparticles via the root system. The accumulation of nanoparticles in the roots and/or the aerial parts of plants may result in transient or long-lasting changes affecting the growth and development of the plant. This property is commonly referred to as phytotoxicity. But apart from their phytotoxicity, a further risk associated with the accumulation of nanoparticles in plants would be the possibility of their entering the food chain, mainly through cultivated plants.

14.4.1 Basic Tools for Studying Nanoparticle Phytotoxicity

At the present time there is little available data regarding the effects of nanoparticles on natural or cultivated plants. The only published studies were carried out under conditions simplified by exposing the plants to nanoparticles dispersed in a hydroponic solution or in water. The guidelines put forward by the Organisation for Economic Cooperation and Development (OECD) for phytotoxicity tests recommend the use of seedling emergence and growth tests [83], growth inhibition assays on duckweed or *Lemna sp.* [84], and vegetative vigour tests [85]. While there are still no guidelines relating to nanoparticle phytotoxicity, these tests could serve as a reference, to a first approximation. In practice the most widely used tests are the germination test, the foliar growth test, and the radicle growth test.

14.4.2 Phytotoxic Effects: Inhibition of Germination and Growth

Germination takes place in several stages, from seed imbibition to the growth of a radicle. The toxic effects of nanoparticles may appear at these two stages. Recent studies have shown that phytotoxicity varies with the type of nanoparticle, its physicochemical characteristics, and the exposed plant. The effects of aluminium (Al), alumina (Al_2O_3), zinc (Zn), and zinc oxide (ZnO) nanoparticles, and multiwall carbon nanotubes have been studied on various plants, including radish, colza, ray grass, lettuce, maize, and cucumber. It seems that the stage most affected is the growth of the radicle rather than imbibition of the seed [86]. Zinc (Zn) and zinc oxide (ZnO) nanoparticles are the most phytotoxic and disturb the root development of all the species studied. Zn nanoparticles delay germination of ray grass while ZnO nanoparticles have the same effect on radish. ZnO nanoparticles also delay the growth of the aerial parts of ray grass [87]. They are detected in the cells of the endodermis and the vascular cylinder of the roots. The nanoparticles presumably cross the epidermis and the root cortex by the apoplastic route, and then the endodermis via the protoplasts, to reach the central part of the root. The hypothesis put forward to explain this is that the nanoparticles create pores in the walls of the plant cells, as they do in bacteria, thereby allowing root uptake [87]. Note that, in this study, the nanoparticles are aggregated in the exposure solution, although some remain isolated. Many nanoparticle clusters are observed at the root surface, and these may mechanically alter their development and restrict the supply of nutrients to the plant.

Another point is that, for a given type of nanoparticle, the surface state governs phytotoxic effects. Al_2O_3 nanoparticles delay root elongation in maize, cucumber, soybean, cabbage, and carrot. On the other hand, when these nanoparticles are first put in contact with phenanthrene, a polycyclic aromatic hydrocarbon, they no longer exhibit phytotoxic effects [88].

14.4.3 Nanoparticle Translocation from Roots to Aerial Parts

Whether they have phytotoxic effects or not, nanoparticles are likely to accumulate in plants and thus be introduced into the food chain. At the present time, the translocation of nanoparticles from the roots to the aerial parts has been reported for several plants, but the phenomenon seems minor. In addition, the concentrations used for laboratory exposure are very high, hence far removed from realistic environmental contamination.

In the study on ray grass described in the last section [87], although ZnO nanoparticles reach the vascular region of the root, few nanoparticles are observed in the aerial parts. The transfer factor (ratio of the zinc concentration in the aerial parts to that in the roots) is very low, viz., 0.01–0.02 compared with 0.03–0.50 in the case of Zn^{2+} ions. On the other hand, another study showed that the aluminium concentration in ray grass leaves increases when the plant is cultivated in soil amended with aluminium nanoparticles [89]. Likewise for the pumpkin, a plant recognised for its ability to absorb a large amount of water, Fe_2O_3 nanoparticles are transferred from the roots to the leaves where they accumulate without inducing phytotoxicity [54]. The ability of the plant to extract water in large amounts from the soil thus seems relevant to the translocation of nanoparticles from the roots to the aerial parts. Presumably the flow of water to regions of transpiration carries the nanoparticles along with it.

14.5 Conclusion

The nanometric size of nanoparticles means they have special properties, quite different from those of the bulk material (see Chap. II). These properties can be exploited to engineer new materials, satisfying constraints of chemical reactivity, electrical conductivity, or optical sensitivity that could not otherwise be achieved. Nanoparticles thus confront us with novel and as yet unknown types of molecular behaviour. However, these new technologies are already being used in many commercial products and will no doubt see a huge development over the coming decades. For this reason, the question of the potential hazards of nanoparticles and the materials incorporating them has already been raised. The stakes are high. The problem is to keep pace with the regulatory measures needed to control the use and dissemination of these objects throughout our environment and our future way of life. The social issues seem enormous, given the vast range of applications expected over the coming years. While the current approach to risk assessment with regard to chemical substances in our environment is organised according to the new European regulation known as Registration, Evaluation and Authorisation of CHemicals (REACH), this does not apply to nanoparticles either directly or simply as-is. So today there is a regulatory vacuum that needs to be filled by risk assessment methods characterised by and tailored for nanoparticles. But the first step here must be to obtain a better understanding of the potential hazards intrinsic to nanoparticles at all the organisational levels of living systems, from the subcellular level to the global level of the ecosystem.

Obtaining this understanding will be a cross-disciplinary exercise in which ecotoxicology must play a dominant role, investigating the potential hazards of these materials for the integrity of our environment. However, the wide range of nanoparticles, the many different forms in which they may turn up in natural environments, and the broad array of responses of living organisms, will make it a very difficult task to analyse their potential effects in this context. The results reviewed in the present chapter already suggest that a simple, fast, or easily generalisable response with regard to all the relevant organisms will probably prove impossible. Several paradigms can nevertheless be discerned:

- The importance of nanoparticle localisation which will dictate those organs or functions potentially affected by them.
- The importance of intrinsic nanoparticle reactivity, and in particular redox activity.
- Nanoparticle-induced oxidative stress seems to be a frequent issue common to many organisms.
- The toxicity and solubility of the chemical elements, e.g., Cd, Zn, making up nanoparticles.

At the present time, few research groups have yet begun to assess the ecotoxicological risks due to the presence of nanoparticles in the environment. Our current understanding of the probable effects of these nanoparticles, even for living beings taken individually, is incomplete to say the least. So what are the potential effects of such particles in the complex natural environments all around us? How will nanoparticles behave in these environments? What will be their distribution? What hazards do they represent for ecosystems?

The problem of hazards is not the only one that must be solved to achieve the overriding objective of managing the risks associated with the presence of nanoparticles in our environment. Understanding the life cycles of these products and the transfer of degradation products within the various environmental compartments, not to mention their behaviour within complex environmental media, are so many challenges that must be met and scientific bottlenecks that must be overcome in order to better control the effects of these nanomaterials within a framework of sustainable development. The answers to such questions could only be obtained by bring together the skills of many different disciplines in order to characterise the effects (if there are such) within organisms, populations, communities, or indeed the ecosystem as a whole.

Acknowledgements

The authors would like to thank the *Centre national de la recherche scientifique* (CNRS) and the *Commissariat à l'énergie atomique* (CEA) for financing the International Consortium for the Environmental Implications of NanoTechnology (GDRi iCEINT) directed by Jean-Yves Bottero, and also the National Science Foundation and the Environmental Protection Agency in the US for financing the Center for the Environmental Implications of NanoTechnology (CEINT) (EF-0830093) directed by Mark R. Wiesner. Some of the results presented in this chapter were obtained by the combined research group NAUTILE (nanotubes and ecotoxicology) CNRS/UPS/INPT–ARKEMA FR (specific collaboration agreement CNRS No. 48361, director L. Gauthier).

References

1. V. Colvin: The potential environmental impact of engineered nanomaterials. Nature Biotechnol. **21**, 1166–1170 (2003)
2. M.R. Wiesner, G.V. Lowry, P.J.J. Alvarez: Assessing the risks of manufactured nanomaterials. Environ. Sci. Technol. **40**, 4337–4345 (2006)
3. E. Oberdörster: Manufactured nanomaterials (fullerene, C_{60}) induce oxidative stress in the brain of juvenile largemouth bass. Environ. Health Perpect. **112**, 1058–1062 (2004)
4. M.M. Huczko, H. Lange: Carbon nanotubes: Experimental evidence of null risk of skin irritation and allergy. Fullerene Sci. Technol. **9**, 247–250 (2001)
5. D.B. Wareight, B.R. Laurence, K.L. Reed, D.H. Roach, G.A. Reynolds, T.R. Webb: Comparative pulmonary toxicity assessment of single-wall carbon nanotubes in rats. Toxicol. Sci. **77**, 117–125 (2004)
6. T.B. Henry, F.N. Menn, J.T. Fleming, J. Wilgus, R.N. Compton, G.S. Sayler: Attributing effects of aqueous C_{60} nano-aggregates to tetrahydrofuran decomposition products in larval zebrafish by assessment of gene expression. Environ. Health Perspect. **115**, 1059–1065 (2007)
7. J.W. Moreau, P.K. Weber, M.C. Martin, B. Gilbert, I.D. Hutcheon, J.F. Banfield: Extracellular proteins limit the dispersal of biogenic nanoparticles. Science **316**, 1600–1603 (2007)
8. L. Sigg, P. Behra, G.N. Stumm: *Chimie des milieux aquatiques, chimie des eaux naturelles et des interfaces dans l'environnement.* (Dunod, Paris 2000) pp. 350–390
9. L.K. Adams, D.Y. Lyon, P.J.J. Alvarez: Comparative eco-toxicity of nanoscale TiO_2, SiO_2, and ZnO water suspensions. Water Research **40**, 3527 (2006)
10. J. Verran, G. Sandoval, N.S. Allen, M. Edge, J. Stratton: Variables affecting, the antibacterial properties of nano and pigmentary titania particles in suspension. Dyes and Pigments **73**, 298–304 (2007)
11. K. Qi, W.A. Daoud, J.H. Xin, C.L. Mak, W. Tang, W.P. Cheung: Self-cleaning cotton. J. Mat. Chem. **16**, 4567–4574 (2006)
12. I. Sondi, B. Salopek-Sondi: Silver nanoparticles as antimicrobial agent: A case study on *E. coli* as a model for Gram-negative bacteria. J. Colloid Interf. Sci. **275**, 177–182 (2004)

13. C.N. Lok, C.M. Ho, R. Chen, Q.Y. He, W.Y. Yu, H. Sun, P. Tam, J.F. Chiu, C.M. Che: Silver nanoparticles: Partial oxidation and antibacterial activities. J. Biol. Inorg. Chem. **12**, 527534 (2007)
14. M. Zheng, A. Jagota, E.D. Semke, B.A. Diner, R.S. McLean, S.R. Lustig, R.E. Richardson, N.G. Tassi: DNA-assisted dispersion and separation of carbon nanotubes. Nat. Mater. **2**, 338–342 (2003)
15. C. Salvador-Morales, P. Townsend, E. Flahaut, C. Venien-Bryan, A. Vlandas, M.L.H. Green, R.B. Sim: Binding of pulmonary surfactant proteins to carbon nanotubes: Potential for damage to lung immune defense mechanisms. Carbon **45**, 607–617 (2007)
16. H. Hyung, J.D. Fortner, J.B. Hughes, J.H. Kim: Natural organic matter stabilizes carbon nanotubes in the aqueous phase. Environ. Sci. Technol. **41**, 179–184 (2007)
17. N.M. Franklin, N.J. Rogers, S.C. Apte, G.E. Batley, G.E. Gadd, P.S. Casey: Comparative toxicity of nanoparticulate ZnO, Bulk ZnO, and $ZnCl_2$ to a freshwater microalga (*Pseudokirchneriella subcapitata*): The importance of particle solubility. Environ. Sci. Technol. **41**, 8484–8490 (2007)
18. L. Zhang, Y. Jiang, Y. Ding, M. Povey, D. York: Investigation into the antibacterial behaviour of suspensions of ZnO nanoparticles (ZnO nanofluids). J. Nanoparticle Res. **9**, 479–489 (2007)
19. R. Brayner, R. Ferrari-Iliou, N. Brivois, S. Djediat, M.F. Benedetti, F. Fievet: Toxicological impact studies based on *Escherichia coli* bacteria in ultrafine ZnO nanoparticles colloidal medium. Nano Lett. **6**, 866–870 (2006)
20. Z. Huang, X. Zheng, D. Yan, G. Yin, X. Liao, Y. Kang, Y. Yao, D. Huang, B. Hao: Toxicological effect of ZnO nanoparticles based on bacteria. Langmuir **24**, 4140–4144 (2008)
21. W.A. Daoud, J.H. Xin, Y.H. Zhang: Surface functionalization of cellulose fibers with titanium dioxide nanoparticles and their combined bactericidal activities. Surface Science **599**, 69–75 (2005)
22. M. Auffan, W. Achouak, J. Rose, C. Chaneac, D.T. Waite, A. Masion, J. Woicik, M.R. Wiesner, J.Y. Bottero: Relation between the redox state of iron-based nanoparticles and their cytotoxicity towards *Escherichia coli*. Environ. Sci. Technol. **42**, 6730–6735 (2008)
23. N. Fauconnier, J.N. Pons, J. Roger, A. Bee: Thiolation of maghemite nanoparticles by dimercaptosuccinic acid. J. Colloid Interf. Sci. **194**, 427 (1997)
24. M. Auffan, L. Decome, J. Rose, T. Orsiere, M. DeMeo, V. Briois, C. Chaneac, L. Olivi, J.L. Berge-Lefranc, A. Botta, M.R. Wiesner, J.Y. Bottero: In vitro interactions between DMSA-coated maghemite nanoparticles and human fibroblasts: A physicochemical and cyto-genotoxical study. Environ. Sci. Technol. **40**, 4367–4373 (2006)
25. R.S. Norman, J.W. Stone, A. Gole, C.J. Murphy, T.L. Sabo-Attwood: Targeted photothermal lysis of the pathogenic bacteria *Pseudomonas aeruginosa* with gold nanorods. Nano Lett. **8**, 302–306 (2008)
26. M. Monthioux: *Introduction to Carbon Nanotubes*. Springer Handbook of Nanotechnology, 2nd edn. (Springer, New York 2007)
27. A. Neal: What can be inferred from bacterium–nanoparticle interactions about the potential consequences of environmental exposure to nanoparticles? Ecotoxicology **17**, 362–371 (2008)

28. G. Reguera, K.D. McCarthy, T. Mehta, J.S. Nicoll, M.T. Tuominen, D.R. Lovley: Extracellular electron transfer via microbial nanowires. Nature **435**, 1098–1101 (2005)
29. K.M. Reddy, K. Feris, J. Bell, D.G. Wingett, C. Hanley, A. Punnoose: Selective toxicity of zinc oxide nanoparticles to prokaryotic and eukaryotic systems. Appl. Phys. Lett. **90**, 213902–213903 (2007)
30. D.Y. Lyon, L.K. Adams, J.C. Falkner, P.J.J. Alvarez: Antibacterial activity of fullerene water suspensions: Effects of preparation method and particle Size. Environ. Sci. Technol. **40**, 4360–4366 (2006)
31. N. Vigneshwaran, R.P. Nachane, R.H. Balasubramanya, P.V. Varadarajan: Functional finishing of cotton fabrics using zinc oxide-soluble starch nanocomposites. Carbohydr. Res. **341**, 2012 (2006)
32. S.H. Joo, A.J. Feitz, D.L. Sedlak, T.D. Waite: Quantification of the oxidizing capacity of nanoparticulate zero-valent iron. Environ. Sci. Technol. **39**, 1263–1268 (2005)
33. K. Sunada, T. Watanabe, K. Hashimoto: Studies on photokilling of bacteria on TiO_2 thin film. J. Photochem. Photobiol. A: Chemistry **156**, 227–233 (2003)
34. M.J. Rorones, J.L. Elechiguerra, A. Camacho, K. Holt, J.B. Kouri, J.T. Ramírez, M.J. Yacaman: The bactericidal effect of silver nanoparticles. Nanotechnology **16**, 2346–2353 (2005)
35. S. Pal, Y.K. Tak, J.M. Song: Does the antibacterial activity of silver nanoparticles depend on the shape of the nanoparticle? A study of the Gram-negative bacterium *Escherichia coli*. Appl. Environ. Microbiol. **73**, 1712–1720 (2007)
36. S. Makhluf, R. Dror, Y. Nitzan, Y. Abramovich, R. Jelinek, A. Gedanken: Microwave assisted synthesis of nanocrystalline MgO and its use as a bactericide. Adv. Function. Mat. **15**, 1708–1715 (2005)
37. L. Huang, D. Li, Y. Lin, D.G. Evans, X. Duan: Influence of nano-MgO particle size on bactericidal action against *Bacillus subtilis var. niger*. Chinese Sci. Bull. **50**, 514–519 (2005)
38. A. Thill, O. Zeyons, O. Spalla, F. Chauvat, J. Rose, M. Auffan, A.M. Flank: Cytotoxicity of CeO_2 nanoparticles for *Escherichia coli*. Physico-chemical insight of the cytotoxicity mechanism. Environ. Sci. Technol. **40**, 6151–6156 (2006)
39. O. Zeyons, A. Thill, F. Chauvat, N. Menguy, C. Cassier-Chauvat, C. Orear, J. Daraspe, M. Auffan, J. Rose, O. Spalla: Direct and indirect CeO_2 nanoparticles toxicity for *E. coli* and *synechocystis*. Nanotoxicology **3**, 284–295 (2009)
40. G. Fu, P.S. Vary, C.T. Lin: Anatase TiO_2 nanocomposites for antimicrobial coatings. J. Phys. Chem. B **109**, 8889–8898 (2005)
41. D. Mitoraj, A. Jańczyk, M. Strus, H. Kisch, G. Stochel, P.B. Heczko, W. Macyk: Visible light inactivation of bacteria and fungi by modified titanium dioxide. Photochem. Photobiol. Sci. **6**, 642–648 (2007)
42. S. Kang, M. Pinault, L.D. Pfefferle, M. Elimelech: Single-walled carbon nanotubes exhibit strong antimicrobial activity. Langmuir **23**, 8670–8673 (2007)
43. D.Y. Lyon, J.D. Fortner, C.M. Sayes, V.L. Colvin, J.B. Hughes: Bacterial cell association and antimicrobial activity of C_{60} water suspension. Environ. Toxicol. Chem. **24**, 2757–2762 (2005)
44. P.K. Stoimenov, R.L. Klinger, G.L. Marchin, K.J. Klabunde: Metal oxide nanoparticles as bactericidal agents. Langmuir **18**, 6679–6686 (2002)

45. Z. Tong, M. Bischoff, L. Nies, B. Applegate, R.F. Turco: Impact of fullerene (C_{60}) on a soil microbial community. Environ. Sci. Technol. **41**, 2985–2991 (2007)
46. M. Auffan, J. Rose, M.R. Wiesner, J.Y. Bottero: Chemical stability of metallic nanoparticles: A parameter controlling their potential toxicity in vitro. Environmental Pollution **157**, 1127–1133 (2009)
47. J.P. Jolivet, C. Chaneac, E. Tronc: Iron oxide chemistry. From molecular clusters to extended solid networks. Chem. Commun. **5**, 481–487 (2004)
48. R.W. Tarnuzzer, J. Colon, S. Patil, S. Seal: Vacancy engineered ceria nanostructures for protection from radiation-induced cellular damage. Nano Lett. **5**, 2573–2577 (2005)
49. J. Kiwi, V. Nadtochenko: Evidence for the mechanism of photocatalytic degradation of the bacterial wall membrane at the TiO_2 interface by ATR-FTIR and laser kinetic spectroscopy. Langmuir **21**, 4631–4641 (2005)
50. J. Fang, D.Y. Lyon, M.R. Wiesner, J. Dong, P.J.J. Alvarez: Effect of a fullerene water suspension on bacterial phospholipids and membrane phase behavior. Environ. Sci. Technol. **41**, 2636–2642 (2007)
51. S.B. Lovern, R. Klaper: *Daphnia magna* mortality when exposed to titanium dioxide and fullerene (C_{60}) nanoparticles. Environ. Toxicol. Chem. **25**, 1132–1137 (2006)
52. S.B. Lovern, J.R. Strickler, R. Klaper: Behavioral and physiological changes in *Daphnia magna* when exposed to nanoparticle suspensions (titanium dioxide, nano-C_{60}, and $C_{60}HxC_{70}Hx$). Environ. Sci. Technol. **41**, 4465–4470 (2007)
53. E. Oberdorster, S. Zhu, T.M. Blickley, P. McClellan-Green, M.L. Haasch: Ecotoxicology of carbon-based engineered nanoparticles: Effects of fullerene (C_{60}) on aquatic organisms. Carbon **44**, 1112–1120 (2006)
54. H. Zhu, J. Han, J.Q. Xiao, Y. Jin: Uptake, translocation, and accumulation of manufactured iron oxide nanoparticles by pumpkin plants. J. Environ. Monit. **10**, 713–717 (2008)
55. S. Zhu, E. Oberdorster, M.L. Haasch: Toxicity of an engineered nanoparticle (fullerene, C_{60}) in two aquatic species, *Daphnia* and fathead minnow. Marine Environmental Research **62** (Suppl. 1), S5–S9 (2006)
56. X. Zhu, L. Zhu, Y. Chen, S. Tian: Acute toxicities of six manufactured nanomaterial suspensions to *Daphnia magna*. J. Nanopart. Res. **11**, 67–75 (2009)
57. X. Zhu, L. Zhu, Y. Li, Z. Duan, W. Chen, P.J.J. Alvarez: Developmental toxicity in zebrafish (*Danio rerio*) embryos after exposure to manufactured nanomaterials: Buckminsterfullerene aggregates (nC_{60}) and fullerol. Environ. Toxicol. Chem. **26**, 976–979 (2007)
58. C.Y. Usenko, S.L. Harper, R.L. Tanguay: Fullerene C_{60} exposure elicits an oxidative stress response in embryonic zebrafish. Toxicol. Appl. Pharmacol. **229**, 44–55 (2008)
59. J.J. Scott-Fordsmand, P.H. Krogh, M. Schaefer, A. Johansen: The toxicity testing of double-walled nanotube-contaminated food to *Eisenia veneta* earthworms. Ecotoxicol. Environ. Safety **71**, 616–619 (2008)
60. E.J. Petersen, Q. Huang, W.J. Weber: Ecological uptake and depuration of carbon nanotubes by *Lumbriculus variegatus*. Environ. Health Perspect. **116**, 496–500 (2008)
61. A.P. Roberts, A.S. Mount, B. Seda, J. Souther, R. Qiao, S. Lin, P.C. Ke, A.M. Rao, S.J. Klaine: In vivo biomodification of lipid-coated carbon nanotubes by *Daphnia magna*. Environ. Sci. Technol. **41**, 3025–3029 (2007)

62. R.C. Templeton, P.L. Ferguson, K.M. Washburn, W.A. Scrivens, G.T. Chandler: Life-cycle effects of single-walled carbon nanotubes (SWNT) on an estuarine meiobenthic copepod. Environ. Sci. Technol. **40**, 7387–7393 (2006)
63. A.J. Kennedy, M.S. Hull, J.A. Steevens, K.M. Dontsova, M.A. Chappell, J.C. Gunter, C.A. Weiss: Factors influencing the partitioning and toxicity of nanotubes in the aquatic environment. Environ. Toxicol. Chem. **27**, 1932–1941 (2008)
64. J. Cheng, E. Flahaut, S.H. Cheng: Effect of carbon nanotubes on developing zebrafish (*Danio rerio*) embryos. Environ. Toxicol. Chem. **26**, 708–716 (2007)
65. J. Cheng, C.M. Chan, L.M. Veca, W.L. Poon, P.K. Chan, L. Qu, Y.P. Sun, S.H. Cheng: Acute and long-term effects after single loading of functionalized multi-walled carbon nanotubes into zebrafish (*Danio rerio*). Toxicol. Appl. Pharmacol. **235**, 216–225 (2009)
66. C.J. Smith, B.J. Shaw, R.D. Handy: Toxicity of single walled carbon nanotubes to rainbow trout (*Oncorhynchus mykiss*): Respiratory toxicity, organ pathologies, and other physiological effects. Aquatic Toxicology **82**, 94–109 (2007)
67. F. Mouchet, P. Landois, E. Flahaut, E. Pinelli, L. Gauthier: Assessment of the potential in vivo ecotoxicity of double-walled carbon nanotubes (DWNTs) in water, using the amphibian *Ambystoma mexicanum*. Nanotoxicology **1**, 149–156 (2007)
68. F. Mouchet, P. Landois, E. Sarremejean, G. Bernard, P. Puech, E. Pinelli, E. Flahaut, L. Gauthier: Characterisation and in vivo ecotoxicity evaluation of double-wall carbon nanotubes in larvae of the amphibian *Xenopus laevis*. Aquatic Toxicology **87**, 127–137 (2008)
69. E.J. Petersen, Q. Huang, J. Weber, J. Walter: Bioaccumulation of radio-labeled carbon nanotubes by *Eisenia foetida*. Environ. Sci. Technol. **42**, 3090–3095 (2008)
70. P.L. Ferguson, G.T. Chandler, R.C. Templeton, A. DeMarco, W.A. Scrivens, B.A. Englehart: Influence of sediment amendment with single-walled carbon nanotubes and diesel soot on bioaccumulation of hydrophobic organic contaminants by benthic invertebrates. Environ. Sci. Technol. **42**, 3879–3885 (2008)
71. P.V. Asharani, N.G.B. Serina, M.H. Nurmawati, Y.L. Wu, Z. Gong, V. Suresh: Impact of multi-walled carbon nanotubes (MWCNTs) on aquatic species. J. Nanosci. Nanotechnol. **8**, 1–7 (2008)
72. R.J. Griffitt, R. Weil, K.A. Hyndman, N.D. Denslow, K. Powers, D. Taylor, D.S. Barber: Exposure to copper nanoparticles causes gill injury and acute lethality in zebrafish (*Danio rerio*). Environ. Sci. Technol. **41**, 8178–8186 (2007)
73. G. Federici, B.J. Shaw, R.D. Handy: Toxicity of titanium dioxide nanoparticles to rainbow trout (*Oncorhynchus mykiss*): Gill injury, oxidative stress, and other physiological effects. Aquatic Toxicology **84**, 415–430 (2007)
74. S. Kashiwada: Distribution of nanoparticles in the see-through medaka (*Oryzias latipes*). Environ. Health Perspect. **114**, 1697–1702 (2006)
75. R.Q. Long, R.T. Yang: Carbon nanotubes as superior sorbent for dioxin removal. J. Am. Chem. Soc. **123**, 2058–2059 (2001)
76. P.A. Gauden, A.P. Terzyk, G. Rychlicki, P. Kowalczyk, E. Raymundo-Pinero, K. Lota, E. Frackowiak, F. Beguin: Thermodynamic properties of benzene adsorbed in activated carbons and multi-walled carbon nanotubes. Chem. Phys. Lett. **421**, 409–414 (2006)

77. Y.H. Li, S. Wang, A. Cao, D. Zhao, X. Zhang, C. Xu, Z. Luan, D. Ruan, J. Liang, D. Wu, B. Wei: Adsorption of fluoride from water by amorphous alumina supported on carbon nanotubes. Chem. Phys. Lett. **350**, 412–416 (2001)
78. X. Peng, Y. Li, Z. Luan, Z. Di, H. Wang, B. Tian, Z. Jia: Adsorption of 1,2-dichlorobenzene from water to carbon nanotubes. Chem. Phys. Lett. **376**, 154–158 (2003)
79. C. Lu, Y.L. Chung, K.F. Chang: Adsorption thermodynamic and kinetic studies of trihalomethanes on multiwalled carbon nanotubes. J. Hazard. Materials **138**, 304–310 (2006)
80. C. Lu, H. Chiu, C. Liu: Removal of zinc(II) from aqueous solution by purified carbon nanotubes: Kinetics and equilibrium studies. Indust. Engineer Chem. Res. **45**, 2850–2855 (2006)
81. G.P. Rao, C. Lu, F. Su: Sorption of divalent metal ions from aqueous solution by carbon nanotubes: A review. Separation and Purification Technology **58**, 224–231 (2007)
82. X. Zhang, H. Sun, Z. Zhang, Q. Niu, Y. Chen, J.C. Crittenden: Enhanced bioaccumulation of cadmium in carp in the presence of titanium dioxide nanoparticles. Chemosphere **67**, 160–166 (2007)
83. OCDE Essai-208, Essai sur plante terrestre: essai d'émergence de plantule et de croissance de plantule. Lignes directrices pour les essais de produits chimiques/section 2: Effets sur les systèmes biologiques. (Edition OCDE 2006)
84. OCDE Essai-221, Lemna sp. Essais d'inhibition de croissance. Lignes directrices pour les essais de produits chimiques/section 2: Effets sur les systèmes biologiques. (Edition OCDE 2006)
85. OCDE Essai-227, Essai sur plante terrestre: Essai de vigueur végétative. Lignes directrices pour les essais de produits chimiques/section 2: Effets sur les systèmes biologiques. (Edition OCDE 2006)
86. D. Lin, B. Xing: Phytotoxicity of nanoparticles: Inhibition of seed germination and root growth. Environmental Pollution **150**, 243–250 (2007)
87. D. Lin, B. Xing: Root uptake and phytotoxicity of ZnO nanoparticles. Environ. Sci. Technol. **42**, 5580–5585 (2008)
88. L. Yang, D.J. Watts: Particle surface characteristics may play an important role in phytotoxicity of alumina nanoparticles. Toxicol. Lett. **158**, 122–132 (2005)
89. R. Doshi, W. Braida, C. Christodoulatos, M. Wazne, G. O'Connor: Nano-aluminum: Transport through sand columns and environmental effects on plants and soil communities. Environ. Res. **106**, 296–303 (2008)

15

Toxicological Models Part A: Toxicological Studies of Nanoparticles on Biological Targets and Attempts to Attenuate Toxicity by Encapsulation Techniques

Roberta Brayner and Fernand Fiévet

Nanotechnology has become a major economic issue today, promising a wide range of innovations and opening up interesting prospects in various areas, such as biomedicine, electronics, computing, transport, and others. At the present time, Europe, the United States, and Japan each devote around a billion euros per year for the development of nanotechnologies. As far as France is concerned, in 2004, a report was drawn up on the funding of nanotechnology and nanoscience at the request of the French Ministry of Youth, Education, and Research [1]. The aim of the report was to identify all forms and all sources of public funding devoted to research and development in the areas of nanotechnologies and nanoscience, identifying also their operational context and distinguishing them as far as possible from the areas of microtechnologies and microelectronics. It transpired that France invested 183.2 MEuros (excluding tax) on average every year between 2001 and 2005, with steady progression over the period (see Fig. 15.1).

Many industrial companies are currently investing in nanotechnologies. Table 15.1 provides a picture of the wide range of activities involved [2]. At the present time, the impacts of engineered nanomaterials on health and the environment cannot be accurately assessed and taken into account, given the absence of specific toxicological data. However, there is considerable concern among scientists and consumers over the potential hazards of these nanomaterials for humans and for the environment. To tackle such questions, many reports have been produced by governmental and non-governmental agencies, such as the AFSSET report published in France in July 2006, entitled *Les nanomatériaux: effets sur la santé de l'homme et sur l'environnement* (Nanomaterials: Effects on human health and on the environment) [3]. These reports generally deal with the following points:

- Definitions, characterisation, and properties of nanomaterials.
- Industrial applications.
- Production methods.
- Detection techniques.

P. Houdy et al. (eds.), *Nanoethics and Nanotoxicology*,
DOI 10.1007/978-3-642-20177-6_15,

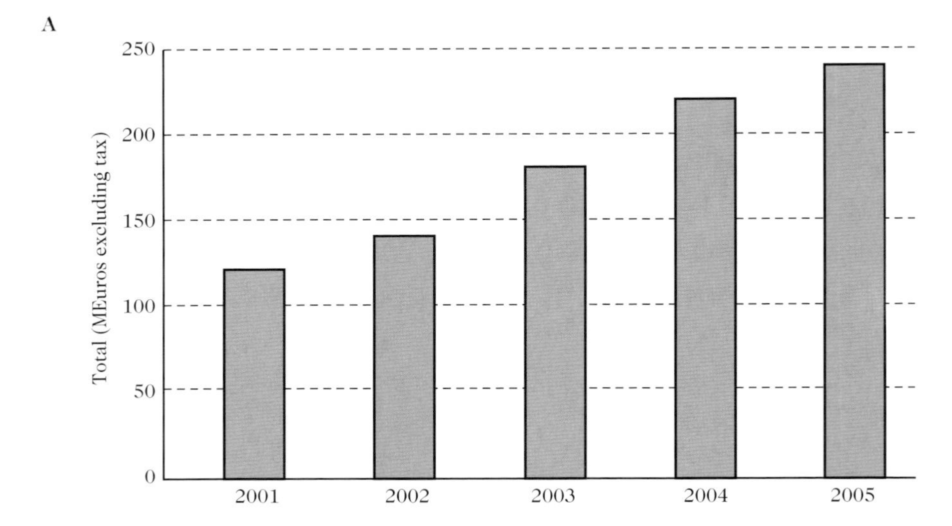

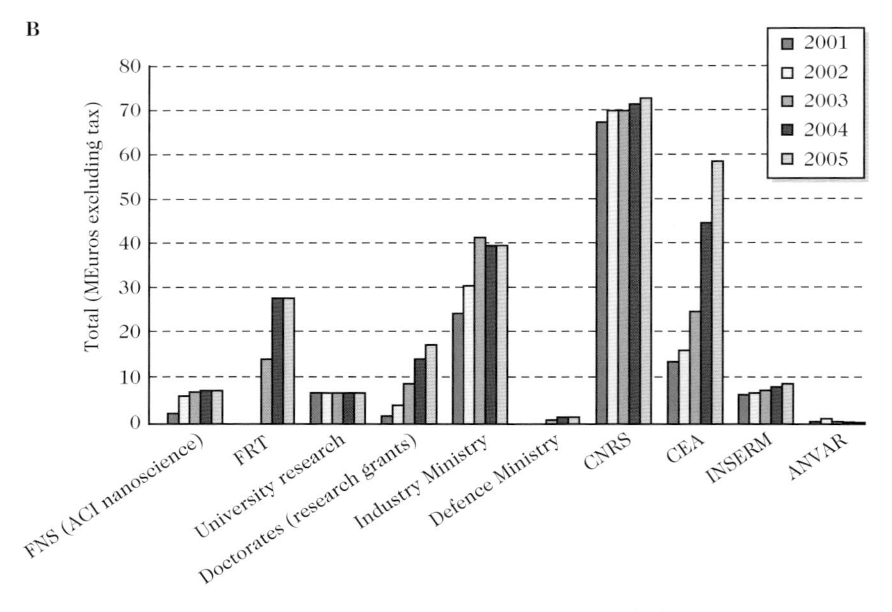

Fig. 15.1. Public funding for nanotechnology in France. (**A**) Overall public spending and (**B**) breakdown across different sectors [1]. Values for 2005 are predictions. FNS *Fonds National de la Science*, ACI *Actions Concertées Incitatives*, FRT *Fonds de la recherche et de la technologie*, CNRS *Centre national de la recherche scientifique*, CEA *Commissariat à l'Energie Atomique*, INSERM *Institut national de la santé et de la recherche médicale*, ANVAR *Agence nationale de valorisation de la recherche*

Table 15.1. Examples of companies active in the field of nanotechnology [2]

Company	Products
Acadia Research Group	Development in genetics, molecular characterisation of disease
Altair Nanotechnologies	Lithium titanate electrodes
Applied Nanofluorescence	Optical instruments for studying nanotubes
Arryx	Nanotweezers for picking up/moving nanoparticles
California Molecular Electronics	Invent and use intellectual property in molecular electronics
Carbon Nanotechnologies	Commercial production of carbon nanotubes
Cima Nanotech	Synthesis of fine and ultrafine powders, metal and alloy nanoparticles
Dendritech	Production of dendrimers
Dendridic Nanotechnologies	Synthesis of dendrimers for pharmaceutical industry
EnviroSystems	Disinfectants for hospitals
eSpin Technologies	Polymer nanofibres
Front Edge Technologies	Ultrathin rechargeable batteries
Hysitron	Instruments for measuring mechanical properties of nanomaterials, e.g., friction, adhesion, elasticity, in industry and research
Intermatix	Electronic materials and catalysts
Kereos	Therapeutic nanoparticles and labels
Lumera	Polymers
Molecular Electronics	Electronic and optoelectronic applications
Molecular Imprints	Nanoimprinting tools for industries making semiconductor and/or electronic equipment
NanoDynamics	Silver, copper, and nickel nanoparticles, nanostructured carbon, and nano-oxides
Nano Electronics	Novel materials, e.g., high K-gate dielectrics, metal gates, silicides
NanoGram	Chips for computers
Nanohorizons	Thin films
NanoInk	Anthrax detection
NanoOpto	Nanostructures for optical systems
Nanophase Technologies	Metal oxide nanopowders
Nanopoint	Equipment for analysing living cells in the infrared and UV–visible
NanoProducts	Metal nanopowders, doped oxides and alloys
NanoSpectra Biosciences	Non-invasive therapies using hollow nanoparticles
Nanosphere	Ultrasensitive detection and analysis of nucleic acids and proteins
Nanosys	Thin films for electronics and biology, solar cells
Nano-Tex	Coatings
Nanotherapeutics	Controlled release of pharmaceutical molecules using nanoparticles as carriers
Neo-Photonics	Nano-optical labels
Novation Environmental Technologies	Nanofiltration/disinfection of water
Ntera	Electronic ink and digital paper

- Risk (exposure) assessment for humans and the environment.
- Risk management.
- Recommendations and lists of funded research projects.

At the same time, scientific research projects are developing and beginning to throw some light on the hazards relating to nanomaterials that are already being produced. For example, Mark Wiesner and coworkers at Duke University (USA) have investigated the toxicological impact of several nanomaterials [2]. They have monitored the risks during industrial production of a series of nanomaterials: carbon nanotubes (SWCNT), fullerenes (C_{60}), quantum dots, alumoxane (alumina gel) nanoparticles, and TiO_2 nanoparticles [2]. Figure 15.2 sums up the relative risks during production of these nanomaterials as compared with those of other widely used products (not necessarily of nanometric dimensions). The estimated risks in using conventional safety equipment appear altogether comparable at every stage of production (temperature, pressure, volatility, inflammability, mobility, etc.).

At the present time there are no regulations specifically referring to the synthesis and manipulation of nanoparticles. Manufacturers working on this type of material react in two ways: either they use safety data for the compounds making up the nanoparticles without further consideration for their specific characteristics, or they consider nanoparticles as potentially dangerous substances and handle them with conventional safety equipment, which is nevertheless not always well suited to the task. In fact, they have not yet realised that, in the case of nanoparticles, certain specific physicochemical parameters, totally absent from conventional toxicity assays, must be taken into account (see Fig. 15.3).

Some parameters are interdependent, e.g., size and solubility. Once these parameters have been identified, one must also take into account those more

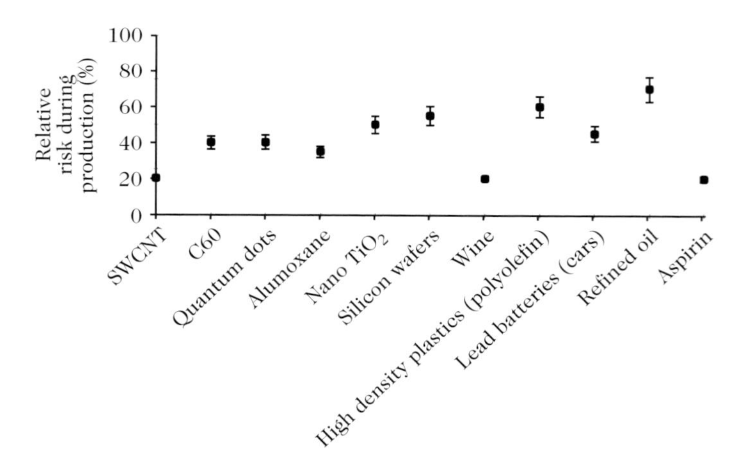

Fig. 15.2. Relative risks during production of nanomaterials compared with risks due to everyday products [2]

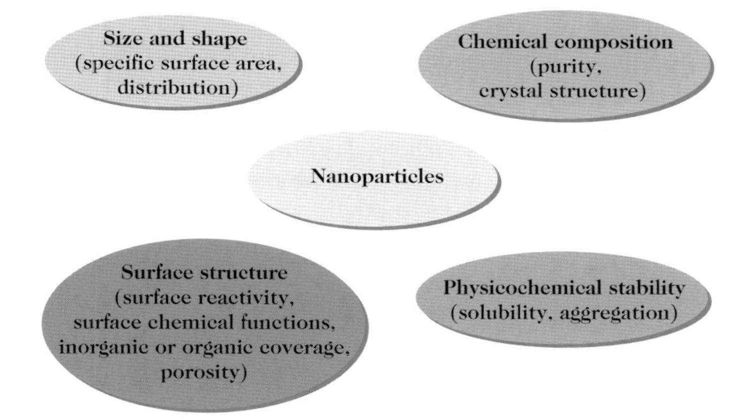

Fig. 15.3. Physicochemical parameters of nanoparticles which may be related to their toxicity

conventional but no less important aspects dealt with in conventional in vitro and in vivo toxicity tests, viz.:

- Tested dose.
- Exposure period.
- Administration route.
- Biological target.
- Composition of the environment in which exposure occurred.

The complexity of the parameter matrix highlights the difficulty in obtaining standardised data for nanoparticles as a whole. Any attempt to uniformise protocols without first building up a detailed understanding of the relevant parameters must therefore be regarded with caution.

In this chapter, we shall illustrate current research with some examples of toxicological studies carried out on nanoparticles already being manufactured. We begin by exemplifying chemical methods of synthesis in solution. We then outline some approaches for minimising the toxicological effects of certain nanoparticles using encapsulation techniques which can be implemented during production and which aim to limit the release of toxic species while preserving the morphological characteristics and physicochemical properties exploited in applications of these nanomaterials.

15.1 Chemical Synthesis of Nanoparticles and Toxicological Studies

It is not easy to make objects of nanometric size. Such a fine division of matter is not thermodynamically the most stable, and steps must be taken to control the kinetics in such a way as to favour these metastable states.

To achieve this, one way is to exploit the nature of the chemical reaction so as to produce an insoluble compound in the medium of synthesis. Another is to control physical aspects of nucleation, such as temperature and the way reagents are introduced, but also growth, e.g., growth in a confined environment; or again to adjust chemical aspects such as the use of protective agents like surfactants, organic molecules, or polymers, controlling the growth of the nanoparticles, so as to obtain nano-objects of uniform size and shape.

15.1.1 Type II–VI Semiconductor Nanoparticles

Over the past few years, many ways have been devised for synthesising semiconductor nanoparticles, usually called quantum dots (QD). In this chapter, we shall mainly focus on type II–VI semiconductors, because nanoparticles are more easily synthesised than with type III–V semiconductors. Type II–VI semiconductors are widely used to make light-emitting diodes (LED), solar and photovoltaic cells, and security inks. They are also used in medicine and biology to label specifically different entities such as cells, proteins, DNA fragments, etc. For efficient labelling, especially in comparison with fluorescent organic molecules (fluorophores), these nanomaterials must produce a strong and lasting signal, while remaining stable in aqueous media and having a sufficiently large surface area to graft biological molecules without perturbing their optical properties (see Fig. 15.4).

Paul Alivisatos and coworkers at Berkeley University, California, have developed a way of synthesising quantum dots (CdS, CdSe, CdSe@ZnS, CdSe@CdS, etc.) with less than 5% size dispersion [4–14]. It has also been possible to synthesise nanoparticles with a range of shapes such as spheres (see Fig. 15.5A), rods (see Fig. 15.5B–D), and tetrapods (see Fig. 15.6).

To clearly separate nucleation and growth stages, the reaction was conducted at high temperature (250–300°C) by fast precursor injection into a complexing organic medium comprising molecules such as trioctylphosphine oxide (TOPO), trioctylphosphine (TOP) and *n*-octadecylphosphonic acid (ODPA). These molecules remain adsorbed at the surface of the nanoparticles after synthesis. For some applications, particularly in biology and medicine, the nanoparticles must remain stable in an aqueous medium. Then they have to be transferred from the organic complexing medium to the aqueous medium. This manipulation must not alter the size and shape distribution which determines the optical properties of the nanoparticles. The simplest method is to replace the organic molecules coating the nanoparticle surface by hydrophilic molecules or by a film of SiO_2 [4]. Since coordinating solvents are harmful and expensive, methods using non-coordinating solvents such as 1-octadecene (ODE) have been proposed [15–17].

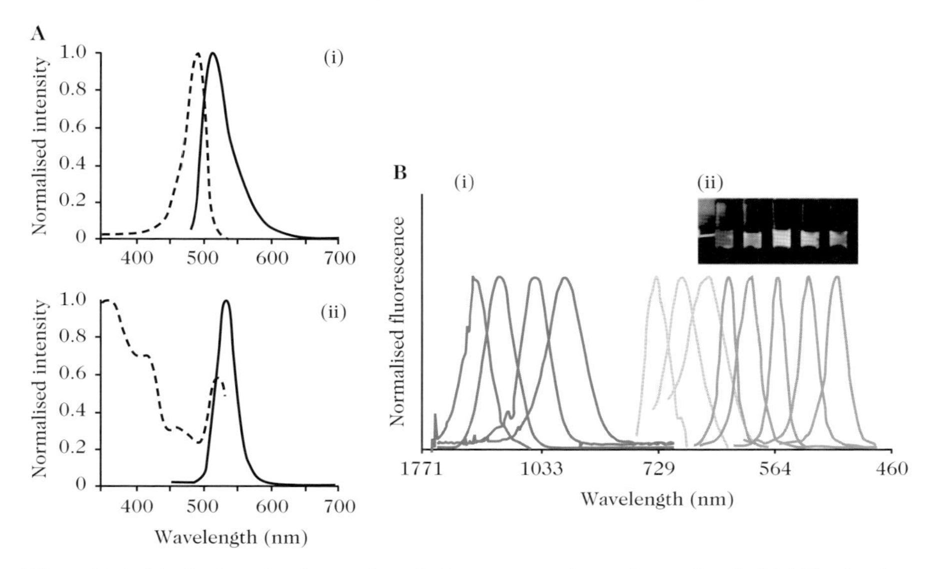

Fig. 15.4. Typical excitation and emission spectra (see colour plate). (**A**) Excitation spectrum (*dashed*) and fluorescence spectrum (*continuous*) (i) of fluorescein and (ii) QDs in a phosphate buffer solution (PBS). (**B**) (i) CdSe nanoparticles of different sizes (*blue spectra*): 2.1, 2.4, 3.1, 3.6, and 4.6 nm. InP nanoparticles (*green spectra*): 3.0, 3.5, and 4.6 nm. InAs nanoparticles (*red spectra*): 2.8, 3.6, 4.6 et 6.0 nm. (ii) True colours of a series of CdSe@ZnS colloids coated with a film of SiO_2. From [8]. Reproduced with the kind permission of Elsevier 2005

The US Environmental Protection Agency used standard aquatic assays to assess the ecotoxicological impact of CdSe@ZnS quantum dots sold on the market under the trade name Qdot 545 ITK carboxyl quantum dots [18]. The target organisms used were the green alga *Pseudokirchneriella subcapitata* and the daphnia *Ceriodaphnia dubia*.

In the case of *Ceriodaphnia dubia*, after contact with 110 ppb of Qdot 545 ITK carboxyl quantum dots for 48 h, no cell death was observed. These QDs are surrounded at the surface by a film of carboxyl groups. For comparison, *Ceriodaphnia dubia* does not survive a concentration 500 times lower of the same quantum dots but without the carboxyl groups at the surface [19]. In the case of *Pseudokirchneriella subcapitata*, cell death was observed for a concentration of 37.1 ppb after 96 h exposure. This is 190 times greater than the concentration of quantum dots without carboxyl groups to which another green alga, *Selenastrum capricornutum*, was exposed [20]. These results show that the carboxyl groups used to stabilise the colloid also play a protective role by limiting the toxicity due to metal released during laboratory exposures. On the other hand, nanoparticles without surface protection may release thetal to concentrations well above allowed toxicity thresholds, and so are likely to have health impacts at higher trophic levels.

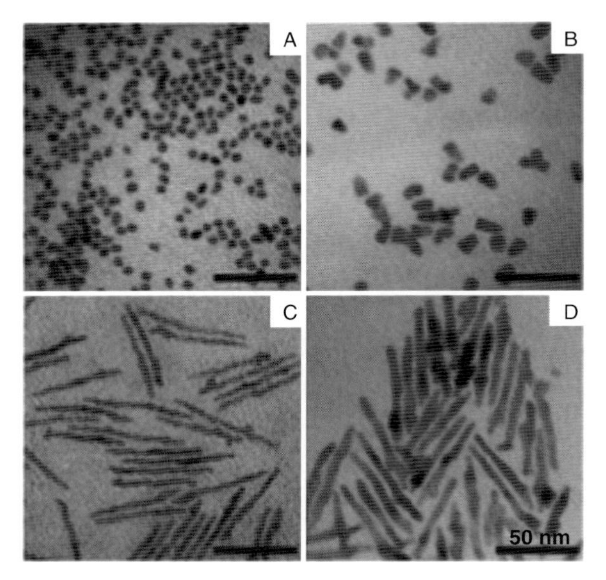

Fig. 15.5. Transmission electron microscope (TEM) images of CdSe nanoparticles: (**A**) 7 nm × 7 nm, (**B**) 8 nm × 13 nm, (**C**) 3 nm × 60 nm, (**D**) 7 nm × 60 nm. From [11]. Copyright Wiley-VCH Verlag GmbH and Co. KGaA. Reproduced with permission

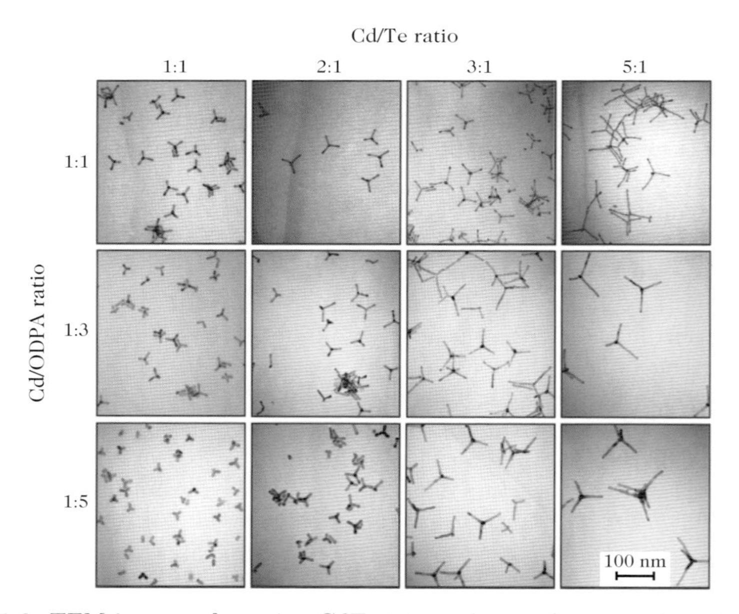

Fig. 15.6. TEM images of growing CdTe tetrapods as a function of reaction conditions. From [10]. Reproduced with the kind permission of Nature Publishing Group

15.1.2 ZnO Nanoparticles

Zinc oxide (ZnO), itself a semiconductor, is used as a pigment in paints, as a filler in rubbers, and to cover certain papers, particularly due to its ability to absorb UV radiation. It also has applications in creams, lotions, and other sunscreen products. It has piezoelectric and light-emitting properties. On the worldwide scale, a million tonnes of ZnO are produced annually.

The technical properties of ZnO are improved by using finer powders. This increases its capacity to absorb UV radiation, and it also raises its specific surface area (up to several tens of m^2/g), an important parameter in the catalysis of certain reactions (antibacterial, antifungal, adherence, etc.). Furthermore, ZnO nanoparticles are transparent, and this is useful for designing cosmetic products such as sunscreen creams. In the laboratory, ZnO nanoparticles have proven to be more efficient absorbers of UVA radiation than TiO_2 nanoparticles. In addition, zinc oxide has the advantage of having a less white appearance [21].

Among the various methods of synthesis that are used, or could be used, we shall discuss here the synthesis by forced hydrolysis in a polyol medium [22–26]. The most widely used polyols are α-diols like ethyleneglycol (ethane-1,2-diol) (EG) and propane-1,2-diol (PEG), or compounds resulting from the condensation of α-diols like diethyleneglycol (DEG). The polyol plays the role of solvent, complexing agent, and growth medium for solid particles. This growth is influenced by the nature of the medium, and in particular its viscosity and surface tension. For example, hydrolysis of zinc acetate in diethyleneglycol yields monodisperse powders of ZnO [19–22]. For reaction temperatures in the range 150–180°C, it is possible to obtain ZnO particles of controlled size and shape for hydrolysis ratios $H = n_{H_2O}/n_{Zn^{2+}}$ in the range between 2 and 300 (see Fig. 15.7).

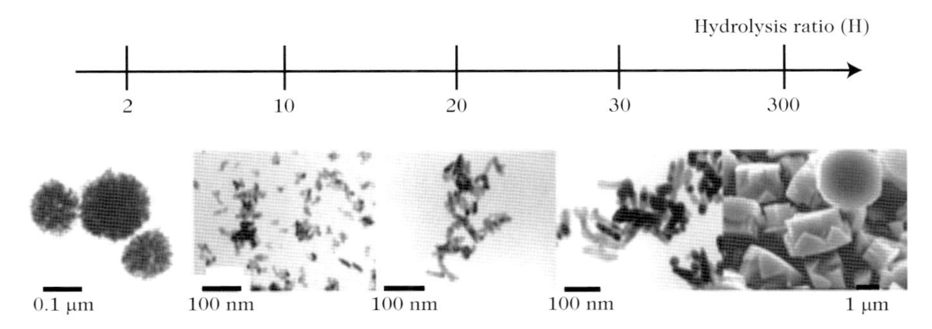

Fig. 15.7. Micrographs of ZnO nanoparticles synthesised in a polyol medium without structuring agent and varying the hydrolysis ratio [24]

As for the synthesis of type II–VI semiconductors, several structuring agents can be used, e.g., trioctylphosphine oxide (TOPO), bovine serum albumin (BSA), sodium dodecylsulfate (SDS), polyoxyethylene-10-stearyl-ether (Brij-76), etc., to control the size and shape of the nanoparticles [25].

The ecotoxicological impact of these nanoparticles has been investigated on two target micro-organisms, namely the cyanobacteria *Anabaena flos-aquae* and the euglena (micro-alga) *Euglena gracilis* [24]. These two micro-organisms were chosen for different reasons. *Anabaena flos-aquae* synthesises large amounts of exopolysaccharides, which could act as a barrier to prevent the internalisation of nanoparticles, while *Euglena gracilis*, able to internalise the nanoparticles by endocytosis, seems a priori to be more exposed to their harmful effects. The photosynthetic activity of these micro-organisms was monitored over a period of one month after contact with ZnO nanoparticles, whereas fixation and resin inclusion were established after 15 days of contact with the nanoparticles (see Fig. 15.8). In every case, the concentration of Zn^{2+} was fixed at 10^{-3} M.

At the beginning, photosynthetic activity is reduced due to the stress caused by pricking out. In general, eukaryotic plankton (green algae, euglenas) exhibit higher photosynthetic activity than prokaryotic plankton (cyanobacteria) (see Fig. 15.8C). ZnO nanoparticles were introduced into the culture in the presence of the micro-organisms, once their photosynthetic activity had stabilised. After adding the nanoparticles, a significant drop in photosynthetic activity was observed in every case (see Fig. 15.8D). For *Anabaena flos-aquae*, following this drop, there was a gradual increase in photosynthetic activity after 15 days of contact with the ZnO nanoparticles. On the other hand, for *Euglena gracilis*, photosynthesis came to a complete halt and cell death was concluded (see Fig. 15.8D). Observing the cell ultrastructure by electron microscopy, it appeared that, for *Anabaena fluos-aquae*, the nanoparticles remain around the outside of the cells without being able to enter them. This is due to the exopolysaccharide synthesised by the cyanobacteria (see Fig. 15.8E). On the other hand, images showed the internalisation of nanoparticles inside vesicles within the cells of *Euglena gracilis* (see Fig. 15.8F). These results agree with observations of photosynthetic activity. We may thus conclude that the exopolysaccharide produced by *Anabaena flos-aquae* really does play the role of barrier, preventing the internalisation of nanoparticles, and hence also cell death. On the other hand, since *Euglena gracilis* can feed by endocytosis, the presence of ZnO nanoparticles within these cells leads to their death.

Any attempt to standardise the protocols for toxicity assays without first building up a detailed understanding (1) of the way the given nanoparticles act and (2) the way the chosen biological targets react to them must therefore be regarded with caution.

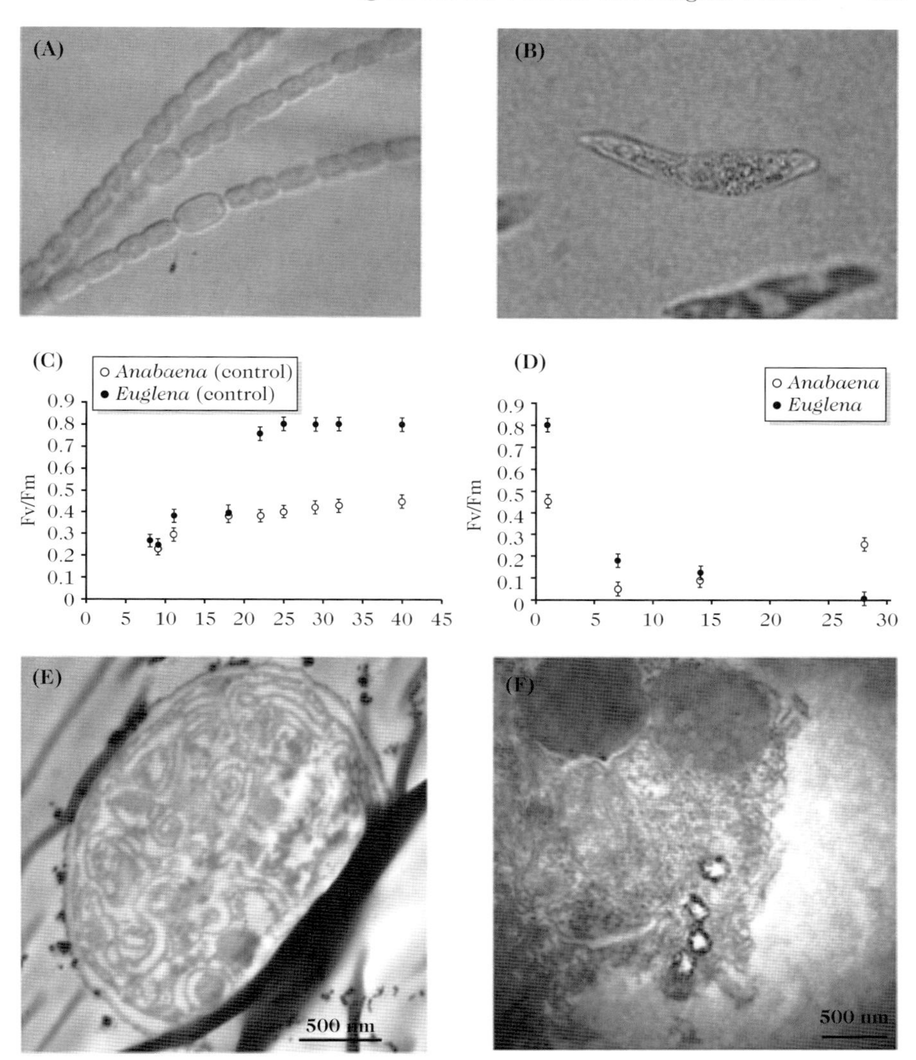

Fig. 15.8. Photon microscope images: (**A**) *Anabaena flos-aquae*, (**B**) *Euglena gracilis*. See colour plates for (**A**) and (**B**). Photosynthetic activity of the microorganisms (Fv/Fm) (**C**) before and (**D**) after adding the ZnO nanoparticles. Cell ultrastructure after contact with ZnO nanoparticles: (**E**) *Anabaena flos-aquae*, (**F**) *Euglena gracilis* [24]. (**A**), (**B**), (**E**), and (**F**) taken from [24], with the kind permission of Langmuir, copyright 2010 American Chemical Society

15.2 New Ways to Synthesise Protected Nanoparticles with Reduced Toxicological Effects

Given the complexity of the parameters describing nanoparticles, biological targets, and the composition of the environment in which exposure takes place, and given the difficulty in obtaining standardised toxicological data for the systems under investigation, despite the rapidly growing interest in nanomaterials in many fields, it makes sense to try to reduce the toxicity of nanoparticles at the very outset, when they are synthesised, using protection techniques such as core–shell, surface functionalisation, etc., without alteration of their physical properties. This protection may also in some cases prevent the partial dissolution of nanoparticles in the natural environment. Let us now illustrate this with a few examples.

15.2.1 Surface Functionalisation and Passivation of Type II–VI Semiconductor Nanoparticles

Cadmium chalcogenide quantum dots are potentially dangerous owing to the toxicity of this metal. The first method for limiting their toxicity is to cover the core of the cadmium chalcogenide nanoparticle by saturating all dangling bonds by a shell of material that does not contain cadmium. The lattice parameter of the shell must be close to that of the core to facilitate epitaxial growth. The shell is usually grown in an organic phase using the core nanoparticles as heterogeneous nucleation sites. It is also possible to functionalise the nanoparticle surface with non-toxic hydrophilic molecules that make these particles stable in an aqueous medium (see Fig. 15.9).

For example, the toxicological impact on human kidney cells (HEK293T) of CdTe, CdTe@CdS, and CdTe@CdS@ZnS nanoparticles functionalised by

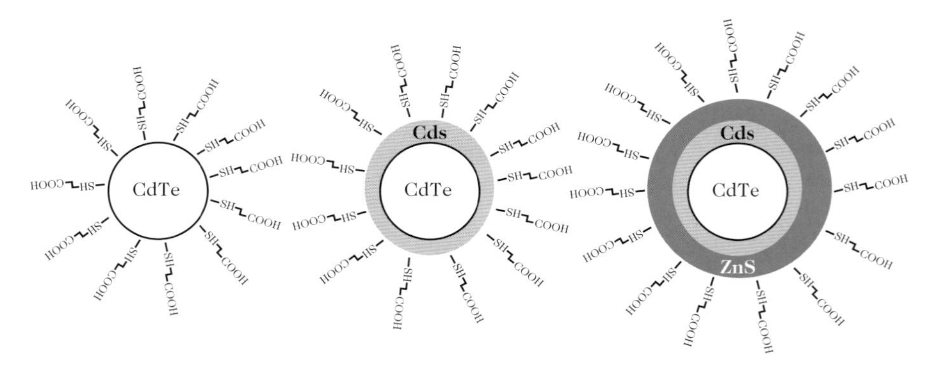

Fig. 15.9. Examples of quantum dots (CdTe, CdTe@CdS, CdTe@CdS@ZnS) functionalised by hydrophilic carboxylated chains [27]

hydrophilic carboxylated chains was investigated by varying the nanoparticle concentration and exposure time (see Fig. 15.10) [27].

These tests show how efficient it can be to passivate the core of semiconductor nanoparticles by a ZnS film. Even after 48 h of incubation, cell viability is very high, whereas viability is estimated at between 60 and 50% for CdTe and CdTe@CdS nanoparticles after 24 h incubation, and it is zero at 48 h (see Fig. 15.10) [27]. Note also that passivating the core of semiconductor nanoparticles by a ZnS film also improves fluorescence yields by confining the exciton within the core.

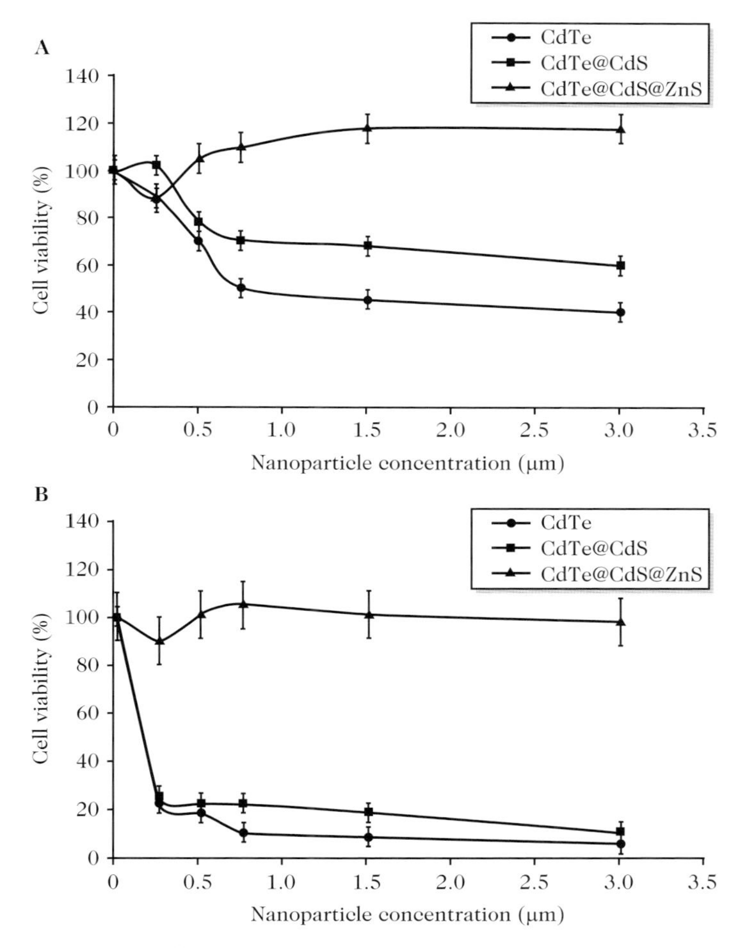

Fig. 15.10. Viability of HEK293T cells after contact with different concentrations of quantum dots at 37°C in a damp atmosphere with 5% CO_2. (**A**) After 24 h contact. (**B**) After 48 h contact. Cell viability is expressed by the percentage of living cells after exposure to the nanoparticles as compared with the living cells in a control sample (without nanoparticles) [27]

15.2.2 Nanoparticle Encapsulation by an SiO_2 Shell

Silica Encapsulation: State of the Art

The first work on the synthesis of silica capsules was published in 1990 by David Avnir and coworkers in Jerusalem [28]. This group showed that enzymes, in this case alkaline phosphatase, can be encapsulated in silica gels. Not only are these enzymes not denatured, but they maintain a non-negligible level of biocatalytic activity. Many other enzymes have since been encapsulated in silica gels with the aim of making biosensors (glucose oxidase) or bioreactors (lipases) [29, 30]. Jacques Livage and his group at LCMCP, Pierre and Marie Curie University and Collège de France, has shown that living cells such as yeasts, bacteria, protozoa, plant cells, etc., can be encapsulated in a silica gel and still retain a high level of viability for more than a month [31].

This same group used this approach to synthesise new hybrid silica–alginate materials [32]. Originally, the most important application of these microcapsules was in the treatment of diabetes. As everyone knows, this disease is related to an insufficient production of insulin in response to glucose. This is normally regulated by the cells of the pancreas, and in particular cells in the islets of Langerhans, clusters of α and β cells. In type I diabetes, these β cells, responsible for the production of insulin, are destroyed by the immune system of the patient (autoimmune disease). One possible solution was to implant immobilised healthy Langerhans islets in alginate microcapsules [33]. These microcapsules are obtained by gelling droplets of alginate in solutions of $CaCl_2$. At contact with this solution, the surface of the droplet is immediately gelled by the Ca^{2+} ions, and the droplet transformed into a bead whose interior will gradually become crosslinked by diffusion of calcium. The beads obtained in this way are then macroporous and hence not very selective with regard to the sizes of the molecules that can diffuse through them. To remedy this, these capsules are put in contact with polylysine. This cationic polymer forms a polyelectrolyte complex with the anionic alginate at the bead surface. Then adding sodium citrate, which complexes the Ca^{2+} ions, the core of the capsule can be liquified, leaving only an outer alginate/polylysine membrane. The diffusion properties of this membrane can be controlled by the thickness of the film and the length of the polylysine chain [34].

To improve the mechanical properties of these membranes, Livage and his group have shown that a silica shell can be deposited around the polylysine film (see Fig. 15.11). This deposit, a few microns thick, is made of very dense and homogeneous silica. After treatment with sodium citrate, the capsules obtained do indeed have much improved mechanical strength, while the diffusion properties of the membrane remain unaffected [32].

Toward Multifunctional Hybrid Nanoparticles

It turns out that the hybrid capsules described above could be used as carriers for transport and controlled delivery of therapeutic agents by incorporating

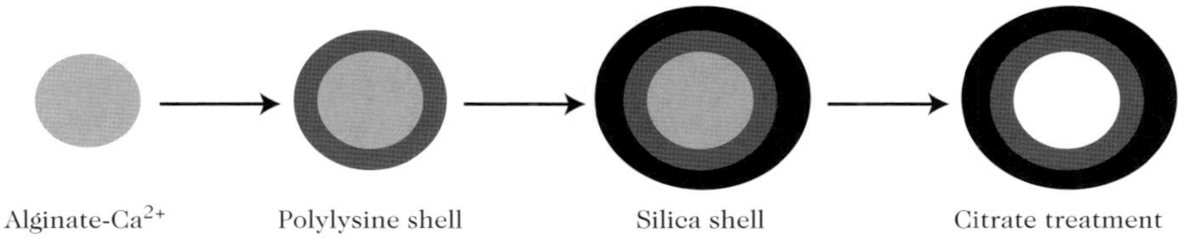

Fig. 15.11. Synthesis of alginate@polylysine@SiO_2 hybrids, followed by treatment with sodium citrate which liquifies the core of the capsule

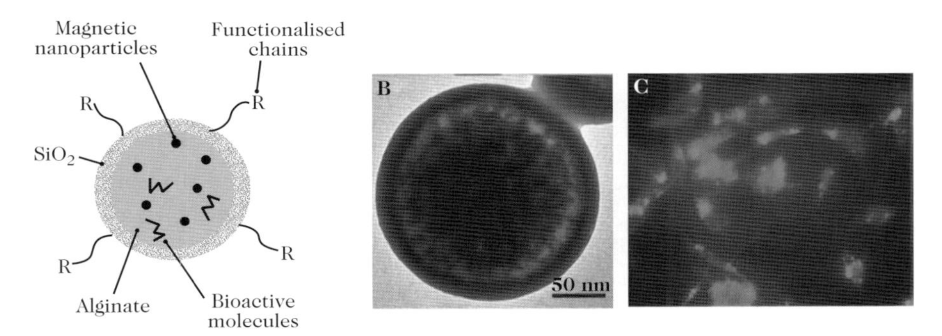

Fig. 15.12. (**A**) Multifunctional hybrid nanoparticle. (**B**) Micrograph of an alginate–Fe_3O_4@SiO_2 nanocapsule. (**C**) Epifluorescence microscopy image of 3T3 fibroblasts following 24 h contact with alginate–Fe_3O_4–carboxyfluorescein@SiO_2 nanocapsules. (**B**) and (**C**) are taken from [36], with the kind permission of Elsevier, copyright 2007. For (**C**), see colour plate

active molecules and magnetic nanoparticles within the capsules. However, the first aim here is to reduce the size of these objects. Micrometric dimensions are compatible with oral administration, but sizes smaller than 50 nm are required for intravenous injections. Thibaud Coradin and coworkers at LCMCP, University of Pierre and Marie Curie in Paris have shown that the size of hybrid alginate–Co^0 and alginate–Fe_3O_4 beads can be reduced to around a hundred nanometers before coating them with silica (see Fig. 15.12) [35, 36]. Hybrid alginate–Fe_3O_4@SiO_2 nanocapsules have superparamagnetic behaviour with blocking temperature 100 K. The Fe_3O_4 nanoparticles remain frozen within the alginate (see Fig. 15.12B). The first toxicity tests were done using 3T3 fibroblasts as biological target and the fluorophore carboxyfluoscein to label the nanocapsules (see Fig. 15.12C). Internalisation of the nanocapsules was observed after a 24 h incubation at 37°C. Viability was maintained at 90% even after 24 h contact with the nanocomposites.

ZnO or TiO_2 nanoparticles used in the composition of solar protection products due to their ability to absorb UV radiation have sizes of the order of 20 nm. Tests have shown that these nanoparticles can, following internalisation by cells, cause modfications to DNA owing to the formation of radical species such as radical oxygen species (ROS) [37]. The authors studied samples

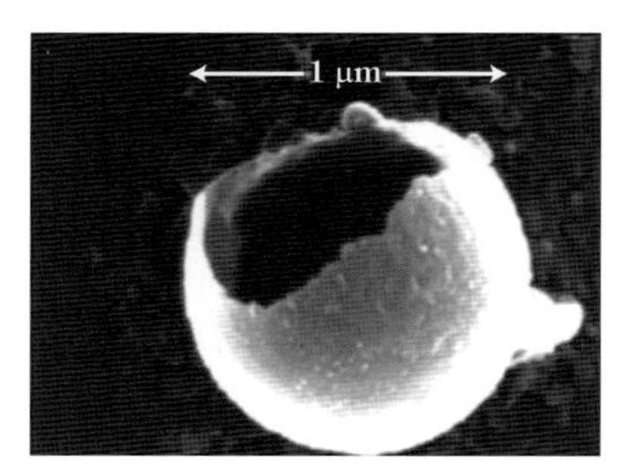

Fig. 15.13. Micronic silica shell

of commercial TiO_2 with diameters in the range 20–50 nm, using varying proportions of anatase and rutile, and also samples containing ZnO. Genotoxicity was monitored in vivo using the comet assay, a technique for detecting and quantifying DNA lesions. This analysis uses samples included in agarose and exposed to an electric field. In these analyses, human cells (MRC-5 fibroblasts) were illuminated with and without the sunscreen cream containing TiO_2 and ZnO nanoparticles. The study showed that human cell DNA was damaged in the presence of these nanoparticles.

To prevent internalisation of these nanoparticles, one possibility would be to encapsulate them in micronic silica shells, following the work by Avnir et al., who patented a new range of sunscreen creams called UV-Pearlsä, where organic molecules able to absorb UV radiation are encapsulated within a silica shell prepared by the sol–gel method (see Fig. 15.13).

Biocompatibility of Colloidal Silica

The biocompatibility of silica has been debated over the past few years. The potential toxicity can in principle depend on several factors relating not only to the physicochemical properties of the nanoparticles but also to the dose, exposure time, administration route, biological target, and composition of the environment in which exposure occurs.

Studies carried out using the green alga *Pseudokirchneriella subcapta* as biological target have shown that the excotoxic effect of colloidal silica is related to the size and specific surface area of the nanoparticles and not to the concentration of nanoparticles in the medium [38]. In this study, a 20% reduction of cell viability was found after 72 h contact with commercial colloidal suspensions of LUDOX (a colloidal silica) with sizes in the range 12.5–27 nm (specific surface area 135–230 m^2/g). No ecotoxic effect was observed after contact with micronic silica particles.

In another study, four groups of 25 rats were exposed to concentrations of colloidal silica (LUDOX, made by DuPont, diameter approximately 22 nm) of 0, 10, 50, and 150 mg/m^3, for 6 h per day, 5 days per week, and over 4 weeks [39,40]. At the end of exposure (day 0), 5 rats were sacrificed, then 10 (day 10), and finally 10 (3 months). For a concentration of 10 mg/m^3, no effect was observed, whereas for concentrations of 50 or 150 mg/m^3, pulmonary modifications arose (macrophages containing silica in the bronchi, infiltrations by neutrophils, hyperplasia of type II pneumocytes), with increasing incidence as the concentration was raised in the test. Most biochemical parameters returned to normal with 3 months of recovery, but there remained tiny nodular lesions in the bronchi and perivascular regions.

15.3 Conclusion

With the intensification of research in nanoscience, the fast development of nanotechnology, and the commercialisation of an ever increasing number of nanomaterials, the scientific community will be called upon to study and assess the toxicological impact of these novel materials throughout every stage of their synthesis and use. The few examples of in vitro studies of nanoparticle toxicity on biological targets discussed in this chapter serve to illustrate just how complex these studies are. They require a cross-disciplinary approach which takes into account the physicochemical characteristics intrinsic to each type of nanoparticle on a case by case basis. This is the price to pay if we are to understand the interactions between cells and nanoparticles, and identify the relevant toxicity mechanisms.

In parallel, research has shown that it is possible to synthesise effective multifunctional hybrid nanomaterials that are made less toxic by encapsulation or surface treatment processes, while their morphological characteristics and physical properties remain unchanged. The implementation of such techniques is still in the exploratory stage, but it already offers interesting prospects, primarily because they are applied during synthesis, i.e., very early in the life cycle of these nanomaterials. Their aim is thus to prevent rather than cure the toxicity of certain nanoparticles with potential applications, or already exploited.

References

1. A. Billon, J.L. Dupont, G. Ghys: Le financement des nanotechnologies et des nanosciences: l'effort des pouvoirs publics en France. Comparaisons internationales. Rapport No. 2004-002, Ministère de la Jeunesse, de l'Education nationale et de la Recherche, Paris (2004)
2. L. Williams, W. Adams: *Nanotechnology Demystified: A Self-Teaching Guide.* McGraw-Hill (2007)
3. AFSSET: Les nanomatériaux: Effets sur la santé de l'homme et sur l'environnement (2006) www.afsset.fr/upload/bibliotheque/587621558014304413168640606286/synthese_nanomateriaux_2006.pdf
4. M. Bruchez, M. Moronne, P. Gin, S. Weiss, A.P. Alivisatos: Semiconductor nanocrystals as fluorescent biological labels. Science **281**, 2013 (1998)

5. G. Dukovic, M.G. Merkle, J.H. Nelson, S.M. Hughes, A.P. Alivisatos: Photodeposition of Pt on colloidal CdS and CdSe@CdS semiconductor nanostructures. Adv. Mater. **20**, 4306 (2008)
6. A. Fu, W. Gu, B. Boussert, K. Koski, D. Gerion, L. Manna, M. Le Gros, C.A. Larabell, A.P. Alivisatos: Semiconductor quantum dots and rods as single-molecule fluorescent biological labels. Nano Lett. **7**, 179 (2007)
7. T. Zhang, J.L. Stilwell, D. Gerion, L. Ding, O. Elboudwarej, P.A. Cooke, J.W. Gray, A.P. Alivisatos, F.F. Chen: Cellular effect of high doses of silica-coated quantum dot profiled with high throughput gene expression analysis and high content cellomics measurements. Nano Lett. **6**, 800 (2006)
8. A. Fu, W. Gu, C. Larabell, A.P. Alivisatos: Semiconductor nanocrystals for biological imaging. Curr. Opin. Neurobiol. **15**, 568 (2005)
9. H. Liu, A.P. Alivisatos: Preparation of asymmetric nanostructures through site selective modification of tetrapods. Nano Lett. **4**, 2397 (2004)
10. L. Manna, D.J. Milliron, A. Meisel, E.C. Scher, A.P. Alivisatos: Controlled growth of tetrapod-branched inorganic nanocrystals. Nature Mater. **2**, 382 (2003)
11. W.U. Huynh, J.J. Dittmer, W.C. Liby, Whiting, A.P. Alivisatos: Controlling the morphology of nanocrystal–polymer composites for solar cells. Adv. Func. Mater. **13**, 73 (2003)
12. L.S. Li, J. Hu, W. Yang, A.P. Alivisatos: Band gap variation of size- and shape-controlled colloidal CdSe quantum rods. Nano Lett. **1**, 349 (2001)
13. D. Gerion, F. Pinaud, S.C. Willians, W.J. Parak, D. Zanchet, S. Weiss, A.P. Alivisatos: Synthesis and properties of biocompatible water-soluble silica-coated CdSe@ZnS semiconductor quantum dots. J. Phys. Chem. B **105**, 8861 (2001)
14. W.U. Huynh, X. Peng, A.P. Alivisatos: CdSe nanocrystal rods/poly (3-hexythiophene) composite photovoltaic devices. Adv. Mater. **11**, 923 (1999)
15. W. Yu, X. Peng: Formation of high quality CdS and other II–VI semiconductor nanocrystals in noncoordinating solvents: Tunable reactivity of monomers. Angew Chem. Int. Ed. **41**, 2368 (2002)
16. M. Protiere, P. Reiss: Facile synthesis of monodisperse ZnS capped CdS nanocrystals exhibiting efficient blue emission. Nanoscale Res. Lett. **1**, 62 (2006)
17. M. Protiere, P. Reiss: Highly luminescent $Cd_{1-x}Zn_xSe/ZnS$ core/shell nanocrystals emitting in the blue–green spectral range. Small **3**, 399 (2007)
18. J.L. Bouldin, T.M. Ingle, A. Sengupta, R. Alexander, R.E. Hannigan, R.A. Buchanan: Aqueous toxicity and food chain transfer of quantum dots in freshwater algae and *Ceriodaphnia dubia*. Environ. Toxicol. Chem. **27**, 1958 (2008)
19. G. Bitton, K. Rhodes, B. Koopman: Cerinofast: An acute toxicity test based on *Ceriodaphinia dubia* feeding behavior. Environ. Toxicol. Chem. **15**, 123 (1996)
20. J.M. Diamond, D.E. Koplish, J. McMahon, R. Rost: Evaluation of the water-effect ratio procedure for metals in a riverine system. Environ. Toxicol. Chem. **16**, 509 (1997)
21. S.R. Pinnell, D. Fairhurst, R. Gilliers, M.A. Mitchnick, N. Kollias: Microfine zinc oxide is a superior sunscreen ingredient to microfine titanium dioxide. Dermatol. Surg. **26**, 309 (2000)
22. D. Jezequel, J. Guenot, N. Jouini, F. Fievet: Submicrometer zinc oxide particles: Elaboration in polyol medium and morphological characteristics. J. Mater. Res. **10**, 77 (1995)

23. C. Feldmann: Polyol-mediated synthesis of nanoscale functional materials. Adv. Func. Mater. **13**, 101 (2003)
24. R. Brayner, S.A. Dahoumane, C. Yéprémian, C. Djediat, M. Meyer, A. Couté, F. Fiévet: ZnO nanoparticles: Synthesis, characterization and ecotoxicological studies. Langmuir **26**, 6522 (2010)
25. R. Brayner, R. Ferrari-Iliou, N. Brivois, C. Djediat, M.F. Benedetti, F. Fievet: Toxicological impact studies based on *Escherichia coli* bacteria in ultrafine ZnO nanoparticles colloidal medium. Nano Lett. **6**, 866 (2006)
26. R. Brayner: The toxicological impact of nanoparticles. Nano Today **3**, 48 (2008)
27. Y. Su, Y. He, H. Lu, L. Sai, Q. Li, W. Li, L. Wang, P. Shen, Q. Huang, C. Fan: The cytotoxicity of cadmium based aqueous phase synthesized quantum dots and its modulation by surface coating. Biomaterials **30**, 19 (2009)
28. S. Braum, S. Rappoport, R. Zusman, D. Avnir, M. Ottolenghi: Biochemically active sol–gel gases: The trapping of enzymes. Mater. Lett. **10**, 1 (1990)
29. D. Avnir, S. Braun, O. Lev, M. Ottolenghi: Enzymes and other proteins entrapped in sol–gel materials. Chem. Mater. **6**, 1605 (1994)
30. B.C. Dave, B. Dunn, J.S. Valentine, J.I. Zink: Sol–gel encapsulation methods for biosensors. Anal. Chem. **66**, 1120A (1994)
31. N. Nassif, O. Bouvet, M.N. Rager, C. Roux, T. Coradin, J. Livage: Living bacteria in silica gels. Nature Mater. **1**, 42 (2002)
32. T. Coradin, E. Mercey, L. Lisnard, J. Livage: Design of silica-coated microcapsules for bioencapsulation. Chem. Commun. **7**, 2496 (2001)
33. P. De Vos, P. Marchetti: Encapsulation of pancreatic islets for transplantation in diabetes: The untouchable islets. Trends Mol. Med. **8**, 363 (2002)
34. H. Uludag, P. De Vos, P.A. Tresco: Technology of mammalian cell encapsulation. Adv. Drug Deliv. Rev. **42**, 29 (2000)
35. M. Boissiere, P.J. Meadows, R. Brayner, C. Helary, J. Livage, T. Coradin: Turning biopolymer particles into hybrid capsules: The example of silica/alginate nanocomposites. J. Mater. Chem. **16**, 1178 (2006)
36. M. Boissiere, J. Allouche, C. Chaneac, R. Brayner, J.M. Devoisselle, J. Livage, T. Coradin: Potentialities of silica/alginate nanoparticles as hybrid magnetic carriers. Int. J. Pharmaceutics **344**, 128 (2007)
37. R. Dunford, A. Salinaro, L. Cai, N. Serpone, S. Horikoshi, H. Hidaka, J. Knowland: Chemical oxidation and DNA damage catalyzed by inorganic sunscreen ingredients. FEBS Lett. **418**, 87 (1997)
38. K. Van Hoecke, K.A.C. De Schamphelaere, P. Van Der Meeren, S. Lucas, C. Janssen: Ecotoxicity of silica nanoparticles to the green alga *Pseudokirchneriella subcapitata*: Importance of surface area. Environ. Toxicol. Chem. **27**, 1948 (2008)
39. K.P. Lee, D.P. Kelly: The pulmonary response and clearance of Ludox colloidal silica after 4-week inhalation exposure in rats. Fund. Appl. Toxicol. **19**, 399 (1992)
40. K.P. Lee, D.P. Kelly: Translocation of particle-laden, alveolar macrophages and intra-alveolar granuloma formation in rats exposed to Ludox colloidal amorphous silica by inhalation. Toxicology **77**, 205 (1993)

16

Toxicological Models Part B: Environmental Models

Jeanne Garric and Eric Thybaud

Assessment of ecotoxicological risks due to chemical substances is based in part on establishing concentration–response relationships for different organisms, including plants, invertebrates, and vertebrates living on land, fresh water, or sea water. European regulations for assessing the risks due to chemical products thus recommend the measurement of toxic effects on at least three taxons (algae, crustacea, fish) [1]. The assessment becomes more relevant when based upon a variety of different organisms, with a range of different biological and ecological features (autotrophic or heterotrophic, benthic or pelagic habitat, and different modes of reproduction, growth, respiration, or feeding, etc.), but also when it describes the effects of contaminants on sensitive physiological functions such as growth and reproduction, which determine the balance of populations of terrestrial and aquatic species in their environment.

This concern over possible ecological hazards due to dispersion of nanoparticles in the environment is very recent. Not only has there been very little work on the assessment of nanoparticle ecotoxicity with regard to terrestrial or aquatic organisms, but it is all relatively new. For example, the first study of nanoparticle toxicity for a non-mammalian vertebrate dates to 2004 and concerns the ecotoxicity of fullerenes for the fish *Micropterus salmoides* [2].

Regarding aquatic organisms, studies have dealt exclusively with planktonic or epibenthic species such as fish, microcrustacea, and algae. There is as yet no published study dealing with the ecotoxicity of nanoparticles for benthic organisms. However, the question of the bioaccumulation of nanoparticles through the sediment and their possible influence on the bioavailability of hydrophobic contaminants has recently received attention [3, 4]. The water compartment and aquatic organisms have been the main subject of investigation, but it should be stressed that there is still almost no data available on the ecotoxic effects for marine organisms.

Taxon. A set of organisms sharing common features. The species is the basic taxon of the systematic classification of life. The higher the rank of the taxon, e.g., family, order, the less the individual organsims, e.g., plants, animals, toadstools, bacteria, in different taxons will resemble one another.

P. Houdy et al. (eds.), *Nanoethics and Nanotoxicology*,
DOI 10.1007/978-3-642-20177-6_16,

The exposure of organisms to nanoparticles depends in part on their ecological characteristics, e.g., their habitat.

Benthic Invertebrates. These are organisms whose life cycle occurs in part or wholly in the sedimentary substrate, e.g., oligochaetes, nematodes, etc.

Epibenthic Invertebrates. Organisms whose life cycle occurs mainly at the interface between water and the sedimentary substrate, such as amphipod crustacea, bivalves, freshwater gastropods, etc.

Pelagic Invertebrates. Organisms living mainly in open water.

16.1 Types of Nanoparticles Investigated

Ecotoxicity studies have been carried out on a relatively narrow range of nanoparticles. This work mainly concerns nanoparticles made from aluminium, copper, zinc, and metal oxides (oxides of zinc, aluminium, and titanium), single-wall or multiwall carbon nanotubes, and quantum dots, such as CdSe or CdTe. These have been favoured owing to the increasing applications of these substances (see Table 16.1).

16.2 Types of Preparation

There are several ways to prepare nanoparticle suspensions for ecotoxicity tests: without dispersion, single or successive ultrasonic dispersions possibly followed by filtration, and mechanical shaking possibly followed by ultrasonic dispersion [5]. For tests in solid media, nanoparticle powders can be mixed directly with the medium, without dilution.

The review by Crane et al. [5] concludes that, among the most important obstacles to be overcome in estimating nanoparticle ecotoxicity, the main

Table 16.1. Types of nanoparticles and associated applications

Type of nanoparticle	Examples of applications
Aluminium	Fuels, especially for the aerospace industry
Copper	Electronics, ceramics, polymers, inks, metal and surface treatment industries
Metal oxides	Combustion catalysts, solar cells, metallurgical industry, treatment of pollution, environmental remediation, disinfection, self-cleaning glass, food additives, pharmaceutical products, and cosmetics
Single- and multiwall carbon nanotubes	Plastics, catalysis, batteries, adhesives, therapeutic implants, car and aeronautics industries, water treatment
Quantum dots, e.g., CdSe or CdTe	Medical imaging, electronics, photovoltaic conversion

ones to be considered at the outset are ways to implement realistic exposure scenarios and robust methods for characterising the exposure of organisms. The latter point is all the more delicate to deal with in that nanoparticles are synthesised with specific features of size, shape, surface, and functionalisation that are tailored to meet the requirements of specific applications, whence their physicochemical properties and reactivities are extremely varied. For example, the kind of coating can play a direct role on the measured effect (possible toxicity of the organic coating molecule), or indeed an indirect role through a better dispersion in solution, for example [6]. The functionalisation of carbon nanotubes (nC_{60}) produces lethal and sublethal toxic effects of differing degrees on daphnia. Unmodified fullerenes and hydroxylated fullerenes have been shown to be toxic at a concentration of 100 ppm, whereas hydrogenated fullerenes and TiO_2 nanoparticles caused no mortality in 24 h on *Daphnia pulex* [7]. Likewise, significant changes in the detoxification enzyme activities of glutathione sulfo-transferase (GST) and catalase (CAT) were measured in daphnia at different concentrations and to different extents depending on the type of nanoparticle to which the organisms were exposed.

The ecotoxicity of nanoparticle solutions is influenced to a certain extent by the way the solution is prepared. For example, Oberdörster et al. [8] have shown that the toxicity of titanium dioxide nanoparticles can vary by a factor of 100 depending on the way the suspension is made (see Table 16.2).

By measuring the mortality of *Daphnia magna* after 48 h exposure to different preparations of TiO_2 and C_{60}, it was shown that toxicity was greater with filtered suspensions. This was attributed to a higher proportion of small particles [9].

The behaviour of nanoparticles, and in particular their state of aggregation in the medium, which conditions the size of the objects to which organisms are exposed, depends on various abiotic factors such as pH, ionic strength, or available organic matter [10], and this varied behaviour will in turn affect concentration–response relationships, as well as repeatability and reproducibility of results. For example, the pH 7–9, the conductivity (100–600 μS), and the hardness of the water (1–5 mM $CaCO_3$) in which organisms are conventionally exposed in toxicity tests tend to favour the aggregation of TiO_2 nanoparticles into clusters with size distributions centered on the micron [11].

Table 16.2. Ecotoxicity (half-maximal effective concentration EC_{50}) of various nanoparticles depending on how the suspension is made

Nanoparticle	Preparation	Daphnia	Algae
TiO_2	Sonication	> 500 mg/L	400 mg/L
TiO_2	Shaking	5.5 mg/L	25 mg/L
Fullerene	THF	0.46 mg/L	–
Fullerene	Shaking	> 35 mg/L	–

The aggregation phenomenon leads to a loss of nanoparticles in suspension and varies in relation to the test medium, but it also depends on the organism tested and the duration of the test [12]. Living organisms such as daphnia, algae, fish, etc., release various biomolecules into their environment, and these may react with the nanoparticles. For example, in bioassays, exposure conditions change with time and this evolution depends in part on the organisms and the test system. This observation confirms the need to define the relevant physical quantities, which must be representative of the exposure and easily monitored throughout the bioassay.

16.3 Compartments

16.3.1 Terrestrial Compartment

Plant Models

There have not been many nanoparticle ecotoxicity studies on higher land plants. Those that have been done were carried out on plants conventionally used in ecotoxicity studies, as recommended by the Organisation for Economic Cooperation and Development (OECD, Paris, France), the US Environmental Protection Agency, the US Food and Drug Administration, or the American Society for Testing and Materials (West Conshohocken, PA, USA) to study the phytotoxicity of chemical substances or pesticides. The main species used are summarised in Table 16.3.

Invertebrate Models

The ecotoxicity of nanoparticles for soil invertebrates has been studied on the isopod *Porcellio scaber* or common woodlouse, which is widely distributed around the world (see Fig. 16.1).

Toxic Effect Criteria

Nanoparticle ecotoxicity for soil micro-organisms has been measured by studying changes in the structure and composition of communities, but also by observing changes in biological functions. The impacts on the structure of microbial soil communities are assessed by counting the various organisms present [13] (the most likely number method for protozoans, or colony forming units for bacteria), or by studying the biomass produced [14]. Regarding the biological functions of soil micro-organisms, these are measured either globally by studying the alteration of respiration [13,14] or the mineralisation of glucose [15], or more specifically by studying perturbations in the functioning of various enzymes such as acid phosphatases like β-glucosidase, urease, or dehydrogenase [14].

Table 16.3. Main plant species used in ecotoxicity tests

Family	Species	Common name
Dicotyledonae		
Amaranthaceae	*Spinacia oleracea*	Spinach
Apiaceae (Umbelliferae)	*Daucus carota*	Carrot
Asteracea (Compositae)	*Lactuca sativa*	Lettuce
Brassicaceae (Cruciferae)	*Brassica napus*	Colza
Brassicaceae (Cruciferae)	*Brassica oleracea*	Cabbage
Brassicaceae (Cruciferae)	*Raphanus sativus*	Radish
Cucurbitaceae	*Cucumis sativa*	Cucumber
Fabaceae (Leguminosaea)	*Glycine max*	Soybean
Fabaceae (Leguminosaea)	*Phaseolus vulgaris*	Common bean
Solanaceae	*Solanum lycopersicon*	Tomato
Monocotyledonae		
Liliaceae (Amarylladaceae)	*Allium cepa*	Onion
Poaceae (Gramineae)	*Lolium perenne*	Ray grass
Poaceae (Gramineae)	*Triticum aestivum*	Wheat
Poaceae (Gramineae)	*Zea mays*	Maize

Fig. 16.1. *Porcellio scaber* or common woodlouse. Model used to study the ecotoxicity of nanoparticles for terrestrial invertebrates. This crustacea, which inhabits damp environments, was originally found in the Atlantic regions of the Iberian peninsula. But by virtue of its adaptability and anthropophilic habits, this species has been carried far and wide around the world

The toxic effect criteria studied in land plants are mainly inhibition of germination or inhibition of root growth in the early stages of development [16–20]. One study supplemented this by observing effects on photosynthesis, chlorophyll content, and activity of the enzyme ribulose bisphosphate carboxylase/oxygenase [20].

For the crustacean *Porcellio scaber*, toxic effect criteria concern either global perturbations to the physiology of the organism, corresponding to global criteria (feed rate, excretion rate, assimilation efficiency, change in weight, and mortality), or toxic action mechanisms in cells measured using biochemical biomarkers involved in the antioxidant response, such as catalase and glutathione S-transferase.

Methodology

In order to investigate the impact of nanoparticles on soil microbial communities, studies are carried out in the laboratory with natural soils sifted to remove particles larger than 2 mm. Studies are usually carried out in a soil column for periods of 7–180 days.

Most protocols recommend artificial substrates such as filter paper or agar-agar, limiting interactions between the given nanoparticles and the test medium. This practice does not contradict the recommendations of the OECD guidelines, which recognise that using natural soils can complicate the interpretation of the results, owing to the variability of the physicochemical properties of soils and the microbial populations living in them. However, it seriously reduces the ecological representativity of the results.

Three main types of protocol have been developed to study the impacts of nanoparticles on the germination and growth of land plants. One method favours a realistic scenario and uses a natural soil, while the others favour exposure of the organisms with tests in agar-agar and on filter paper.

- *Test in Natural Soil.* The plants are cultivated in natural soils sifted at 2 mm to remove stones, roots, and animal and plant organisms. They are watered each day and the effects on growth are noted over the 2 months of the test. The main aim of this test, developed by Doshi et al. [15], was to study the transport of nanoparticles in the soil and their accumulation in the plant tissues.
- *Test in Agar Medium.* Since many nanoparticles are insoluble in water, Lee et al. [17] developed a test using a culture medium based on agar-agar. On the one hand, this allows a good homogenisation of the nanoparticles within the culture medium, and on the other, it is not necessary to use a solvent to solubilise the nanoparticles. After sterilising by immersing in a bath of sodium hypochlorite at 5% for 10 min and rinsing in distilled water, the seeds are placed in darkness at 25°C for 24 h, wrapped in damp cotton. After this period, only those seeds that have germinated are used in the following. The test is carried out in 87×18 mm Petri dishes containing 30 mL of agar-agar culture medium including the nanoparticles, distributed in 20 mL of the medium at 2.5% agar-agar and 10 mL of a 1% agar-agar medium, the latter being placed in layers with the medium at 2.5% being placed below. At the beginning of the test, 10 pregerminated seeds are introduced just under the surface of the upper layer of agar-agar in each dish. These dishes are then placed in darkness at 25°C for 48 h. After this incubation period, the plants are separated from the agar medium and the length of the roots is measured.
- *Filter Paper Assay.* After sterilising by soaking in a sodium hypochlorite solution and reimbibing overnight in darkness at 25°C, the seeds are placed

on filter paper in Petri dishes (100 × 15 mm). Two different protocols have been developed for the following:

1. The nanoparticle solution is applied in concentric circles in such a way as to completely imbibe the filter paper, and the Petri dishes are transferred to culture chambers at 25°C for 48 h. The length of the roots is then measured after 24 and 48 h [16, 19].
2. The seeds are exposed by soaking them for 2 h, then transferred to filter paper in Petri dishes. The latter are subsequently placed in the culture chamber for 5 days, in darkness and at room temperature [18].

A variant proposed by Zheng et al. [20] is to incubate the seeds between two layers of expanded perlite. This highly absorbent substrate is strongly discouraged by the OECD [21, 22].

Jemec et al. [23] studied nanoparticle ecotoxicity for soil invertebrates on the isopod crustacean *Porcellio scaber*. The tests were carried out on adult individuals of both sexes, with weights of 30–80 mg. Each animal was placed in its own Petri dish and fed with disks of hazel leaf on which the nanoparticles had been deposited. Observations were made after 3 days of exposure by the alimentary route.

16.3.2 Aquatic Compartment

Concerning the aquatic compartment, the review by Klaine et al. [24] assembles a large amount of data on the effects of various types of nanoparticles on aquatic organisms. The information and results available refer to the most widely used biological models in aqueous phase bioassays: green algae, the cladoceran microcrustaceans *Daphnia magna* and *Daphnia pulex*, and fish like the rainbow trout (*Oncorhynchus mykiss*), the zebrafish (*Danio rerio*), the fathead minnow (*Pimephales promelas*), or again the see-through medaka (*Oryzia latipes*). Other models used in freshwater ecotoxicology have also been the subject of nanoparticle toxicity tests: the amphipod crustacean *Hyalella azteca*, the pulmonate gastropod *Lymnaea stagnalis*, or the freshwater mussel (*Elliptio complanata*).

Various toxic effects are sought, including lethal effects, but also sublethal, biochemical, transcriptional, physiological, or behavioural effects.

Alga Models

The toxicity of nanoparticle suspensions of different kinds are also studied for their toxic effects on single-cell freshwater algae, mainly green algae (see Fig. 16.2) [25–28]. The main global toxic effect criterion concerns the growth of algal cultures, innoculated in a culture medium in the presence of a nanoparticle suspension. Physiological effects such as inhibition of photosynthesis, but also biochemical or genomic effects, can also be observed to elucidate toxicity mechanisms, and in particular the activity of free radicals and oxidative stress induced by the presence of some kinds of nanoparticles.

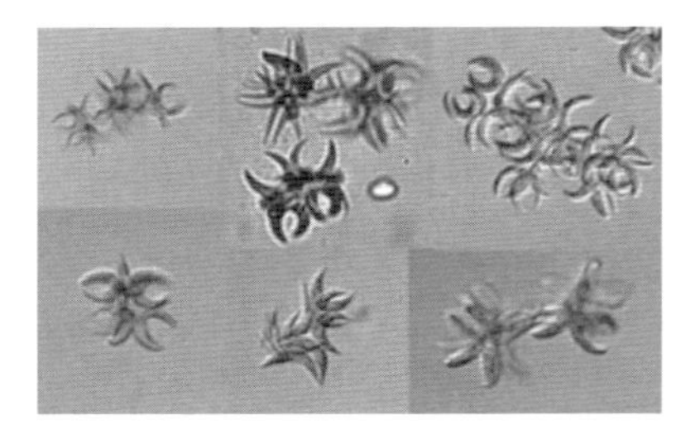

Fig. 16.2. The green alga *Pseudokirchneriella subcapitata* is a model traditionally used to assess the ecotoxicity of chemical substances for single-cell algae

For example, the effects of TiO_2 nanoparticles and CdTe quantum dots (QD) coated with thioglycolate have been studied by Wang et al. [26] on *Chlamydomonas reinhardtii* using various toxic effect criteria such as the number of cells and photosynthetic activity. These global effects were complemented by more specific measurements. The production of reactive oxygen species (ROS) by algal cells was investigated via a biochemical marker (induction of malondialdehyde) and by observing the expression of genes coding for the antioxidant enzymes superoxide dismutase (SOD), glutathione peroxidase (GPOx), and plastid terminal oxidase (PTOx). A 5 day exposure to suspensions of TiO_2 nanoparticles or QDs, present in the form of aggregates (median size 700–800 nm) in the algal culture medium, causes a significant alteration of algal growth at the highest concentrations, viz., 100 mg/L and 10 mg/L, respectively. The 50% inhibition concentrations (EC_{50}) for algal growth were 10 mg/L (TiO_2) and 5 mg/L (QD). It was established that this algal growth inhibition was in part caused by oxidative stress mechanisms occurring earlier on, with increased production of malondialdehyde and transient overexpression of genes for defensive enzyme activities from the first few hours of exposure.

Invertebrate Models

Most toxicity tests use daphnia (see, e.g., [7–9, 29, 30]) and standardised procedures [31] to carry out short term tests of variable length from 24 to 48 h, during which the survival rate of the organisms is measured (see Fig. 16.3).

Sublethal effects can also be sought. Such perturbations usually occur at lower exposure concentrations than those inducing mortality, and may signal more serious risks. Behavioural and physiological modifications affecting displacement, the beat rate of feeding appendages, and heart rate have been studied on daphnia exposed to nanoparticle suspensions, along with oxidative stress biomarkers [7, 30]. Some results have shown that, for 24 h exposure to fullerene suspensions at concentrations lower than 100 mg/L, there was no significant effect on the survival of adult *Daphnia pulex* [7]. However,

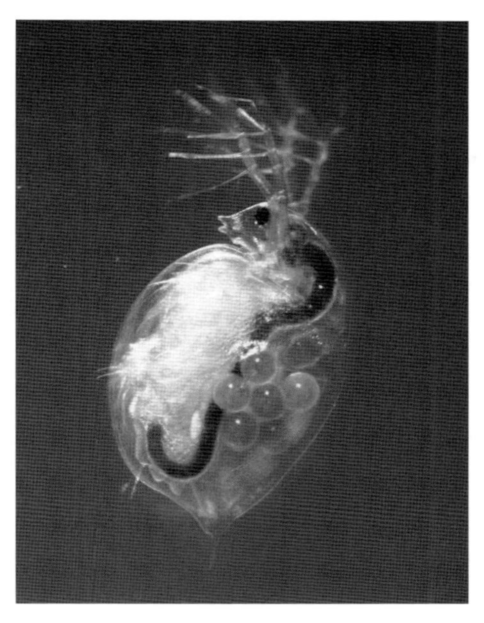

Fig. 16.3. The cladoceran crustacean *Daphnia magna* used to assess acute and chronic toxicity of chemical substances for aquatic invertebrates

effects on sublethal biochemical parameters (catalase and glutathione sulfotransferase or GST) were detected at lower concentrations than those inducing mortality. These early signals of perturbations to behaviour or metabolism confirm that studies need to be done to investigate the long term hazards of exposure to nanoparticles, and in particular their effects on survival, growth, and reproduction, these conditioning the dynamics of exposed populations.

Some studies have also investigated the ingestion and accumulation of nanoparticles in daphnia. Transmission electron microscope (TEM) observation has revealed the internalisation and accumulation of gold nanoparticles in the gastro-intestinal tract of *Daphnia magna*, exposed for 1 to 12 h to suspensions at sublethal concentrations, and the possible elimination of these particles as time goes by. Although there would not appear to be any obvious uptake by the microvilli in the tract, the internalisation of nanoparticles by daphnia nevertheless constitutes a potential source of contamination for their predators [32].

Other taxons have also been studied for the effects of nanoparticles on their development. The nematode *Caenorhabditis elegans*, exposed in an aqueous suspension, has thus been used to measure the effects of zinc oxide, aluminium oxide, and titanium dioxide nanoparticles [33]. This model is useful because lethal and sublethal effects can be reached with short exposure times (24 h).

Finally, original work has recently been done to study the uptake and depuration of single-wall and multiwall carbon nanotubes by benthic organisms. The natural sediments were enriched with ^{14}C-tagged carbon nanotubes. Accumulation by the oligochaete *Lumbriculus variegatus* and subsequent depuration of the nanoparticles were measured after 7, 14, and 28 days of exposure, in parallel with a similar study of the accumulation and depuration of pyrene, a polycyclic aromatic hydrocarbon (PAH) [3]. The results showed that the accumulation of carbon nanotubes in the organism was ten times less than the accumulation of pyrene, with a bioaccumulation factor (the equilibrium ratio of the concentration in the organism and the concentration in the exposure medium) less than unity. Moreover, the relative speed of elimination of nanotubes by the organism, compared with the elimination of pyrene, confirms the low uptake of nanotubes by the organism. The nanoparticles probably remain bound to the sediment in the gastro-intestinal tract and are then rapidly eliminated with the sediment particles when the organism is returned to a clean environment.

Similar studies have also been carried out on a benthic polychaete and an epibenthic amphipod [4] to investigate the possible effects of the presence of nanoparticles (carbon nanotubes) on the accumulation of hydrophobic organic contaminants (HOC), such as PAHs, polychlorobiphenyls (PCB), or polybrominated diphenyl ethers (PBDE). The natural sediments were enriched with different carbon sources (nanotubes or diesel soots). The joint addition of carbon nanoparticles and HOCs in enriched natural sediments significantly reduced the accumulation of HOCs by the polychaetes after 14 days of exposure, confirming the considerable adsorption capacity of carbon nanotubes for organic molecules. On the other hand, the presence of nanoparticles did not modify the uptake of HOCs by the copepod, perhaps due to differences in feeding habits and/or digestive capacity between the two taxons. In contrast to the effect of carbon nanotubes, the addition of diesel soots to the sediment led to an increase in the bioaccumulation of hydrophobic contaminants in the polychaete, without significantly changing the contamination of the amphipods.

These results illustrate the need to develop work on animal and plant models with different biological and ecological characteristics, in order to make an ecologically relevant and multispecific assessment of the impact of nanoparticles on ecosystems.

Fish Models

Embyronic, larval, and adult fish are the most widely used aquatic vertebrate models for investigating the effects of nanoparticles on individuals, and in particular lethal effects [12, 29, 34–36], but also for observing the accumulation of particles in tissues, or measuring sublethal parameters such as biochemical reactions, e.g., signatures of oxidative stress, tissue alterations, or gene expression [34–41].

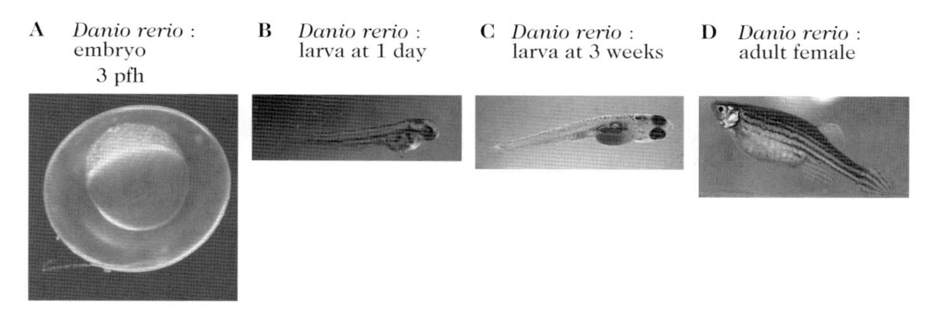

Fig. 16.4. The zebrafish (*Danio rerio*). Fish model for assessing nanoparticle toxicity. pfh post-fertilization hours

Nanoparticle toxicity mechanisms and effects can be analysed at different biological length scales (genes, cells, organisms) using the zebrafish (*Danio rerio*) (see Fig. 16.4), thanks to their transparent eggs and larvae, which are easy to manipulate and can be observed directly with the naked eye (for example, it is easy to remove the chorion, which partly protects the egg from the external medium, without raising embryonic mortality), but also the fast development of the embryo and the availability of known gene sequences. This model is also used as a vertebrate model in vivo to investigate the behaviour (real-time internal transport) and cell biocompatibility [38] of nanoparticle probes used in particular for cell imaging.

Exposures are carried out a few hours after fertilization and last for different lengths of time (24–120 h), depending on the effects under investigation (physiological, biochemical, or transcriptional). Eggs and larvae are incubated in natural or synthetic water contaminated by nanoparticle suspensions of varying concentrations. As far as possible and as for other types of bioassay, the nanoparticle suspension is usually cleared of any solvent by evaporation to limit the possible residual toxic effects of solvents after their use.

Changes in the usual characteristics of the fish, in particular the zebrafish, such as survival rates of embryos and larvae, embryo hatch time and rate, embryonic development (presence and type of caudal deformations, and so on) are also widely used criteria for characterising the effects of nanoparticles in bioassays, standardised or otherwise. Other effects may be observed, such as subcellular responses, enzyme activity (osmoregulation, oxidative stress), and gene expression.

Other work has used a trout model (rainbow trout) for short-term lethality tests [29] and more detailed assessment of physiological perturbations [41,42]. These studies investigate biochemical parameters relating to branchial and intestinal ion regulation (Na/K-ATPase), oxidative stress biomarkers (GST and catalase activities), and gill histology. They showed that 14 days of exposure of juvenile trout to a suspension of TiO_2 nanoparticles induced lesions in gill tissue and intestinal tissue, as well as stimulating antioxidant defence mechanisms.

Amphibian Models

Protocols developed to assess nanoparticle ecotoxicity for amphibians have been based on two species, the axolotl (*Ambystoma mexicanum*) and the xenopus (*Xenopus laevis*) (see Fig. 16.5). The first belongs to the urodeles, while the second is an anuran. These are species traditionally used in laboratory ecotoxicity tests [43–45].

The methods so far proposed for nanoparticle ecotoxicity assessments on amphibians adopt the procedures standardised by the *Association française de normalisation* in France [43, 44] or the International Standards Organization (ISO) [45], except that contamination is not achieved by periodically renewing the test medium, but rather the medium is kept the same throughout the test in such a way, according to the authors, as to provide a more realistic imitation of environmental exposure conditions [46, 47].

In this method, xenopus larvae at stage 50 of the development table or axolotl at the stage of indentation on the hind limb bud are exposed for 12 days [43, 44] to different nanoparticle concentrations in the test medium (distilled water plus $CaCl_2$, $MgSO_4$, $NaHCO_3$, and KCl). A negative control (test medium) and a positive control (cyclophosphamide) are carried out in parallel.

There are two types of toxicity effect criteria for amphibians, concerning acute toxicity and genotoxicity. Acute toxicity (mortality and abnormal development) is examined after 12 days' exposure of the organisms. The parameters characterising abnormal development are reduction or cessation of growth, reduced feeding, and altered mobility. Regarding genotoxicity, the parameter measured (appearance of micronuclei in circulating erythrocytes) indicates whether the substance studied is clastogenic (breaking chromosomes) or aneugenic (modifying the number of chromosomes) [46, 47].

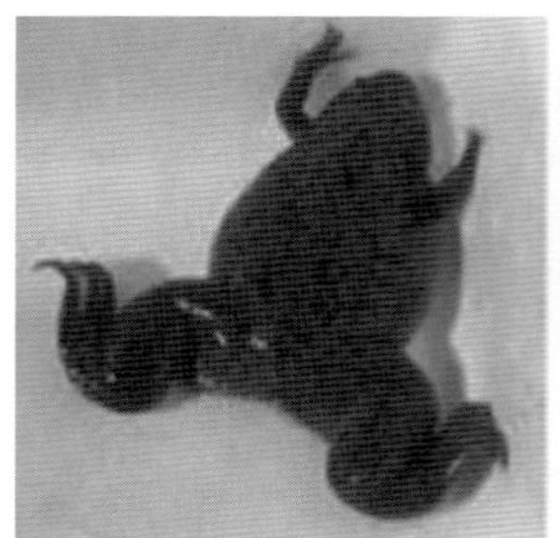

Fig. 16.5. *Xenopus laevis* and *Ambystoma mexicanum*, amphibians traditionally used to assess the genotoxic potential of chemical substances

16.4 Limitations of these Tests

The sometimes contradictory results obtained with ecotoxicity tests can be explained in part by the influence of experimental conditions. For example, the toxicity of fullerenes is affected by the composition of the test medium. The exposure of *Escherichia coli* and *Bacillus subtilis* to different fullerene concentrations under various environmental conditions leads to contradictory results [48, 49]: growth and respiration are inhibited in a minimal medium (Davis medium containing 10% of the recommended concentration of potassium phosphate and 1 g/L of glucose), whereas these parameters remain unaltered in a richer culture medium (Luria medium).

Furthermore, it is important to monitor the concentration of metal dissolved in the test medium, resulting from the solubilisation of particles in suspension, which can in some cases explain the observed toxicity [33]. One study monitored the dissolution of zinc in suspensions of zinc nanoparticles using dialysis, and showed that the concentration of metal dissolved in the test media was sufficient to explain the toxicity observed in algae (*Pseudokirchneriella subcapitata*) [25]. These results were confirmed by other groups [28].

Other work investigating the action of silver nanoparticles (Ag NP) on photosynthesis in *Chlamydomonas reinhardtii* also demonstrated that dissolved Ag^{2+} ions in the suspension play an important role in the algal response, resulting from the interaction of the Ag NPs with the algae [27].

Given that many nanoparticles tend to agglomerate, and that these agglomerates exhibit different bioavailability and toxic potential to non-agglomerated nanoparticles, combined with the fact that the physicochemical properties of these substances vary with the nature of the medium and its physicochemical properties, it is crucial to obtain a precise characterisation of the properties and behaviour of the nanoparticles in order to carry out ecotoxicity tests that are fully relevant for assessing the risks inherent in these compounds [50]. Indeed, any uncertainty regarding the state of the particles may lead to a wide range of different results. Due to the physicochemical properties of the nanoparticles, their toxic potential and bioavailability depend not only on their exposure concentration, but also on their number, surface area, and size [50].

For example, silver nanoparticles with average diameter 13 nm can cause root growth inhibition in higher plants, whereas particles with average diameters in the range 200–300 nm have no effect on this parameter [19]. Likewise, the toxic effect is maximal on *Daphnia magna* exposed to titanium dioxide nanoparticles of smaller diameters (25 nm) [51, 52].

16.5 Conclusion and Prospects

Considering the current development and use of nanoparticles, and given the likelihood of their being released into the environment, it is essential to obtain a better understanding of their acute and chronic ecotoxicity using a broad range of organisms on different trophic levels. These organisms must have biological characteristics (e.g., type of feeding, such as filter feeders, detritivores, or predators) and ecological characteristics (e.g., type of habitat, such as water, sediment, or soil) that are representative of the full range of organisms and their living environments.

Different exposure routes (respiratory and trophic) must be considered, in order to evaluate possible bioaccumulation and biomagnification processes. In addition, toxic effect criteria need to be extended to cover sublethal parameters, in particular biochemical ones, such as those considered to be important in mammals, especially oxidative stress. And as for other pollutants in the natural environment, exposure assessments should take into account the total exposure to this type of entity, whatever its chemical composition [53].

Particle synthesis is a key stage in organising reliable experimentation to assess the hazards in environmental matrices. Without clearly characterised protocols, it will always be a difficult matter to interpret the results of ecotoxicology studies, but also even to carry out routine bioassays with the possibility of interlaboratory comparisons [53].

In order to develop realistic exposure scenarios, the behaviour of nanoparticles must first be described in laboratory test media and in the aquatic environment (water and sediment) [12,54]. In particular, in the face of nanoparticle aggregation processes in natural environments and the possible deposition of these aggregates on substrates, benthic species must be used to assess the biological consequences.

References

1. Technical Guidance Document on Risk Assessment in support of Commission Directive 93/67/EEC on Risk Assessment for new notified substances, Commission Regulation (EC) No. 1488/94 on Risk Assessment for existing substances, Directive 98/8/EC of the European Parliament and of the Council concerning the placing of biocidal products on the market. Office for Official Publications of the European Communities, Luxembourg (2003)
2. E. Oberdöster: Manufactured nanomaterials (fullerenes, C_{60}) induce oxidative stress in the brain of juvenile largemouth bass. Environ. Health Perspect. **112**, 1058–1062 (2004)
3. E.J. Petersen, Q. Huang, W.J. Weber: Ecological uptake and depuration of carbon nanotubes by *Lumbriculus variegatus*. Environ. Health Perspect. **116**, 496–500 (2008)
4. P.L. Ferguson, G.T. Chandler, R.C. Templeton, A. DeMarco, W.A. Scrivens, B.A. Englehart: Influence of sediment amendment with single-walled carbon nanotubes and diesel soot on bio-accumulation of hydrophobic organic contaminants by benthic invertebrates. Environ. Sci. Technol. **42**, 3879–3884 (2008)
5. M. Crane, R.D. Handy, J. Garrod, R. Owen: Ecotoxicity test methods and environmental hazard assessment for engineered nanoparticles. Ecotoxicology **5**, 421–437 (2008)

6. T.C. King-Heiden, P.N. Wiecinski, A.N. Mangham, M. Metz Kevin, D. Nesbit, J.A. Pedersen, R.J. Hamers, W. Heideman, R.E. Peterson: Quantum dot nanotoxicity assessment using the zebrafish embryo. Environ. Sci. Technol. **43**, 1605–1611 (2009)
7. R. Klaper, J. Crago, J. Barr, D. Arndt, K. Setyowati, J. Chen: Toxicity biomarker expression in daphnids exposed to manufactured nanoparticles: Changes in toxicity with functionalization. Environmental Pollution **157**, 1152–1156 (2009)
8. E. Oberdöster, S. Zhu, T.M. Blickley, P. McClellan-Green, M.L. Haasch: Ecotoxicology of carbon-based engineered nanoparticles: Effects of fullerene (C_{60}) on aquatic organisms. Carbon **44**, 1112–1120 (2006)
9. S.B. Lovern, R. Klaper: *Daphnia magna* mortality when exposed to titanium dioxide and fullerene (C_{60}) nanoparticles. Environ. Toxicol. Chem. **25**, 1132–1137 (2006)
10. R.D. Handy, R. Owen, E. Valsami-Jones: The ecotoxicology of nanoparticles and nanomaterials: Current status, knowledge gaps, challenges and future needs. Ecotoxicology **17**, 315–325 (2008)
11. R.A. French, A.R. Jacobson, B. Kim, S.L. Isley, R.L. Penn, P.C. Baveye: Influence of ionic strength, pH, and cation valence on aggregation kinetics of titanium dioxide nanoparticles. Environ. Sci. Technol. **43**, 1354–1359 (2009)
12. R.J. Griffitt, J. Luo, J. Gao, J.C. Bonzongo, D.S. Barber: Effects of particle composition and species on toxicity of metallic nanomaterials in aquatic organisms. Environ. Toxicol. Chem. **27**, 1972–1978 (2008)
13. A. Johansen, A.L. Pedresen, K.A. Jensen, U. Karlson, B.M. Hansen, J.J. Scott-Fordsmand, A. Windind: Effects of C_{60} fullerene nanoparticles on soil bacteria and protozoans. Environ. Toxicol. Chem. **27**, 1895–1903 (2008)
14. Z. Tong, M. Bischoff, L. Nies, B. Applegate, R.F. Turco: Impact of fullerene (C_{60}) on a soil microbial community. Environ. Sci. Technol. **41**, 2985–2991 (2007)
15. R. Doshi, W. Braida, C. Christodoulatos, M. Wazne, G. O'Connor: Nano-aluminium: Transport through sand columns and environmental effects on plants and soil communities. Environ. Res. **106**, 296–303 (2008)
16. J.E. Canas, M. Long, S. Nations, R. Vadan, L. Dai, M. Luo, R. Ambikapathi, E.H. Lee, D. Olszyk: Effects of functionalized and nonfunctionalized single-walled carbon nanotubes on root elongation of selected crop species. Environ. Toxicol. Chem. **27**, 1922–1931 (2008)
17. W.M. Lee, Y.J. An, H. Yoon, H.S. Kweon: Toxicity and bioavailability of copper nanoparticles to the terrestrial plants mung bean (*Phaseolus radiatus*) and wheat (*Triticum aestivum*): Plant agar test for water insoluble nanoparticles. Environ. Toxicol. Chem. **27**, 1915–1921 (2008)
18. D. Lin, B. Xing: Phytotoxicity of nanoparticles: Inhibition of seed germination and root growth. Environmental Pollution **150**, 243–250 (2007)
19. L. Yang, D.J. Watts: Particle surface characteristics may play an important role in phytotoxicity of alumina nanoparticles. Toxicol. Lett. **158**, 122–132 (2005)
20. L. Zheng, F. Hong, S. Lu, C. Liu: Effect of nano-TiO_2 on strength of natural aged seeds and growth of spinach. Biol. Trace Element Res. **104**, 83–91 (2005)
21. OCDE: Ligne directrice pour les essais de produits chimiques: essai sur plante terrestre; essai d'émergence de plantules et de croissance de plantules, LD 208. OCDE, Paris, 27 p. (2006)

22. OCDE: Ligne directrice pour les essais de produits chimiques: essai sur plante terrestre; essai de vigueur végétative, LD 227. OCDE, Paris, 27 p. (2006)
23. A. Jemec, D. Drobne, M. Remskar, K. Sepcic, T. Tisler: Effects of ingested nano-sized titanium dioxide on terrestrial isopods (*Porcelio scaber*). Environ. Toxicol. Chem. **27**, 1904–1914 (2008)
24. S.J. Klaine, P.J. Alvarez, G.E. Batley, T.F. Fernandes, R.D. Handy, D.Y. Lyon, S. Mahendra, M.J. McLaughlin, J.R. Lead: Nanomaterials in the environment: Behavior, fate, bioavailability, and effects. Environ. Toxicol. Chem. **27**, 1825–1851 (2008)
25. N.M. Franklin, N.J. Rogers, S.C. Apte, G.E. Batley, G.E. Gadd,P.S. Casey: Comparative toxicity of nanoparticulate ZnO, bulk ZnO and $ZnCl_2$ to a freshwater microalga (*Pseudokirchneriella subcapitata*): The importance of particle solubility. Environ. Sci. Technol. **41**, 8484–8490 (2007)
26. J. Wang, X. Zhang, Y. Chen, M. Sommerfeld, Q. Hu: Toxicity assessment of manufactured nanomaterials using the unicellular green alga *Chlamydomonas reinhardtii*. Chemosphere **73**, 1121–1128 (2008)
27. E. Navarro, F. Piccapietra, B. Wagner, F. Marconi, R. Kaegi, N. Odzak, L. Sigg, R. Behra: Toxicity of silver nanoparticles to *Chlamydomonas reinhardtii*. Environ. Sci. Technol. **42**, 8959–8964 (2008)
28. V. Aruoja, H.C. Dubourguier, K. Kasemets, A. Kahru: Toxicity of nanoparticles of CuO, ZnO and TiO_2 to microalgae *Pseudokirchneriella subcapitata*. Science of the Total Environment **407**, 1461–1468 (2009)
29. D.B. Warheit, R.A. Hoke, C. Finlay, E.M. Donner, K.L. Reed, C.M. Sayes: Development of a base set of toxicity tests using ultrafine TiO_2 particles as a component of nanoparticle risk management. Toxicol. Lett. **171**, 99 (2007)
30. S.B. Lovern, J.R. Strickler, R. Klaper: Behavioral and physiological changes in *Daphnia magna* when exposed to nanoparticle suspensions (titanium dioxide, nano-C_{60}, and $C_{60}H_xC_{70}H_x$). Environ. Sci. Technol. **41**, 4465–4470 (2007)
31. OCDE: Ligne directrice pour les essais de produits chimiques: *Daphnia sp.*, essai d'immobilisation immédiate et essai de reproduction, LD 202. OCDE, Paris, 17 p. (1984)
32. S.B. Lovern, H.A. Owen, R. Klaper: Electron microscopy of gold nanoparticle intake in the gut of *Daphnia magna*. Nanotoxicology **2**, 43–48 (2008)
33. H. Wang, R.L. Wick, B. Xing: Toxicity of nanoparticulate and bulk ZnO, Al_2O_3 and TiO_2 to the nematode *Caenorhabditis elegans*. Environmental Pollution **157**, 1171–1177 (2009)
34. R.J. Griffitt, R. Weil, K. Hyndman, N.D. Denslow, K. Powers, D. Taylor, D.S. Barber: Exposure to copper nanoparticles causes gill injury and acute lethality in zebrafish (*Danio rerio*). Environ. Sci. Technol. **23**, 8178–8186 (2007)
35. X.S. Zhu, L. Zhu, Y. Li, Z.H. Duan, W. Chen, P.J. Alvarez: Developmental toxicity in zebrafish (*Danio rerio*) embryos after exposure to manufactured nanomaterials: Buckminsterfullerene aggregates (nC_{60}) and fullerol. Environ. Toxicol. Chem. **26**, 976–979 (2007)
36. X.S. Zhu, L. Zhu, Z. Duan, R. Qi, Y. Li, Y. Lang: Comparative toxicity of several metal oxide nanoparticle aqueous suspensions to zebrafish (*Danio rerio*) early developmental stage. Journal of Environmental Science and Health Part A **43**, 278–284 (2008)
37. S. Kashiwada: Distribution of nanoparticles in the see-through medaka (*Oryzias latipes*). Environ. Health Perspect. **114**, 1697–1702 (2006)

38. K.J. Lee, P.D. Nallathamby, L.M. Browning, C.J. Osgood, X.H.N. Xu: In vivo imaging of transport and biocompatibility of single silver nanoparticles in early development of zebrafish embryo. ACSNAno **1**, 133–143 (2007)
39. C.Y. Usenko, S.L. Harper, R.L. Tanguay: Fullerene C_{60} exposure elicits an oxidative stress response in embryonic zebrafish. Toxicology and Applied Pharmacology **229**, 44–55 (2008)
40. J. Cheng, E. Flahaut, S.H. Cheng: Effect of carbon nanotubes on developing zebrafish (*Danio Rerio*) embryos. Environ. Toxicol. Chem. **26**, 708–716 (2007)
41. G. Federici, B.J. Shaw, R.D. Handy: Toxicity of titanium dioxide nanoparticles to rainbow trout (*Oncorhynchus mykiss*): Gill injury, oxidative stress, and other physiological effects. Aquatic Toxicology **4**, 415 (2007)
42. C.J. Smith, B.J. Shaw, R.D. Handy: Toxicity of single walled carbon nanotubes on rainbow trout (*Oncorhynchus mykiss*): Respiratory, organ pathologies, and other physiological effects. Aquatic Toxicology **82**, 94–109 (2007)
43. AFNOR: Norme NFT 90-325 Essais des eaux. Détection en milieu aquatique de la génotoxicité d'une substance vis-à-vis des larves de batraciens (*Pleurodeles waltl* and *Ambystoma mexicanum*). Essais des micronoyaux. AFNOR Monograph, Paris, 12 p. (1987)
44. AFNOR: Norme 90-325. Qualité de l'eau. Evaluation de la génotoxicité au moyen des larves d'amphibien (*Xenopus laevis, Pleurodele waltl*). AFNOR, Paris, 17 p. (2000)
45. ISO: Water quality. Evaluation of genotoxicity by measurement of the induction of micronuclei. Part 1. Evaluation of genotoxicity using amphibian larvae. ISO 21427-1, ICS: 13.060.70, Geneva, 15 p. (2006)
46. F. Mouchet, P. Landois, E. Flahaut, E. Pinelli, L. Gauthier: Assessment of the potential in vivo ecotoxicity of double-walled carbon nanotubes (DWNTs) in water, using the amphibian *Ambystoma mexicanum*. Nanotoxicology **1**, 149–156 (2007)
47. F. Mouchet, P. Landois, E. Sarremejean, G. Bernard, P. Puech, E. Pinelli, E. Flahaut, L. Gauthier: Characterisation and in vivo ecotoxicity evaluation of double-wall carbon nanotubes in larvae of the amphibian *Xenopus laevis*. Aquatic Toxicology **87**, 127–137 (2008)
48. J.D. Fortner, D.Y. Lyon, C.M. Sayes, A.M. Boyd, J.C. Falkner, E.M. Hotze, L.B. Alemany, Y.J. Tao, W. Guo, K.D. Ausman, V.L. Colvin, J.B. Huges: C_{60} in water: Nanocrystal formation and microbial response. Environ. Sci. Technol. **39**, 4307–4316 (2005)
49. D.Y. Lyon, J.D. Fortner, C.M. Sayes, V.L. Colvin, J.B. Huges: Bacterial cell association and antimicrobial activity of a C_{60} water suspension. Environ. Sci. Technol. **24**, 2757–2762 (2005)
50. Scientific committee on emerging and newly-identified health risks (SCENIHR): Opinion on the appropriateness of the risk assessment methodology in accordance with the technical guidance documents for new and existing substances for assessing the risks of nanomaterials. European Commission, Health & Consumer Protection, 68 p. (2007)
51. K. Hund-Rinke, S. Markus: Ecotoxic effect of photocatalytic active nanoparticles (TiO_2) on algae and daphnids. Environ. Sci. Poll. Res. **13**, 225–232 (2006)
52. A. Jemec, D. Drobne, M. Remskar, K. Sepcic, T. Tisler: Effects of ingested nano-sized titanium dioxide on terrestrial isopods (*Porcelio scaber*). Environ. Toxicol. Chem. **27**, 1904–1914 (2008)

53. G. Oberdöster, V. Stone, K. Donaldson: Toxicology of nanoparticles: A historical perspective. Nanotoxicology **1**, 2–25 (2007)
54. P. Christian, F. Von der Kammer, M. Baalousha, T.H. Hofmann: Nanoparticles: Structure, properties, preparation and behaviour in environmental media. Ecotoxicology **17**, 326–343 (2008)

17

Life Cycle Models and Risk Assessment

Jérôme Labille, Christine O. Hendren, Armand Masion,
and Mark R. Wiesner

Nanomaterials are incorporated into more and more products. There can be little doubt that they will end up in the natural environment, by different pathways and at different stages right through their life cycle. In this respect, they do not differ from other manufactured substances. However, nanomaterials, that is, objects with at least one dimension measuring less than 100 nm, are likely to display novel characteristics and behaviour due to their small size. And as the size of these particles decreases, so the ratio of their surface area to volume increases, thereby altering fundamental characteristics such as reactivity and magnetic and/or optical properties. Indeed, it is precisely these modifications that make nanotechnology so promising. They can result in useful features, such as increased physical strength, better electron transport, or better control of the response to an incident energy in terms of colour or photoreactivity. Many of these novel properties that make nanomaterials so promising will be retained right through their life cycle, and may therefore induce new responses from organisms and the environment.

This chapter will describe the state of the art in the study of the life cycles of nanomaterials and their derivatives in the environment. In the first instance, this will be grounded upon the huge knowledge base and acquired experience built up from previous studies of the life cycles of well established conventional products such as chemicals, coarse materials, and so on, asking each time to what extent they might apply to the special case of nanomaterials. Then, in response to the limitations of those earlier studies, we shall put forward several ideas for experimental methods or models, emphasising what is still lacking for a good understanding of the overall risks due to nanomaterials throughout their life cycle.

17.1 Potential and Risks of Nanotechnologies

Every year many new applications of nanotechnology come to light, continually pushing back the frontiers of efficiency and performance by exploiting the novel features and behaviour of these highly reactive materials. From

P. Houdy et al. (eds.), *Nanoethics and Nanotoxicology*,

DOI 10.1007/978-3-642-20177-6_17,

industrial uses, through decontamination processes, to products for the general public, everyone agrees today that nanotechnologies have the potential to revolutionise many aspects of our technological world. However, given that the benefits of these new materials will probably come along with just as revolutionary effects on and unexpected reactions with environment, it is essential to conduct research on the impacts of these materials in parallel with research on their applications. A balance must be found between the potential for major innovations and the potential for environmental damage. And to achieve this, an adequate assessment must be made of the risks for the environment and human health.

In this context, it is interesting to recall several examples from previous experience where, carried away by enthusiasm for their discovery and a genuine belief in the benefits they could bring, several new products ended up having serious repercussions on the environment. For example, the powerful insecticide dichloro-diphenyl-trichloroethane (DDT) promised to rid the planet of the insect vectors of malaria and typhus, thereby significantly reducing the number of people afflicted by these diseases. However, its widespread use before obtaining a full understanding of its impacts resulted in a cascade of negative effects on the environment. DDT was condemned for ever as a major error in the chemical revolution. In the same way, methyl terbutyl ether (MTBE) was widely used to improve the octane index in fuels, in order to improve combustion and thereby reduce particle emissions. But in the United States, leaks from storage tanks into the water table made huge amounts of water improper for human consumption, owing to its strong and unpleasant taste. Its resistance to treatment and the broad geographical extent of the contamination made this fuel additive a very costly mistake indeed, even several years after it had been banned in the United States. Finally, asbestos, whose fireproofing properties were supposed to make buildings safer, turned out to be extremely dangerous when inhaled, resulting in many deaths, and some of the most expensive decontamination projects in history.

Risk can be viewed as the product of the magnitude of the negative consequences of an event and some measure of the degree of exposure to this event. In this way, the hazard is combined with the exposure to define the risk due to a given substance or process [1]. In general, hazards are identified, then characterised, exposure routes and levels of exposure are evaluated, and the two are somehow quantified. These two risk determinants are then combined to characterise a relative risk [1, 2]:

$$\text{risk} = \text{hazard} \times \text{exposure} \, .$$

This chapter is mainly concerned with the mathematical evaluation of risk. However, while examining the subject of risk assessment during the life cycle of nanomaterials, it is important to bear in mind the role of risk perception and public opinion with regard to decision-making processes.

17.1.1 Hazards

For nanomaterials, the hazards are mainly associated with toxicity and effects on the environment. This may be toxicity for bacteria, protozoa, or any other animal in the trophic chain, up to and including effects on human health. Other hazards, or effects, may be environmental changes induced by the presence of nanomaterials or their byproducts, such as interruption of the nutrient cycle. Today, these potential impacts are more and more often studied on the laboratory scale, where scientists try to determine the effects and mechanisms involved, and also to understand the dose–response relationships for a large panel of biological targets [3].

This kind of information informs as to the magnitude of the negative consequences of nanomaterials entering into contact with the environment or with human beings. However, taken alone, the toxicity of a nanomaterial cannot tell us about the global risk it represents. Although a tidal wave is extremely dangerous, the risk associated with this type of disaster is very low for people living hundreds of kilometers inland. This is why the other factor in the definition of risk is exposure, which tells us who and what will be likely to enter into contact with the nanomaterial.

17.1.2 Exposure

It is a major task to understand the ways in which a nanomaterial can enter into contact with the environment and with human beings. The amounts and the forms in which they are released into the environment must first be established. The physicochemical characteristics of the nanomaterial must be determined, along with its alterations, coatings, functionalisations, and matrices within which it is released. The transformations and interactions that modify these materials, along with the kinds of intermediate phases and their effects, must also be identified. These and many other factors controlling the fate of nanomaterials in the environment determine not only the hazards, but who and what will run the risk of being exposed to them.

A full determination of all the exposure routes may look like a formidable task. In order to achieve it, risk specialists break the problem down into smaller elements. Monitoring a nanomaterial right through its life cycle provides a systematic method for elucidating the potential exposure routes. Figure 17.1 represents the global risk from the life cycle angle, illustrating the different stages in the life cycle of a nanomaterial, the exposure routes taking the nanomaterial into the various environmental compartments in each stage, and the impacts these materials may then have.

There is an urgent need to assess the risks due to nanomaterials. Even now, the world of industry and the general public are forging their opinions on this matter, whether they are based on an adequate assessment or not [4]. Although there are significant gaps in our ability to characterise

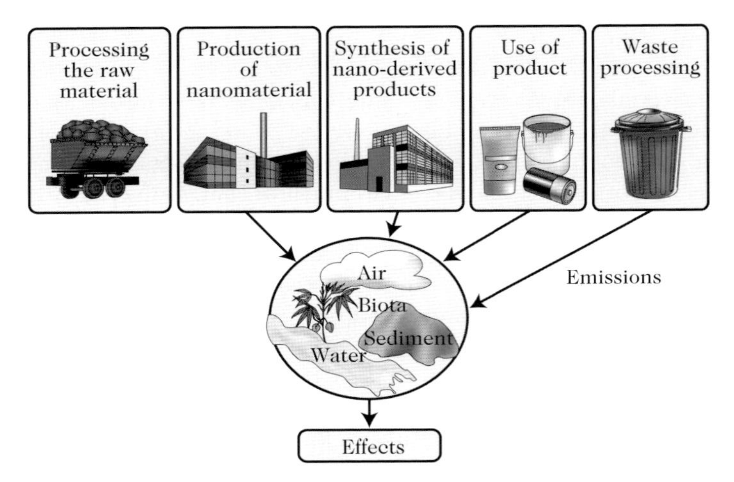

Fig. 17.1. Exposure risks during the life cycle of a nanomaterial. This can be visualised as a sequence of stages, from extraction of the basic raw material for synthesising the nanomaterial right up to waste management at the end of its life, going through each stage of production, synthesis, and use

these risks [5, 6], a lack of information will not prevent stakeholders from introducing measures to define future orientations of what they consider to be the least risky strategies, whether they are right or wrong about the matter. An editorial of the Marine Pollution Bulletin in 2005 [7] summed up the importance of having research scientists participate in the decision-making process. According to this source, environmental research scientists need urgently to meet the decision-makers to discuss the impacts of the manufacture and use of nanomaterials. If they do not, it is almost certain that the media will influence the understanding and trust of the public with regard to nanotechnologies, spreading unjustified stories and unreasonable conjectures, which may then be used as the basis for an ill-informed debate whose consequences could be out of all proportion with regard to the actual risks.

17.2 Monitoring Nanomaterials Throughout Their Life Cycle to Predict Emissions into the Environment

It is important to take into account all exposure routes represented by the various stages in the life cycle of a nanomaterial: extraction and processing of the raw material, manufacture of the nanomaterial, manufacture of the product based on the nanomaterial, use of the product, and disposal or recycling of the product when it reaches the end of its useful life. In each of these stages, the nanomaterial, its constituents, or its byproducts may be released into the

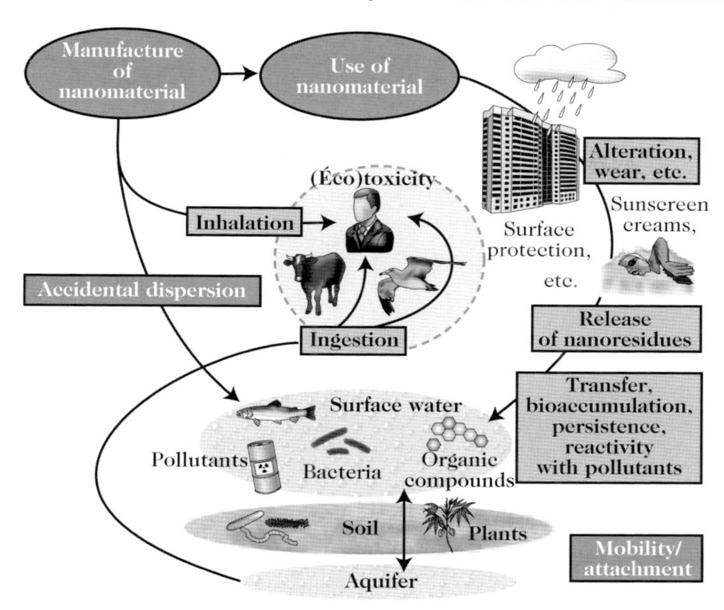

Fig. 17.2. Life cycle of a nanomaterial in the different environmental reservoirs. Nanomaterials rejected into the environment accidentally, professionally, or in the form of altered residues following their normal use distribute themselves between the aquifer, the soil, surface waters, and organisms living on land or in the sea

environment in different forms, and by different mechanisms (see Fig. 17.2). The emission scenarios must be studied systematically in order to identify and if possible quantify the substances released into the environment. Table 17.1 presents some examples of the variety of substances and exposure routes that must be taken into account at each stage of the life cycle, in order to assess all possible forms of exposure.

When we examine this table, it becomes clear that the kind of nanomaterial is only the starting point on a long list of related substances and exposure routes by which the material in question may eventually have an impact. However, this way of thinking does provide a list of potential forms of exposure, although it does not specify which routes are the most likely. In practice, it would be difficult to take into account each gram of nanomaterial, its precursors, and its byproducts. On the other hand, detailed data regarding the amounts produced, e.g., the number of kilos of nanometric silver produced annually in Europe, and the likely fate of a material, e.g., the percentage of sewage sludge incinerated, are important data for filtering and classifying the most significant materials and exposure routes that need to be monitored. Once these emission scenarios have been outlined, they constitute a good starting point, but it is only a starting point when investigating the interactions between a material and the environment.

Table 17.1. Nanomaterial exposure routes during their life cycle, depending on their properties and original function

Emitted phase	Airborne particle Aqueous phase Solid phase
State of substance	Raw material Byproduct Nanomaterial Nanomaterial incorporated in product: • functionalised • coated • in solution • in a solid matrix
Uses of the product (may inform as to possible exposure mechanisms)	Fixed use indoors Dispersed use indoors Fixed use outdoors Dispersed use outdoors Mobile consumer product
Emission mechanism	Abrasion, seepage Waste water effluent Sewage sludge Volatization by incineration Lixiviation from rubbish dump

17.3 Nanomaterial Interactions with the Environment

By taking into consideration all possible ways nanomaterials or their constituents may be released into the environment during their life cycle, it becomes possible to build up a better understanding of, or even predict, the ways in which they are likely to interact with the natural environment. The intrinsic physicochemical properties of the nanomaterial, along with many environmental parameters such as the pH, the ionic strength, the redox potential, the wind, or hydrodynamic flows, to mention only the most important, determine the fate of nanoparticles in the environment, that is, their transport and the ways in which they interact and change, or fractionate within the various media they encounter. By studying these developments, we can obtain a better comprehension of the ways in which nanoparticles may come into contact with living organisms, and of the consequences that may ensue. This is a multifaceted and complex task which may be approached from different standpoints.

17.3.1 Models for Estimating Existing Risks

Some models for estimating and building up a hierarchy of risks are based on simple algorithms, taking into account a few key parameters describing the substance in order to calculate a relative classification of the various types of risk, such as the risk of explosion or the risk of persistence in water. Such models are used by insurance companies to calculate their prices, for example [8]. Other more accurate approaches are based on more detailed models, e.g., considering multiple transport in the oil program of the US Environmental Protection Agency, run by the Mineral Management Service (MMS), to monitor the fate and transport of non-nanometric materials [9]. The latter also depends on the input parameters describing the characteristics of the materials in question and the surrounding medium, but it is based on more detailed mathematical relations in order to monitor all the transformations and associations that characterise the fate of the material as it transfers through the various environmental compartments.

Whichever model is used from the two described above, whether its predictions refer to the relative risk or the contaminant concentration in some environmental reservoir, its accuracy will always depend on the veracity of the input parameters.

17.3.2 The Situation for Nanomaterials

In the case of non-nanometric, homogeneous materials, the mass concentration of the byproducts released can be used to predict the scale of the consequences in terms of reactivity and interactions. But with a nanomaterial of the same composition, the release of byproducts per unit mass is likely to be greater owing to the higher specific surface area. The same model would make the incorrect prediction that the two materials would behave in the same way, unless some further characteristic were input to better account for the specific features of the nanomaterial, such as the number of particles or the specific surface area.

When the behaviour of a material is modified by some nanoscale effect, specific experimental research is required to better characterise the intrinsic product, its reactivity, and its behaviour. A great deal of scientific effort is currently being expended on precisely this point (see Chap. II). Measurements of nanoparticle aggregation and surface charge can be used to predict the stability of their dispersion, and how likely they are to aggregate or to associate with other components in suspension, and hence modify their availability for other reactions or internalisation [10–14].

For example, nanoparticles in suspension are potentially bioaccessible to pelagic flora and fauna, but when environmental conditions favour the aggregation of nanoparticles, the latter are no longer stable in suspension and tend to sediment out. However, this does not mean that the process of transfer between the various environmental compartments has reached an end, because benthic organisms are then likely to take over from pelagic organisms, and these in turn will play their role in the global life cycle of the nanoparticles, for instance, by internalising or transforming them. In this context, investigations of the deposition of metal- or carbon-based nanomaterials can provide a better understanding of their affinity for sediments, minerals, or the biotope [15–17].

Furthermore, in parallel with this succession of environmental or biological repositories, there will be transfer of internalised nanoparticles through the food chain. This will very likely lead to bioaccumulation of nanoparticles, something that needs to be taken into account when estimating the risks due to chronic toxicity.

On the other hand, certain general scientific models for the behaviour of materials can sometimes be applied to nanomaterials. When this is the case, it is important to take advantage of already acquired data, thereby saving resources and time in the overall approach to estimating the final risk [18]. For example, for silver nanoparticles, part of their toxicity comes from the dissolved ionic form of silver. But the fact that these silver ions come from nanometric particles does not make them different from other silver ions, and since there are already many publications dealing with the behaviour of these ions [19, 20], this literature should be exploited whenever appropriate, and integrated into the panoply of tools for predicting the impact of silver nanoparticles.

17.3.3 Degradation of Nanomaterials

In their everyday use, some products incorporating nanomaterials already on the market are more prone to degradation or alteration than others, e.g., tyres, paints, cosmetics, window glasses, cements, textiles, surface coatings, etc. On the other hand, for other such products designed for short term use, e.g., in electronics or new technologies, it is the unsuitable storage of waste materials that can result in deterioration. In every case, the degradation of nanomaterials as time goes by may lead to the introduction of adventitious nanometric byproducts into the environment, sometimes in considerable quantities.

Physicochemical surface properties of nanomaterials play the dominant role in their fate, reactivity, and toxicity within the environment. But in fact, the nanomaterials implemented in existing industrial processes are usually functionalised in order to bestow them with the new surface properties required to incorporate them in the finished commercial product. This is why the evolution of the nanomaterial coating during the life cycle of the product must be taken into consideration in order to make realistic assessments. The size and surface properties of adventitious byproducts depend on the physicochemical process and the degree of alteration of the original material. For example, depending on whether the smallest such byproducts are elementary nanoparticles or micrometric fragments of the finished product containing thousands of nanoparticles imprisoned in a matrix, their surface properties and hence their fate and ecotoxicity will be totally different.

In this context of indirect exposure to the biomass, the investigation of risk must necessarily begin with an understanding of degradation, dispersion, and absorption by organisms.

17.3.4 Compiling Data

The problem here is to seek out those physicochemical parameters and environmental conditions that can serve as the best indicators of the impact of the nanomaterial. This goes hand in hand with the study of the relationships between these different parameters. When the data on aggregation, transport, attachment, degradation, and dissolution have been acquired, there still remains the difficult task of exploiting them to draw conclusions about the life cycle of the material. To achieve this, the data must be processed in one way rather than another. Relative classification models allow us to build up a hierarchy of risks in boxes assigned with distinct values, while multiple transport models can be used to calculate and predict the fate of the material over a long period of time.

17.4 Characterising the Hazards of Nanotechnologies: Toxicity

Once the various known scenarios and the behaviour in each compartment of the life cycle have been sized up, all this information about exposure and the parameters taking into account the relevance of the given hazard must be put together. To achieve this, toxicological studies provide ways of putting these considerations into perspective and making a global estimate of the risk.

Regarding nanoparticle toxicology studies, most effort is expended on very specific interactions between well defined nanomaterials and well characterised biological models. This kind of research has generated a huge amount of toxicity data in the form of dose–response relationships [18]. However, there are still major gaps in our understanding of the mechanisms underlying these effects. A good review of the state of our knowledge of nanomaterial toxicity as of 2009 can be found in Chap. 14. The data resulting from early studies are highly contradictory, but it can at least be concluded that the toxicity of nanoparticles depends to a large extent on their surface chemistry (degree of oxidation for metal oxides, biocompatibility of coatings, etc.), whereas size controls mainly the mobility and bioaccessibility of the nanoparticles.

In the end, toxicity mechanisms must be properly assimilated if there is to be any hope of predicting the effects of a nanomaterial purely on the basis of its chemical and morphological characteristics, without having recourse to a whole series of controlled biological tests on different targets for each new material, which would be inconceivable. In the meantime, research on exposure to and toxicity of nanomaterials must attempt to pinpoint certain products or environments that can be considered as potentially more risky, and thus favour further, more relevant investigations.

17.5 Present State of Knowledge Regarding Nano Risk

The final aim is to integrate exposure and toxicity data together into a consistent estimate of the overall risk. However, while the first notions have already been obtained on the basis of a great deal of work using different approaches, much remains to be done before we can make precise and rigorous assessments of the risks due to nanomaterials.

An early approach aiming to study the environmental impact of nanomaterials focused on collateral damage related to manufacture, corresponding to the first two stages in Fig. 17.1. Using a simple insurance company algorithm applied to 1 200 manufacturing processes for non-nanomaterials, five nanomaterial production processes were classified according to their own relative risks [8]. The nanomaterials produced were themselves excluded from the study, partly to be able to focus on pollution and environmental risks induced by the fabrication process, and partly because the algorithm was unable to account for the nanoscale effects of the nanomaterial as compared with those of a coarser material (as discussed earlier). The results of the study are shown in Fig. 17.3. It transpires that the nanomaterial production processes considered (I) exhibit equivalent or lower levels of risk on the whole as compared with other production processes for well established chemical products (II).

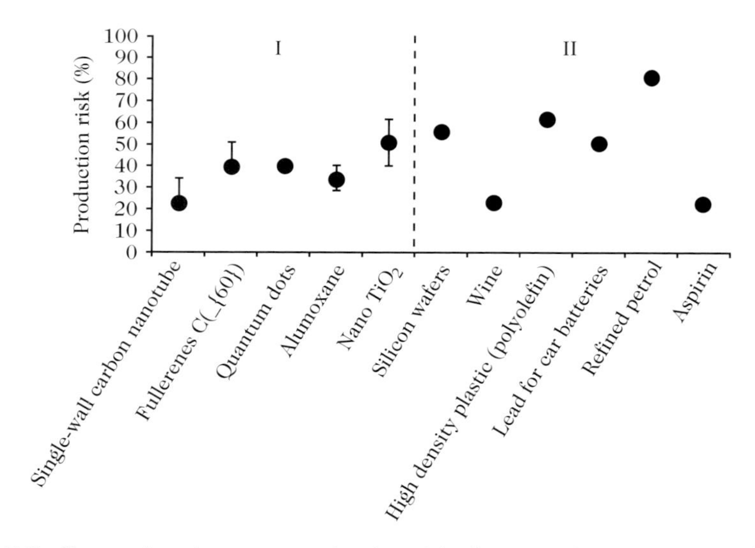

Fig. 17.3. Comparison between production risks for several nanomaterials (I) and for some well-established standard products (II). Adapted from [8]

A second way to represent the risks due to nanomaterials is to take into account the characteristics of the nanomaterial in the model to generate an influence diagram which accounts for cause and effect relationships between the parameters and the respective relations which link them together. This is a preliminary method, providing a structural organisation of the way the risk is viewed. The enormous amount of data acquired, which is difficult to synthesise, can then be reduced to a shorter list by updating this data and by clarifying the mathematical relationships governing their interconnections. A study by Morgan (2005) [21] was based on the acquisition of expert data, a method for interviewing research professionals, and the recording of their opinions on the interparameter relations to create just such a structural organisation in the form of interdependent influence diagrams. Figure 17.4 provides an illustration on several levels of the risks due to nanomaterials, obtained using this approach.

This kind of diagrams and their more detailed subdiagrams seek to elucidate the mathematical relationships between the parameters. These can then be used to predict the fate and transport of nanoparticles in risk models. They can also serve as a concrete foundation for supporting and guiding discussions and decisions relating to the risks of nanoproducts. In particular, being in possession of such a mass of data, it will be possible to guide immediate scientific research toward the most crucial issues, and also to specify standard procedures within the given scientific community.

A third approach has also been widely followed. It consists in characterising and quantifying the potential emissions of specific nanomaterials in well defined systems. This is used to make a first predictive estimate of the concentrations of nanomaterials one might expect to find in the environment. Indeed, by combining production data with environmental partitioning models, they can predict the scale and destination of emissions into the environment [22,23].

For example, across the territory of Switzerland, Mueller and Nowack [22] considered three nanomaterials and were able to follow the products through

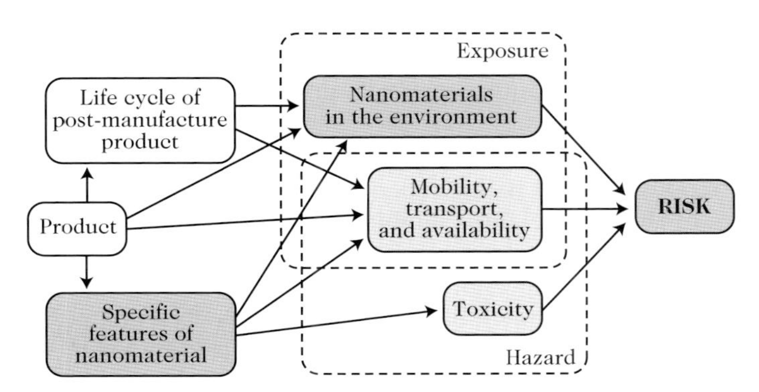

Fig. 17.4. Interdependent influence diagram for the parameters controlling exposure and toxicity of nanomaterials in a risk assessment. Adapted from [21]

the various stages of their life cycle and produce arguments concerning the potential exposure vectors. Production emissions were monitored through the municipal waste disposal and water treatment processes up to their final discharge into the environment. Although this represents a very limited amount of data and a distinctly preliminary view of the global risk, it is a necessary contribution to the general study of ways of monitoring nanomaterials and their life cycles in order to predict their impact on the environment and human health.

17.6 Titanium Dioxide Nanoparticles in Sunscreen Creams

Any study of the life cycle of a nanomaterial must necessarily take into account all the intrinsic and environmental parameters that determine its fate and its toxicity. This requires a well-defined, rigorous, and cross-disciplinary approach. Below we outline an approach that could serve as the basis for a wide range of nanomaterials. The example discussed and illustrated here is the life cycle of products incorporating nanometric titanium dioxide of the kind now used in most sunscreen creams with a high index of protection [24, 25].

The mineral form of titanium dioxide (TiO_2) exhibits specific UV photosensitivity and absorption characteristics which makes it extremely useful for many industrial applications, such as photocatalysis or sunscreen creams. Derivative products of TiO_2-based nanomaterials are currently flooding the cosmetics market (sun creams), but they are also commonly used in self-cleaning surfaces (window glass, cements, etc.). However, the classification of this mineral in class 2B of potentially carcinogenic products by the International Agency for Research on Cancer (IARC) in 2006 [26] has led to controversy. It means that it will be necessary to reexamine its safety of use, in particular in nanoparticulate form, since this involves higher reactivity for an equivalent mass. Sunscreen creams using rutile TiO_2 as a UV filter are particularly relevant here, since most incorporate this mineral in nanoparticulate form for aesthetic reasons. Indeed, nanometric TiO_2 can be used to formulate colourless creams, favouring more frequent cutaneous application by the consumer and therefore optimising its efficiency.

The TiO_2 nanoparticles used to formulate sunscreen creams are always coated to modify some of their properties. A film of aluminium hydroxide at the surface of the nanoparticles inhibits the photocatalytic properties of the TiO_2 by preventing the release of radical oxygen species during activation by UV radiation. The dispersion of these hydrophilic inorganic nanocomposites in the lipophilic organic phase of the cream is then enabled by coating with a surfactant dispersing agent such as stearic acid, polydimethylsiloxane (PDMS), dimethicone, etc. (see Fig. 17.5).

The aging of this nanomaterial integrated into the sun cream when it is subjected to various sources of deterioration during its life cycle (sweat,

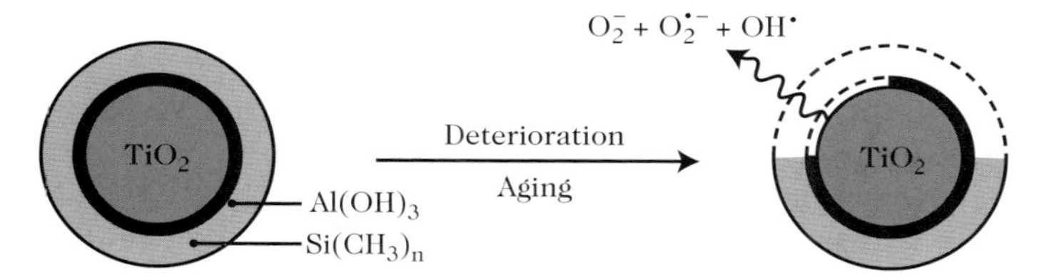

Fig. 17.5. Formulation of the TiO_2-based nanocomposite, coated with $Al(OH)_3$ and PDMS, and used as a UV absorber in sunscreen creams. Differing degrees of deterioration of its coating depending on environmental conditions. The extreme case is a total stripping of the nanoparticle coatings with direct exposure of TiO_2 at the surface, which would favour the production of radical oxygen species

soap, sea water, shaking, etc.) is likely to degrade all or part of its coating (see Fig. 17.5). To begin with, degradation of the lipophilic dispersant would facilitate a stable dispersion of the nanoparticles in the aqueous phase, hence favouring its transfer into various environmental compartments. In the case of more serious deterioration, the aluminium hydroxide film might dissolve in places, exposing mineral TiO_2 at the surface of the nanomaterial. This would strongly favour the production of radical oxygen species under UV exposure, and might have serious consequences for living organisms coming into contact with the product.

17.6.1 Protocol for Laboratory Reconstitution of Deterioration

In order to estimate the various stages of deterioration, fractioning, and internalisation during the life cycle of such a nanomaterial, aging conditions must be reconstituted in the laboratory and the adventitious byproducts investigated. A protocol was followed for simulating aging by shaking the nanomaterial under controlled conditions of light, acidity, salinity, and time (see Fig. 17.6) [24, 25]. Indeed sunscreen creams constitute a ductile nanomaterial, designed for prolonged immersion, and the protocol for deterioration by shaking seems the most appropriate. On the other hand, for most other commercialised nanomaterials, alternative protocols may be better suited, using an environmental test chamber and/or simulating specific abrasion conditions, for example.

The visual changes observed in the behaviour of the nanomaterial during the aging process suggest a certain degree of alteration of its coating. Originally hydrophobic due to its lipophilic dispersing agent, the nanomaterial does not wet and is retained at the surface of the aqueous phase (see Fig. 17.6A). However, after just 30 min of shaking, the product begins to disperse in the water, indicating a likely deterioration of the lipophilic agent (see Fig. 17.6B). When shaking is stopped, the nanomaterial forms three fractions with different behaviour (see Fig. 17.6C): a hydrophobic layer remains at the surface of

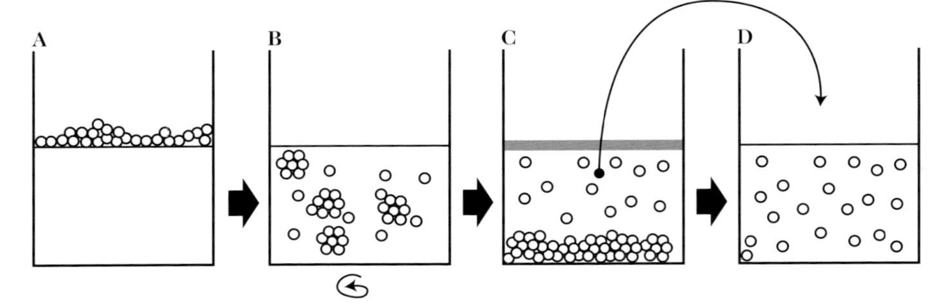

Fig. 17.6. Nanocomposite deterioration protocol reconstituted in the laboratory. (**A**) 100 mg of nanocomposite powder +250 mL of water. (**B**) The system is shaken up under controlled conditions of lighting, pH, salinity, and time. (**C**) Deterioration byproducts are classified in three fractions: hydrophobic, fine colloidal, and sedimented coarse products. (**D**) Fine colloidal fraction recovered for characterisation and ecotoxicological study

the water, a fraction of large aggregates sediments out at the bottom of the reaction vessel, and a finer, colloidal fraction remains in stable suspension for several months.

17.6.2 Nanomaterial Dispersion Kinetics in an Aqueous Medium

This colloidal fraction of adventitious byproducts is of particular interest for investigations into the life cycle of the material since it is well placed for transfer into the environment and highly bioaccessible. Its size distribution measured by laser diffraction (see Fig. 17.7) extends over 50–700 nm. This corresponds to single, perfectly dispersed nanoparticles (50 nm) and small, less well dispersed aggregates of these, but which are nevertheless small enough to remain in stable suspension. The sedimentation cutoff threshold around 1 000 nm is a classic consequence of Stokes' law, which governs the regime, depending on their size and density, where the particle motion is dominated by gravitational attraction. Above this size, aggregates of the adventitious byproducts, reaching sizes up to 1 000 μm in diameter, make up the coarse sedimented fraction noted earlier. Although less mobile, this fraction of the adventitious byproducts nevertheless remains accessible to benthic organisms in the aquatic environment.

During this deterioration, the proportion of fine colloidal particles released from the nanomaterial tends to increase almost immediately. They appear after only 30 minutes and reach a maximum of about 20% of the volume after just two hours (see Figs. 17.7 and 17.8). Note that such a volume percentage of fine particles in fact represents a very large majority of the particles, since a 50 nm particle occupies a volume 10^6 times smaller than a 50 μm particle.

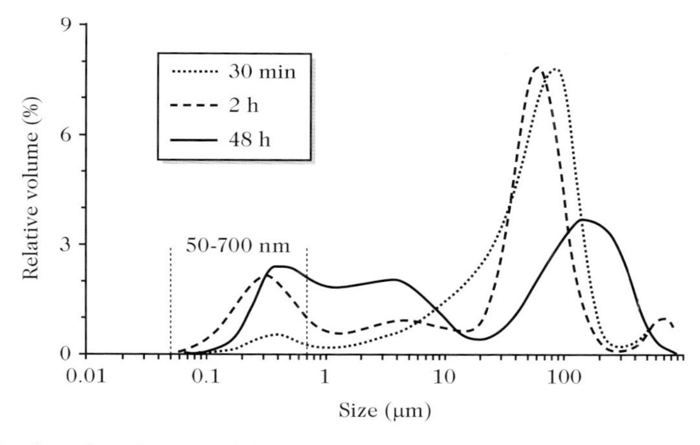

Fig. 17.7. Size distribution of the nanocomposite dispersed in water as a function of time. The proportion of submicronic particles increases very quickly after 30 min shaking. Measurements were made by laser diffraction under mechanical shaking to allow suspension of the aggregates. Adapted from [25]

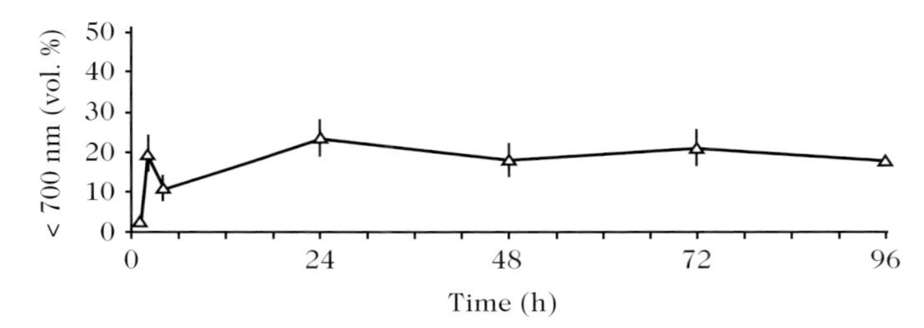

Fig. 17.8. Time dependence of the volume proportion of particles or aggregates with diameters smaller than 700 nm, characteristic of the stable colloidal fraction in suspension, measured by laser diffraction. Adapted from [25]

17.6.3 Characterising Deterioration Reactions

In order to get a better idea of how an initially hydrophobic nanomaterial can be so quickly stabilised in suspension in an aqueous phase, one must study the physicochemical reactions causing this alteration of the surface. For example, deterioration of the lipophilic dispersing agent polydimethylsiloxane (PDMS) $[\text{Si-O}(\text{CH}_3)_2]_n$ has been studied by measuring the release of silicon dissolved in the surrounding medium (see Fig. 17.9A) as a function of time. Not only did this confirm that the polymer started dissolving in the first hour of shaking, but it also provided a way of quantifying the reaction. For example, the proportion of silicon released into the solution relative to the total amount initially in the nanomaterial was 30% under conditions of illumination. However, there is very probably a third speciation of silicon, first released by the decomposition of PDMS, then remobilised for example by readsorbing onto

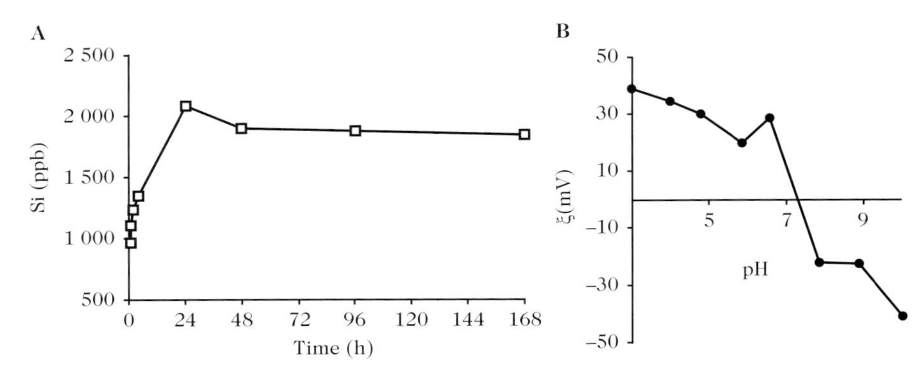

Fig. 17.9. Characterising the chemical reactions involved in deterioration of the nanomaterial. Adapted from [25]. (**A**) Release of dissolved silicon as a function of time, produced by deterioration of the PDMS coating and measured by atomic emission spectroscopy (ICP AES). (**B**) Zeta potential (ζ) of adventitious byproducts (stable fraction) as a function of pH. The isoelectric point close to 7.5 is characteristic of an $Al(OH)_3$ surface

the nanomaterial, something which would not be taken into account in these measurements.

The pH dependence of the ζ potential of the particles making up the stable colloidal fraction is shown in Fig. 17.9B. This also confirms the total or partial disappearance of the lipophilic coating. Indeed, the pH value for which the ζ potential of the colloids vanishes, called the isoelectric point (IEP), found here to be about 7.5, is typical of an aluminium hydroxide surface. On either side of this isoelectric point, the ζ potential increases in absolute value, favouring the stabilisation of the colloids in suspension. On the other hand, in the vicinity of pH 7.5, non-charged colloids tend to aggregate by coagulation and sediment out.

The salinity of the medium is also a decisive factor for the stability of colloids in suspension. A critical concentration of each salt, beyond which the colloids aggregate by coagulation, can be specified by measuring the aggregation of the particles or the turbidity of the suspension as a function of the ionic strength. These measurements, not presented here, are essential for estimating the fate and stability of nanoparticles in suspension during their life cycle.

17.6.4 Life Cycle in a Biotic Medium and Introduction into the Food Chain

When the conditions of degradation of the nanomaterial and the relevant chemical reactions causing this degradation are established, and when the resulting byproducts are perfectly characterised in terms of size, surface chemistry, and stability in suspension, only then does it become possible to trace the life cycle of the byproducts in more complex environmental compartments such as those represented by biotic media.

The example presented in Figs. 17.10 and 17.11 illustrates a possible route for the uptake of adventitious byproducts by the trophic network via the food chain, following these substances through the first two links in the chain. They first enter the micrometric freshwater green alga *Selenastrum capricornutum* (see Fig. 17.10A), and then the freshwater microcrustacean *Daphnia magna* (see Fig. 17.11A) which feeds on these algae.

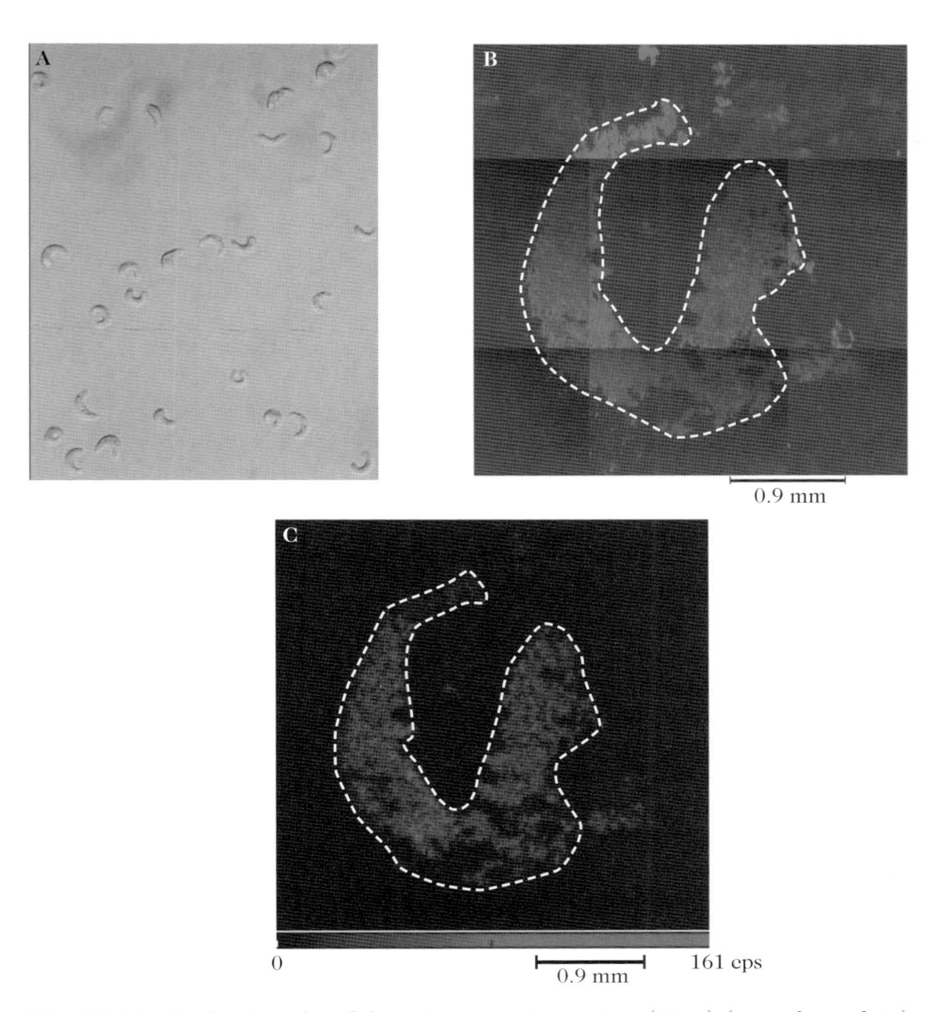

Fig. 17.10. Freshwater alga *Selenastrum capricornutum* (5 μm) (see colour plate). (**A**) Photograph of a culture through a binocular magnifying glass (Cemagref, B. Vollat). (**B**) Photograph of an alga cell (*dashed contour*) isolated after incubation in contact with adventitious byproducts. (**C**) X-ray microfluorescence map of titanium (*pink*) for the same alga cell. The titanium distribution is uniform throughout the cell

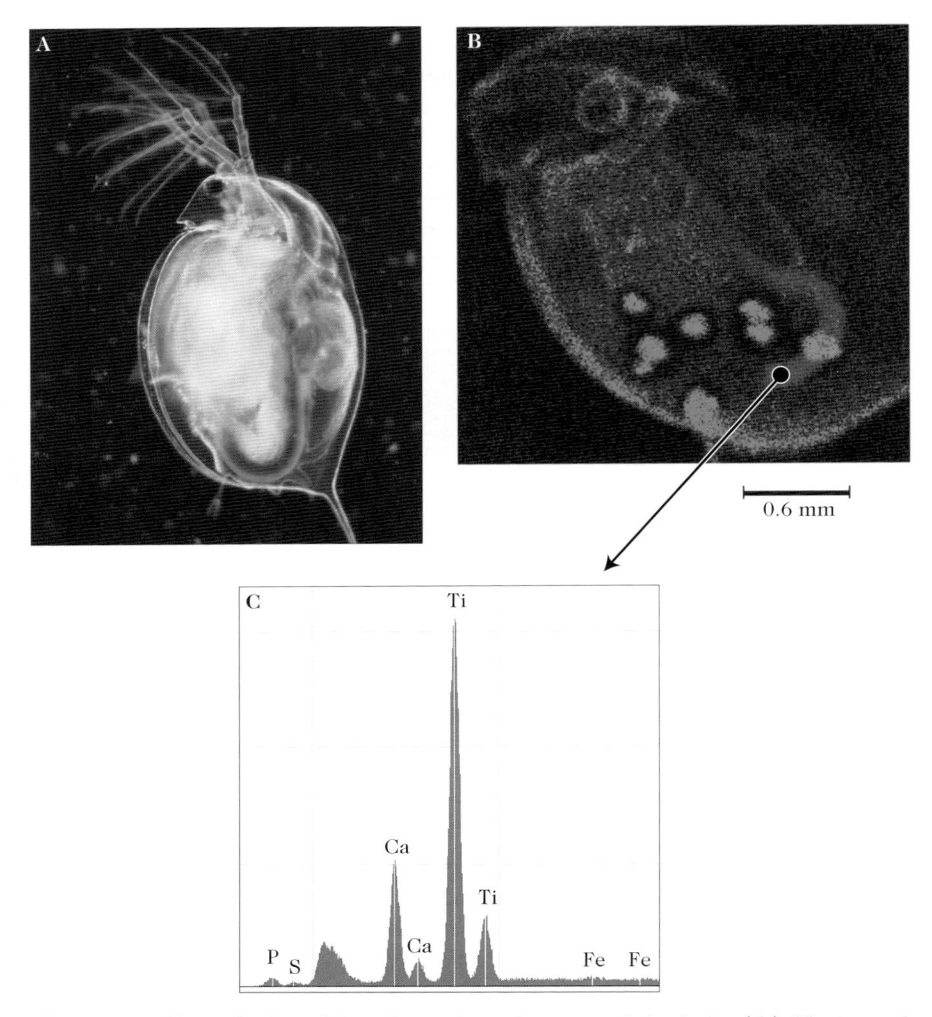

Fig. 17.11. Fate of adventitious byproducts in exposed daphnia. (**A**) Photograph of *Daphnia magna* through a binocular magnifying glass (×10) (Cemagref, B. Vollat). (**B**) X-ray microfluorescence mapping of calcium (*red*) and titanium (*blue*) in an adult *Daphnia magna* that has been in contact with adventitious byproducts (10 mg/L, 9 days). (**C**) Energy dispersive X-ray (EDX) spectroscopy at the point indicated (size 10 μm)

The first incubation experiments on these organisms in the presence of adventitious byproducts show that these are indeed transferred from the alga to the daphnia. Juvenile and adult daphnia were put in contact with these byproducts for a period of 11 days, during which the regular intake of alga was their sole source of food. Ecotoxicological measurements carried out during these experiments showed normal growth of the juvenile daphnia, implying normal feeding. On the other hand, the rate of reproduction dropped by 25% in the adults at the highest incubation concentration of 10 mg/L, and by 67% in the juveniles for all incubation concentrations (0.1, 1, and 10 mg/L).

X-ray fluorescence microscopy of alga cells taken from the incubation medium showed a clear affinity for the adventitious byproducts. Indeed, titanium mapping of the alga cells shows that these byproducts accumulate there in significant amounts by adsorption or absorption (see Figs. 17.10B and 17.10C). There is a clear link between the freshwater trophic network and the life cycle of this nanomaterial.

The same approach is shown in Fig. 17.11 to monitor in the incubation medium the continued fate of the adventitious byproducts accumulated by the alga, and in particular to detect their presence in the daphnia. The titanium map, superposed on the calcium map which provides the image of the daphnia, reveals an accumulation of adventitious byproducts in the digestive tube of the individual.

These studies demonstrate that adventitious byproducts are transferred from the aqueous medium to algae, and from there to the primary consumers, viz., the daphnia, via the food chain. The next step in this study might be to investigate the possible transfer of adventitious byproducts to the secondary consumers, viz., fish, and the consequent ecotoxicological effects. In the long term, considering the final link in the food chain, it would be interesting to study the fate and toxicity of these byproducts in the human gastric system.

17.7 Conclusion

The risks due to the emission of a nanomaterial into the environment are assessed by considering it global life cycle, considering in particular its interactions, transformations, and fractionation during the various stages of its life. There is no standardised approach to study nanomaterial life cycles, just as there is no such approach for chemical products in general. Although there is a set of normalised methods, each study must take into account the specific features of the given product. Regarding nanomaterials and their derived products, since the environmental impact may occur at various stages throughout their life cycle, all the stages in the life cycle must necessarily be taken into account when making global risk assessments.

The main problem encountered at the present time is the lack of specific data concerning certain points, limiting our knowledge of the global life cycle and making it difficult to assess the risk. Examples are the toxicity due to the transport and fate of nanomaterials and adventitious products in the environment. In future research, it will be important to focus in priority on these points, developing standardised methods and protocols to be followed by the whole of the relevant international scientific community. This is the motivation behind the *Consortium international pour l'étude des implications environnementales des nanotechnologies* (iCEINT, *Groupement de recherche international* GDRI), founded in 2008 by the *Centre national de la recherche scientifique* (CNRS) and the *Commissariat à l'énergie atomique* (CEA) in France and the National Science Foundation (NSF) in the United States, which aims to improve coordination of research on either side of the Atlantic.

These approaches will require investigation of several key and often interdependent parameters regarding nanomaterial life cycles. On the one hand, intrinsic properties, and in particular surface properties, of nanoparticles and nanomaterials determine their behaviour in the various environmental compartments they encounter during their life cycle (hydrophilicity/hydrophobicity, reactivity, charge, etc.). On the other hand, the environmental conditions specific to each of these compartments (pH, ionic strength, mechanical conditions, etc.) determine the behaviour of the nanomaterials in relation to their own intrinsic properties. And finally, the latter are sure to evolve during the life cycle, by degradation of the nanomaterial or interaction with the medium, thereby engendering new characteristics in terms of behaviour and potential toxicity.

Combining a toxicological study with knowledge of these parameters which control dispersion, transport, and bioaccessibility of the nanomaterial in the environment, it should be possible to improve our understanding of the conditions and reasons for its toxicity, and hence to better assess the related risks. But this can only be done through a cross-disciplinary approach, seeking to measure all the various parameters and decisive factors discussed above. This is the aim of the French national programme ANR AGING NANO & TROPH [27], which brings together chemists, physical chemists, ecotoxicologists, toxicologists, and manufacturers to carry out a global study of the life cycle of commercialised nanomaterials and their degradation residues released into the environment.

References

1. C.L. Tran et al.: A scoping study to identify hazard data needs for addressing the risks presented by nanoparticles and nanotubes. Institute of Occupational Medicine (2005)
2. SCENIHR: Opinion on the appropriateness of existing methodologies to assess the potential risks associated with engineered and adventitious products of nanotechnologies. European Commission, Scientific Committee on Emerging and Newly Identified Health Risks (2005)
3. AFSSET: Les nanomatériaux. Effets sur la santé de l'homme et sur l'environnement. Saisine AFSSET 2005/010 (2006)
4. M.R. Wiesner, G.V. Lowry, P. Alvarez, D. Dionysiou, P. Biswas: Environmental Science and Technology **40**, 4336 (2006)
5. Nanotechnology White Paper: US Environmental Protection Agency (2005)
6. Nanotechnologies, a preliminary risk analysis on the basis of a workshop. European Commission: Health and Consumer Protection Directorate General, Brussels (2004)

7. R. Owen, M. Depledge: Marine Pollution Bulletin **50**, 609–612 (2005)
8. C.O. Robichaud, D. Tanzil, U. Weilenmann, M.R. Wiesner: Environmental Science and Technology **39**, 8985 (2005)
9. US EPA: Oil program, Mineral Management Service (2008)
10. N. Saleh et al.: Environmental Engineering Science **24**, 45 (2007)
11. J. Brant, H. Lecoanet, M.R. Wiesner: Journal of Nanoparticle Research **7**, 545 (2005)
12. J. Brant, H.F. Lecoanet, M. Hotze, M.R. Wiesner: Environmental Science and Technology **39**, 6343 (2005)
13. J. Brant, J. Labille, C.O. Robichaud, M. Wiesner: Journal of Colloid and Interface Science **314**, 281 (2007)
14. K.L. Chen, M. Elimelech: Langmuir **22**, 10994 (2006)
15. H.F. Lecoanet, J.Y. Bottero, M.R. Wiesner: Environmental Science and Technology **38**, 5164 (2004)
16. H.F. Lecoanet, M.R. Wiesner: Environmental Science and Technology **38**, 4377 (2004)
17. AQUANANO: Programme national ANR Pnano, www.ineris.fr/aquanano (2008–2010)
18. S.J. Klaine et al.: Environmental Toxicology and Chemistry **27**, 1825 (2008)
19. S.N. Luoma, Y.B. Ho, G.W. Bryan: Marine Pollution Bulletin **31**, 44 (1995)
20. C.M. Wood, R.C. Playle, C. Hogstrand: Environmental Toxicology and Chemistry **18**, 71 (1999)
21. K. Morgan: Risk Analysis **25**, 1621 (2005)
22. N.C. Mueller, B. Nowack: Environmental Science and Technology **42**, 4447 (2008)
23. S.A. Blaser, M. Scheringer, M. MacLeod, K. Hungerbühler: Science of the Total Environment **390**, 396 (2008)
24. M. Auffan et al.: Environmental Science and Technology **44**, 2689–2694 (2010)
25. J. Labille et al.: Environmental Pollution, DOI: 10.1016/j.envpol.2010.02.012 (2010)
26. CIRC: Titanium dioxide (group 2B). Centre international de recherche sur le cancer (2006)
27. AGING NANO & TROPH: Programme national ANR CES, cerege.fr/aging nano (2009–2011)

Part III

Nanoethics: Ethical Questions Raised by Nanotechnology and Scientific Discovery on the Nanoscale

The instrumentation and software now available for observation, interpretation, and synthesis of nanoscale entities at the beginning of the twenty-first century are already having a significant impact on the rate and extent of scientific discovery.[1] These advances are being made simultaneously in laboratories located all round the world, even though clusters of worldwide dimensions remain tightly concentrated, if we are to judge by the traditional indicators of publications and patent applications.

Scientific excellence is based on the organisation and transmission of the body of scientific knowledge in the traditional disciplines of physics, chemistry, and biology. However, there is the prospect of new disciplines, bringing logical and ontological solutions to the growing complexity generated by the convergence of the traditional disciplines at the nanometric scale. Information science, together with the many possibilities today for sharing knowledge and creating understanding by computer simulation, but also the models provided by computer archives and open innovation, are unprecedented resources for promoting initiative, especially at the frontier between life and the creation of hybrid entities.

The need to find solutions that suit the market, but which take into account current challenges on the planetary scale, such as poverty, climate change, pollution, water treatment, agricultural production in the face of fast-increasing populations, and the availability of novel nanostructured materials and systems, has led whole sectors of industry to contemplate the incorporation of artefacts with radically different properties into their product range.

The changes brought about by nanotechnology are not at the present time accompanied by, nor therefore preceded by, any hierarchical organisation of the emergent knowledge resulting from nanotechnology. Attempts to name and define, in order to find a terminology that could be easily adopted by all stakeholders, are the result of fragmentary initiatives or standardisation organisations, while ontological questions and the taxonomic approach are by their very nature a scientific prerogative.

The resistance shown by traditional disciplines when required to produce a suitable taxonomy for coping with the nanoscale convergence of their fields of expertise, the roadmap of synthetic biology providing an illustration of this convergence, has led to the generation of a semantic fog (or nanosmog!) that has done little to encourage high quality communication or sharing of points of view between the various components of society. If we can invest a little in understanding the state of the art on the nanoscale, it may be possible to overcome our purely emotional reactions and turn to a more reasoned dialogue with some historical, if not philosophical, depth.

In order to rise above these difficulties, it would be useful to mobilise international scientific cooperation, in order to set up a large scale project that

[1] Introduction by Françoise Roure, *Conseil général de l'industrie, de l'énergie et des technologies*.

could establish a serviceable taxonomy, capable of describing the evolutions of our knowledge base at the intersection of the relevant scientific disciplines.

Scientists sometimes accuse politicians and their entourage of raising ethical questions that they believe do not exist (e.g., improvement of human performance, uncontrollable and irreversible forms of pollution), while politicians do not yet have at their disposal the means of understanding and expressing the state of the art when they need to base policy on methods for assessing the changes induced by convergent technologies at the nanoscale using ethical, legal, and social criteria. The most likely situation is then a lasting dialogue between parties that hear nothing of the other's point of view, leaving only the scientists' codes of conduct to guide the free will and soothe the consciences of those engaged in the scientific aspects of the problem, while attempts to organise intergovernmental cooperation are left to handle all the difficulties involved in promoting a responsible development of nanoscience and nanotechnology, without being provided with any truly appropriate methodological tools.

The contributions in the following part of the book deal with just these ethical issues, on the basis of complementary analyses, bringing in questions and subsidiary issues from a broad international range of origins, with a view to raising doubt, compelling those involved to seek a meaning for the notion of scientific progress, and to find a balance between the effort expended, social innovation, and fairness with respect to sharing of profits, by means of a systemic and prospective approach, before taking any major decision about the large scale implementation of nanotechnological applications.

18

Nanoethics: Challenges and Opportunities

Alain Pompidou

Nanoscience and its technological consequences constitute a relatively recent field of knowledge. Their rapid development around the world is characterised by an absence of specific norms and standards. Industrial applications, already promising, are not without risk, and this risk deserves to be fully and rigorously assessed. On the other hand, this vigilance and anticipation of risk does not need to deprive society as a whole of new opportunities, whether they be for developed countries and the emerging economies, or for less developed nations.

The challenge here concerns not only the research community, but also the political decision-makers and the general public. An integrated approach involving all actors needs to be set up as quickly as possible. The aim is to avoid making mistakes on the basis of false interpretations and unjustifiable hopes. Indeed, science fiction tends to exaggerate the extent of both expectation and apprehension.

With the consequences of Hiroshima, Chernobyl, and Bhopal in mind, and following the sanitary disasters of 'mad cow' disease and food contamination by melamine, while controversy rages over genetically modified organisms (GMO), the rapid development of nanoscience and its technological spinoffs should give rise to another debate, but built upon a solid foundation, and well before releasing onto the market products that may be promising but whose potential risks have been covered up.

The questions raised are often new and relate to problems of managing complexity. This is yet another reason for anticipating them and learning from past experience. Indeed, the most promising applications are in medicine (in particular, targeted drug delivery), cosmetics, energy and freshwater management, information and communication technologies, and more generally manufacturing processes for high-tech products.

It is indeed at this stage that we need to adopt a cross-disciplinary approach to assess the trends in terms of risk and benefits. The idea behind this fundamentally ethical step is to build up a dialogue between those involved in research and development and those members of society, such as philosophers, legal experts, historians, sociologists, economists, and

P. Houdy et al. (eds.), *Nanoethics and Nanotoxicology*,
DOI 10.1007/978-3-642-20177-6_18,

perspectivists, who enjoy close relations with political decision-makers and elected representatives. The aim is to clarify ethical, social, and legal aspects of the use of nanotechnological research, raising public awareness in a rational way by introducing an appropriate policy of norms and labelling.

At the present time, our approach is inadequately structured, leaving room for potential conflicts of interests between those concerned, i.e., researchers, producers, and consumers. The main challenge we must meet, beyond the purely scientific and technical issues, is to establish a decision-making process based on the responsibility of the stakeholders, transparency, and objective assessment. To achieve this it will be necessary to reassess and set up mechanisms for normative and legal regulation, adapting them to suit the specific features of novel materials (and notably, nanoparticles), with variable, adjustable, and hence often somewhat complex characteristics.

The present regulatory environment only exists in the form of an outline. While certain principles have been fixed, both in Europe and in the United States, an integrated and worldwide approach is essential, given the rapid evolution of knowledge and associated technology developed in the present context of globalisation.

As an example, consider the environmental risks, health risks, or infringement of privacy threatened by new integrated surveillance systems, not to mention the other potential risks, not always clearly identified and a possible subject of controversy.

In this area, uncharted in so many respects, it will be essential to revise not only the regulatory framework, but also the technical and legal systems, organising concerted action on both regional and international levels, and involving not only the most developed countries, but also the developing nations, and those that are currently the least developed but who may become consumers, and who may suffer the consequences of potential risks.

Whatever is organised on the level of technical, social, and in particular political experts must be backed up by suitable training or preparation for the general population. The solution here is to educate teachers and trainers, and to inform the media and public opinion as objectively as possible. Hence the importance of the authorities involved in technological and scientific ethics, on all levels, from national, through regional, to international. These are institutions like the International Council for Science (ICSU) or UNESCO with its World Commission on the Ethics of Scientific Knowledge (or *Comité mondial sur l'éthique de la science et de la technologie* COMEST). The cross-disciplinary nature and independence of such bodies must be emphasised, since their credibility depends on it.

Certain basic principles should also be stressed:

- Emotional reactions, often encouraged by the media, tend to have the edge on a rational approach, which can be difficult to explain and share with others, through lack of knowledge more than through any lack of mutual understanding.

- The fact of having explained does not mean that one has been understood.
- The more complex the investigations and the ensuing discussion, the longer it will take for ideas to mature in people's minds.

The integrative approach becomes less accessible when it requires a flexible outlook owing to the fast-evolving nature of our knowledge and related techniques, and this is very much the case with nanoscience and nanotechnology.

If we are to engage upon a reasoned approach, based on a principle of precaution, an attitude of vigilance, the will to learn from experience, and the step-by-step and prospective assessment of what is known, it must be borne in mind that public opinion is generally torn between false hopes and unfounded rejection. Hence the importance of favouring a scientific, therefore rational approach, based on demonstration and a critical spirit, among experts committed to expressing their views in full independence.

Given the risks of not obtaining adequate expert advice, this reminds us that the expert must publicly declare any possible conflict of interests and commit himself or herself to full independence. Without this proviso, specialist advice might easily be monopolised, hence biased, by stakeholders of debatable objectivity, despite the fact that they may address the task in good faith.

Before concluding, it is important to point out some approaches that actually have nothing to do with ethics, if we define the latter to deal with principles of good conduct and anticipation of risk:

- The so-called grey goo scenario, based on the fear of uncontrolled self-replication of nanoparticles. It might be defended from a purely theoretical standpoint, but completely neglects the reactive and changing capacities of the environment.
- Discussion of the artificial potentiation of the human being and its consequences. It is true that improvement of daytime or nighttime sight, or heightened athletic performance are possible and even probable. But these are highly specialised and complex areas which remain for the moment in the minds of the perspectivists. Their spectacular, even anxiogenic, nature might make one lose sight of the ethical approach as it should be defined and adapted to the problem of nanoscience, nanotechnology, and specifically, nanoparticles.

We must look beyond these two ways of thinking which clearly fall outside a reasoned ethical approach.

The international community must therefore focus its attentions on national resources that may lead to nanoproducts by the optimal exploitation of results obtained from the nanoscience and nanotechnology that arise from them.

To conclude, despite the fact that nanotechnology looks very promising, the ethical and political consequences of such research are not radically different from those with which we have already been confronted in the past, so here perhaps is an opportunity to tackle them with full knowledge of the facts and with greater success than in previous instances. The resulting rules can be put together by a constructive process in such a way as to transform these challenges into opportunities for all the inhabitants of our planet.

References

1. Technology assessment on converging technologies, European Parliament (IP/A/STOA/SC/2005-183); www.europarl.europa.eu/stoa/publications/studie
2. A.M.J. Henk (Ed.): Nanotechnologies, ethics and politics, UNESCO Publishing, publishing.unesco.org/details.aspx?Code_Livre=4
3. Nanotechnologies and ethics: Policies and actions, COMEST Policy Recommendations, 2007, unesdoc.unesco.org/images/0015/001521/15214
4. The ethics and politics of nanotechnology, UNESCO, 2006, unesdoc.unesco.org/images/0014/001459/14595

19

Ethics and Medicine: Philosophical Guidelines for a Responsible Use of Nanotechnology

Corine Pelluchon

19.1 Definition of Ethics

Ethics is not an isolated discipline, standing aloof from science, economics, and politics. And neither is it an authority devoted to censure, for it is not the philosopher's role to set up as an authority of any kind, nor to dictate to others what is good or bad in itself on the basis of some personal morality. Ethics is that part of philosophy that allows us to acquire the tools that serve to elucidate actions and assess them critically. The aim is to identify principles, that is, notions that are taken as fundamental and must guide our actions in medicine, in business, or in the application of biotechnology. However, these principles are not empty of content, and part of the philosopher's work in the field of applied ethics is to elucidate the values underlying the notion of autonomy and distributive justice, and to determine the relationship between the latter and the notion of equality. Likewise, the ethicist must consider the implicit and explicit norms belonging to some narrowly defined community (a group of professionals) or a broader community (a country), or even the international community.

Finally, there are three levels of judgement in ethics, according to Ricœur [1]:

- The first level deals with the relationship between particulars. It refers to those qualities or virtues which help one to make the right decision in a completely novel situation which is not without uncertainty. Ethics exists precisely because the right action and the sensible decision are not obvious and do not follow from some simple rule, as they might in mathematics. As Aristotle reminds us in Book II of *Nicomachean Ethics*, angles are not straight lines. This is why prudence, or practical wisdom, is the virtue of deliberation, and involves a particular management of risk.
- The second level concerns norms, that is, the universalisation of maxims or precepts discovered by practitioners in the individual pursuit of their profession. This is the deontological level of ethical judgement. Here one finds the main tools of applied ethics (medical ethics, bioethics, environmental

P. Houdy et al. (eds.), *Nanoethics and Nanotoxicology*,
DOI 10.1007/978-3-642-20177-6_19, © Springer-Verlag Berlin Heidelberg 2011

ethics, business ethics). These norms, validated by professional associations and set down in deontological codes, charters, or declarations, give content to the principles used, and at the same time reaffirm the importance of human rights.
- The third level of ethical judgement is the teleological level. This concerns society's ends and choices. The philosopher's task here is to articulate the first two levels at the third, which means to say that the questions raised by the various fields of applied ethics require a move from moral philosophy to political philosophy, whence one may pose the problem of what kind of society, and even what kind of human being, we wish to advocate.

However, such an investigation presupposes that we first ask whether nanotechnology raises *specific* problems. Are these problems radically different from those encountered in medicine when we reflect upon the use of biotechnology, or nuclear energy?

19.2 Exacerbation of Problems Inherent in Conventional Techniques

Nanotechnologies are based on the physics, chemistry, and physicochemistry of matter, but what is specific about them is the length scale used here by researchers. As discussed in the previous chapters, nanotechnologies manipulate matter on the scale of the nanometer. They thus involve characteristics that are specific to this range of sizes, and that bestow particular properties upon them, and even a certain unpredictability. This type of technology therefore necessitates an assessment of sanitary and environmental risks relating to the use of nanoparticles. The problem is all the more important in that some nanoparticles, as has been demonstrated, are able to cross biological barriers in living organisms and can cause cancers, in an analogous way to those induced by asbestos particles. We must therefore give careful consideration to their potential risks and carry out adequate tests before accepting large volumes of materials or objects containing nanoparticles on the open market. The political authorities thus have a duty to organise detailed studies of their potential impacts, and to inform the public of the results obtained.

However, the central issue is not the question of risk. The specificity of nanotechnologies is that they can be combined with other 'sensitive' technologies, such as biotechnologies, in the context of genetic manipulation. Likewise, they boost the potential of any interaction between living organism and machine. Nanotechnologies can increase our control over matter, life forms, and even the human brain, and their use is accompanied by a degree of uncertainty which makes risk assessment unavoidable. On the other hand, the ethical and political problems that can be attributed to them are not completely new. Some of the problems are raised quite generally by all forms of contemporary technology. To be precise, their future potentialities, both positive and negative, are likely to *exacerbate* the problems we encounter when we reflect upon the relationship between science and current technology.

In other words, just like science and technology in general, nanotechnologies have come about in a given social context. What is important is to identify the problems characterising this context, such as unequal access to technology, information, or health care. Nanotechnologies, so promising in the field of reparative and predictive medicine, will clearly aggravate these problems or, at the very least, make them more acute. The question of environmental justice, which refers to the equitable access to a high quality environment, and which requires us to ask how the benefits and burdens of manufacturing technologies and product recycling will be shared out among the world's populations, also lies at the heart of any philosophical reflection on the use of nanotechnologies. In order to answer these questions, we need to formulate our priorities and decide what kind of society we wish to live in. We must also reflect upon the decision-making authorities, and the place of ordinary citizens in public deliberation. Not only must we establish the very meaning of a responsible use of nanotechnologies, but in addition this process of reflection must be carried out upstream. This implies that citizens must be properly informed and trained, and they must in all respects be given the means to take part in the decision-making process.

We already possess the means to pose the ethical and political questions relating to these forms of technology and to take them into account in public policy. We can set up guidelines for the use of nanotechnologies and for the promotion of policies that privilege one kind of research over another. This does not mean that policy should have *total* control over research, which, since it concerns knowledge, is an end in itself, but what is at issue is to decide what we want to do and what we do not want to do, and why. Decisions must be linked to our choice of society and assessed in the light of the ends and ideals that we continue to honour. Now, one of the main problems lies in the fact that such an investigation is ruled out from the start. It is said to be impossible or vain. Ethics becomes a trapped authority, a mere guarantee, or conversely, it is taken as an instrument of censure, as though its purpose were to introduce virtue in a world that did not want it. But, on the contrary, the task of the political philosopher is to identify ways of posing the central question: what constitutes a responsible use of nanotechnologies?

19.3 The Use of Nanotechnologies and Society's Purpose

This question requires an investigation of society's final causes. As noted by Ronald Sandler [2], professor at the Northwestern University in Massachusetts, technology must contribute to human happiness and social progress in a fair, realistic, and sustainable way as regards the environment. Now many would agree that technology should have this aim, and yet it is not clear that all forms of technology currently on the market or benefiting from huge investment on the part of public or private organisations will allow us to achieve it. In any case, such a claim presupposes that ethics has meaning insofar as the idea of a responsible use of nanotechnologies is accepted. Indeed, the problem here is to identify the obstacles to achieving these objectives (and there are plenty, especially when one considers the inequitable access to technology and environmental justice). Finally, we need to specify what is meant by human happiness and social progress.

If we talk about human happiness and the question of sustainable development, then we are compelled to clarify the responsibilities of current generations with regard to future generations who may be required to pay for decisions which in some cases may have dramatic and irreversible consequences. In this context, the issues relating to nanotechnologies are not without parallel in the questions raised by nuclear energy, or indeed by any technique which confers such tremendous power on the decision-makers, and as a consequence, a much greater responsibility. Likewise, going back to the statement made above, the relationship with other species must be taken into account. This broadening of the scope of our responsibility to future generations and other species suggests a move from a negative definition of freedom (freedom from) which is still that of human rights to a consideration centered on the capabilities that allow humans to use certain goods and resources [3], and even to a reflection on the limits of our rights. Can the source of what we consider to be legitimate refer exclusively to the person who thinks of this right as an instrument of his own power [4]? Should we continue to base human rights on the moral agent and on the individual considered as an empire within an empire? This question was asked by Claude Lévi-Strauss, who suggested a reformulation of political principles wherein humans would be treated as a species whose rights come to an end at the precise moment when their exercise puts the existence of another species in danger. When we think of the responsible use of technology and everything that comes under the heading of sustainable development, if it is not just to be a pious hope, does this not presuppose a reflection on the relationship between peoples of different cultures, between humans today and future generations, and between humans and other species, or nature as a whole?

19.4 What Criterion Can Distinguish Between Legitimate and Illegitimate Uses of Bionanotechnologies?

The philosopher does not pronounce on which technique is good or bad in itself, but instead will examine its impact on institutions, the family, the arrangements, and the traditions which up to now have made democracy possible [5]. If we consider the example of the interaction between nanotechnology and gene therapy with a view to improving the sensorial, physical, intellectual, and cognitive abilities of a human being, then it belongs to the philosopher to ask whether this application is compatible with the values upon which our institutions are founded. Likewise, the question as to whether there is a contradiction between certain practices and the ideals underlying our institutions can serve as a guideline for the philosopher's enquiry. For example, one may ask whether it is acceptable to manipulate an individual's genotype, that is, the genetic heritage specific to that individual, while at the same time claiming the equality of all individuals. The freedom of those who wish to endow their future offspring with superior capabilities threatens the freedom of individuals whose children have not been 'improved' and who will thus find themselves bottom of the class at school, last in competitive sports, and so on. Likewise, one should stress the contradiction between this ideal of total control, which finds an ally in bionanotechnology, and the worship of singularity in culture and art.

So in contrast with what might be thought at first glance, the criterion whereby one may distinguish a legitimate use from an illegitimate one is not simply a distinction between therapeutic use and one which aims to 'improve' the individual. For example, there are predictive tests for the prevention of cancer which are perfectly legitimate, showing that the aim of medicine is not merely to cure. Furthermore, this distinction presupposes a fixed definition of what is normal, considered as an average to be attained with regard to size, IQ, or behaviour. But it is hard to distinguish between hyperactivity and being dynamic, social anxiety and being shy, as Leon Kass has reminded us [6].

A practice or usage is illegitimate when it debases the very meaning of an activity. Doping in sport is a good example. It is contemptible because it corrupts the meaning of competition. The doped athlete reduces the race to its outcome alone. In addition, he uses his body as a machine and debases the intrinsically human meaning of physical effort which manifests the phenomenological unity of mind of body. So the discriminating criterion we seek must not be based on any rigidifying vision of nature, an ideology banishing artifice and technology. The problem here is *to question the impact of science on our social practices and to examine the compatibility or incompatibility between habits (among which there are induced habits) produced by certain technologies and values underpinning the way we live together and the exercise of democracy.*

The emergence of nanotechnologies and other contemporary forms of technology compel us to ask just how far we are ready to evolve, and why. To make this enquiry, we must first clarify the content of certain notions often used as principles, such as autonomy, solidarity, and justice. But we must also reflect upon the human condition, the meaning of mortality and birth, and unpredictability. As pointed out by H. Arendt in *The Human Condition*, the newly born introduces something new into the world, and this is an essential safeguard for the creativity of a society. But this creativity becomes less obvious if, by constantly extending the human lifespan, we keep the same people in power and maintain young adults in a state of adolescence [7]. And nor can we disregard the need for an ontological consideration of the relationship between man himself and what is not man, or of responsibility which is, even more than the possession of reason, what is specific to humans as compared with other living beings, a responsibility that scientific knowledge emphasises.

19.5 International Norms and the Political Community

One of the tasks of the political philosopher would be to identify the common values which underlie a political community and are expressed through its institutions. These values are also bound to its traditions and its moral stances, what Rousseau called "this fourth kind of law which maintains a people in the spirit of its institutions" in *The Social Contract*. The results of this description of the 'strong evaluations' which reflect the sources of morality and politics in a community should also be subjected to deliberation. The philosopher, by contextualising notions and reflecting upon the contents attributed to the notions of solidarity and justice, would make explicit the implicit values that govern our practices and are reflected in our laws. This would be a task of translation, with all that must remain unfinished about that. This attempt to view a community in its own terms and to express rights and morals in a relatively immanent way, by basing itself on what constitutes the narrative identity of the community [8], implies that there cannot be a valid international ethics for all problems and in all contexts.

For sure, there are international norms that could serve as points of reference to contain or even prohibit certain practices. However, a political community cannot escape the need to undertake this reflexive examination of itself, because words do not have the same meaning from one country to another, and the content of principles serving as ethical guidelines must be specified. While the working rules of procedural justice are common to all liberal democracies (transparency, publicity, revisability and rationality of norms, participation), the question of usage begs the question of ends, which themselves depend on the sources of morality and politics, the traditions, and the ipseity of a country.

In this sense, we may say that the industrial emergence of nanotechnology is an opportunity to go beyond the post-modern credo which required political philosophy to abstain from any substantial vision of what is good, and even to refrain from any reflection upon the common good in order to abide by the procedural rules. It might even be thought that the national and international commissions set up to consider these technologies and the proliferation of public information meetings on their potential and their risks are a sign that we are aware of the urgent need to pose these questions and to find new forms of governance that are more rigorously democratic.

References

1. P. Ricœur: Les trois niveaux du jugement médical (1996). In: *Le Juste 2*, Le Seuil, Paris (2001) pp. 227–243
2. R. Sandler: Nanotechnologies: The social and ethical issues. In: *Project on Emerging Nanotechnologies* (PEN) 16 (January 2009) pp. 4–63
3. A. Sen: *Poverty and Famine. An Essay on Entitlement and Deprivation.* Oxford University Press, Oxford (1987)
4. C. Lévi-Strauss: Réflexions sur la liberté. In: *Le regard éloigné*, Plon, Paris (1983) pp. 376–377
5. C. Pelluchon: *L'Autonomie brisée. Bioéthique et philosophie*, PUF, Paris, p. 134 (2009)
6. L. Kass: *Beyond Therapy.* The Report of the President's Council on Bioethics, Washington (DC), (October 2003) pp. 13–20
7. H. Jonas: Le fardeau et la bénédiction de la mortalité (1992). In: *Evolution et liberté*, transl. by S. Cornille and P. Ivernel. Payot/Rivages, Paris (2000) pp. 129–157
8. C. Pelluchon: *La raison du sensible. Entretiens autour de la bioéthique*, Chap. I, Artège, Perpignan (2009)

Part IV

Nanoethics and Regulation: The Situation in France

A new form of technology has rarely aroused so much debate and controversy as nanoscience and nanotechnology. Indeed, since 1990 in the United States and since the beginning of the 2000s in Europe, scientific communities touching upon these areas have often been faced with aggressive polemic and contestation regarding the well-foundedness, or even the dangerousness, of their work, and accused of scientific irresponsibility. Scientists are suspected of worrying only about understanding the mechanisms of the physical or chemical phenomena of the objects they study or manipulate, without consideration of the potential risks involved in implementing and disseminating them. The aim of the next five chapters is to provide an overview of the positions adopted by the various scientific bodies, institutions, and consumer societies on the subject, and the evolution of the notion of precautionary principle in society.

The main points to come out are naturally, at the first link in the chain, an appeal to scientists to remain vigilant about the goals of their research, but then an appeal to the authorities to require scientists to provide a better anticipation of the risks, while consumer societies insist on the need for the relevant authorities to promote citizen deliberation in connection with the major changes resulting from these new technologies.

Chapter 4 discusses the principle of precaution and the way its interpretation has evolved in a more general context than the one provided by the nanosciences. One such change is the introduction of a principle of the ethical dangers of innovation, to be taken into account before new technologies are launched on the market, going beyond a simple risk–benefit analysis.

20

Situation in France: Ethical Reflection on Research in Nanoscience and Nanotechnology

Jacques Bordé

20.1 Awareness of Nanotechnological Risk in North America

By the end of the 1990s, the possible impacts of nanotechnologies on humans and the environment had already come under the spotlight. Owing to the tremendous promise of the development programmes for these technologies, it seemed important to reflect upon a responsible way of implementing them. Even in 1989, as part of the MIT course on *Law, Technology and Public Policy*, David Forrest spoke in particular on the subject of regulating nanotechnology development, and in February 1999 a seminar was organised in California to devise guidelines on how to control the new risks associated with these recent technological possibilities [1].

The surge of interest in nanotechnology in the US at the beginning of the 2000s led to a proliferation of meetings and publications on the subject over the following years (see, for example, the seminar run by the National Science Foundation (NSF) [2] or the report entitled *The big down, from genomes to atoms*, by the Canadian group ETC [3]). The 'nano' topic had suddenly become a dominant concern among those involved in the societal impacts of scientific research.

20.2 Reaction of the European Union

Europe followed a short way behind, having already set up a European excellence network by the name of Nanoforum in the Fifth Framework Programme. This network soon became interested in research into the ethical, legal, and social aspects (ELSA) of nanotechnology [4]. In parallel, the ideas of the NSF (which had already recognised the need to finance ELSA research) were quickly transmitted to the European Commission, for example during the third EC–NSF seminar at the beginning of 2002, bearing largely on social aspects of the problem [5]. Then the European Parliament was directly

P. Houdy et al. (eds.), *Nanoethics and Nanotoxicology*,

DOI 10.1007/978-3-642-20177-6_20,

influenced by the ETC Group mentioned above, during a meeting it organised in June 2003 with the relevant NGOs, just when the European Commission was devising its strategic plan for the nanotechnologies [6].

The EC soon grasped the need to look more deeply into these issues, and they included a session on this at the nanotechnology forum in Trieste in December 2003 [7]. In addition, the Health and Consumer Protection Directorate General organised a seminar to examine the risks of nanotechnologies [8] at the beginning of March 2004, and research into related health problems was stimulated rather early on by the EC, in particular through the activities of the European project Nanosafe. The Research DG also convened a social and human sciences seminar in Brussels in April 2004 [9] with the aim of deciding on suitable lines of action.

Finally, the EC report [10] entitled *Converging technologies: Shaping the future of European societies* in August 2004 was intended to be Europe's answer to the NSF's 500 page report [11], published in 2002 under the title *Converging technologies for improving human performance.* It was a response that sought to contrast starkly with the objectives stated in the American document. The EC's various initiatives and actions on the subject of the nanotechnologies were then substantially supported by a mobilisation of the human sciences to anticipate the ethical, legal, and social impacts of these new technologies.

20.3 Mobilisation of Research in the Human Sciences

An international scientific community springing from several disciplines of the human and social sciences was soon up and running. The departments of philosophy of the universities of Darmstadt (Germany) and South Carolina (US) brought it together for the first time on the theme of nanotechnologies in the United States in 2003 (giving rise to an important book entitled *Discovering the Nanoscale* [12]), then a second time for the NanoEthics congress in March 2005, once again in South Carolina.

In Europe, human science research in the area of nanotechnology has grown, not only in the philosophy of science, but also in the sociology and economics of science. In 2003, in the Netherlands, where there are well known research centers on the problems of innovation, Arie Rip of the University of Twente suggested that the Dutch consortium NanoNed should adopt a method called constructive technology assessment whose aim was to encourage scientists to evaluate innovation in real time.

In the United Kingdom, a country that has long been interested in communication between the scientific community and the general public, research on this interface was promoted in such a way as to establish a form of communication that was no longer always one-way, encouraging the public not to simply accept information passively, but to make their voice heard to scientists. The ultimate aim was, in the case of scientific priorities that might strongly affect the future of the whole of society, to make the corresponding decision-making processes more democratic.

As a result, the European academic environment in human and social sciences that reflects upon the nanotechnologies is immersed in this three-point approach, based on philosophy, sociology, and politics, and fully aware of the ethical questions raised by these innovations.

20.4 Motivation for the Ethics Committee of the CNRS

It was from about 2003 that France was really confronted with an explosion of public debate on nanotechnology. It was sparked off in Grenoble by a group who contested the technological development of our society across the board [13], but who were particularly worked up about the huge investments devoted to nanotechnologies (which they nicknamed necrotechnologies) by the CEA and by the city of Grenoble. This contestation led members of the *Institut national polytechnique de Grenoble* (INPG) to reflect upon a suitable way to respond to this situation. They were already concerned about technology ethics, and had produced a manifesto in 2000 for the consideration of future engineers [14]. At the end of 2003, they contacted the *Comité d'éthique du CNRS* (COMETS). In the March 2004 session, COMETS heard a delegation from the INPG. Then, before taking action, they had discussions with two others, Jean-Yves Marzin (in June 2004), who coordinated the CNRS contribution to the ministerial programme for nanotechnologies, and Jean-Pierre Dupuy (in January 2005), whose views on the ethical aspects of the nanotechnologies were already well known [15, 16].

As a result of these three meetings, and given the breadth and scope of the debate in France, COMETS reached the following conclusions:

- Research in nanotechnologies was likely to fuel society's mistrust of scientists.
- It was important to examine the goals of the research, given the considerable impact it promised to have on society.

Now, COMETS was created in 1994 on the initiative of its then Director General, François Kourilsky, but by 2003 it was made up of new members and had obtained a new statute, since the decree of 2000 concerning the CNRS had conferred a more institutional existence upon it. The changed status of COMETS provided it with an opportunity to redefine its objectives. These now explicitly include deliberation on the final purpose of research. Among other things, the charter established by the modernised COMETS now clearly states its intention to "pursue the reflection on ethical issues, a reflection focusing on the way research is carried out, and taking into account its ends and its consequences" (COMETS Charter, July 2003).

So one of the tasks of the new COMETS was to reflect upon the consequences of research in nanotechnology and to promote a strong and lasting relationship of trust between society and scientists. This was particularly relevant for the CNRS, an important actor in the field of nanoscience research. Its ethics committee could not therefore reasonably refrain from reflecting upon the responsibility of CNRS researchers and clarifying the ethical issues surrounding future research and expected progress. In 2005, at its own instigation, COMETS set up a working group to pronounce on these questions.

The spokesperson for the working group was a philosopher and historian of science, Bernadette Bensaude Vincent, specialising in the history of chemistry and the relationships between science and society. Moreover, her research theme in 2004 already focused upon nanotechnologies and their nature, philosophical significance, and social impact. She had contributed to a special issue of the journals HYLE (International Journal for Philosophy of Chemistry) and TECHNE (Research in Philosophy and Technology) with an article entitled *Two Cultures of Nanotechnology* [17]. She thus already belonged to the incipient scientific community mentioned above. There was thus a good level of synergy between the questions raised by COMETS and the data gathered by Bernadette Bensaude Vincent when she had been questioning scientists in nanotechnology research centers for her own research (supported in part by the CNRS during a sabbatical year).

This contribution allowed COMETS to remain in direct relation with the grass roots of scientific research during its reflection upon the ethical aspects of nanotechnology and its consequences for society as a whole. The working group was thus able to adopt a bottom-up approach, rather than just imposing some 'ready-made' ethics to the problems of nanotechnologies. We shall see that, in the final declaration, this resulted in recommendations for raising the awareness of scientists about the impacts of their research, not only with regard to health and environmental risks, but also concerning their social, ethical, and more broadly cultural consequences.

20.5 A State of Turmoil in France

There was a somewhat contrasting situation among the European institutions in 2004, and especially between the United Kingdom and France. In the UK, the Royal Society report of November 2003 [18] had caused something of a stir by its open-minded attitude with regard to the human sciences and society. (This report was in fact preceded in July 2003 by an ESRC report [19] entitled *Social and Economic Challenges of Nanotechnology*.) In France, the report [20] by the *Académie des Sciences* and the *Académie des Technologies* looked in comparison much more like a case of the exact sciences withdrawing into their own inscrutable world, and the relevance of nanotechnology to biology was not even included. However, it should be said that OPECST had for its part organised a session on nanoscience and progress in medicine in May 2004 [21], in which the need for ethical reflection was clearly specified, but only to announce future deliberation by the *Comité consultatif national d'éthique* (CCNE).

The action of the CCNE did not originate solely in France, but was largely encouraged by the European Commission, which as we have seen was closely following events in North America. Indeed, the EC had set up a network of national ethics committees across Europe, including the CCNE, in Rome in December 2003, and had decided by June 2003 to put a session entitled *Ethical Implications of Nanotechnologies* on the agenda, with a presentation on nanomedicine and an address by Jean-Pierre Dupuy [15,16]. The CCNE drew its conclusions [22] a few months after COMETS. Although the two committees never really worked together, they had maintained quite close contacts, e.g., a CCNE observer attended COMETS meetings, so their declarations differed somewhat but were otherwise globally consistent.

In France, from 2004, institutional initiatives, scenes of debate and reflection, and reports began to snowball, organised by a broad range of different bodies, from associations, through agencies and organisations to ministries, including AFSSA, AFFSET, CPP, ECRIN, INERIS, INRS, *Conseil général des Mines*, *Conseil général des technologies de l'information*, CEA, INSERM, OPECST, *Cité des sciences et de l'industrie*, the Ministry of Health, *Association entreprises pour l'environnement* (EPE), and *Association pour la prévention de la pollution atmosphérique*, *Fondation Sciences citoyennes*, *France Nature Environnement*, *Mission agrobiosciences*, *Fondation Internet nouvelle génération* (FING), *Commission nationale informatique et libertés* (CNIL), *Direction générale de l'armement* (DGA), *Conseil économique et social*, the city hall of Paris , the region Ile-de-France, *Mouvement universel pour la responsabilité scientifique* (MURS), the political movement CAP21, the parliamentary association PACTE, the association Vivagora, and the list continues, with all this activity reverberating through the press in the form of commentaries and special reports.

20.6 Position of the CNRS

As an institution (noting that some of its research scientists took part in these debates of their own accord), the CNRS tended to remain rather in the background with regard to the issue of science and society (prior to the declarations of COMETS), but it did not remain silent about its involvement in nanoscience and technology, because from a strategic point of view it was important to show the decision-makers (the *Agence nationale de la recherche* or ANR had just listed it as one of its priorities) that it was one of the main actors in this sector. In actual fact, and it is not widely known, the CNRS was one of the very early actors in this field, with a heading 'Nanotechnologies' which already featured in the cross-disciplinary programme Ultimatech as early as 1991 [23]. So nanotechnology was already an issue at the CNRS well before 2000, and this shows to what extent the origins of this movement emanate from scientists rather than from some demand by society (as is often suggested to qualify certain research priorities).

What changed in the 2000s was the general propagation of the term 'nanoscience' (until then rarely used in France) to rename a whole range of scientific fields located upstream of the nanotechnologies and which had existed previously under different appellations. The CNRS thus advertised its strong presence in these 'nanosciences' on several occasions. The first was through a dossier of the *Journal du CNRS* in July 2002 [24], although there was more there on nanotechnology than on the question of nanoscience. Then the website of the department of physical sciences and mathematics gave a panorama of the nanosciences which presented only the expected benefits. Note also that the 2004 situation report of the *Comité national du CNRS* included a chapter on nanoscience and nanotechnology written by a group containing only members involved in the exact sciences, and thus not mentioning social issues. In that respect, it followed the tone of the guide produced by the ministry responsible for research in 2003, entitled *A la découverte du nanomonde* (Discovering the Nanoworld). In May 2005, the same ministry published a prospective reflection on nanoscience and nanotechnology [25], where impacts on society are simply presented as a problem of risks, and according to which it is simply the ignorance of protesters that has led to a completely unjustifiable mistrust on the part of the general public.

Subsequently, in September 2005, the CNRS published a glossy 40 page brochure called *Les Nanosciences* in the series *Focus* [26], which presented the full extent of the CNRS's commitment, but this time also in the human and social sciences. Consequently, they now mentioned the social impact as well as risks, and described the work underway at COMETS, but without entering into the ethical debate. (The press conference convened to present the brochure at the CNRS headquarters was the scene of a lively intervention on the part of the PMO group with regard to the final purpose of the nanosciences.) Dealing more explicitly with the ethical issues, the subsequent special report in the *Journal du CNRS* no. 189 of October 2005, entitled *La Déferlante Nano* (The Nano Tidal Wave) [27] contained an interview with Pierre Léna, the then president of COMETS, and another with Jean-Pierre Dupuy.

For its part the brochure on nanoscience in the *Focus* series was designed largely to demonstrate the major role played by the CNRS in this area, the role of European leader that it seemed important to assert at a time (2004) when France was proposing a European network NanoSci-ERA to complement the existing ERA-Net on nanotechnology [28]. It is instructive to note that this ERA-Net NanoSci-ERA did indeed include a brief discussion of the interactions between nanoscience and society, although the accent was still placed on communication with the public rather than the real ethical questions relating to nanoscience. At this time, most researchers were still convinced that ethical questions concerned only nanotechnology and not nanoscience.

Finally, for the sake of completeness, it should be mentioned that, in 2004, INIST put together a bulky dossier on nanoscience and nanotechnology, but excluding human and social sciences, while the information and public relations office of the CNRS produced *Sagasciences* [29], which gives ethics its full due. Note also that, in 2008, CNRS Images published a DVD which includes an interview with Bernadette Bensaude Vincent, providing numerous references on the societal consequences of nanoscience and nanotechnology [30]. It is thus possible to trace the evolution of ethical questions concerning nanoscience and nanotechnology within the CNRS, through the concurrent evolution in its style of communication.

20.7 The Position of Other French Institutions

The French atomic energy authority, the CEA, was also inspired to improve its communications on these issues, especially since it was so heavily involved in nano–bio convergence as coordinator of the European excellence network Nano2Life, inaugurated in the Savoy region of France in February 2004. It is notable that this network included an ethics committee which stimulated a significant debate among researchers belonging to it through seminars specifically organised to this effect. However, it is also notable that, in the *Biofutur* dossier devoted to nanobiotechnologies in November 2004 [31], the coordinator Patrick Boisseau wrote an article on Nano2Life in which there was no mention of ethics.

The CEA changed the orientation of its communications drive on the nanotechnologies slightly later, in the summer of 2005, when it published the special issue of *Clefs du CEA* (no. 52) entitled *Le nanomonde, de la science aux applications* (The Nanoworld: From Science to Applications) [32], where it began to adjust its image by including a paper by Louis Laurent, a CEA researcher. This paper contained a deep reflection on the good and bad consequences that might be engendered by nanotechnologies, and at the same time demonstrated that there was a real debate going on. Louis Laurent is a talented speaker and writer, attending many public meetings and cowriting popular accounts with Jean-Claude Petit, with evocative titles like *Nanosciences: nouvel âge d'or ou apocalypse?* (Nanoscience: New Golden Age or Apocalypse?) (July 2004 at the CEA website and then published in English in the journal HYLE [33]), or *Les nanotechnologies doivent-elles nous faire peur?* (Should We Be Afraid of the Nanotechnologies?) [34].

At the CEA, they share this talent for communication with Etienne Klein who, in a little book intended for the general public, had already raised the question of the potential threats of science [35]. And later on, Etienne Klein would produce a report with two other colleagues on the importance of nanotechnology for the CEA, under the title *Le débat sur les nanosciences: enjeux pour le CEA* (The Nanoscience Debate: What Is at Stake for the CEA) [36].

The CEA had thus evolved, and in issue no. 18 of the series *CEA Jeunes*, entitled *Le nanomonde* (The Nanoworld) in May 2008 [37], there are three pages describing a development of nanoscience and nanotechnology that is based upon a public-spirited attitude. It even mentions the code of conduct of the European Commission on research in nanoscience and nanotechnology [38], although this has in fact received a great deal of criticism.

Among the other French institutions that should be mentioned for their reflection on the way nanotechnology is controlled, there is the *Conseil général des technologies de l'information* (CGTI), where Françoise Roure has been deeply involved in the debate over the future promised to us by nanotechnology, and suggested various ways to remain in control of it. We could cite her report [39] with Jean-Pierre Dupuy which, among other things, recommended the creation of a societal observatory of nanoscience and nanotechnology, urged organisations to launch ELSA research, and called upon the French ethics committees to state their opinions on the subject. We could also cite Françoise Roure's article entitled *New Ethics for Nanosciences and the Future of Information Technology? Let the Limits Move*, published by the European Science Foundation in 2006 [40], which is remarkable for drawing the parallels with ethical questions in information technologies and, to a lesser extent, those in neuroscience (neuroethics).

20.8 Further Developments Within the CNRS

The CNRS had also made decisions in line with several reports advocating an emphasis on ELSA research in the context of nanoscience and nanotechnology. We have already mentioned that this type of research had been strongly promoted in the United States, and significantly but to a lesser extent by the European Commission. (It was advocated as early as 2001 by Roger Strand of the University of Bergen in Norway, in the European programme COST [41].) However, in France, prior to 2006, the research ministry did not support this type of research in its national nanoscience and nanotechnology programme PNANO. As a consequence, and unfortunately, the CNRS component of this PNANO programme did not involve researchers in the human and social science department. In order to develop ELSA research, the CNRS created a cross-disciplinary section of its national committee devoted to the social impacts of nanoscience and nanotechnology in 2004 (CID 43), and this meant that researchers were recruited from the human and social sciences, two in law and one in political science, who could work on research subjects relating to the regulatory issues raised in this area.

It is regrettable that philosophers of science were not represented in this section, despite the available posts. It no longer exists, but for four years it allowed a genuine cross-disciplinary dialogue between its members.

20.9 Recommendations by COMETS

The opinion of COMETS was published in October 2006 [42], at the same time as the deliberations of several of the other groups mentioned above were also coming to an end, and this blossoming of reports and declarations triggered a number of meetings to debate these issues. The conclusions drawn by COMETS occupied a special place because they concerned fundamental research, considered as being the source of this industrial revolution due to nanoscience and nanotechnology, and because they were addressed to a research organisation encompassing all disciplines. Furthermore, the scope of the opinion was not restricted to the risk aspect alone, but is in keeping with Jean-Pierre Dupuy's idea that one should not confuse ethics with risk. In that respect, it differs significantly from concomitant conclusions drawn for example by AFSSET [43] and CPP [44], which for their part address the issues of environmental and sanitary risk.

The full text of the COMETS opinion can be found at the CNRS website [42], but we can summarise the main points. COMETS suggests to the establishment and its researchers certain lines of investigation and recommendations that fall into three parts:

1. The opinion takes into account several specific features of research in nanoscience and nanotechnology: the difficulty in separating science from technology; the tension between opposing aspirations which underpin research (the will to control nature on the one hand, and the desire to discover unknown and unexpected properties on the other); the aura of fiction which accompanies the launch of new initiatives, and which must be taken seriously because the fiction makes research part of an economy of promise, creating expectations as well as fears in society; and finally, the generic nature of the nanotechnologies which will affect production across the board and create a situation of uncertainty.
   It also takes into account the competitive context in which research in nanoscience and nanotechnology is carried out. Naturally, 'nano' has become a convenient slogan, but three new features need to be considered: the scientific context with the nano-bio-info-cognitive (NBIC) convergence, the political context of globalisation and competition, and the social context with an ever more demanding public.
   It gives an overview of the rather contrasting ethical and social initiatives developed in other European countries: Constructive Technology Assessment in the Netherlands, Public Engagement in Science in Great Britain, and *Pour une symbiose entre science et culture* (Symbiosis Between Science and Culture) in Germany and France.
   Finally, it lays the foundations of an applied ethics for nanoscience, stressing the fact that ethics concerns not only good practice in research, the prevention of risks, and precautionary measures in the face of uncertainty, but also a reflection on values and purpose.

2. The report states the importance of the freedom of research, but also stresses the social responsibility of researchers, urging all parties to reestablish a high level of trust between the world of research and the general public.
3. The report makes 8 recommendations which emphasise the following three aims: (i) to raise the awareness of researchers about ethical considerations by creating places and opportunities to get them involved; (ii) to develop ELSA research within the organisation; and (iii) to favour a coevolution of science and society through an appropriate dialogue involving all concerned.

One might criticize these recommendations for being too general, and for being applicable to the whole of scientific research. Of course, it is perfectly true that what is said about nanoscience and nanotechnology on the social impact and the responsibility of researchers could sometimes be transposed *mutatis mutandis* to many other areas of research, but that should not diminish the force of the recommendations. Indeed, any form of research should be accompanied by this kind of reflection, but nanoscience and nanotechnology have in a sense brought this upon themselves, partly through what is commonly known as 'hype' today, and partly because they are generic technologies for other technologies that are themselves generic (e.g., computing), underpinning NBIC convergence, and they thus cover a very broad field of research. In fact, the programming of nanoscience and nanotechnology research constitutes an archetypal problem of scientific policy, with all the accompanying doubts and debates, including the ethical questions, and it would be a mistake not to carry out this reflection in their particular case on the grounds that it is not specific to nanoscience and nanotechnology.

20.10 Impact of the COMETS Conclusions and the Role of the CNRS in the Debate

It was the first time that a declaration by the ethics committee of the CNRS was not simply dealing with some deontological question to do with the inner workings of the scientific community, but instead with a specific sector of research and its impact on the future of society. It was thus noticed outside the CNRS, even in the press and televised debates [45–48], but above all by the organisers of the debates mentioned earlier, which prospered in 2006 and 2007. In particular, the CNRS was invited to talk about the ethics of nanoscience at the seminar [49] organised by the ministry of health in October 2006 (which gave the Directorate General for Health the opportunity to compare the conclusions announced by the CNRS, the CPP, AFSSET, AFSSAPS, and the future conclusion of the CCNE), but also at the public hearing organised by OPECST in November 2006 [50], at the national debate orchestrated by the *Cité des sciences et de l'industrie* [51] in March 2007, and at the workshop organised by the European Commission in May 2007 to prepare the code of conduct in nanoscience and nanotechnology [38] of February 2008.

The CNRS could be present at these meetings and debates in three guises: through the people responsible for its scientific policy with regard to nanoscience and nanotechnology (e.g., Michel Lannoo at OPECST), through the people representing COMETS, which drew up this ethical statement (although this was not strictly the CNRS's statement, but that of an independent committee, and confusion was commonplace), and finally through its research members who were scientific experts on questions relating to nanoscience and nanotechnology. Among these experts, it is interesting to mention Eric Gaffet, research director in chemistry at the CNRS, who played an important role for many authorities assessing the toxicological risks due to nanomaterials, and in particular AFSSET, ANR, the association ECRIN, and the OECD (which set up several working groups on nanotechnologies).

Among the decision-makers, Jean-Claude André should also be noted. Former assistant director at the CNRS department of engineering sciences, he became interested in nanoscience and nanotechnology when occupying the post of scientific director he subsequently took up at the INRS [52]. He then returned to the CNRS as consultant to the ST2I department, where he proposed the concept of socially responsible research (SRR) [53]. He represented the CNRS at interministerial meetings which discussed the code of conduct proposed in 2008 by Brussels for responsible research in nanoscience and nanotechnology [38]. He then piloted the implementation of a 'Wiki' CNRS devoted to this area. This shows that CNRS influence on the public debate in this area exceeds that of its COMETS committee or CNRS management, and that it is more difficult to identify exactly.

As a further example, Stéphanie Lacour, one of the two researchers in law recruited at the CNRS by the CID 43, attended the ministry of health seminar and the *Conférence de Citoyens* (Citizens' Conference) organised by the Ile-de-France region of France in January 2007, before being assigned to the OECD to support its activities in nanoscience and nanotechnology.

20.11 Dialogue with Civil Society

A common factor in many reports making recommendations is that it is important to change the form of the dialogue between scientists and the public. This particular recommendation, which features in the COMETS statement, was already at the heart of the report [54] made in Grenoble in 2005 (whose first recommendation was to organise a citizens' conference). It materialised in different forms and on many separate occasions in France and in other European countries (see, in [55–57], a list of meetings with public participation, or indeed the final report of the Nanotechnology Engagement Group, or again the OECD seminar of 30 October 2008 in Delft to explore methods of communication and participation).

In order to implement this recommendation, it is now possible to exploit the results of research in the science of communication, but also to refer to a great deal of experience gained in the field, like the evening debates organised by *Association Vivagora* in Paris and Grenoble, the *Conférence de Citoyens* organised in January 2007 by the Ile-de-France region of France, or the event organised at the *Cité des Sciences* in March 2007 [51] (to fulfill a promise by Dominique de Villepin, who had earlier announced a national debate for the inauguration of MINATEC).

Looking more closely at these public debates, one can classify them to some extent by distinguishing the following features:

- Those intended primarily to persuade the public of the social acceptability of choices that have already been made, and generally to obtain public trust and support.
- Those, more profound in aim, that seek to convey a full and transparent description of the risks, so that the democratic choice can be made by enlightened members of the public.
- Those, more ambitious in scope, that seek to provide a genuine opportunity for joint decisions about priorities, moving toward a new way of governing scientific and technological priorities (consider the seminar on public debates organised at INRA in January 2008 by the *Risques, sciences, société* network).

The CNRS can adapt to this new situation and avoid the many pitfalls. At a time when the *Commission nationale du débat public* (National Committee for Public Debate or CNDP) is launching a national public debate on nanoscience and nanotechnology, the CNRS has many assets, for example in the person of Daniel Boy, a researcher in sociology, who fully understands the two essential components of scientific debate: the public's perception of science and what constitutes a genuine public debate [58, 59].

20.12 Preparing Researchers for a New Form of Communication

Without going so far as to suggest joint governance of research with the civil society, the very fact of encouraging scientists to engage directly in a balanced dialogue with society on the ethical, legal, and social impacts of their research (the proactive dialogue that is also advocated by the Swiss academies [60]) requires training for researchers. In March 2007, the COMETS committee of the CNRS organised a training school at Les Houches for researchers in nanoscience and nanotechnology on the theme of freedom and responsibility with regard to this kind of research (*Entre liberté et responsabilité: la recherche en nanosciences et technologies*). The aim was that these researchers could learn to view their work, and be able to speak about it, not only purely from the angle of a scientific challenge, but also in the light of the ethical issues it raises for our society. Indeed, it is also with regard to these aspects that their civil society interlocutors expect to hear their position in the debate today.

The real difficulty with nanotechnologies is that they mainly concern developments that do not yet exist, so that the discussion is really about a future vision of our society. Now researchers do not particularly like to discuss science fiction, especially in public, while recommendation 7 of the COMETS conclusion states that they must "rise to the challenge of considering very long term issues, even if they may have the appearance of science fiction" and also that they should "give their dreams explicit form".

The fact of presenting science fiction as a thought experiment that might throw light on the debate is somewhat original compared with the statements made by other organisations, which only attempt to deal with the immediate problems. The UNESCO document of 2006 [61] even presents this kind of science fiction as dangerous, because its authors consider that it distracts attention from the true, more imminent problems. This is not completely wrong if we consider the immediate risks relating to products already on the market, or the ethical and geopolitical problems which are already with us and which may soon grow in importance. But this does not mean that the goals hidden in dream and fiction should not serve as a useful, even inescapable, guide in the dialogue, since literature has long since taken possession of nanotechnology [62] and the common myth it creates is not always totally disconnected from the final aims of current research (as we say, reality sometimes exceeds fiction). Since research prepares the world of tomorrow, it would be difficult to disconnect the ethics of research from a prospective approach, and many reports have understood this, in particular those of the Science and Technological Options Assessment (STOA) [63], which is to the European parliament what the OPECST is to the French parliament.

20.13 Conclusion

In conclusion, this statement by the CNRS is ambitious, because it does not try to offer either researchers or decision-makers a ready-made applied ethics, a ready-to-use ethics so to speak, but urges a general vigilance built upon experience in the field, with regard to the final aims of research. It is much more difficult, but the cross-disciplinary assets of the CNRS are largely sufficient to bring this reflection to fruition by getting researchers of different backgrounds into the discussion (as was observed at Les Houches).

This cross-disciplinary contact and the reflection on final purpose provide a way out of the highly blinkered view that threatens any given researcher working often largely alone in some narrow area, a situation that could easily diminish a scientist's feeling of responsibility. Hopefully, researchers will realise that they are not just a cog in the machine of an evolving society (as research is often described), and that they too can be the copilot, that this will not necessarily reduce their creativity, but may even stimulate it, for ethics should certainly not be seen as a curb on research.

The recommendations of the ethics committee of the CNRS (COMETS) aim to implement an ethical approach within the organisation, and the stakes are high, because this approach could equally apply to all emerging technologies where, as with nanoscience and nanotechnology, the frontier between science and technology is growing ever fuzzier.

Appendix: Table of Acronyms

AFFSET	Agence française de sécurité sanitaire de l'environnement et du travail
AFSSA	Agence française de sécurité sanitaire des aliments
ANR	Agence nationale de la recherche
CCNE	Comité consultatif national d'éthique
CEA	Commissariat à l'énergie atomique
CGTI	Conseil général des technologies de l'information
CNDP	Commission nationale du débat public
CNIL	Commission nationale informatique et libertés
CNRS	Centre national de la recherche scientifique
COMETS	Comité d'éthique du CNRS
CPP	Comité de la prévention et de la précaution
DGA	Direction générale de l'armement
ECRIN	Association échange et coordination recherche-industrie
ELSA	Ethical Legal and Social Aspects
EPE	Association entreprises pour l'environnement
ERA	European Research Area
ESRC	Economic and Social Research Council
ETC Group	Action Group on Erosion, Technology and Concentration
FING	Fondation Internet nouvelle génération
INERIS	Institut national de l'environnement industriel et des risques
INIST	Institut de l'information scientifique et technique
INPG	Institut national polytechnique de Grenoble
INRA	Institut national de la recherche agronomique
INRS	Institut national de recherche et de sécurité
INSERM	Institut national de la santé et de la recherche médicale
MINATEC	Campus d'innovation pour les micro- et nanotechnologies
MIT	Massachusetts Institute of Technology
MURS	Mouvement universel pour la responsabilité scientifique
NBIC	Nano-bio-info-cognitive
NSF	National Science Foundation
OECD	Organisation of Economic Cooperation and Development
NGO	Non-governmental organisation
OPECST	Office parlementaire des choix scientifiques et technologiques

PMO	Pièces et Main d'œuvre
PNANO	Programme en nanosciences et nanotechnologies
SRR	Socially responsible research
STOA	Science and Technological Options Assessment

References

1. Foresight Institute: www.foresight.org/guidelines/2004oct.html. Foresight Institute's Worskshop on Molecular Nanotechnology Research Policy Guidelines, Monterey workshop, 19–21 February 1999. Environmental Regulation of Nanotechnology, G.H. Reynolds, Environmental Law Reporter, 6-2001
2. Nanotechnology: Societal implications, maximizing benefit for humanity, December 2003, NSF, United States
3. The big down, from genomes to atoms, ETC Group, January 2003
4. Benefits, risks, ethical and social aspects of nanotechnology, 4th Nanoforum Report, June 2004
5. Nanotechnology: Revolutionary opportunities & societal implications, 3rd Joint EC-NST Workshop on Nanotechnology, 31 January 2002, Italy
6. Vers une stratégie européenne en faveur des nanotechnologies, Communication de la Commission COM, 338 (2004)
7. S. Fantechi, R. Tomellini (eds.): Proceedings of the EuroNanoForum 2003, DG Research, European Commission, www.euronanoforum2007.de/ENF2003/index.htm
8. Nanotechnologies: A preliminary risk analysis, 1–2 March 2004 Workshop, Health and Consumer Protection Directorate General of the European Commission
9. P. Healey, H. Glimell (eds): European workshop on social and economic issues of nanotechnologies and nanosciences, 14–15 April 2004
10. Converging technologies: Shaping the future of European societies, European Community, August 2004
11. Converging technologies for improving human performance, NSF, United States (2002)
12. D. Baird, A. Nordmann, J. Schummer: *Discovering the Nanoscale.* IOS Press, Amsterdam (2004)
13. Nanotechnologies/Maxiservitude, January 2003, Pièces et Main d'œuvre (PMO)
14. Manifeste pour la technologie au service de l'homme, INPG, Grenoble, 12 October 2000
15. J.P. Dupuy: Pour une évaluation normative du programme nanotechnologique. Communication au Forum européen des comités d'éthique nationaux, Rome, December 2003 and in a special issue of Les Nanotechnologies, Réalités Industrielles, Annales des Mines, February 2004
16. J.P. Dupuy: Complexity and uncertainty: A prudential approach to nanotechnology. Contribution to the workshop Nanotechnology: A preliminary risk analysis, European Commission, March 2004
17. Nanotech Challenges, Vol. 10, no. 2 (2004), HYLE (International Journal for Philosophy of Chemistry) and TECHNE (Research in Philosophy and Technology)

18. Nanoscience and nanotechnologies: Opportunities and uncertainties. The Royal Society, July 2004
19. The Social and Economic Challenges of Nanotechnology. ESRC, United Kingdom, July 2003
20. Nanosciences, nanotechnologies. Rapport de l'Académie des Sciences et de l'Académie des Technologies, April 2004
21. Nanosciences et progrès médical. Rapport de l'OPECST, J.L. Lorrain and D. Raoul, May 2004
22. Questions éthiques posées par les nanosciences, les nanotechnologies et la santé, 1 February 2007, Avis no. 96, CCNE
23. Lettre des programmes interdisciplinaires de recherche du CNRS. Ultimatech, no. 2, August 1991
24. 0.000 000 001 mètre sous la matière. Le Journal du CNRS no. 151–152, July–August 2002
25. Nanosciences et nanotechnologies: Une réflexion prospective, May 2005, Mission scientifique, technique et pédagogique, ministère délégué à la Recherche
26. Les Nanosciences, collection Focus, CNRS, September 2005
27. La Déferlante nano. Le Journal du CNRS, no. 189, October 2005; Riding the nano wave. CNRS International Magazine no. 2, Spring 2006
28. NanoSci-ERA. NanoScience in the European Research Area, ERA-Net of the Research DG of the European Commission (2004)
29. www.cnrs.fr/cw/dossiers/dosnano
30. www.cnrs.fr/cnrs-images/nano/ressources_web.html
31. Nanobiotechnologies: des avancées pour les sciences du vivant. Dossier de Biofutur 249, November 2004
32. Le nanomonde, de la science aux applications. Clefs CEA, no. 52, summer 2005
33. L. Laurent, J.C. Petit: Nanosciences and its convergence with other technologies: New Golden Age or Apocalypse? HYLE **11**, 45–76 (2005)
34. L. Laurent, J.C. Petit: *Les nanotechnologies doivent-elles nous faire peur?* Editions Le Pommier Paris (2005)
35. E. Klein: *La science nous menace-t-elle?* Editions Le Pommier, Paris (2003)
36. E. Klein, A. Grinbaum, V. Bontems: Le débat sur les nanosciences: enjeux pour le CEA. DSM-LARSIM, CEA, June 2007
37. Le nanomonde, CEA Jeunes no. 18 (2008)
38. Code de bonne conduite pour une recherche responsable en nanosciences et nanotechnologies, Recommandation de la Commission, 7 February 2008
39. J.P. Dupuy, F. Roure: Les nanotechnologies: éthique et prospective industrielle. CGM and CGTI (2004)
40. F. Roure: New ethics for nanosciences and the future of information technology? Let the limits move. In: *Nanosciences and the long-term future of information technology*, European Science Foundation, 2006
41. ELSA studies of nanosciences and nanotechnology, MEMO to the COST NST Advisory Group, Roger Strand, 16 November 2001
42. Enjeux éthiques des nanosciences et des nanotechnologies, avis du COMETS du CNRS, 12 October 2006 www.cnrs.fr/fr/organisme/ethique/comets/docs/ethique_nanos_061013.pdf
43. Les Nanomatériaux. Effets sur la santé de l'homme et sur l'environnement, AFSSET, July 2006
44. Nanotechnologies, nanoparticules. Quels dangers, quels risques? Report by the CPP, May 2006

45. Y. de Kerorguen: Le débat sur les nanotechnologies prend de l'ampleur. La Tribune, 20 October 2006
46. G. Toulouse: Ethique des sciences: une étape majeure? La Croix, 28 November 2006
47. S. Hasendahl: Faut-il avoir peur des nanosciences? Le Quotidien du Médecin, 15 January 2007
48. Touche pas ma planète, Debates on nanotechnologies in January 2006 and December 2006 with Corinne Lepage on Direct 8
49. Séminaire interministériel sur les risques liés aux nanomatériaux et nanotechnologies, Direction Générale de la santé, Paris, 19 October 2006
50. Les nanotechnologies: risques potentiels, enjeux éthiques, compte rendu de l'audition publique, OPECST, 7 November 2006
51. Nanotechnologies: le point sur les débats, des orientations pour demain, Cahier d'Acteurs, Cité des sciences et de l'industrie, March 2007
52. J.C. André: Réflexions autour de la nano-éthique et de la nano-normalisation, INRS, September 2005
53. J.C. André: Vers le développement d'une recherche durable ... ou vers une (ré)humanisation des sciences et des artefacts. Environnement, Risques & Santé **7**, 47–54 (2008)
54. P.B. Joly (ed.): Démocratie locale et maîtrise sociale des nanotechnologies, report for La métro de Grenoble, September 2005
55. N. Baya Laffite, P.B. Joly: Nanotechnology and society: Where do we stand in the ladder of citizen participation? CIPAST Newsletter, March 2008
56. Democratic technologies? Final report of the Nanotechnology Engagement Group, Involve 2007, www.nanowerk.com/nanotechnology/reports/reportpdf/report105.pdf
57. How to best engage the general public in nanotechnology? OECD Conference, Delft, The Netherlands, 30 October 2008
58. D. Boy: Vers une démocratisation des choix scientifiques et techniques? Deux mille cinquante, PUF, **7**, April 2008
59. D. Boy: *Pourquoi avons-nous peur de la technologie?* Sciences PO Les Presses, Paris (2007)
60. Thèses sur les nanotechnologies, a+, Académies suisses des Sciences (2008)
61. The ethics and politics of nanotechnology, UNESCO (2006)
62. G. Collins: Nanosciences et science-fiction. Pour la Science, no. 290, December 2001
63. Technology assessment on converging technologies. Literature and vision assessment, Document for the workshop: Converging technologies in the 21st century: heaven, hell or down to earth? STOA, European Parliament, June 2006

21

Situation in France: Nanoparticles in the Grenelle Environment Forum

Philippe Hubert

The aims of the round table talks in France known as the grenelle environment forum were to implement plans of action with regard to sustainable development and reach a consensus between the various economic and social players. It lasted 2 years, from the first meetings of the working groups in the summer of 2007 until the law referred to as *Grenelle I* was voted after a second reading to parliament in the spring of 2009. Over the two years, the various positions adopted with regard to nanomaterials, nanotechnologies, and nanoparticules (the three lines of attack were used) gradually began to consolidate.

21.1 Health–Environment Working Group and Conclusions of the Round Tables

The first stage of the environment forum took place in the summer to autumn of 2007. Working groups were set up and their conclusions were reformulated by a round table. The health–environment working group tackled this question in the summer of 2007, and its proposals were established at three levels: summary, text, and detailed text.

In the summary[1] of the health–environment working group, the main lines were already identified. The group thus proposed the following five points:

- Organisation of a consensus conference by the *Commission nationale du débat public*.
- Establishment of a consultative committee involving all stakeholders.
- Improvement of understanding regarding manufactured nanoparticles.
- Maximum reduction of employee exposure (applying a principle of precaution).

[1] See Summary 3.3 in the Appendix.

P. Houdy et al. (eds.), *Nanoethics and Nanotoxicology*,

DOI 10.1007/978-3-642-20177-6_21,

- Establishment of a compulsory declaration with rules for labelling products and informing users to be determined after consultation.

There was a consensus on these points.

The group organisation of the forum allowed the expression of dissent concerning two issues. The ONG representatives proposed a ban on all commercialisation in food applications and personal hygiene, cosmetic, and clothing products, whereas employers' representatives preferred a case by case approach. Between the option of national regulation and recourse to a process of authorisation through European regulations, the group had still not come to a conclusion.

Moving ahead to the full text, the group allowed 2 years to set up a regulatory framework in order to assess, then regulate products involving nanomaterials. The notion of compulsory assessment was thus advocated, while the debate on the regulatory framework was just beginning, for not all participants shared the idea that further regulation was necessary. Doubts were expressed over the ability of REACH (Registration, Evaluation, Authorization and restriction of CHemicals) to integrate substances in the nanoparticulate state within its field of application.

The detailed text contained even more precise recommendations. Those recommendations not mentioned in the summary and text are detailed below. Inventories were requested, both for industrial activity and for the safety and precautionary measures adopted. The French government was requested to take steps to improve REACH.

The detailed text was also more prescriptive with regard to research, and required, apart from fundamental developments, an applied research programme so that current regulatory tests for toxicity and ecotoxicity could be rapidly adapted to nanoparticles, leading within 2 years to a proposal for a regulatory framework along with relevant criteria of classification and dangerousness. Finally, it suggested that assessment should include assessment of alternative solutions, i.e., substitution possibilities, and this in the context of a pluralistic process. In addition, the group requested that nanomaterials be taken into account in the *Plan national santé-environnement 2* (Second National Health and Environment Plan), expected to get underway in 2010.

In conclusion, the final outcome of the round tables was a four-point recommendation (Action 159):

- Anticipation of risks due to nanomaterials. The *Commission nationale du débat public* will organise a debate on the risks related to nanoparticles and nanomaterials.
- The presence of nanoparticles in products destined for the general public must be clearly stated, from 2008.
- Systematic cost–benefit analysis before commercialising products containing nanoparticles or nanomaterials, from 2008.

- Ensure that employees are adequately informed and protected, on the basis of the AFSSET study.

The deadlines fixed by the round table were more ambitious than those of the working group, with commitments for 2008. The more operational arrangements (the need for an assessment process and adequate tools, texts, coordination of some kind with REACH) were not spelt out in this recommendation. They were postponed to subsequent work by the operational committees.

21.2 Recommendations of the Three Operational Committees

The second phase of the grenelle environment forum was conducted in the form of operational committees. The subject was thus taken up again by three operational committees: research (Comop 30), health surveillance (Comop 19), and the *Plan national santé-environnement* (National Health and Environment Plan) (Comop 20).

The operational committee for research dealt with nanomaterials among other innovations relating to the building industry, and nanotechnologies for solar energy and aeronautics. It considered nanoparticles from the standpoint of the development of research on health risks among other emerging factors (e.g., vector-borne infectious diseases, climate change, emerging pollutants, radiofrequencies), and thus integrated them into the research it proposed on toxicology and ecotoxicology.

The committee relating to health watch and emerging risks focused a large part of its work on this subject. It restricted attention to the area of deliberately manufactured nanoparticles, excluding nanoparticles that are naturally present or emitted unintentionally.

The range of themes on which it requested research development with regard to the harmful aspects of nanotechnologies was rather broad, since it included methods for measurement and testing, toxicology and ecotoxicology, the fate of nanomaterials in the environment, safety, toxic action mechanisms, and epidemiology (with priority for exposed workers). Concerning research, after examining the relationships with toxicological approaches to chemical substances, proposals were made to establish a structure, and it was noted that competent bodies would need to be created. Regarding communications, the idea of public debate was favoured, in association with many initiatives (nanoforums, regular opportunities for dialogue).

The reflection was extended to proposals for changes in the regulatory framework dealing with occupational and environmental safety, with proposals for specific texts on declaration. The committee suggested taking the issue to Europe by supporting a project in the context of the French presidency. One interesting detail is that labelling must appear on finished products containing manufactured nanoparticles. The committee drew up texts in the form of legislative bills, including one article specifically dealing with the obligation of declaration:

> Article L. 523-1 makes it compulsory to declare substances thereby produced in the nanoparticulate state. This declaration includes features allowing the precise identification of the relevant substances, as well as the uses and the amounts on the market. The obligation to declare these contents applies to manufacturers, importers, and those responsible for their release on the market, both on an industrial level and in research.

Such precise use of language may seem exaggerated at this stage, but it was essential in order to leave the realm of principle and test the realism of the proposals.

It was proposed to run this through an interdepartmental group on nanotechnologies, bringing together the relevant ministries, in relation with all the stakeholders. The associated coordination was to include public–private partnerships.

The third committee on this subject, the committee for the *Plan national santé-environnement*, announced its conclusions somewhat later, in the spring of 2009. It proposed to act in the following five ways (Action 47: To strengthen regulation, surveillance, and expertise with regard to nanomaterials and risk prevention):

- To reinforce surveillance and expertise with regard to nanomaterials following the public debate to take place from September 2009.
- To strengthen regulations on nanomaterials by making it compulsory to declare their use in commercialised products, by studying the possibilities for changing the regulation of classified installations to ensure that they take into account activities relative to the manufacture of nanomaterials, and their possible impacts on humans and the environment, and by setting up a programme of specific controls to check as soon as possible that the new regulations are implemented.
- To strengthen efforts to inform and dialogue with the public.
- To reinforce safety measures in the workplace with regard to nanomaterials, in conformity with the recommendations made by the *Agence française de sécurité sanitaire de l'environnement et du travail* (AFSSET) and the *Haut Conseil de santé publique* (HCSP).
- To develop and validate the relevant tests.

The committee also integrated nanomaterials among emergent risks, for which it requested support for research (Action 54). The developments planned for research picked up some of the points made by the committee with regard to surveillance, but emphasised other aspects, such as the need to coordinate research on the European level and work with European institutions, e.g., the Health and Consumer Protection Directorate General of the European Commission.

21.3 Law and Prospects Opened by the Grenelle Environment Forum

The law *Grenelle 1* of 3 August 2009 took up the main recommendations in its article 42.[2] This begins with a point concerning European government:

> France will promote, on the European level, a renewal of expertise and the evaluation of emerging technologies, especially with regard to nanotechnologies and biotechnologies, in order to update the knowledge used in all disciplines.

This proposal did not come directly from the working groups, which had discussed national organisation in greater depth, and focused assessment rather on emergent risks than on emerging technologies. It is important to stress this double discrepancy.

The themes raised by practically all the groups are then taken up:

- A public debate before the end of 2009.
- Compulsory declaration for substances in the nanoparticulate state or materials destined to reject such under certain conditions, to be introduced within 2 years. The notion of rejection "under normal or reasonably predictable conditions of use" is one that deserves attention, and probably a certain amount of research. However, it might exempt some materials containing substances in the nanoparticulate state.
- A campaign to inform the public and consumers.
- A methodology for assessing risks and benefits relating to these substances and products.
- Improved information from employers to employees, regarding risks and measures to be taken to guarantee worker safety.

21.4 Conclusion

The evolution of ideas in going from a working group via a round table and three committees to a bill of law is rather complex, but it was achieved in under 2 years. During the process as a whole, diverse recommendations tended to consolidate rather than to diverge. The different stages made it possible to include a growing number of players, in such a way that the steps taken today can claim a broadened level of support.

The evolution of the debate brought out certain themes, while others tended to be played down. For example, there was no initial consensus on demands for a ban or a moratorium, and these ideas finally disappeared from the proposals.

[2] See Article 42 in the Appendix.

The initial alternative between national regulation and an evolution of international regulation was finally decided in the bill, which elected for the national option, without completely rejecting the possibility of European measures. The relationship with REACH remains an open issue.

While the law does not pick up on them, the proposals for research were consensual and should be integrated into the *Plan national santé-environnement*. It is important to note that they refer to two different levels: research upstream, more or less finalised, but which complements highly applied research on the development of measurement tools and operational test systems.

The example of operational test systems pinpoints the main weakness of this set of proposals. They are based on two levels of operationality and temporality. One series of proposals is very short term, and no date beyond 2 years is mentioned. Another series involves no schedule at all. Intermediate horizons at 5 to 10 years are not mentioned.

In the final analysis, the main difficulty to be overcome will be the contradiction between requests for assessment, labelling, etc., to be imposed in the very short term, and the fact that usable practical tools are neither ready nor validated.

Appendix

Summary 3.3. To anticipate the risks due to nanomaterials. The group suggests the organisation of a scientific consensus conference followed by a public debate, e.g., by the *Commission nationale du débat public*, in 2008. A consultative committee will be set up, bringing together all stakeholders. A research campaign will be carried out to improve understanding of manufactured nanoparticles. Given the lack of knowledge concerning their health impacts, employee exposure to manufactured nanomaterials or nanoparticles should be reduced as far as possible as a matter of precaution. The NGO representatives propose to ban all commercialisation in food applications, personal hygiene products, cosmetic products, and clothing. Employers' representatives prefer a case by case approach. A compulsory declaration will be set up and the information will be made transparent for users of nanomaterials in ways (e.g., labelling) to be defined by consultation. Opinions are divided between the implementation of specific national regulations and recourse to an authorisation process specified by European regulations.

Article 42. The surveillance of emergent risks for the environment and health will be intensified by strengthening coordination and modernising all existing health watch networks.

France will promote, on the European level, a renewal of expertise and the evaluation of emerging technologies, especially with regard to nanotechnologies and biotechnologies, in order to update the knowledge used in all disciplines.

The use of substances in the nanoparticulate state or materials containing nanoparticles will be the subject of a public debate organised on a national level before the end of 2009. The State sets the objective that, within two years of the promulgation of the present law, the manufacture, importation, or commercialisation of substances in the nanoparticulate state or materials destined to reject such substances under normal or reasonably predictable conditions of use, must be the subject of a compulsory declaration to the administrative authority, relating in particular to the amounts and uses, and must also be the subject of a campaign to inform the public and consumers. A methodology will be devised for assessing risks and benefits relating to these substances and products. The State will ensure that the information due to employees by employers will be improved regarding risks and measures to be taken to guarantee worker safety.

The government will set up a body to watch over and measure electromagnetic waves, run by officially authorised independent organisations. This will be financed by an independent fund provided for by the contributions of network operators emitting electromagnetic waves. The results of these measurements will be sent to the *Agence française de sécurité sanitaire de l'environnement et du travail* and the *Agence nationale des fréquences*, who will make them public. A decree by the Council of State will specify the way these bodies will operate, together with a list of legal entities approved to request measurements and the conditions under which such measurements can be requested. Parishes and municipalities will be involved in any decision to allow operators to set up emitters in the context of local charters or new procedures of consultation on the parish or county level. A synopsis of scientific studies relating to the effects of electromagnetic fields on health will be presented by the government to parliament before the end of 2009.

A nationwide plan of climate adaptation will be prepared by 2011 for the various sectors of activity.

22

Situation in France: The Position of a Federation of Environmental Protection NGOs

José Cambou and Dominique Proy

France Nature Environnement (www.fne.asso.fr) is the French federation of associations for the protection of nature and the environment, and a member of the European Environmental Bureau. Founded in 1968, it has explicitly been in charge of environmental health issues since 1997.

Would the study of potential risks due to nanotechnologies by researchers at the University of British Colombia and the University of Minnesota constitute an objective endorsement of the urgent need to reconsider French legislation on this subject? Is France once again going to drag its heels in comparison with the United States? The study concludes that the current American regulatory system for commercialised nanotechnological products needs to be modified. And do the regulations of the European Union not also attest to the urgent need for legislation tailored to the real risks of nanoparticles in the food chain and nanocosmetics? And therefore, is *France Nature Environnement* not justified in devoting its attentions to such a vast subject, poorly understood by the general public and barely acknowledged until recently by the players in the economic and political world?

Are the scientific studies around the world of sufficient quality to demand monitoring of production processes involving nanoparticles, a form of labelling, and an information campaign for public authorities and consumers?

From an ethical standpoint, the response of *France Nature Environnement* (FNE) is clearly affirmative. The hazards relating to nanoproducts seem at this stage to be even more pernicious than those of asbestos, because the potential economic profits and technological spinoffs are extremely attractive. If the fascination for money and science are not to lead to risks that could have serious impacts on human health on a transgenerational scale, and on the equilibrium of the natural environment, it is essential that scientific options should be clearly understood by the different stakeholders.

France Nature Environnement, a federation of associations for the protection of nature and the environment, has maintained the same clear stance since 2006, specifically regarding nanotechnology, nanoparticles, and nanomaterials, validated by its board of directors.

P. Houdy et al. (eds.), *Nanoethics and Nanotoxicology*,

DOI 10.1007/978-3-642-20177-6_22,

Nanotechnology encompasses many different areas whose common denominator is the use of nano-objects[1] (nanoparticles have linear dimensions less than 100 nanometers, and one nanometer is one billionth of a meter). Given the novel physical, chemical, and biological properties induced on this length scale, experts predict exponential growth over the next few years. These nanomaterials, and more generally nano-objects, constitute a new source of exposure that involves specific features due to the size of the nanoparticle. It is thus both legitimate and urgent to address the question of their possible toxicity for humans and ecosystems.

Certain particles, including carbon nanotubes (CNT) and silver nanoparticles, have already diffused into the natural environment. The latter are beginning to hamper the correct functioning of sewage treatment stations.

The main objectives of *France Nature Environnement* are:

- To encourage a policy of precaution and prevention, including appropriate regulation, as soon as possible.
- To request an exhaustive inventory of everything manufactured and commercialised, to be kept permanently up to date, in order to establish effective traceability.
- To take appropriate measures depending on use, exposure, and sanitary and environmental risks.
- To encourage scientific understanding to obtain a better grasp of the hazards and risks and inform the public.
- To involve all stakeholders, including environmental protection NGOs.

The main expectations and demands of *France Nature Environnement* regarding the nanotechnologies feature in a dedicated platform.[2]

Nanomaterials provide an illustration of potentially risk-bearing emerging technologies. Our demands involve a specific platform which simultaneously addresses manufacturers, researchers, and public authorities. The positions adopted there were developed by the board of FNE's health–environment network. They took into account observations received from correspondents within the member groups of the federation.

In 2006, we made an official presentation in France, and also to the European parliament, of a first version of this platform. The second version was presented to stakeholders in the Grenelle Environment Consultation in 2007, and was considered positively. Scientific publications since 2006 have validated the hypotheses cited by *France Nature Environnement* and confirmed

[1] While there is no consensual terminology on the international level, the generic term 'nano-object' is used in the recommendation of the European Commission of 7 February 2008 for a code of good conduct for responsible research in nanoscience and nanotechnology to refer to products arising from such research. It covers nanoparticles and their aggregation on the nanometric scale, nanosystems, nanomaterials, nanostructured materials, and nanoproducts. Sometimes the term 'nanoform' is used.

[2] www.fne.asso.fr/fr/nos-dossiers/sante-environnement/nanotechnologies.html

the relevance of the stances adopted. A third version was produced at the beginning of June 2009. It begins with a set of preambles before presenting demands targeting each stakeholder profile.

The potential risks of exposure to nanotechnologies and nanomaterials are considered by the representatives of *France Nature Environnement* to be extremely worrying, and this for precise reasons. It is already officially and clearly established that the nanometric scale of substances and materials leads to radically different physical, chemical, electromagnetic, and biological properties compared with what was previously known, forbidding any extrapolation with regard to their potential risks, and therefore that only case-by-case studies can provide reliable scientific data.

Certain specific physicochemical properties of synthetic nanoparticles can also induce unexpected physical and chemical risks for human safety, such as fire hazards, a risk of explosion, or unforeseen catalytic activity. New parameters must be taken into consideration, such as the surface area to volume ratio and specific form, rather than traditional data and threshold values relating to mass.

Nanotechnology will soon concern most sectors of activity.[3] Indeed, in August 2008, 803 nanoproducts for use by the general public were identified by the Woodrow Wilson Institute.[4] Nanotechnology does not constitute a new product, a new sector, or even a new process, but rather a completely new way to handle matter, involving many innovative processes.

Nanoparticles induce unknown environmental risks. Of the 803 nanoproducts listed by the Woodrow Wilson Institute in August 2008, 56% are produced from nanosilver, a bactericide according to French legislation but listed as a pesticide in the USA. So what will be its effects in the air, the water, and the soil, in environmental terms? What will be its effects on wildlife? Or on human health? For example, through drinking water obtained from contaminated raw water, even in a postponed form? Nanosilver is only one example of nanoparticles that have been released into the environment, but it attests to the wide range of possible negative interactions for other types of nano-object.

Nanoparticles are produced either by reducing the size of existing microsystems (the top-down method), or by creating structures on the atomic or molecular scale (the bottom-up method). Bottom-up methods are less costly in energy and produce less waste than top-down methods.

Health risks concern both workers and consumers, and this at various stages in the life cycles of the products.

---

[3] Even now, nanotechnology is all around us, in tyres (silica), the formulation of certain cosmetics, sports equipment, cloths, etc., that are already commercialised. We should thus expect a very fast development of the use of nanoparticles across a wide range of different industrial sectors, notably including foods, aeronautics, cars, chemistry, construction, cosmetics, defence, electronics, energy production, optics, pharmaceuticals, textiles, and others.

[4] www.nanotechproject.org/inventories/consumer/analysis_draft/

Studies made in 2009 show that the vital organs are rapidly contaminated by nanoparticles, some within 24 h, including the brain, the neuronal system, the liver and digestive system, and the reproductive system, implying transgenerational risks and of course a highly negative impact on the immune system.

A Swiss study[5] of cell cultures has shown that certain nanoparticles easily absorbed by cells have harmful effects depending on their chemical composition. They appear to cause inflammatory reactions and even modify tissues, including brain tissues. The impacts of nanoparticles on the cardiovascular system, the blood, the heart, the spleen, the liver, and the reticulo-endothelial system have been established. It has been confirmed by toxicokinetic studies that the lungs and digestive system suffer major effects due to nanoparticle deposits.

It has been shown that nanoparticles in the blood can be deposited in mouse bone marrow,[6] where they may have consequences on the immune system and hematopoiesis. The Swiss study shows that it is possible for nanoparticles present in the blood to reach ovocytes or spermatozoids, thereby generating a potential risk for future births.

The information available[7] shows that some insoluble nanoparticles can be disseminated through the body and accumulate in organs (lungs, heart, kidneys, intestine, stomach, liver, and spleen), and that they can even cross protective barriers (placental, blood–brain, etc.). Once in the organism, nanoparticles are difficult to eliminate. The way nanoparticles translocate in the organism is still poorly understood. The hair follicles are considered by the scientific community as being a possible uptake route for nanoparticles.[8]

The impact of nanozinc on the brain has been published recently, and attests once again to the high level of potential dangerousness. A publication in 2009 suggests that neurovegetative diseases may appear very quickly, with nerve cells being permanently asphyxiated within 24 h.[9]

France Nature Environnement appeals to the scientific community to obtain further understanding. The little we know at present is already very worrying. The total lack of data for whole ranges of nanomaterials only exacerbates these worries, whence the importance of acting quickly. The current

---

[5] Full report in German (286 pages): *Aktion plan: Risikobeurteilung und Risikomanagement synthetischer Nanopartikel.*

[6] Banerjee et al., 2002; Oberdoerster et al., 2005.

[7] See the documents produced by the FNE's health–environment network: *Nanotechnologies, opinions and recommendations returned in 2006.*

[8] Report by the *Centre d'analyse stratégique.* Preparation for the grenelle environment forum. Inventory of issues regarding nanotechnologies and their relationships with the environment, 24 July 2007, www.strategie.gouv.fr/IMG/pdf/GRENELLE_nanos.pdf.

[9] Toxicity of nanoparticles: Zinc oxide on the brain, published online, 11 May 2009 at www.natureasia.com/asia-materials/highlight.php?id=438.

risks take us into the unknown, both in terms of the seriousness of their consequences and their likelihood. Indeed, nanoparticles are divided into 5 groups[10] and our knowledge of them varies enormously from one group to the next. The warning signs given by several scientific publications in May 2008 concern certain carbon nanotubes. At the present time, data on the behaviour of nanomaterials in the environment and also on their toxicity are extremely limited. The rapid development of techniques for nanosilvers to exploit their bactericidal attributes raises many questions, to which some Chinese, Australian, and American scientists have responded in the form of scientific papers, but also publicly on the Internet: the French scientific community must therefore decide upon its own position as clearly and quickly as possible. It is essential to assess the hazards and risks for ecosystems, and to assess the hazards and risks of human exposure, whether it concern voluntary consumers, employees, or unwittingly exposed citizens. Complete kinetic models for each type of nanoparticle or nanomaterial, often described as ADME (absorption, distribution, metabolism, excretion) models, should be the main objective of quantitative studies and deal with uptake mechanisms and target organs.

The whole issue of risks must be taken into account transversely in scientific work. It is important to ensure that hypotheses respect quality procedures for international scientific work and allow cross-disciplinary teams to cooperate together using compatible methodologies. Measurement tools must be devised to suit nanoparticle characteristics, along with test systems to assess the risks due to nanomaterials, including information about acceptable exposure limits. Transmission paths from one (eco)system to another must be investigated. The disturbance of aquatic environments is difficult to quantify and nothing is yet known about the impact on soils.

It must be made possible to assess the validity of scientific theses. Decision-makers, e.g., in administration and public bodies, do not have sufficient specific training on nanotechnologies and await clarification by the scientific community. It is essential that the methodologies and hypotheses selected by scientists to support their thesis should respect quality procedures and that comparison between them should be straightforward, especially with regard to the procedures, measurement instruments, symptoms, interactions, and so on. The capacity to carry out studies and publish the results in countries like China and the United States has introduced a degree of uncertainty regarding the validity of scientific theses in certain countries, like France, which do not have the means to keep up with this rate of publication. It is important that French theses should be able to remain consistent with what is published internationally.

---

[10] Fullerenes, carbon nanotubes, inorganic nanoparticles (made from pure metals or different inorganic products or alloys), organic nanoparticles (made from various organic substances, often insoluble polymers onto which various organic radicals have been grafted), and quantum dots (semiconductor nanoparticles).

The scientific community must remain vigilant regarding conflicts of interest. Research work requires funding, and this fact may result in genuine conflicts of interest which a researcher may, in perfectly good faith, not be aware of, since for him or her they are not real, whereas for other members of society, and in particular the general public, the situation may be viewed quite differently.

France Nature Environnement appeals to industry to reinforce management practices for anticipating new and emerging risks with a whole set of precautionary measures for employees in the relevant sectors, manufacturers or users of nanoparticles, accompanied by information concerning possible risks, legible and clear labelling, with compulsory traceability at all stages of handling and transformation, storage and elimination. In the framework of Occupational Health and Safety Groups, employees should be fully informed and trained. Medical observation of exposed workers should be strengthened and integrated into national or European cohorts.

Naturally, *France Nature Environnement* expects industry to develop production in such a way as to integrate the notion of new and emerging risks, with appropriate measuring equipment on site to indicate the presence of nanoparticles, and if so, to characterise and/or measure their concentration, to know under what conditions specific intervention would be necessary. They must also set up systems for managing residues and waste from production in order to exclude all dispersion of nanoparticles into the environment. And they must accompany any programme for the development of new products with tests regarding their harmful properties, notably tests on cells, since the international community has doubts about the reliability of systematic animal testing.

The cost of nanomaterials does of course depend on production costs, but also on the level of reliability (the electronics industry has observed that the lifetime and reliability of a product decrease in proportion with size reduction). Are the resulting additional costs justified when the cost of risk management is included, along with the full cost over the whole life cycle, including the storage of production waste? Would conventional substitution products not be more profitable, even though the marketing effects may be less spectacular? It is important to integrate external costs in economic evaluation of nanoparticle products.

France Nature Environnement expects industry to inform consumers and public authorities. To begin with, product labels should provide clear and exhaustive information on the presence of nanomaterials, traceability being an essential feature at all stages. As 'nanoparticle' alone is too vague to be useful in this context, the concentration, size, and shape of nanoparticles must be indicated, due to their specific properties. An inventory of all bodies involved in production, use, and elimination of nanoparticles must be established officially at a rate consistent with the exponential development and commercialisation of products, including imported products. Nanomaterials already commercialised or in the pipeline, and all products containing them with their characteristics, must be identified and the list made accessible to the public.

France Nature Environnement appeals to the administrators and decision-makers of the French State to do everything possible to improve our knowledge of these issues, and hence to contribute to international research and development programmes. The State must introduce suitable regulation urgently. France must encourage the European Union to adopt specific legislation regarding the risks relating to nanomaterials. However, without awaiting the finalisation of European regulations, the French government must anticipate and set up its own legislation on these issues. This should be possible on the basis of recommendations made by the *Comité de la prévention et de la précaution* (CPP) and the *Agence française de sécurité sanitaire de l'environnement et du travail* (AFSSET).

France Nature Environnement urgently requests the French State to impose a strict moratorium regarding the commercialisation in France of food products, food packaging products, all personal hygiene products, including cosmetics, sunscreen creams, and all other products entering into direct contact with the human body (excepting medical products, submitted to a specific regulation) in the context of normal use, whenever they contain nanoparticles,[11] and a scheduled withdrawal on the shortest possible timescale of all such products currently on the market. This partial moratorium is justified by the fact that it is urgent to find out more about possible health risks before large scale commercialisation gets underway, and these products become commonplace. Even if the risk per person is very small, but real, it could induce a genuine health problem on the scale of the population as a whole.

Precise labelling of products containing nanoparticles must be made compulsory and standardised. Consumers must not be able to purchase them unwittingly, without being fully informed. The specific features of nanomaterials must be taken into account in regulation concerning waste.

Finally, advertising aimed at non-professional consumers should be forbidden in all forms of media and as soon as possible, once it has been shown that a product contains nanoparticles likely to migrate into the environment and come into contact with humans.

Systems for maintaining vigilance and surveillance must be set up. The surveillance of installations of all kinds involved in the manufacture or use of nanoparticles by specially appointed inspectors must become a reality. Systems for monitoring ambient air and indoor air must also monitor nanoparticles. It is necessary to set up suitable training for factory doctors and emergency services, including personal safety procedures for rescue officers required to intervene in atmospheres contaminated by nanoparticles, but also for hospital staff who may be called upon to treat victims.

[11] For example, the titanium dioxides used by certain manufacturers have proinflammatory effects.

The inventory of all bodies involved in production, use, and elimination of nanoparticles must be compulsory. An inventory of all nanomaterials already commercialised or expected to be launched on the market, together with the products containing them and their characteristics, must be made available to the public and kept up to date. This is allowed for by Article 42 of law no. 2009-967 of 3 August 2009 regarding the implementation of the results of the Grenelle Environment Consultation.

The *Plan national santé-environnement* (PNSE) and its regional counterpart the *Plan regional santé-environnement* (PRSE) provide an opportunity to integrate the whole issue of nanotechnologies in an explicit way. This undertaking will involve several different facets: information, research, observation, regulation, and consultation.

A broad and sincere information campaign is expected with regard to the general public. A website containing available public information about nanomaterials should be opened, providing public access to all this information in the French language. At the same site, in French, all results of international research should be made accessible in order to fuel a high level debate, with the translation of some documents originally published in German, Swedish, and so on. Naturally, information made available in the context of the public debate devoted to nanotechnologies should be online.

Finally, *France Nature Environnement* expects the public authorities to set up an appropriate government control of nanomaterials. This must make full allowance for the fact that we are currently in a context where the principle of precaution must apply. In this situation of uncertainty, an iterative decision-making model must be used. It should also be carried out in the spirit of the Aarhus convention of 25 June 1998, bearing in mind that two of the three themes of this agreement concern access to information and public participation in the decision-making process.

All members of society should be involved in decision-making. The FNE demands that authorities be set up at different geographical levels, i.e., national and regional, but with local focal points in geographical sites like Saclay, Grenoble, and Toulouse, which contain active centers of research and development. On the regional level, Regional Councils, which have the role of supporting the economic sector, but which are also involved in support for research, higher education, training, etc., should take the initiative of creating consultative opportunities devoted to nano issues. On the national level, two responses need to be set up in parallel: opportunities for negotiation as in the Grenelle Environment Consultation involving the full range of stakeholders; and opportunities for consultation, information, exchange of ideas, and free speech, taking inspiration from the Nanoforum organised by the *Conservatoire National des Arts et Métiers* (CNAM), which did a remarkable job in the circumstances, but remedying certain weak points in order to increase accessibility to as many people as possible.

In order to generate responsible decisions, a set of values needs to be placed at the center of the debate, and not only those relating to competitivity and the importance of this or that market, the aim being to take economically, ecologically, and sociologically acceptable decisions.

The position adopted by *France Nature Environnement* has always been to encourage public debate and consultation. For this reason, the FNE took an active part in the Grenelle Environment Consultation. Working Group no. 3 in the summer of 2007 carried the heading *Instaurer un environnement respectueux de la santé* (For an Environment Respectful of Health). In the context of this group, *France Nature Environnement* introduced the nanotechnologies theme which featured among its proposals under the title *Faire mieux prendre en compte les risques sanitaires associés aux nanotechnologies* (Promote a Better Consideration of Health Risks Associated with Nanotechnologies), using its dedicated platform as support. During the discussions, our demands mainly concerned three points. First of all, a moratorium on all products targeting the general public when these products come into contact with the body during normal use, e.g., clothing, foods, cooking products, food packaging, cosmetics, sunscreen creams, etc., but with the exception of medicinal products which are subject to specific protocols. We also requested a form of labelling that would allow consumers the freedom to make purchases in full possession of the facts. Finally, we requested that opportunities be organised for stakeholders to continue their consultation on the nanotechnologies.

The following extract[12] from the report by this working group (pp. 7–8) attests to the positive evolution in the consideration of risk thanks to the efforts made by *France Nature Environnement*:

> The group suggests the organisation of a scientific consensus conference followed by a public debate, e.g., by the *Commission nationale du débat public*, in 2008. A consultative committee will be set up, bringing together all stakeholders. A research campaign will be carried out to improve understanding of manufactured nanoparticles.
> Given the lack of knowledge concerning their health impacts, employee exposure to manufactured nanomaterials or nanoparticles should be reduced as far as possible as a matter of precaution. The NGO representatives propose to ban all commercialisation in food applications, personal hygiene products, cosmetic products, and clothing. Employers' representatives prefer a case by case approach.
> A compulsory declaration will be set up and the information will be made transparent for users of nanomaterials in ways (e.g., labelling) to be defined by consultation. Opinions are divided between the implementation of specific national regulations and recourse to an authorisation process specified by European regulations.

Other longer texts feature in the body of the report on pp. 27, 28, and 72–74.

[12] The report can be consulted online at www.legrenelle-environnement.fr/grenelle-environnement/IMG/pdf/G3_Synthese_Rapport.pdf

From the round table negotiations of 25 October 2007,[13] two commitments refer directly or indirectly to nano issues. Commitment 159 reads as follows:

> Anticipation of risks relating to nanomaterials: The *Commission nationale du débat public* will organise a debate on the risks relating to nanoparticles and nanomaterials. The presence of nanoparticles in products destined for the general public must be clearly stated, from 2008. Systematic cost–benefit analysis will be carried out before commercialising products containing nanoparticles or nanomaterials, from 2008. Guarantees will be laid down to ensure that employees are adequately informed and protected, on the basis of the AFSSET study.

Commitment 138 concerning the second *Plan national santé-environnement* (PNSE-2) specifies the following:

> After the *Plan national santé-environnement* (PNSE) planned for the period 2004–2008, a new PNSE beginning in 2008 will bring together all stakeholders, but extending the field of action of the first to include new technologies, new pathologies, environmental equity, and so on.

In 2008, *France Nature Environnement* contributed to Workshop 19, devoted to emerging risks, and set up by an operational committee called Comop 19. The initiatives of *France Nature Environnement* led to clear measures in the Comop 19 final report,[14] referring to nanotechnologies and manufactured nanoparticles.

Within the framework of the Grenelle Environment Consultation, we made it quite clear that we would have liked these statements to go further. But progress has already been made. Now there remains one essential point that cannot be neglected: certain negotiations, decisions, and regulations can only be handled on a Europe-wide scale, or even on a worldwide scale.

At the time of writing, that is, in June 2010, nano issues feature in the legislative consequences of the Grenelle, both in the planning law to implement the conclusions of *Grenelle 1* and in the bill for national environmental commitment known as *Grenelle 2*, still under examination by the parliament.

France Nature Environnement contributed to these discussions with regard to the organisation of a public debated piloted by the *Commission nationale du débat public* in 2009, and produced a contributor's report. Even after the end of the public debate, all the documentation brought together, full minutes of public meetings, all the contributor reports, etc., will remain accessible at the website www.debatpublic-nano.org. This will thus constitute a very useful source of information.

[13] These commitments are available online at www.legrenelle-environnement.fr/grenelle-environnement/IMG/pdf/GE_engagements.pdf

[14] The Comop 19 report is available online at www.legrenelle-environnement.fr/grenelle-environnement/IMG/pdf/Rapport_ Comop_19_veille_sanitaire_et_risques_emergents.pdf

France Nature Environnement contributed to the public consultation on nanotechnology risk assessment in the framework of the Scientific Committee on Emerging and Newly Identified Health Risks (SCENIHR)[15] in June 2009. Its views can be found online at the website of the Health and Consumer Protection Directorate General. *France Nature Environnement* insists upon an immediate and efficient implementation of the principle of precaution and appropriate safety measures.

Owing to the wide variety of different sizes and shapes of nanoparticles, with the resulting diversity in their potential health and environmental impacts, it is no longer possible to speak about nano-objects in general, and an in-depth investigation of suitable measures is required. Moreover, this is something we already suspected in 2006, when we used the title *The World of Nanos* for one of our commentaries!

France Nature Environnement has a crucial ethical role to play in challenging all stakeholders to acknowledge their responsibilities right away, at whatever level they operate. We cannot discuss nanotechnologies without tackling the ethical questions. This is clearly illustrated by the following non-exhaustive list of points:

- There is a tremendous imbalance between the state of our knowledge of sanitary and environmental risks and the production and commercialisation of applications.
- Certain practices, like carrying out inquiries about people and ideas, are inacceptable as regards the protection of individuals and their private lives. The fact that the president of the *Commission nationale de l'informatique et des libertés* (CNIL) intervened personally to express his concern lends support to this conviction.
- The development of these technologies may widen the gulf between countries in the north and south.
- NBIC convergence (bringing together nanotechnologies, biotechnologies, information technology, and cognitive sciences) raises real ethical problems. The possibility of developing human capacities, or 'repairing' what are considered as weak points, by technological means, may eventually be taken up in the military sector, sports, and health. Even in the health sector, ethical limits need to be set up.

The public authorities must produce appropriate regulations and as quickly as possible. Why wait until the negative effects show up in the statistics before legislating? Let us learn from the bitter experience of asbestos. When sickness and death come years after exposure, the product has already been widely disseminated, and decontamination becomes an extremely difficult and costly operation.

[15] To find out more, consult ec.europa.eu/health/ph_risk/committees/04_scenihr/04_scenihr_fr.htm

By raising the awareness of user industries, they will be able to devise manufacturing procedures that protect both employees in contact with the products and consumers downstream of the production process. Waste processing is another major issue that must be treated immediately, because the impact on the environment is likely to be considerable.

And above all, it is not conceivable from an ethical point of view that the consumer should be the only one to bear the weight of this responsibility. Product labelling is essential, but the consumer should not even have access to products whose consequences are beyond his or her control, whether they be cosmetics (sunscreen creams), DIY products (paints containing nanosilvers, which can be rinsed with water and which will pollute the environment on a very short timescale indeed, disturbing or even preventing the operation of sewage treatment systems), but also medical and food products (ingested substances).

Two questions are fundamental for *France Nature Environnement*. What regulations would we like to see for this technology? What can this technology usefully do for us, and what uses would be acceptable? *France Nature Environnement* would like to see France become a driving force on the European and international level, so that in the country of human rights, science really does remain in the service of the human being and respect the environment.

23

Situation in France: The Position of a Consumer Protection Group

Christian Huard and Bernard Umbrecht

Our association ADEIC for consumer protection, education, and information (*Association de défense, d'éducation et d'information du consommateur*) has made its name in the world of consumerism through its attempts to foresee the questions that will be raised by consumers of ever faster moving innovative techniques. Nanotechnologies fall well inside the latter category. And it will not be the least challenge we have faced since our beginnings in 1983, not by a long way!

Indeed, the problems raised by the introduction of nanotechnologies into everyday consumer products will revive all the issues that came up about previous innovations, and there will be other questions too in this case: sanitary, ecological, and ethical risk–benefit analysis, the problem of protecting personal data and privacy, economic dependence, largely unpredictable social evolution, and so on. Here we have food for thought to animate many a debate for some time to come, although it must be said that, for the time being, few seem to be much involved.

23.1 French National Consumer Council

The introduction of nanotechnologies, combined with an explosion in the information and communication technologies (ICT), has happened in a situation where the consequences of many disasters (financial, economic, social, moral, ecological, sanitary, demographic, etc.) are breaking over us faster than the high tides. They are in the process of generating a serious crisis of mutual mistrust between citizen and public authority. The powers that be are ever more distrustful of private citizens and their organisations, seeking to short-circuit consultative bodies in favour of small (carefully selected?) and shortlived ad hoc groups or committees. Conversely, those running independent organisations try to experiment with other consultative regulatory channels than those set up by political decision-makers.

P. Houdy et al. (eds.), *Nanoethics and Nanotoxicology*,
DOI 10.1007/978-3-642-20177-6_23,

In the face of this situation, ADEIC pushed for the matter of the introduction of nanotechnology into everyday consumer products to be referred to the French National Consumer Council (*Conseil national de la consommation* or CNC). A mandate for this was proposed and a working group was able to begin its work in September 2008. It should be noted that, throughout the whole history of the CNC, this group brought together a record number of representatives of consumer societies, professionals, experts, and, quite exceptionally, directors of state administrative bodies. To our knowledge, such a broad-based consultation has no equivalent anywhere in the world. For the moment, this form of consultative regulation under the aegis of the public authorities exists only in France.

Here are the main clauses of the mandate:

> While still largely unknown to the general public, nanotechnologies are likely to lead to far-reaching changes which are expected to have repercussions in practically all sectors of the economy, bringing many new solutions to satisfy the needs of the consumer, whose everyday life will thus be transformed. They would therefore appear to be the key technologies of the twenty-first century.
>
> Due to their enormous potential, nanotechnologies have become an essential consideration for the competitivity of national economies. Worldwide public and private spending on nanoscience and nanotechnology over the period 2004–2006 amounted to some 24 billion euros, of which one quarter was in Europe. According to the forecast by the European Commission: "The world nanotechnology market should represent between 750 and 2 000 billion euros by 2015 and the job creation potential could be as high as 10 million in areas relating to nanotechnologies by 2014, i.e., 10% of all jobs in manufacturing industries worldwide."
>
> Now, a poorly controlled use of nanotechnologies would cause serious risks for the health and safety of consumers, or for the environment. Poorly controlled, these risks could impair the trust of consumers, as noted by the statement of the *Conseil économique et social* of 25 June 2008 on nanotechnology: "It is known that fears are born from ignorance of phenomena, but also from the feeling that concerns expressed have not been taken into account."
>
> At the request of several consumer societies, the Minister[1] thus decided to organise a working group, within the framework of the *Conseil national de la consommation*, with the following aims:
>
> - To gather information about the use of nanotechnologies for the production of goods and services destined for consumers (type of application, economic weight per sector and overall, number of consumers concerned, potential benefits, etc.).

[1] At this time, Luc Châtel.

- To consult experts on the risks associated with this kind of production, and also on the steps already taken or to be taken to understand, reduce, and eliminate these risks, and in particular the studies that should be conducted.
- To organise an economic dialogue between professionals and consumers to examine in a reactive manner the novel problems that nanotechnology may raise.
- To assess the effectiveness of existing legal rules and control systems to protect the consumer when he/she acquires or uses products incorporating nanotechnologies.
- To identify regulatory needs brought up by nanotechnology and, where necessary, to formulate proposals for extending legislation and regulation on both the national and European levels, in order to adapt them to the context of nanotechnological developments.
- To set up a simple and understandable source of information for the consumer on questions regarding the nanotechnologies, the benefits they may bring, but also the risks they may involve, and the precautions they occasion. This reflection will bear not only upon general information about these technologies but also upon the information that should accompany each free or paying product put on the market.

The group worked intensively on these issues. The work was temporarily suspended in April 2009 for reasons totally beyond the control of those involved, but should be resumed soon.

23.2 Consumer Information: Failings and Modest Steps Forward

Launched in Germany in 2006, with huge advertising support, the bathroom cleaning product *Magic Nano* was down to revolutionise the housekeeper's life by projecting an invisible film on the tiling that could repel dirt and bacteria. But alas! The miraculous product was withdrawn hastily from the shops after triggering breathing problems in 97 consumers, after just 3 days. Several were even hospitalised with a pulmonary edema (accumulation of fluid in the lungs). Nanotechnology was off to a bad start. Except that the problem was not strictly speaking the nanoparticles. The only thing nano about the product was the thickness of the film it deposited on surfaces. The pathological problems came from the extremely small size of the droplets projected by the spray, and which had allowed inhalation of the chemical product. But the affair brought back the asbestos syndrome, and reminded everyone that it is not because it is small that it is not dangerous.

On the packaging it said 'nano' in big letters. It was even the music-hall version. But the fact that it was written there changes nothing. In this precise case, we are dealing with – or so the manufacturer thought – a positive perception which was clearly shared by the consumers, at least at the outset, until the incident occurred. Soon afterwards, a major manufacturer of cosmetic products removed the term 'nano' from its labelling. There too the product was not a true example of nanotechnology, or so they claimed later on.

In this kind of situation, is the label really such a good thing? If it is written there, does that mean it could be dangerous? If there are risks, the label will teach us nothing about the precautions the user should take. The idea that it is important to know whether the nanomaterials used involve risks and that it then belongs to the consumer to choose can hardly be taken seriously.

Naturally, that does not mean that there is no need or obligation to specify the kind of ingredients contained in the products. There is nothing fundamentally different there with regard to legal obligations. The information on the label does in fact have some virtues. If the presence of components in nano form is indicated, then it can be checked whether they satisfy the general safety obligations required by the law, and in particular, article 221-1 of the French consumer code: "products and services [...] must fall within the safety requirements that can be legitimately expected of them and not affect the health of users."

Obviously, this assumes the existence of a precise nomenclature which remains to be established, together with an inventory of manufacturing companies. In the case of beauty products, the cosmetics regulation adopted by the European parliament in March 2009 extends the rule about mentioning the presence of nanomaterials on the label by stressing the need to improve monitoring systems, and also stipulating that the public authorities be informed and a data base updated. Indeed, the main changes concern these three points:

- Improvement of systems for monitoring the market. The manufacturer must now notify the European Commission of any product launched on the market. This notification will allow the monitoring authorities in each member state to increase their visibility of commercialised products.
- Creation of a shared data base of undesirable effects across all member states. The principle of cosmetics surveillance will be applied in all countries of the European Union.
- Consideration of the specific features of nanomaterials. Any manufacturer wishing to incorporate nanomaterials in one of their products must inform the European Commission 6 months prior to commercialisation. The EC reserves the right to consult an expert committee.

In addition, the manufacturer must indicate the presence of these nanomaterials in the list of ingredients which already features by obligation on all products. A labelling rule has been laid down to this effect, in the form: Name of ingredient [nano], e.g., Titanium dioxide [nano].

But why should we retain a conventional form of labelling? Is this not an area where the very notion of label could be enriched with further innovations, such as those provided by radiofrequency identification (RFID) tags?

Could these not provide the consumer with more complete information? Rather than considering intelligent chips solely for checking up on consumers, why not use them to allow consumers to check up on what they purchase?

Even in the construction mentioned above by the European parliament, we see that consumers are left with only a strict minimum of information. This does not seem satisfactory, because it gets things out of all proportion, neglecting to emphasise that we are dealing with a much broader technological revolution. The trick with labelling has already been pulled for the problem of GMOs, with the success everyone knows.

In our opinion, for each product incorporating nanomaterials, those responsible for launching it on the market must provide consumers with all the useful information, and in particular:

- The nanomaterial(s) present and their various structures.
- The reason for their presence and/or expected benefits.
- Safety precautions, especially under certain conditions of use.
- Conditions of storage and conservation.
- Conditions for disposal by destruction or recycling.

This information must be honest, and not in any way misleading (e.g., stressing the small quantity included). It must also be possible to check it, with sanctions when necessary.

The means for providing full information remains a subject of discussion at the CNC working group mentioned at the beginning of the chapter. But a rather obvious point, often played down, should be borne in mind here: consumers do not constitute a homogeneous group, but quite the opposite, a broad range of people with extremely varied, even contradictory needs.

The scientific innovations and techniques currently under development will probably lead to deep changes in producer–consumer relations which it would be wrong to treat as somehow immutable. But whatever else happens, one fact will always stand: confidence will always be a key feature of economic relations. And confidence is a fragile entity, rather like a house of cards in which many different elements need to support one another. We would all feel reassured, for example, if we had guarantees that sufficient effort was being made in toxicological and ecotoxicological research, but this is far from being the case.

In addition, there can be no satisfactory consumer information if we do not at the same time deploy sufficient means to educate, train, and inform. This aspect of the problem alone deserves more attention, when we observe the almost underdeveloped state of our nation when it comes to consumer education.

The whole range of tools at our disposal should also be made available to consumers, and at the same time, we should stop treating consumers as a different category of persons, but just as much concerned by the toxicological risks they run, along with their environment. It should also be remembered that they are just as much concerned by the ethical questions raised by the possibilities for manipulating living matter. Although they can be justly concerned by the suject, environmental organisations do not seem to be, any more than consumer societies would seem to be. A debate with one group and a label for the other is an untenable position in the long term.

23.3 The Need to Go Beyond Labels. A New Form of Governance

However hesitant one may be about a word which is too often used to describe the notion of government, we will nevertheless adopt 'governance' to refer to the complex problems raised by nanotechnologies, in the sense of facilitating, even organising, a public deliberation.

Confronted with the present tidal wave of technical innovations, how could we just sit back and observe the breakdown of democratic innovation? We must assemble the means, both human and financial, to set up new ways of dealing with this kind of situation. We must make sure that the ever-present discrepancy between the rate of technical transformations and the rate at which habits build up and innovations are adopted socially does not get transformed into an unfathomable gulf for many people, generating pointless tensions.

Adoption of these new techniques and the acceptance of the products containing them cannot simply be based on marketing and communications strategies. Indeed, these caused one of the first difficulties in public opinion with regard to products carrying the prefix 'nano'. To quote J.-P. Dupuy:[2]

> The future of nanotechnologies is real enough. The only uncertainty lies in the way people and their governments will react to the major changes they will produce.

This issue requires innovation in governance, given its complexity and the fact that it sits at a crossroads of scientific disciplines (the whole of science and its technical spinoffs seem to want to be 'nano', or move towards it), engendering risks of all kinds (health, freedom, environment), and affecting social and societal issues relevant to a variety of public policies.

These issues go beyond the ones covered by the grenelle environment forum, important though they may have been. And no specific authority has exclusive rights here.

Nanotechnology is not just a mere domain of activity, and it is not just a matter for a single public policy, but several, and then on the national, European, and worldlwide scales. At each level, we need to set up a governing

[2] J.-P. Dupuy: Impact du développement futur des nanotechnologies sur l'économie, la société, la culture et les conditions de la paix mondiale. Projet de mission, Paris, Conseil général des Mines (2002).

authority to reduce this dispersion. On the worldwide scale, a suitable structure might be similar in methods and powers to the International Atomic Energy Agency (IAEA).

The nano question lies at the heart of the challenges of the twenty-first century. In terms of risks, nanotechnologies seem to condense and cumulate all previous crises, from asbestos, through mad cow disease, to GMOs, and the invisible threat of radiation. But an approach that only considered the balance of risks and benefits would be too simplistic, because what the nanotechnologies impose are social choices. The problem here is to specify our future, and the society in which we would like to live.

As stressed by the philosopher Marcel Gauchet, there is a risk that the construction of science-run societies will go hand in hand with a 'de-intellectualisation' of their citizens, i.e., government by pure foolishness and folly. The old arrangement wherein science seeks and sometimes finds in the secrecy of its laboratory, while industry develops, then by marketing, designs new consumer behaviour, and the State watches over this and controls what happens to a greater or lesser extent, this tried and tested configuration has reached its limits.

What we need today is a form of governance which deals in a new relationship between science and society, one which remains to be established, and not merely governance of the research itself. Citizens must be treated not only as coproducers of new habits, but as coproducers of knowledge.

This governance must be guided by a principle of openness toward all stakeholders, respect of the right to information (as stated, for example, in the Aarhus convention), respect of transparency, and respect of each and everyone's position. And it must itself be transparent and represent all stakeholders, from citizen to researcher, expert, politician, industrialist, militant, and so on. This collective governance must also be maintained on a permanent basis.

The regulatory authority must be able to influence research orientations and should employ scientifically trained specialists so that industrialists, citizens, consumer organisations, environmental groups, and public authorities would be in a position to set up controls and testing. It should also be equipped to assess the social utility and social risks, guiding towards sustainable development, and able to decide whether or not legislation is required.

In short, it should allow us to recover the meaning of public deliberation.

24

Situation in France: The Principle of Precaution

François Ewald

The 'nano' paradigm is revolutionising science, technology, and their industrial implementations, just as chemistry did in the years following World War II.

Today this kind of progress cannot simply concern researcher and industrialist, under the benevolent supervision of the state. Today, scientific research, together with technical developments like the manufacture and commercialisation of new products, have become social issues in principle, not by the nationalisation of the players, but because specifications and requirements are imposed in the name of society as it is now and as we hope it will become. Any refusal to comply would amount before long to a ban, or at least a moratorium.

One of the key instruments in this socialisation of technical and industrial projects is the principle of precaution, although unfortunately, in France, its use seems to be poorly understood. It should be applied to nanotechnologies insofar as these operate at a scale where there are many uncertainties about the possible consequences with regard to health and the environment, even though it might be thought that the debate about the risks would give way to the ethical question of the problems raised by the 'benefits' in terms of plant, animal, or human performance.

24.1 Definition

The term 'principle of precaution' comes from a somewhat simplistic translation of the German *Vorsorgeprinzip*, coined at the beginning of the 1970s. (The German *Vorsorge* means something more like foresight.) In its *Directives for precautionary measures with regard to the environment* (1986), the German government makes the following statement:

> The word 'precaution' refers to all measures designed either to preclude specific damage to the environment, or, from the safety point of view, to reduce and limit the risks for the environment, or again,

P. Houdy et al. (eds.), *Nanoethics and Nanotoxicology*,
DOI 10.1007/978-3-642-20177-6_24, © Springer-Verlag Berlin Heidelberg 2011

> by anticipating the future state of the environment, to protect and improve the natural conditions for life, these aims being closely related.

So precaution refers to a temporal hierarchy of three kinds of action ranging from short-term safeguards against imminent risk to long-term management of natural resources. The precautionary perspective integrates safety measures. It involves three essential conditions: to reduce risks and avoid emissions even when no effects are immediately observed; to formulate goals with regard to environmental quality; and to lay down an ecological approach to environmental management.

At the beginning of the 1980s, the German government, concerned about the state of the North Sea, requested a report by a team of independent experts who proposed to base its protection on the principle of precaution and to organise this in the form of an international cooperation. A first meeting of the ministers of the neighbouring countries was held at Bremen in 1984, and it was declared that states should not wait until damage to the environment had been proven before taking action. This was followed by a second meeting in London in 1987, where ministers formally accepted that, in order to protect the North Sea from the possible harmful effects of the most dangerous substances, a precautionary approach was necessary, which might require the adoption of controls over these substances even before any relation of cause and effect had been formally established scientifically. This approach was spelt out with the declaration that, if the state of knowledge was insufficient, a strict limitation would be imposed at source on pollutant emissions for safety reasons.

At the Rio Earth Summit organised by the United Nations Organization in June 1992, the precautionary principle was built into a complete system of principles for specifying a new relationship between humans, and between humans and the Earth, alongside the principles of participation, cooperation, and accountability:

> In order to protect the environment, the precautionary approach shall be widely applied by States according to their capabilities. Where there are threats of serious or irreversible damage, lack of full scientific certainty shall not be used as a reason for postponing cost-effective measures to prevent environmental degradation. (Principle 15)

In keeping with the philosophy of sustainable development, the principle of precaution was soon integrated in one form or another into a whole series of conventions bearing upon the management of natural resources (biodiversity, fishing, forests) and environmental protection, either in a regional context (Mediterranean, North-East Atlantic, Baltic Sea), or with regard to the handling of problems like waste, climate change, and the ozone layer. There are now countless international conventions referring to the precautionary principle. An inventory of applications of this principle is beginning to look a bit like the narrative of *Genesis* in the Bible: the principle protects the sea and the oceans, the rivers, the atmosphere, the land, its fauna, and its flora.

The European Community was soon involved in this vast movement to enforce environmental protection. It began to develop an active precautionary policy which we now describe:

1. The statement (Articles 2 and 3 of the EC treaty) that the EC must set up a high level of protection and improvement of the quality of the environment, and achieve a high level of health protection.
2. The objective as regards the environment is laid down in Article 174R (health is dealt with in Article 152) in the following terms:
   - EC environmental policy pursues the following objectives: preservation, protection, and improvement of the quality of the environment; protection of human health; cautious and rational use of natural resources; promotion on the international level of measures designed to face regional or global environmental problems.
   - EC environmental policy aims at a high level of protection, taking into consideration the wide range of different situations prevailing in the various regions of Europe. It is based upon the principles of precaution and preventive action, the principle of correction, preferably at source, of damage to the environment, and the polluter-pays principle.
   - When formulating its environmental policy, the EC takes into consideration available scientific and technical data, environmental conditions in the various regions of Europe, the benefits and costs that may result from action or absence of action, the economic and social development of the community as a whole, and the balanced development of its regions.
3. The desire to maintain a high level of protection with regard to food, as attested by directives concerning the use of hormones in cattle feed, the embargo upheld against US beef imports, and also the 1996 embargo on exports of British beef.
4. The *Communication* of the European Commission on 2 February 2000, approved by the European Parliament and taken up again by the Council, is an important document for understanding the principle of precaution. It is less concerned with providing an impossible definition than with setting down, in a kind of vade mecum, the rules to be observed at the different stages of its implementation (risk assessment, risk management, risk communication), together with the principles that should guide any decisions, i.e., principles of proportionality, non-discrimination, consistency, and transparency, taking into account benefits and costs, scientific progress, and burden of proof. On this occasion, the Commission proposed different formulations of the principle, which applies:

   > [...] when a preliminary objective scientific evaluation indicates reasonable grounds for concern that the potentially dangerous effects on the environment, plant, animal, or human health may be inconsistent with the high level of protection chosen for the Community.

> This amounts to linking the principle of precaution with the definition of a certain level of protection, on the one hand, and with a scientific evaluation on the other. The text states that deciding to take steps without having in one's possession all the necessary scientific information clearly conforms to an approach based on the precautionary principle. It then goes on to say that the appeal or otherwise to the principle of precaution is a decision taken when information remains incomplete, inconclusive, or uncertain, and when all the indications are that the possible effects on the environment or plant, animal, or human health might be dangerous and incompatible with the chosen safety level. Having laid down the rules for a preliminary risk assessment, such as the rules for expertise, the *Communication* emphasises the political nature of the precautionary decision, stating that the choice of response appropriate to a certain situation is an eminently political responsibility, depending on what constitutes an *acceptable* level of risk for the society that is about to suffer the consequences. This amounts to introducing a new parameter, namely the acceptability, independent of both the level of protection and the scientific evaluation.

The recent Lisbon treaty pursues the same orientation. In Chap. 4 of the *Charter of Fundamental Rights of the European Union* entitled *Solidarity*, it is stipulated that (Article 37):

> A high level of environmental protection and the improvement of the quality of the environment must be integrated into the policies of the Union and ensured in accordance with the principle of sustainable development.

Article 35 specifies that:

> A high level of human health protection shall be ensured in the definition and implementation of all Union policies and activities.

Article 174 of the treaty on the functioning of the European Union uses practically the same wording as Article 174R of the previous treaty.

In France, the precautionary principle was first introduced into national law by the Barnier Act of 2 February 1995, which defines it as the principle [art. L. 200-1 of the new *Code rural* – drawn up in Book II (new), *Protection of Nature* of the *Code de l'environnement*, in a slightly modified version which includes air quality (Sect. 1) and health (Sect. 2)]:

> [...] whereby a lack of certainty, given the scientific and technical knowledge of the day, should not be allowed to delay the adoption of effective and proportionate measures aiming to prevent a risk of serious and irreversible damage to the environment at an economically acceptable cost.

The precautionary principle was also written into the preamble of the Constitution of the Fifth Republic when the *Charte de l'environnement* was constitutionalised. Article 5 of this charter states that:

> When the realisation of some form of damage, although uncertain in the present state of scientific knowledge, might seriously and irreversibly affect the environment, the public authorities will ensure by application of the precautionary principle and in their various fields of action, that risk assessment procedures are implemented and temporary and proportionate measures adopted with a view to staving off the realisation of this damage.

This article must be understood in conjunction with Article 7 of the same text which states:

> [...] the right of each citizen to have access, under the conditions and within the limits laid down by the law, to the information about the environment held by the public authorities, and also to take part in the formulation of public decisions affecting the environment.

Clearly, these two texts refer to the environment. But this does not mean that the principle does not apply to health issues, only that the procedure spelt out in Article 5 of the *Charte de l'environnement* might be different in this area.

24.2 Legal Status of the Precautionary Principle

On the international, European, and national levels, this principle may have come into being for environmental issues, but now applies to the protection of human, animal, and plant health, too.

With its ever-expanding field of application, the precautionary principle is destined for the public authorities, and not directly for private individuals. In other words, while an individual can criticise the action of the state in the name of the precautionary principle (because they are not implementing it, or apply it inadequately – the judge sitting in the administrative courts has had plenty of opportunity to build up an abundant jurisprudence in this area), he cannot use the notion directly against another individual (civil action), no more than a company can sue an individual for not having respected it *proprio motu* (penal action). This state of affairs, the subject of continual debate, would be hard to contest on the basis of the cited texts. This point, namely that the principle of precaution concerns the public authorities, was stressed by the resolution of the European Council of Nice and taken up again in Article 5 of the *Charte de l'environnement*. But naturally, as soon as the public authorities of the State have specified precautionary measures, private individuals are bound to comply.

The precautionary principle habilitates the public authorities to impose a certain number of measures, expertise, and management, when a precautionary situation is identified, i.e., with the potential for serious and irreversible damage, in a context of scientific uncertainty. Better, under these conditions, the precautionary principle compels these same public authorities to take action. In short, the precautionary principle extends and strengthens the policing powers of the administration whenever health and environmental issues come up. In France, the use of the principle is verified by the administrative judge.

Some French civil courts of first instance called upon to deal with conflicts between mobile phone operators and people living near relay stations sought to apply the principle of precaution to condemn the operators. Such decisions are systematically rejected upon appeal, the appeal judge requiring that the party invoking the principle of precaution supply sufficient reason for doing so. This explains why plaintiffs referred to the notion of neighbourhood disturbance, insisting on disorders like worry and stress caused by a controversial technology. In a way, this is rather like a double application of the principle of precaution. The public authorities establish guidelines for an activity invoking the precautionary principle. The private individuals concerned would then be justified in arguing that there was a disturbance, in the name of a sort of right to tranquility. The French Supreme Court of Appeal has not yet pronounced a verdict on this practice derived from the precautionary principle.

24.3 Decisional Aspects of the Precautionary Principle

Strictly speaking, the principle of precaution, expresses the State's determination, either in its relationship with another State or with regard to internal affairs, to conduct a certain kind of resource management and risk protection policy.

When applying the principle, a state implements a precautionary policy. Such policies, always specific to the relevant sector, e.g., medicines, food safety, climate change, preservation of biodiversity, fishing resources, nuclear energy, GMOs, nanotechnologies, etc., become concrete when precautionary systems are devised, which themselves involve precautionary techniques.

The government and the press often invoke the principle of precaution when required to manage crisis situations like mad cow disease, bird flu, or swine flu. The principle is then supposed to authorize the government to take stiff measures. However, the principle of precaution was never intended simply to manage crises. It also serves to manage very long term situations like global warming. The texts cited in Sect. 24.1 remind us that the problem is not to manage precautionary situations as a matter of urgency in the face of a ban, but rather to engineer improvements over time relative to a context which may itself evolve and which must always be taken into account. However, through a sort of inner logic which is not necessarily favourable to the way of working of a state governed by laws, some consider the precautionary principle as a reason for treating the management of collective risk as a sort of state of emergency, even when the issue is very long term.

Precautionary techniques fall into three main categories: (1) risk assessment techniques, (2) risk management techniques; (3) communication techniques.

24.3.1 Risk Assessment

The principle of precaution begins with a fundamental obligation to find out, i.e., to establish a more or less detailed inventory of the risks involved in an activity. They must be 'produced', revealed by a scientific method. This is done by mobilising scientific and technical research, setting up expert systems, and other systems for early warning, monitoring, and traceability. In the 1990s, France set up a fully comprehensive system of institutions for expertise, health watch, and environmental surveillance: the *Agence de sécurité sanitaire des aliments* (food safety, AFFSA), the *Agence de sécurité sanitaire des aliments et des produits de santé* (food and drug safety, AFSSAPS), the *Agence de sécurité sanitaire de l'environnement et du travail* (environmental health and safety at work, AFSSET), and the *Institut de veille sanitaire* (health watch).

24.3.2 Risk Management

Knowledge of the risks thereby produced should lead politicians to take measures to manage them, although these measures will in principle be temporary, since they depend on a certain state of knowledge that is expected to evolve. They will not be understood in the same way depending on whether one adopts a substantive or procedural view of the precautionary principle.

In its substantive version, the principle of precaution enforces a certain type of decision. In a situation of uncertainty, i.e., in a situation where conflicting arguments cannot be objectively settled, it favours certain arguments over others even though it is not possible to decide between them on a scientific level. It always goes in favour of a ban, a restriction, or some form of caution, and never condones risk-taking. Indeed, by precaution, one must give extra weight to the argument that favours the value to be protected, even though there may be less scientific basis for this than the opposing argument. In this substantive version, the principle of precaution requires one to support the argument that favours the protected value, e.g., health or the environment. The substantive dimension of the precautionary principle invites us, in a sense, mechanically, to take a certain decision.

But this somewhat mechanistic vision of the principle does not fully accord with the texts cited earlier, which for their part plead in favour of a proportionality principle, so that the debate may take into account scientific arguments (it is a scientific assessment that must underlie our understanding of the production of serious and irreversible risks), whence the decision will very likely aim to reduce the possible damage, but in a way appropriate to the situation. These reminders privilege a procedural vision of the principle of precaution. The principle then tells us that, when a problem situation comes about and there is some level of doubt, a cautious and considered approach should be observed, without prejudice regarding the solution that should be adopted to overcome it. The solution only has to be tailored to the situation, suitable, and in just proportion. This dimension of the principle is particularly clearly asserted in the description of the principle adopted by the European Council of Nice in December 2000. It also appears in Article 5 of the *Charte de l'environnement*.

These two versions of the precautionary principle are often confused. For example, we may hear that such and such a governmental decision was taken 'in application of the principle of precaution', as though it had somehow constrained the decision by the authorities. The systematicity inherent in the substantive dimension of the precautionary principle leads to ill-suited decisions, even when the principle is implemented by the public authorities. This is the case in the medical field, where the authorities are led to take certain decisions by application of this principle because their own responsibility might be called into question if they did not apply it. This is why there is a certain consensus to go beyond the substantive dimension of the precautionary principle.

24.3.3 Communication

The final facet of a precautionary policy concerns communication of the understanding of the risks that one has been compelled by precaution to obtain. Indeed, this knowledge cannot be confined to those concerned and the administration, but must be transmitted to all those who are potentially exposed to the risks. And this all the more so since Article 7 of the *Charte de l'environnement* declares that every citizen has the right not only to be fully informed but also to participate in the public decision. This aspect of precautionary policies is not yet completely settled. And yet it is decisive. It leads to the question of participatory democracy, something which has been steadily developing since the Public Debate Commission, citizens' conferences, and more recently, the various *grenelles* (round-table talks in France involving all stakeholders).

24.4 Beyond the Principle of Precaution

The principle of precaution to which the various texts aspire remains a decision-making principle belonging to the universe of scientific knowledge. The aim is to transform threats, doubts, and suspicions into 'risks', then to measure, assess, and quantify them, and finally to make a decision on the basis of the models of scientific rationality that the economists of decision have put forward and formalised by building sophisticated forms of conventional risk–benefit analysis. The best known example of such reasoning is no doubt to be found in the Stern Report on the political conclusions to be drawn regarding global warming.

The principle of public participation, at least when it concerns those in the affected neighbourhood, if not the participation of the citizen in general, brings in new forms of assessment through the presence of associations, and a debate that is no longer so much concerned with risks as with the legitimacy, utility, and necessity of a given activity, whatever the risks it may happen to involve. The debate on GMOs is a case in point. In effect, in the name of the principle of precaution, the problem here is no longer really an assessment of the risks, where benefits must be weighed up against costs.

We may note to begin with that scientific expertise has been gradually disqualified through a form of dialectic particular to the principle of precaution; then that other principles have been built in around it: the principle of anticipation upstream, and the principle of attention downstream; finally that, in the face of the new technological paradigm – bio and nano – the question of risks tends to become marginal in political decision-making, giving way to philosophical and ethical debate.

The Disqualification of Expertise

The following stages can be identified in the dialectic of expertise, where all the conflicts tend to be focused in precautionary circumstances. The first was to contest the experts, their competence, their impartiality, and their independence. Consider the question of competence. A book appeared in the 1990s with the French title *Les experts sont formels*, the word 'formal' carrying a double meaning of formality but also certainty. It stigmatised the inability of 'experts' to anticipate risks and disasters, denounced their role in the poor assessments which had led to several major health crises, and criticised their tendency to minimise, and even to normalise, the abnormal. One example among many has become a point of reference: the Chernobyl cloud stopping at the French border. The expert was said to have given biased information.

Consider also the expert's impartiality. Before any expert assessment, an expert must first present a CV indicating all the circumstances which might affect his or her opinion. The effect is disastrous: there are very few eligible experts in any given discipline. Scientists in university laboratories or public research centers have almost always worked at some point with a private company. Indeed, this practice is encouraged by the government in the name of public–private partnership, competitive clusters, the Lisbon strategy, and the knowledge-based society. And there they are, disqualified because biased. In a sense this amounts to denying the idea of autonomous scientific reasoning, an idea which justifies talk of the scholar's probity. This in turn amounts to denying the very idea of science as an activity wherein all players are detached from their emotions and personal interests. The scientific discourse itself would never be more than the expression of a more or less conscious or acknowledged interest; quite possibly at the service of this scientific and technological 'enframing' characteristic of the West in Heidegger's view.

Into the breach opened up in this way came the idea of surveillance, monitoring, warning, and counter-expertise. Now we must listen to the 'weak signals', wherever they come from, even if they come from beyond the horizon of science. Every form of 'noise' must be recorded, listened to, and examined. All the voices of dissent with regard to the science of the academies. The principle for selection of hypotheses, the filter that underpins each scientific discipline, must be suspended. Very soon, every hypothesis will be worth every other hypothesis. No science will escape the threat of controversy. Expertise will be relativised, for it will lack the expertise of expertise once trusted to the discipline of science, brought together in an academy. And in the relativism thereby created, a new principle of credibility will triumph: the more often the disaster is announced, the more likely it is to be believed. Since there is no longer any foundation on which to build, we must restrain ourselves from everything, starting with the worst. Doom-mongering will replace veracity.

In this epistemological conjuncture, conspiracy theory flourishes: it was the Americans themselves who destroyed the Twin Towers, and Neil Amstrong never set foot on the Moon. The film and photographs which attest to the latter are all bogus. The whole thing is a web of lies, a sham. Everything deserves its own controversy.

Every form of opposition, contradiction, or scientific dissent is esteemed and exploited. It is not because some statement is scientifically false that it should be ignored. Whence the various systems of scientific democracy (sic), like citizens' conferences, where a public randomly selected to represent a whole population is asked to give its opinion on scientific controversies cleverly staged for its consideration.

So this is the scientific revolution: the learned are no longer the judge but the judged. There can be two kinds of controversy: either mainstream science versus dissident science, or judgement against expertise. Beyond the question of what is true and what is false, one can imagine the importance of rhetoric in such encounters. The person we listen to may not be the most learned, and hence the most expert, but simply the best speaker, the most mediagenic, and in the best case, the person who seems to be wisest. Experience will have the edge over expertise. We will listen to the person we would trust in a universe where knowledge is no longer sufficient to establish its own legitimacy.

From this movement there springs a multitude of different kinds of expertise, as attested by the debates on GMOs and mobile phones. Quantity takes precedence over quality. No-one can ever claim to have the last word. We may always be accused of having neglected something. We can be sure there will be another expertise tomorrow or the day after, which must be taken into account no matter what its origins. And this implies several lines of conduct. To begin with, they must all be gathered in without any form of selection. Then they will spread anxiety, turning the most everyday act into a problem or a difficulty. On the one hand, some experts tell us we must eat citrus fruits, while others claim that they are dangerous because they are full of pesticides. At this point, recourse to expertise is no longer useful when we must make a decision. Recourse to expertise results in disagreements and clashes between experts which cannot reach any conclusion and which in the end even rule out any decision. This is what we have seen with GMOs. And this is why we have seen the government organise a round table on relay antennas (*Grenelle des antennes*) where, for the first time in such a process, the experts were not even invited.

On top of that, the precautionary principle has considerably complicated the notion of risk. For example, in their report on this principle, Philippe Kourilsky and Geneviève Viney introduce the idea that the scientific mexpertise should be supplemented by a social, or societal expertise. The consequences are significant. The notion of risk is broken into its scientific components, but also its economic, social, and soon also its ethical components. The points of view that may be put forward on a given theme by a group of scientists may become totally divergent, depending on what each considers to be a good society. We no longer inquire only about the value of an activity, but about the value we attribute to our own values. These are discussions about principles, where the only result is to clarify one's differences, and where the social aspect will always carry more weight than the scientific. The decision must necessarily become political. For expertise, we substitute debate, and now round-table negotiations (*grenelles*).

The Precautionary Principle Complemented and Contested by the Principle of Anticipation and the Principle of Attention

Anticipation covers all the processes of vigilance and monitoring that industry must observe, or expose itself to increasing legal sanctions. They must be aware of everything that is known worldwide, and admit everything they know without reticence. They have the particular responsibility of being the ones that must concentrate all known information, bringing it together for the benefit of everyone.

The principle of attention appeared in 2003 in an AFSSET report on mobile telephones. It was then reformulated as one of the guiding principles of the recent round table on waves and relay stations (*Grenelle des ondes ou des antennes*). The idea is that, even if we are scientifically certain that the WHO standard with regard to emission power is without risk, it is not sufficient reason to ignore the sufferings of those who complain or the demands of the protection groups. We must not reject them, excluding them and treating them like the sick, mocking their psychological fragility. We must take care of them, make available some suitable medical and psychological support that can help them to overcome their difficulty and adapt to technological change. We thus see that the principle of attention stands independently of the principle of precaution insofar as it seeks to build up a policy precisely on the basis of what is not scientific. It is contemporary with the recognition that victims have a peculiar status: the primary damage is not physical and material, but psychological and intangible.

The Principle of Precaution Will Soon Be, or Already Is, Completely Overthrown in Its Original Sense, in the Context of Discussions on How to Implement the New Technical Paradigm of Nano and Bio

Naturally, at a first level, these technologies involve hazards that raise classic problems of risk. What will happen when we get down to the ultimate components of matter, when we begin to manipulate individual atoms. How do they behave? How do they communicate amongst themselves? On the one hand, there are new technical capabilities, generating great promise and hope, while on the other there is great ignorance. We shall say that for this reason we must apply the principle of precaution.

Now these techniques are presented in a somewhat ambiguous manner. On the one hand we proclaim gains in efficiency, new products, and heightened performance, while on the other we speak less of reducing risk and suffering than of the somewhat troubling possibility of 'improving' the performance of natural species – plant (GMO), animal, or human. Here we encounter the problem of the post-human. From this standpoint, technology is no longer this aid, this crutch, this expedient that intelligence has devised to help humans fight against the precariousness of their existence. It is no longer a necessary supplement to assist a human nature that is too limited to organise its own survival.

The nano paradigm seems to inaugurate a new programme: it is no longer there to compensate for a natural weakness, written somehow into nature's project, but to release us from nature for an existence that would become technological and thoroughly artificial. On the one hand, such a release has been associated since the Bible with disaster. On the other, it looks like a manifestation of an exorbitant power, implemented for what may be selfish reasons. This was already the subject of discussion with the advent of the GMO, but here it is amplified. It is understandable that, faced with these questions, the debate over levels of hazard becomes secondary compared with the debate which is no longer about health or economics, but concerns philosophy and ethics. The principle of precaution may here be extended in a new direction, bearing upon the ethical hazards of technological innovation. But then we enter a different universe to the one that dealt in the traditional comparison between risks and benefits.

Part V

Nanoethics and Regulation: The Situation in Europe and the World

If we extend the debate from the national level, as discussed in the last five chapters, to Europe and the rest of the world, nanoscience and nanotechnology have stimulated a previously unknown level of reflection on the international scale, adding ethical concepts to the previously dominant economic issues.

In this context, in 2008, the European Commission drafted a code of good conduct for responsible research in nanoscience and nanotechnology. However, this code depends solely on the good will of each member state to apply the seven principles specified in it, making them responsible for accepting or refusing to implement research into or synthesis of nanometric objects on the basis of expected benefits or potential risks.

There are a huge number of possible applications for nanotechnologies. To confirm this, one only has to consult the data base set up by the European Observatory on Nanotechnologies, where one can sift through hundreds of thousands of references in an almost infinite range of possibilities and in every area of industry.

It is well to ask whether this universal availability of scientific knowledge, combined with the current facility for disseminating it, will really be of benefit to developing countries, allowing them to devise a scientific policy in phase with their specific needs in health, education, and so on. This is the aim of the *Millenium Development Goal*. But will these countries have the means to achieve it?

The next four chapters provide a glimpse of a very important debate, regarding individual and collective awareness of responsibilities in research, innovation, industrialisation, and globalisation, not to mention the crucial need to "control risk without hindering progress" (*Le Monde*, 26 February 2010, p. 15).

25

Situation in Europe and the World: A Code of Conduct for Responsible European Research in Nanoscience and Nanotechnology

Philippe Galiay

25.1 Introduction

The code of conduct for responsible research in nanoscience and nanotechnology adopted and proposed to the Member States of the European Union by the European Commission on 7 February 2008 is the only one of its kind in the world. It results from an approach that is unusual enough to deserve deeper analysis.[1]

Indeed, one need only consider the combination of different logics (scientific, political, economic, and societal) that went into the drafting of this document to begin to grasp its scope. Too narrow for some, who criticise a European Commission under the sway of the markets, a landmark on the road to future (and hence disgraced) regulation for the champions of a free and unconstrained development of nanotechnologies. A happy medium for others, who see this code as a useful guide for the progress to be made in nanoscience and nanotechnology, in Europe and beyond. But no-one could be indifferent to it in a world in crisis, torn between opposing factions, preface to the advent of a new worldwide society based on knowledge rather than material goods.

It will thus be useful to look back quickly over the motivations that underlie the actions of the European Commission in the field of research in general and in the areas of nanoscience and nanotechnology in particular. Along the way, we shall also consider the motives behind the EC's Science and Society Action Plan drawn up in 2001 in the context of a broad reflection on European governance and the European research scene. We can then understand what persuaded the Commission to draw up this code of conduct for responsible research on nanoscience and nanotechnology at the beginning of 2007.

Having understood this, we shall discuss the way the code of conduct is organised, how it is perceived and used, in particular by the European

[1] The opinions expressed in this chapter are those of the author and should not in any way be considered to represent the official position of the European Commission.

P. Houdy et al. (eds.), *Nanoethics and Nanotoxicology*,
DOI 10.1007/978-3-642-20177-6_25, © Springer-Verlag Berlin Heidelberg 2011

Commission in its ethical review of projects competing for EC funds, how it ties in with existing EC regulations, how it relates to international dialogue, and finally, how it is likely to evolve in the future.

25.2 European Research

Research in Europe is not a new thing, and Europe has at several times in history even been the center of the scientific world, although it has never had exclusive rights to scientific curiosity. But few are those who realise that European research has been around for a certain time. On the other hand, if we wish to date it, we only need to agree on what is meant by 'European research'.

If we understand it to mean an initiative jointly run by the governments of several European countries, we come up with 1954, the date when the *Centre européen de recherche nucléaire* (CERN) was created. But if we take it to mean an initiative jointly run by the research organisations of various European countries, we would probably come up with 1974, date marking the beginning of the European Science Foundation. If we think of Europe in its broadest geographical extent, the date would be rather 1984, with the resolutions of the Council of Europe calling for the creation of a European scientific and technical area.

Finally, if we should mistakenly refer to EC research, several important dates come to mind, the oldest being the creation of the *Communauté européenne du charbon et de l'acier* (CECA) in 1951, which sought in particular to encourage research in these areas. However, the launch of the series of framework programmes for research and technological development in 1984 was no doubt the decisive step, marking the birth of a financial tool that has proven itself, in the sense that it has lasted for more than a quarter of a century. The Single European Act of 1986, followed by the Maastricht Treaty of 1992 (and subsequently Amsterdam and Nice) set about laying down, extending, and consolidating the legal foundations of a genuine EC research policy.

The debate was extended in 2000 with the re-emergence of the idea of a European research area (ERA) [1]. It was promoted by Philippe Busquin, who advocated the Lisbon strategy,[2] from the first year of his mandate as European Commissioner for Research, then by Janez Potocnik, Research Commissioner from 2005 to 2009, notably through the Ljubljana process and efforts to introduce a fifth European freedom, namely, the free circulation of knowledge within the ERA.

[2] In March 2000, the European Council of Lisbon set the European Community the target of becoming, within the decade that followed the founding of the knowledge-based economy, the most competitive and dynamic in the world, capable of sustainable economic growth with more jobs and better quality employment, together with greater social cohesion.

The history of European research, while it has no precise definition or unified political theory, is thus studded with initiatives, concrete realisations, and discussions which are too briefly and incompletely described here,[3] and which fall into two distinct categories. One is rather utilitarian, oriented toward economics and politics, while the other aims purely at the advancement of knowledge.

Up to now, EC research actions have been mainly utilitarian, since devoted to the construction of a community through the policies of the European Union. With the introduction of the notion of a knowledge-based society and the ERA in the Lisbon Treaty, the perspective broadened and became more systemic. It was now the whole of European society that was invited to transform itself and better direct its efforts toward a finer understanding of the world in which we live, in all its complexity, and toward fairer realisations, more respectful of human beings, their values, and their environment.

By their very nature, nanoscience and nanotechnology can bring a key contribution to this knowledge-based European society of tomorrow, and it is with this ambition that their governance should be conceived.

25.3 EC Research and Nanotechnology

Today, nanoscience and nanotechnology are primarily characterised by their potential to revolutionise all other scientific and technological activities, whether these be in materials science, medicine, energy technologies, information and communication technologies, and so on. Objects can be manufactured differently, with material and energy gains, while new functions can be invented and others improved. The prospects for nanoscience and nanotechnology are thus full of promise and, in a competitive world environment, it is vital to master them in order to maintain European competitiveness.

However, the idea of creating and using nanometric particles and systems brings up new questions about their fate, i.e., their toxicological effects, when they come into contact with humans and the environment. And it also raises certain ethical issues peculiar to this kind of entity.

This is why the European Commission began to pay particular attention to nanoscience and nanotechnology in its Sixth Framework Programme for Research and Technological Development (2002–2006), when 1 400 Meuro was invested in more than 550 projects.[4] This financing was accompanied by a move toward better coordination with the adoption of a strategy and action plan for Europe over the period 2005–2009.

---

[3] For more detail, see in particular the Seventh Framework Programme in the history of European research, RTD Info Special, June 2007.

[4] This funding should be increased in the Seventh Framework Programme (2007–2013). The budget for nanoscience, nanotechnology, materials and new production technologies was set at 3 450 Meuro.

25.3.1 EC Strategy and Action Plan for Nanoscience and Nanotechnology

The EC directorates most directly concerned by these issues[5] soon realised that the development of nanoscience and nanotechnology would require an integrated approach, linking not only the various EC policies but also the different Member States, in order to ensure responsible and sustainable development. In 2004, the European Commission thus presented a European strategy in favour of nanotechnology to the Member States of the European Union [2], and in 2005, following the favourable reception of this European strategy by the Council, a plan of action [3] specifying a series of interrelated actions to implement the strategy.

The plan of action aims not only to support research, development, and innovation, to favour the inception of infrastructures and competitive clusters in Europe, and to increase human resources particularly in cross-disciplinary areas, but also to integrate the social aspect into nanoscience and nanotechnology by taking into consideration the preoccupations of European citizens with regard to the environment, safety, and public health.

To implement the strategy and action, a new mode of governance is needed. In particular, in its plan of action, the European Commission indicates that a better dialogue between researchers, public and private decision-makers, other stakeholders, and the public can only help to develop a better understanding of the possible concerns, to tackle them from the point of view of science and governance, and to favour a judgement and a commitment in full possession of the facts.

In its resolution of 15 December 2006, the European Parliament, then the Council in its conclusions of 23 November 2007, supported this approach by recognising the need to favour synergy and cooperation between all stakeholders.

25.3.2 Support for Research in Nanoscience and Nanotechnology

Even more noticeable than the integration of the social dimension into nanoscience and nanotechnology is the gradual integration of nanoscience and nanotechnology into society. Indeed, nanoscience and nanotechnology came onto the scene in an environment that had been seriously marked by several scientific and technological controversies toward the end of the twentieth century. Nuclear power, genetics, and human and animal foodstuffs provide a few examples, and in order to allay fears of some connection with GMOs, it was often advisable to stress the fact that nanosciences and nanotechnology had absolutely nothing to do with them. But this was no doubt a little optimistic.

[5] The Directorate Generals responsible for industry, the environment, health, and consumers.

Indeed, parameters had already been set. In February 2000, the European Commission had adopted a communication referring to the principle of precaution [4], proposing a common form of risk governance in situations of scientific uncertainty. In June 2001, the Council had adopted a strategy in favour of sustainable development,[6] which aimed to protect the environment and health. Although not in any way legally binding, it would be difficult for the Commission itself or the Member States who adopted these political orientations to ignore or transgress them. So it was in this soft law context that nanoscience and nanotechnology were destined to come of age.

Furthermore, the opinions of the advisory committees helping the Commission to prepare its policies were also made public. The Scientific Committee on Emerging and Newly Identified Health Risks (SCENIHR) made two statements, the last in March 2006, on the relevance of assessment methods for risks generated by the products of nanoscience and nanotechnology. In January 2007, the European Group on Ethics in Science and New Technologies (EGE) made a statement on the ethical questions relating to nanomedicine. These statements were themselves based upon principles announced in the Charter of fundamental rights of the European Union [5], jointly signed and proclaimed in Nice by the European Parliament, Council, and Commission on 7 December 2000.

Finally, nanoscience and nanotechnology had to face up to more general reflections upon the relationship between science and society.

25.4 Science in Society

Shortly after its communication entitled *Towards a European Research Area* [6, 7] in January 2000, the services of the Commission put forward a paper on *Science, Society and the Citizen in Europe* [9] for the consideration of its Member States. This paper observed that science was appreciated by European citizens for the progress it could bring and for points of reference it could provide in a crisis situation. But at the same time, this science was raising more and more questions and turned out to have little attraction for young people.

In December 2001, at the request of its Member States, the Commission adopted a Science and Society Action Plan, and the Sixth Framework Programme (2002–2006) put aside 88 Meuro to carry out its actions in the fields of communication, education, equality for women, ethics, and governance. This effort was reinforced in the Seventh Framework Programme, with 330 Meuro allocated to Science and Society for the 7 years of this programme (2007–2013).

[6] A strategy revised in Gothenburg in June 2006.

The projects financed by these framework programmes in the areas of governance and ethics served to provide better coordination for reflection carried out in the different European countries, to draw conclusions, and to set up specific orientations and tools on a Europe-wide scale to deal in particular with the issues of nanoscience and nanotechnology.

25.4.1 Toward a Governance That Makes More Allowance for Scientific Knowledge

As regards the governance of research, the main lesson concerns inclusiveness, i.e., increased exchange between the worlds of research and civil society. These exchanges serve to improve mutual comprehension, either with regard to the specific concerns of each of these worlds, or with regard to the values that preside over the choices to be made.

The Commission thus set up new instruments (Research for the Benefit of Specific Groups – Civil Society Organizations, or BSG-CSO) which would allow civil society and research organisations to interact and generate cooperative research schedules specific to this type of relationship. Calls for proposals for public commitment to research were launched in 2009 and more ambitious action plans (Mobilization and Mutual Learning Action Plans) were initiated in 2010.

Besides this, the pilot project *Meeting of Minds – European Citizens' Deliberations* [8], financially supported by the Commission, had shown between 2005 and 2007 that it was possible to extend citizens' deliberations with scientific or technological themes, carried out so successfully on the regional and national levels in certain EU Member States such as Denmark, to the European level.

Several governance projects made reference to nanoscience and nanotechnology, some concerning citizen participation (CIPAST, DECIDE), some risk governance (RISKBRIDGE, CARGO, MIDIR), and some ethics (NANOCAP, DEEPEN). Instruments of governance were subsequently developed, taking into account what was learnt from these projects in order to adjust them to the future development of nanoscience and nanotechnology.

25.4.2 Elaboration, Dissemination, and Application of Ethical Rules

Europe was built on the basis of shared values that need to be protected. These values, laid down in the Charter of fundamental rights of the European Union [5], together with the decision to adopt the Seventh Framework Programme, form the basis for current EC actions in the field of ethics.

The ethical activities of the Sixth and Seventh Framework Programmes contributed not only to coordinating ethical research in Europe, but also to coordinating the reflections of national ethical committees and promoting their implantation in Member States and third party countries.

In addition, the European Commission set up an ethical review procedure [10] for the proposals submitted in answer to calls made in the framework programmes. This ethical review is carried out whenever the proposals identify activities raising ethical questions, such as the informed consent of patients, research on embryo or foetus, privacy and data protection, dual use of technology, animal experimentation, or research involving developing countries.

We shall see later how questions that touch upon nanoscience and nanotechnology are now taken into consideration by the EC's ethical review.

25.5 EC Recommendation for a Code of Conduct

The idea of a code of conduct for nanoscience and nanotechnology was already mentioned in the EC's 2004 strategy, but at the time, the suggestion was to initiate this reflection on the international rather than simply European level. However, the international dialogue on nanoscience and nanotechnology quickly revealed its limitations and the idea disappeared from the agendas. This is why, faced with persistent scientific uncertainties, Janez Potocnik, the European Commissioner for Research, took the initiative in 2007 of preparing a code of conduct for Member States of the European Union.

25.5.1 Choice of Recommendation

The instrument selected to attract the attention of Member States to the code of conduct was not a communication by the Commission, as would be common practice for research policy, but rather a recommendation. Indeed, according to Article 211 of the treaty founding the European Community, in order to ensure the good working and development of the common market, the Commission can formulate recommendations or opinions on issues covered by the treaty, if the latter explicitly allows for that or if it deems it necessary.

Following a public consultation, the Commission thus chose to submit a recommendation to its Member States integrating a code of conduct which advocates safe and responsible research in nanoscience and nanotechnology, based on general principles and suggesting certain guidelines. This recommendation does not in any sense replace existing regulations, but aims to complement them by facilitating the management of uncertainties arising because nanoscience and nanotechnology are still in their infancy.

25.5.2 Content of Recommendation

The body of the EC recommendation includes suggestions for Member States intending to set up a common approach for using the code of conduct in the European Research Area. The recommendation suggests that Member States should adopt the general principles and guidelines of the code of conduct in their own research strategy for nanoscience and nanotechnology. These principles and guidelines should be used to devise the regulations and norms applying to this area of research. They can also be used as a reference in order to determine funding criteria or audit procedures for research organisations.

The Commission recommends disseminating and encouraging the adoption of the code of conduct by national and regional authorities, employers, research funding organisations, researchers, and any person or civil society organisation involved in or interested in research and other activities in the field of nanoscience and nanotechnology. Furthermore, the code can become an effective tool for coordination between Member States, but also at the European and international level. It is an instrument for dialogue at all levels of governance.

25.5.3 The Code of Conduct

In conformity with the good practices of the European Commission, the principles and proposals for action were the subject of an open public consultation through Internet, whose results were taken into consideration when the code of conduct was drawn up. It thus lays down a series of simple principles and guidelines in the form of proposals for action. It also establishes the priorities for and restrictions upon research in nanoscience and nanotechnology.

Principles

The code of conduct is based on the voluntary adoption of seven principles by stakeholders, and covers all research activities in nanoscience and nanotechnology undertaken in the European Research Area. Of the three principles put forward at the public consultation, viz., precaution, inclusiveness, and integrity, the first two were retained and five others were added to the list, viz., meaning, sustainability, excellence, innovation, and accountability:

- *Meaning.* Research activities in nanoscience and nanotechnology[7] must be comprehensible to the public, and the latter must be helped to understand how they serve the well-being of people and society.
- *Sustainability.* These research activities must accord with EC sustainability objectives as well as the Millennium Development Goals of the United Nations.

[7] In the broadest sense, as understood in the present chapter, research in nanoscience and nanotechnology encompasses all forms of research dealing with matter on the nanometric scale (1–100 nm). It includes all nano-objects produced by humans either intentionally or accidentally. Naturally produced nano-objects do not come within the scope of the code of conduct. Research in nanoscience and nanotechnology includes all activities from the most fundamental research to applied research, together with the technological development and pre- and conormative research underpinning scientific opinions, standards, and regulations.

- *Precaution.* Research activities in nanoscience and nanotechnology must anticipate their potential environmental, health, and safety impacts, taking every useful precaution in proportion with the required level of protection.
- *Inclusiveness.* The governance of research activities in nanoscience and nanotechnology should be guided by the principles of openness to all stakeholders,[8] transparency, and respect for the legitimate right of access to information.
- *Excellence.* Nanoscience and nanotechnology research activities should meet the best scientific standards, including standards underpinning the integrity of research and standards relating to good laboratory practices.
- *Innovation.* Governance of nanoscience and nanotechnology research activities should encourage maximum creativity, flexibility, and planning ability for innovation and growth.
- *Accountability.* Researchers and research organisations should remain accountable for the social, environmental, and human health impacts that their nanoscience and nanotechnology research may impose on present and future generations.

Proposed Actions

The proposed actions are mainly concerned with good governance of nanoscience and nanotechnology research and respect of the precautionary principle. They are supplemented by proposals to disseminate and monitor the code of conduct.

Good governance of nanoscience and nanotechnology research is achieved by providing access to high quality information for all concerned and developing a culture of responsibility among all stakeholders, notably in the respect of existing laws. It also requires implementation of the principle of inclusiveness in decision-making processes and in the establishment of research agendas or in the research itself.

In the field of good governance, priority is given to fixing standards and thereby harmonising practices, and also to risk assessment methods. Priority must be give to research aiming to protect the public, the environment, consumers, and workers, and to improve, reduce, or replace animal experimentations.

The code also introduces a small number of restrictions. For example, it advises against financing any research likely to lead to violations of fundamental rights or ethical principles, such as a fraudulent 'improvement' of the performance of the human body. It also stresses that, in the absence of risk assessment, the deliberate introduction of nano-objects into the human body for research purposes should be banned.

---

[8] Member States, employers, research funders, researchers, and more generally all individuals and civil society organisations involved in or interested in nanoscience and nanotechnology research.

The code of conduct calls explicitly for respect of the precautionary principle, notably for the benefit of those required to work in contact with nanoproducts, but also more generally for consumers and the environment. Indeed, in February 2007, the Commission made a firm commitment on this point to set up guidelines for the prevention of pathologies due to nano-objects.[9] Here it encourages the pursuit of research on information systems for researchers and others likely to come into contact with nano-objects.

The precautionary approach also involves analysing risks in proposals submitted for funding and monitoring the impact of programmes. Funding and research organisations should devote an adequate part of their efforts to a better understanding of the risks, and more broadly, to analysing the ethical, legal, and societal consequences of nanoscience and nanotechnology.

25.5.4 Code of Conduct and Ethical Review

We saw earlier how the science and society consultations led the Commission to incorporate in their Framework Programme an ethical review of projects submitted in response to calls for proposals. Indeed, EC funding now involves obligations to respect the fundamental values of the European Union. And nanoscience and nanotechnology funding is no exception to this rule. In fact, quite the opposite, since it is only by respecting these fundamental values that researchers will enjoy the maximum possible freedom for their research in nanoscience and nanotechnology.

From this point, the Commission decided to integrate nanoscience and nanotechnology into its ethical review. Any research proposal involving nanoscience and nanotechnology is now subjected to assessment by an ethics committee, whose role is to detect any violation of fundamental rights and fundamental ethical principles (at the research and development stage), any breach in the informed consent of patients or in the rules for publication of results relating to human health, and any dual use of the research results.

These ethicists must also report research activities aiming to obtain a better understanding of impacts in areas opened up by research in nanoscience and nanotechnology and the way in which these impacts could be predicted, in particular through participative processes also involving ethics committees. Furthermore, this ethical review is used to measure the extent to which the code of conduct and the opinion of the European Ethics Group on the ethical aspects of nanomedicine has penetrated the world of research.

[9] See the EC Strategy 2007–2012 on Health and Safety at Work [11]. In the absence of an internationally accepted terminology, the generic term 'nano-object' refers throughout the code of conduct to all products resulting from research into nanoscience and nanotechnology. It covers nanoparticles and their aggregates on the nanometric length scale, nanosystems, nanomaterials, nanostructured materials, and nanoproducts.

25.5.5 Code of Conduct and Regulation

In June 2008, the Commission adopted a communication [12] aiming to clarify the way in which EC regulations meet the needs of nanoscience and nanotechnology development, occupational safety requirements, and consumer and environmental protection. This communication reviews existing legislation in these different areas, together with the systems for implementing it, and concludes that current legislation covers in principle the potential risks of nanomaterials for health, safety, and the environment. It also states that it is through improved implementation that the protection of health, safety, and the environment will have to be reinforced.

For their part, authorities and bodies responsible for implementing legislation must continue to monitor the market and employ existing EC mechanisms for dealing with cases where it turns out that products already commercialised could induce risks.

The Commission also promises to broaden understanding of nanoscience and nanotechnology, especially with regard to their characterisation and the assessment of associated risks, and this in a coordinated way on the international scale with other stakeholders in organisations like the OECD or ISO. Finally, the Commission promises to report on the progress in these areas within 3 years of presenting its communication, i.e., before June 2011.

25.5.6 Code of Conduct and International Dialogue

The first international discussions were held in Alexandria (USA) on 17 and 18 June 2004,[10] at the invitation of the US National Science Foundation, only one month after the adoption of the EC strategy by the Commission. However, it must be said that these discussions did not come up to expectations in terms of coordination. Indeed, public and private initiatives aiming to supervise and control nanoscience and nanotechnology research have proliferated since 2007 when the Commission undertook broad consultations to prepare its recommendation.

The Royal Society in the United Kingdom joined up with private partners to draw up a Responsible Nano Code [13]. Discussions were held in France on nanoscience and nanotechnology in the context of the Grenelle Environment Forum, and led to the formulation of a law in 2009 (*Loi Grenelle 1*), in the form of a compulsory declaration when products result from any form of nanotechnology. Such an obligation should also be imposed in Canada.

As a result, in just a few months, the code of conduct adopted by the European Commission became a natural reference in the dialogue with third party countries and with international organisations. Today, Argentina is

[10] The next discussions took place in Tokyo (26–28 June 2006), then Brussels (11–12 March 2008).

considering the adoption of a code of conduct with regard to nanoscience and nanotechnology and South Africa is also interested in the EC code.

As invited by the Commission, Member States should take due account of the recommendation in the context of their bilateral agreements on nanoscience and nanotechnology research strategies and activities with third party countries, and in their capacity as members of international organisations, in order to give the EC code of conduct its full scope.

25.6 Conclusion

It is difficult today to predict what will become of the code of conduct in the years to come, nor what will be its impact. Indeed, this is for European Member States to decide. For it is up to them to disseminate it and promote its use by national and regional authorities, or by organisations funding nanoscience and nanotechnology research.[11] Europe's universities and research organisations will also have to make their own decisions concerning the use of these principles and guidelines.

For its part the Commission will use the code of conduct for its own involvement in the funding and supervision of nanoscience and nanotechnology research. It will also monitor its use in the European Union and in countries associated with the Framework Programme for Research and Technological Development, and see to its revision, first in 2010 and every two years thereafter, and this in concert with the Member States.

The code of conduct will also be discussed with third party countries and international organisations.

To sum up, after this brief history of the birth of the code of conduct, it turns out that, rather than being just a unique entity, it is an element of a unique process, an open process, whose calling requires it to integrate opposites, fears and hopes, haste and caution, in an action that should be less troubled and more efficient in the long term for European society as a whole.

References

1. M. André: L'Espace européen de la recherche: histoire d'une idée. Revue d'histoire de l'intégration européenne, **12**, no. 2 (2006)
2. European Commission: Towards a European Strategy for Nanotechnology. COM(2004) 338, 12.5.2004
3. European Commission: Nanosciences and Nanotechnologies: An Action Plan for Europe 2005–2009. COM(2005) 243, 7.6.2005

[11] According to the conclusions of the Competitivity Council of 26 September 2008 (13672/08), Member States showed that they thoroughly supported the Commission's commitments with regard to responsible research in nanoscience and nanotechnology.

4. European Commission: Communication from the Commission on the Precautionary Principle. COM(2000) 1, 2.2.2000
5. European Commission: Charter of fundamental rights of the European Union (2000/C 364/01) FR.18.12.2000
6. European Commission: Towards a European Research Area. COM(2000) 6, 18.1.2000
7. European Commission: Towards a European Research Area: New Perspectives. COM(2000) 161, 4.4.2007
8. www.meetingmindseurope.org
9. European Commission: Science, Society and the Citizen in Europe. SEC(2000) 1973, 14.11.2000
10. European Commission: Ethics for Researchers: Facilitating Research Excellence in FP7. Eléonore Pauwels, 2007
11. European Commission: EC Strategy 2007–2012 on Health and Safety at Work. COM(2007) 62, 21.2.2007
12. European Commission: Regulatory Aspects of Nanomaterials. COM(2008) 366, 17.6.2008
13. www.responsiblenanocode.org

26

Situation in Europe and the World: Societal Risks and Benefits of New Nanometric Products

Jean-Marc Brignon

Nanometric products promise a wide range of applications which should bring benefits to society in many vital areas, including energy, drinking water, health, environmental protection, and others. At the same time, these products involve risks, some due to there use as-is, some due to applications in which they are combined with other materials. In order to avoid the often excessive fears these new technologies inspire (just as enthusiasm for them is often exaggerated), it is important to carry out as objective an assessment of the risks and benefits as possible.

The aim in this chapter is to cast a glance over the regulatory context for assessing the risks and benefits of chemical substances in Europe, and to try to determine how well it is currently adapted to the case of nanotechnological products, not only from a regulatory standpoint, but also with regard to the relevance and sufficiency of the conceptual tools that are actually employed in practice.

26.1 Socio-Economic Assessment of Chemical Substances in Europe

In the REACH regulatory framework (Registration, Evaluation, Authorization and restriction of CHemicals), extremely hazardous chemical substances can only be authorised if the health or environmental risks involved in their use are adequately controlled [1]. If this is not the case, their use may still be authorised if it can be demonstrated that the socio-economic benefits justify taking the risks involved in their use, and if there are no alternative, economically and technically viable substances or technologies that could replace them [2].

In a complementary way, the authorities can demand the partial or total restriction of a chemical product, but they must then check that the benefits of this restriction in terms of avoided impacts remain 'proportionate' when compared with the economic and social costs it induces.

P. Houdy et al. (eds.), *Nanoethics and Nanotoxicology*,
DOI 10.1007/978-3-642-20177-6_26, © Springer-Verlag Berlin Heidelberg 2011

The theoretical framework underlying these assessments is the cost–benefit analysis [3], which is a classic economic tool for evaluating a project and its alternatives. It attempts to identify all the advantages and disadvantages of the different scenarios, if necessary qualitatively, but in the most exhaustive way possible.

The cost–benefit analysis, originally applied in the USA for infrastructure projects, has been extended to public policy assessments in the field of health and the environment. While leaving Member States considerable freedom regarding the exact choice of measures they implement to achieve the objectives for health and environmental protection, several European directives (water framework directive, directive on industrial installations, REACH regulation, etc.) stipulate or recommend that these choices should be based upon socio-economic or cost–benefit analyses.

The interest in such an approach is without doubt to try to rationalise and obtain a balanced global view of the advantages and disadvantages associated with introducing new products into society. This is even more necessary in the context of nanotechnology, where one sometimes encounters an almost mindless optimism regarding their benefits, along with a systematically pessimistic view of their risks.

Socio-economic analysis goes beyond the probabilistic risk analysis, not only in that it brings in the economic aspects, but also that it is concerned with collective impacts, e.g., possible numbers and types of pathologies in the case of health impacts, and types, distribution, and seriousness of damage that might be inflicted upon ecosystems.

But rationality soon reaches its limits, and one of the main difficulties is then to weigh up all these impacts as a whole. Multi-criteria analysis and costing of impacts are classic tools here, but they suffer from both practical and theoretical shortcomings. How can one attribute a value to an ecosystem? How can one add up or weight impacts on the health of the population and damage to fish stocks?

Despite these open questions, the European Chemicals Agency, responsible for implementing the REACH regulation, has taken the initiative of editing a guide to socio-economic analysis and setting up a Socio-Economic Analysis Committee which, on the basis of studies prepared by industry and Member States, will give an opinion on whether certain hazardous chemical substances should be authorised in Europe.

26.2 Does the REACH Regulation Apply to Nanometric Products?

Insofar as they are not explicitly excluded, the REACH regulation applies to nanometric products. However, the conditions of application remain poorly determined, and the effectiveness of this regulation for managing the risks due to nanometric products is still being debated by stakeholders [4].

This situation will probably not be clarified until this regulation is revised in 2012.

A first element of uncertainty is the fact that there are thresholds in the REACH regulation. The threshold of one tonne below which a substance does not have to be registered, and the threshold of ten tonnes which must be exceeded before assessment of exposure to the substance is required, these raise problems in the case of nanoproducts. Given that it is likely that many nanometric products are manufactured in or imported into the European Union in small amounts, all such products would escape registering or exposure assessment. However, it should be said that the safety net of authorisation procedures and restrictions applies without these notions of threshold.

A second element is a certain fuzziness in the REACH regulation concerning the notion of substance or preparation. For example, two nanometric products with the same structure but carrying a different functional coating might not be considered as two distinct substances, or two distinct preparations, even though they may exhibit very different toxicological properties.

Finally, as stressed by the SCENIHR experts, risk analysis methods used for applications of the REACH regulation to decide the regulatory fate of a conventional substance might in fact be ill-suited to nanotechnological products [5].

Going beyond these questions, it is clear that the REACH regulation applies on a case by case basis, adopting no overview of the nanotechnologies and their applications and without considering the interactions between them or with other technologies. It is thus also clear that it is not up to the task of managing these risks.

26.3 Limitations of Cost–Benefit Analysis in the Case of Nanotechnology

We have already mentioned some of the difficulties inherent in socio-economic analyses, and in particular those pertaining to cost–benefit analyses, but which were not specific to nanotechnologies. On the other hand, some limitations can be put down to the fact that nanotechnology potentially involves technological breakthroughs, and hence radical change.

To begin with, one postulate of a cost–benefit analysis is that it is legitimate to seek a global balance between risks and benefits: a compromise leaving winners and losers will nevertheless be acceptable, provided that the gains are greater than the losses. This assumption is based on the possibility of winners compensating losers, but it leaves two problems.

In practice, do the losers not run the risk, for want of a better solution, of accepting inadequate compensation? In other words, the amount of compensation demanded by the losers may be much lower than the amount that the winners would accept to pay in indemnities, and the disparity would then never be corrected. Is this principle realistic and reasonable when the stakes are very high, as might well be the case with nanotechnology? The question of the unfair distribution of benefits from nanotechnologies is not just an internal problem of our European societies, but may even be exacerbated worldwide. The gulf could widen between those countries able to benefit and the rest, notably owing to the absence of any real compensation mechanism. It is important to bear in mind in this respect that the REACH regulation focuses on the protection of workers and consumers residing in Europe.

A second point is that one can contest the legitimacy of taking into account a comparison of benefits and risks when making a decision about authorizing nanometric products if, independently of the benefits, which may be very high, the risks are also very high. When the risks run by society are very significant, who could take the decision to commit themselves on the basis of benefits which, even though they may well exceed the risks, would not be able to remove them? It should be added that the methods used to estimate risks and benefits, which may be valid when the latter are marginal, might be quite unreasonable when the benefits and the risks are no longer marginal. Several official documents, notably in the United States, make it an objective of nanotechnology development policies to maximise the benefits and minimise the risks [6]: this is indeed necessary, but it is not always sufficient, if the risks remain too high.

Apart from questions of method, the relevance of socio-economic analysis depends more concretely on the relevance of exposure scenarios, dose–effect relationships, and socio-economic responses to nanotechnologies. Experience gained in implementing European environmental regulations has taught us that, when these regulations are devised by stakeholders (authorities, companies), the scenarios often represent a compromise between several opposing visions, supplying plausible pictures of the future, but avoiding more extreme pictures, and in particular ones that envisage disaster [7]. Now one of the fears over the release of products in nanometric form is precisely the possible occurrence of health crises like the one created by asbestos.

The socio-economic responses taken into account in socio-economic studies relating to European environmental regulations are limited to conventional questions, e.g., the impact on employment, training levels, growth, and so on. However, in the case of technologies that might radically alter our points of reference, the response may require consideration of other fields, such as the use of nanoproducts for medication or for weapons, the impact on private life, and so on. Nanotechnological products are likely to change our lives, so their impact does not merely result from the characteristics of these products, but just as much from the way our society receives and transforms them. Research is important and legitimate, along with methodological and scientific questions such as specifying dose–effect relationships, identifying the fate of nanoparticles in aquatic media, etc., but they should not allow us to forget the importance of using them in the context of socio-economic analyses in which the scenarios are also the subject of detailed investigation, and we should not be content to merely repeat the automatic responses deriving from previous studies. To this end, it could be profitable to use so-called backcasting techniques, which identify undesirable futures to be averted, and seek the conditions that would lead to them, precisely in order to be able to detect and avoid them.

26.4 Using the Results of Socio-Economic Analysis. The Precautionary Principle

In the case of nanotechnological products, the results of socio-economic assessments are likely to be hard to translate into decisions and actions: high uncertainties and important consequences may all add up to leave the decision-maker somewhat paralysed. The result then would be that nanotechnology would develop anyway, but without proper controls. Under these conditions, the precautionary principle may well be able to bring solutions, despite the fact that it is often seen as an obstacle to innovation and development.

Indeed, the precautionary principle can be conceived of as a dynamical process, adjusting the rate of approach toward new technologies to match the rate of acquisition of information about their risks. Economists consider that the principle of precaution is not necessarily in conflict with cost–benefit analyses, or the principle of optimisation of well-being, but that on the contrary it can be justified in this context [8, 9]: in an adaptive approach, reducing exposure to risk today provides a way of limiting future damage. The point is not therefore to give in simply to an aversion for risk, but rather to formulate public policies based on caution. The principle of precaution is a way of correcting, by attributing more weight to future damage or costs, the tendency of socio-economic analyses to base themselves upon best possible but somewhat unrealistic estimates, when uncertainties are high and when very unlikely events can lead to very serious consequences. In this sense, applying the precautionary principle can be compared with using low depreciation rates when assessing future damage due to environmental problems or new technologies.

26.5 Beyond Risk and Precaution

Using the precautionary principle is a response to the problems of uncertainty and irreversibility of risks and their consequences. But what about our ignorance? What about deficiencies, gaps, and possible errors in our scientific knowledge? And what about crisis situations in which the lack of information no longer concerns the probability of occurrence (the notion of risk) but rather the nature of the phenomena which might occur? In this case, the potential damage is such that the notion of economic efficiency of safety measures hardly seems to adequate to the situation [7].

Nanotechnological products may well lead to radical technological, social, and economic transformations, and even more so when combined with other new technologies in the fields of biology, information, and communication (the so-called convergent technologies). Regarding the benefits, the stakes are incredibly high, e.g., in health, nutrition, etc., but the same is also true of the risks, which may be disastrous, involving the very notion of humanity (surveillance of private individuals, modifications of the human body, relationship with nature, etc.). Nanotechnology is thus liable to generate extreme scenarios with impacts on health and ecosystems, in fact crisis scenarios, that is, by definition, events that are difficult to imagine and even more difficult to predict. By their very nature, nanotechnologies create entities with entirely novel, hence largely unpredictable functions, that might be misused in quite unthinkable ways.

In this context, socio-economic analysis and a conventional and purely economic interpretation of the precautionary principle would be inoperative. The principle of precaution should here mean rather an attitude of caution in the face of the lack of scientific knowledge and the complexity of human behaviour. The problem here is no longer simply to assess risks and the issues they raise between toxicologists and economists. It encompasses also the whole issue of scientific discourse, the roles of all stakeholders, and the introduction of ethics. The precautionary principle then extends explicitly to moral issues, as in the definition given by UNESCO [10], which defines the notion of a morally unacceptable threat and refers to the human rights of those who suffer the impacts. This desire not to limit the precautionary principle to a simple economic logic of well-being by introducing an accountability principle [11], to use the term coined by Hans Jonas, is particularly relevant in the case of nanotechnologies.

References

1. Regulation (CE) no. 1907/2006 (REACH), consideration no. 22
2. Regulation (CE) no. 1907/2006 (REACH), consideration no. 69
3. *Guidance on socio-economic analysis*, European Chemicals Agency (2008)
4. European Commission: Doc. CA/59/2008 rev. 1. Follow-up to the 6th Meeting of the REACH Competent Authorities for the Implementation of Regulation (EC) 1907/2006 (REACH), 15–16 December 2008
5. Scientific Committee on Emerging and Newly Identified Risks, Modified Opinion (after public consultation) on The appropriateness of existing methodologies to assess the potential risks associated with engineered and adventitious products of nanotechnologies, 10 March 2006: http://ec.europa.eu/health/ph_risk/committees/04_scenihr/docs/ scenihr_o_003b.pdf
6. M.C. Roco, W. Sims Bainbridge: Nanotechnology: Societal implications. Maximizing benefits for humanity. Report of the National Nanotechnology Initiative Workshop 3–5 December 2003, Arlington, Virginia: www.nano.gov/nni_societal_implications.pdf

7. J.P. Dupuy, A. Grinbaum: Living with uncertainty: Toward the ongoing normative assessment of nanotechnology. Techné **8**, 4–25 (2004)
8. C. Gollier et al.: Scientific progress and irreversibility: An economic interpretation of the 'Precautionary Principle'. J. Publ. Econ. **75**, 229–253 (2000)
9. K. Kuntz-Duriseti: Evaluating the economic value of the precautionary principle: Using cost benefit analysis to place a value on precaution. Environmental Science & Policy **7**, 291–301 (2004)
10. Nanotechnologies and ethics, policies and actions, World Commission on the Ethics of Scientific Knowledge and Technology (COMEST), UNESCO, Paris (2007)
11. H. Jonas: *Le principe de responsabilité. Une éthique pour la civilisation technologique*. Editions du Cerf (1990)

27

Situation in Europe and the World: The European Nanotechnology Observatory

M. Morrison

27.1 The Role of the ObservatoryNANO Project

Nanotechnology is a complex and rapidly changing field, which is often difficult to assess in terms of opportunities, challenges and risks. Due to its strong interdisciplinary nature and rapid evolution, nanotechnology has wide-spread and fragmented impacts. Thus, the creation of a reliable source of data and analysis, which is continuously monitored and updated, is critical to provide comprehensive information to decision makers. Governments and businesses are interested in the market potential of nanotechnology enabled products and processes (estimated to underpin a total market of up to one trillion US dollars by 2015); and the possibility of these products contributing significantly to alleviating global problems such as major diseases, energy, clean drinking water, and environmental pollution. However, it is also recognized that the socio-economic impacts of nanotechnologies are often exaggerated or placed in an over-optimistic time-frame. Furthermore, nanotechnologies, as with any new technology, have potential risks (socio-economic, to human health, and the environment), and it is important that these are identified early on and appropriate actions taken, to ensure that development occurs in a safe and responsible manner. Even more important are the 'unknowns' which will inevitably remain by the time nano-enhanced products reach the market, and will require more advanced forms of risk communication and corporate responsiveness to ensure market success.

The observatoryNANO project is funded by the EC for 4 years under FP7 to address this. It is assessing all aspects of the value chain from basic research to market applications in terms of scientific, technological and socio-economic developments and prospects. At the same time it is assessing ethical and societal aspects; potential environment, health and safety issues; and developments in regulations and standards.The project employs a combination of literature review; trend analysis of patents and peer-reviewed publications; and engagement with experts from different fields through interviews, workshops, and questionnaires.

P. Houdy et al. (eds.), *Nanoethics and Nanotoxicology*,
DOI 10.1007/978-3-642-20177-6_27, © Springer-Verlag Berlin Heidelberg 2011

The overall strategy is to create a European Observatory on Nanotechnologies to present reliable, complete and responsible science-based and economic expert analysis, across technology sectors, establish dialogue with decision makers and others regarding the benefits and opportunities, balanced against barriers and risks, and allow them to take action to ensure that scientific and technological developments are realized as socio-economic benefits. The key customer for these analyses is the European Commission, with which the consortium interacts closely to ensure that it continues to meet the Commission's needs within its original remit. All reports issuing from the observatoryNANO are reviewed by external experts before being made publicly available.

Summary of Project Objectives

- To observe nanotechnology developments in ten broad sectors: aerospace, automotive, and transport; agrifood; chemistry and materials; construction; energy; environment; health, medicine, and nanobio; ICT; security; and textiles.
- To engage with the expert communities to discuss and review scientific and technical developments, and relate these to socio-economic impacts and wider issues.
- To consolidate this analysis to produce an online database of concise reports, which clearly identify developments, opportunities, challenges and risks in each of these sectors.
- To provide information and tools for the scientific and business communities to support the responsible development of nanotechnologies.
- To support policy and decision makers by providing validated information on the current and forecasted state of nanotechnology development.

All information from the project is freely available through a dedicated website (www.observatory-nano.eu). Users can search for specific topics or browse through the catalogue of reports and articles, and select items in an online 'briefcase' to store for easier access later or to download/print as required (see Fig. 27.1).

27.2 Approach Taken by the ObservatoryNANO

The observatoryNANO combines analyses of different aspects of nanoscience and nanotechnology (N&N) development into concise reports to support policy and decision makers, and at the same time provides tools to assist researchers and business to assess the wider societal implications of their work (Fig. 27.2).

The main emphasis of the project is on science and technology (ST) analysis and linking this to an analysis of economic impacts.

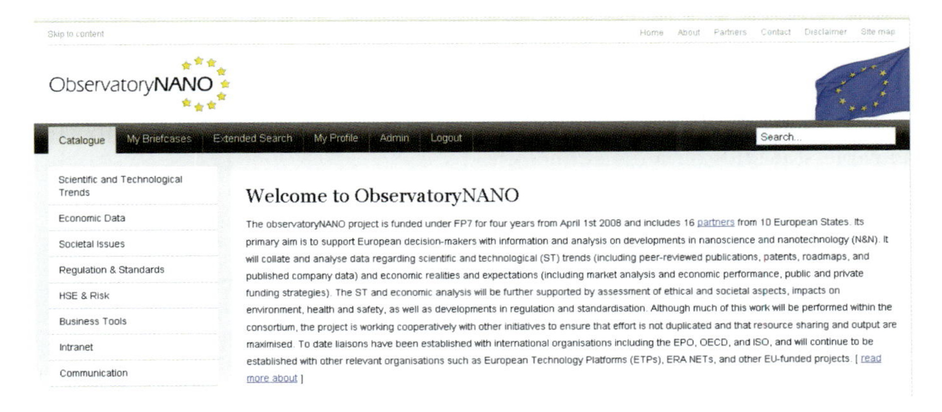

Fig. 27.1. Homepage of the observatoryNANO project

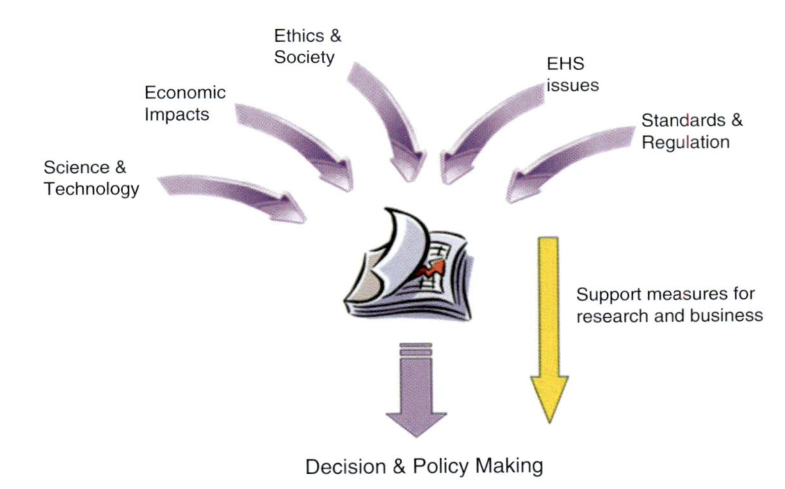

Fig. 27.2. Overall approach taken by the observatoryNANO project

27.3 Interaction with Other Organizations

A critical aspect of the work of the observatoryNANO project is to liaise with other organizations and analyze the implications of their output in the context of other activities. The number of such liaisons and reviews is extensive, however the key ones are indicated below.

27.3.1 Organisation for Economic Co-operation and Development (OECD)

The OECD has established two Working Parties related to nanotechnologies. The first, on Manufactured Nanomaterials (WPMN), has several initiatives that are significant to the work of the observatoryNANO, including the investigation of current research into EHS aspects of nanomaterials, regulation and standards. The second working party, on Nanotechnology (WPN), works together with another OECD group, the National Experts of Science and Technology Indicators to foster the development of internationally comparable statistics on nanotechnology.

27.3.2 International Activities in Standards

The International Standards Organization (ISO) has a technical committee (TC 229) that is responsible for developing standards for nanotechnologies. Several of the partners of observatoryNANO participate in working groups of ISO TC 229, the parallel European standards organization (CEN) technical committee (TC 352) or national standards committees. The observatoryNANO integrates output from these organizations with analysis performed within the project.

27.3.3 European Technology Platforms (ETPs) and Joint Technology Initiatives (JTIs)

European Technology Platforms (ETPs) are industrially led networks that include SMEs, and were established to develop strategic research agendas (SRAs) in specific industrial sectors. Joint Technology Initiatives (JTIs) have evolved from some of these ETPs and are public-private partnerships (PPP) which are eligible for FP7 funding. The purpose of these PPPs is to implement the SRAs of the ETP. There are currently 5 JTIs: Fuel Cells and Hydrogen, Innovative Medicines Initiative, Embedded Computing Systems, Aeronautics and Air Transport, and ENIAC. Of the 40 ETPs and JTIs, 16 explicitly mention nanotechnology within their vision statements or SRAs, and a further 16 by their nature are likely to make use of nanotechnology enabled advances (see Table 27.1).

27.3.4 Manufacturing Initiatives

Manufacturing plays a strategic role in the future of the EU economy and is implicit in wealth generation from each of the sectors in which nanotechnology will have an impact. There are several pan-European initiatives funded to network and support manufacturing industries and coordinate the development of strategies for the future of the EU manufacturing industry. These include, but are not limited to, the micro and nanomanufacturing and MANUFUTURE ETPs described in the table above, and roadmapping projects such as microsapient [which produced roadmaps to "prepare the European industry for a move from designing MST-based products for specific materials and technologies (platform and technology push products) to adopting new disruptive processes/process chains to satisfy specific functional and technical requirements of new emerging multi-material products"] and IPMMAN which supported

Table 27.1. List of current ETPs and JTIs. (1) Explicitly involves nanotechnology, (2) implicitly involves nanotechnology, (3) probably does not involve nanotechnology. *Asterisk* indicates JTI

ETP	Acronym	1	2	3
Advanced Engineering Materials and Technologies	EuMaT	1		
Advisory Council for Aeronautics Research in Europe	ACARE		1	
Aeronautics and Air Transport	Clean Sky*		1	
Embedded Computing Systems	ARTEMIS*		1	
European Biofuels Technology Platform	Biofuels		1	
European Construction Technology Platform	ECTP		1	
European Nanoelectronics Initiative Advisory Council	ENIAC*	1		
European Rail Research Advisory Council	ERRAC			1
European Road Transport Research Advisory Council	ERTRAC			1
European Space Technology Platform	ESTP		1	
European Steel Technology Platform	ESTEP	1		
ETP for the Electricity Networks of the Future	SmartGrids			1
ETP for Wind Energy	TPWind			1
ETP on Smart Systems Integration	EPoSS	1		
ETP on Sustainable Mineral Resources	SMR	1		
Farm Animal Breeding and Reproduction Technology Platform	FABRE TP			1
Food for Life	Food	1		
Forest based sector Technology Platform	Forestry	1		
Fuel Cells and Hydrogen	FCH*		1	
Future Manufacturing Technologies	MANUFUTURE	1		
Future Textiles and Clothing	FTC		1	
Global Animal Health	GAH			1
Industrial Safety ETP	IndustrialSafety		1	
Innovative Medicines Initiative	IMI*	1		
Integral Satcom Initiative	ISI		1	
Micro- and NanoManufacturing	MINAM	1		
Mobile and Wireless Communications	eMobility	1		
Nanotechnologies for Medical Applications	NanoMedicine	1		
Networked and Electronic Media	NEM		1	
Networked European Software and Services Initiative	NESSI			1
Photonics21	Photonics	1		
Photovoltaics	Photovoltaics	1		
Plants for the Future	Plants		1	
Renewable Heating & Cooling	RHC		1	
Robotics	EUROP	1		
Sustainable Nuclear Technology Platform	SNETP		1	
Sustainable Chemistry	SusChem	1		
Water Supply and Sanitation Technology Platform	WSSTP		1	
Waterborne ETP	Waterborne			1
Zero Emission Fossil Fuel Power Plants	ZEP		1	

the improvement of industrial production through the integration of macro-, micro- and nanotechnologies for more flexible and efficient manufacturing support. Both of these contributed to the development of the MINAM ETP.

27.4 Science and Technology Assessment

Research and Innovation are integral to the vision of a knowledge-based and low carbon society (as proposed in the Lisbon strategy). Such a vision requires significant investment in science and technology, and a target of 3% of GDP of Member States has been set for 2010. However, the outcomes of such investments are not always easy to determine.

A major goal of the project is therefore to develop suitable methodologies for the identification and validation of ST and economic indicators in different technology sectors and allow government, funding agencies, investors, and industry to make strategic decisions regarding the potential of the sector, and obstacles to the full realization of this potential; and the relative position of EU RTD with regards to the global market. This will allow the EU to focus its energies on the technologies of most relevance to EU society and to take appropriate actions to ensure that such technology development is capitalized by EU industry and not lost to another region. It will also give clear indications of whether the EU is competing effectively with other regions in growth areas, and if not, the reasons why this could be so and recommended actions. This analysis requires two types of data: quantitative (such as numbers of publications, patents, and initial public offerings (IPOs), and funding levels) and qualitative (such as expert opinion of trends, opportunities, gaps). However, quantitative data is rarely complete (or accurate) and qualitative data can be highly subjective (dependent on the pool of experts sampled). A major objective of this project is therefore to develop suitable and robust methodologies to marry quantitative and qualitative analysis of ST and economic data to provide a clear understanding of the potential socio-economic opportunities, limits and risks posed by N&N, to ensure that key decision-makers are presented with enough facts, analysis and recommendations to make informed decisions on future development strategies. It is envisaged that these methodologies will continue to be developed throughout the duration of the project in light of work done within the project, external feedback, and collaboration with other organizations.

To put developments in a market context, the realm of nanotechnology has been divided into ten broad sectors: aerospace, automotive, and transport; agrifood; chemistry and materials; construction; energy; environment; health, medicine, and nanobio; ICT; security; and textiles. In turn each of these is sub-divided into a number of topic areas. The ST analysis takes a broad view encompassing major breakthroughs in basic research that could have potential future impacts on the EU economy, applications of current RTD in the short and medium term, potential barriers to these achievements, and implications for the EU's manufacturing base. The output from this work is used as a basis for activities in other work packages. The approach involves the analysis of

- Basic research that has the potential to have major societal and economic impacts.
- The technological impact of a new development on its own sector, for example, will it displace existing technologies, will it allow or lead to new processes.
- The technological impact on other sectors, for example, will the development and associated IP have potential broad, cross-industry applications or is it specific to one sector.
- Gaps in knowledge or technology, including access to infrastructure; appropriate workforce training and education; lack of appropriate regulations, legislation, and standards.
- Implications for the manufacturing industry in the EU.
- Products, applications, and companies.
- New developments of national funding programmes and their thematic and application focus.
- Final reports of EU projects with the aim to identify gaps.
- Patent trends using the 'Worldwide Patent Statistical database' (PATSTAT) which includes data from 76 patent authorities (national patent offices and the super national authorities of the EPO and the World Intellectual Property Organization (WIPO).
- Available roadmaps from organizations and other projects.

Activities are divided into 3 phases (which have an annual cycle, see Fig. 27.3):

- Phase 1 is a review phase. Existing literature (peer-reviewed scientific publications, company reports, reports from other projects and initiatives, such as the ETPs) is collated and analyzed. Nanoscience and nanotechnology (N&N) publications in peer-reviewed journals between 1998 and 2007 have been identified using published algorithms (Porter et al., Georgia Institute of Technology) to create a database of 544 440 records (derived from Web of Science). These have been analyzed to observe: trends in publications over the ten year period, and trends in each of the ten technology sectors (using sets of keywords developed from other relevant databases, journals, thesauri, and professional associations, such as Inspec, Compendex and EMTREE). Output includes country, institute, and co-authorship data. Patent analysis has been performed in collaboration with the European Patent Office (EPO) using the PATSTAT database, in which there are over 130 000 entries that are relevant to nanotechnology. The output from this analysis shows trends in total nanotechnology patenting per year and per country, as well as trends in the ten different technology sectors (using either sets of keywords, or the EPO's existing categorization). The interim reports produced from all of this work are used as the basis for analysis in other areas (economic, ethical, EHS, standards and regulation-see below).

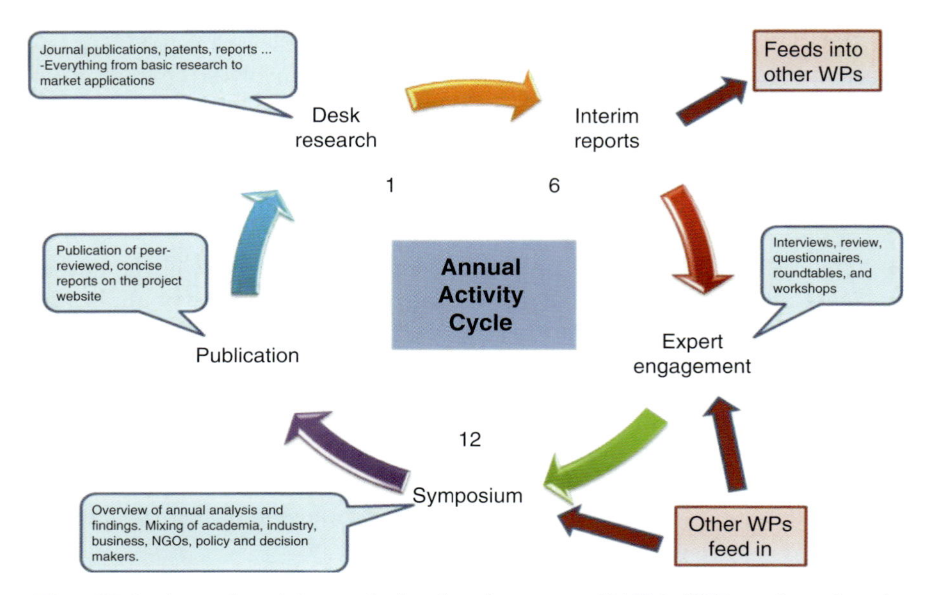

Fig. 27.3. Annual activity cycle for the observatoryNANO (WP work package)

- Phase 2 focuses on expert consultation and analysis. Suitable academic researchers and industrialists are identified based on objective and quantifiable criteria such as authorship of publications and patents, level of activity within a given research area, position in organization, international status, etc. These individuals are invited to participate in a structured discussion on the technical implications of new N&N developments identified in phase 1. The participative process includes interviews, questionnaires, written submissions, roundtables and workshops (both on and off-line). This information is used to enrich the interim reports which are then presented at an annual symposium(phase 3).
- Phase 3 consolidates the data produced during the expert consultation and incorporates analysis from other areas. This is then presented during an annual symposium, to which a selection of experts is invited. This allows for final discussion (particularly of open issues), which is included in reports that are subject to a final round of peer-review before publication on the observatoryNANO website.

In the first year of the project a total of 61 reports have been published describing scientific and technical advances in the ten technology sectors [1]:

- *Aerospace, Automotive & Transport.* Technologies to produce bulk nanostructured metals; technologies to produce polymer nanocomposites; technologies to produce and apply tribological nano-coatings.

- *Agrifood.* Agricultural production; food processing and functional food; food packaging and distribution.
- *Chemistry & Materials.* Carbon based nanomaterials; nanocomposites; nanostuctured metals and alloys; nano-polymers; nano-ceramics; nano-fabrication technologies.
- *Construction.* Cement based materials; coatings; living comfort and building safety; sustainability and environment; civil and underground construction.
- *Energy.* Photovoltaic; thermoelectricity; fossil fuel; energy harvesting; nuclear; renewable energies; fuel cells; hydrogen production and storage; batteries and supercapacitors.
- *Environment.* Air purification; wastewater purification; drinking water treatment; groundwater remediation; soil remediation.
- *Health, Medicine & Nanobio.* Cosmetics; diagnostics; novel bionanostructures; implants, surgery and coatings; therapeutics; regenerative medicine.
- *Information & Communication.* Integrated circuits; memory; displays; manufacturing; photonics; beyond CMOS.
- *Security.* Chemical weapons and industrial toxins detection; biological threat agent detection; radiological-nuclear weapon detection; explosives detection; narcotics detection; neutralising CBRNE effect; decontamination; forensics; personnel protection; equipment and infrastructure protection; condition monitoring of civilian zones; anti-counterfeiting; authentication; positioning and localisation.
- *Textiles.* Nanostructures; fibre production; finishing treatments; textile products.

27.4.1 Publication Analysis

Analysis of publication trends indicates that numbers of N&N publications are increasing each year and that the EU accounts for approximately one-third of these (see Fig. 27.4).In the EU Germany has published the most N&N publications each year, followed by France and the UK (Fig. 27.5). In recent years China has overtaken Japan as the second most prolific country (Fig. 27.4).

The situation is different when considering publication citation indexes, with the US accounting for 28 of the 40 most highly cited organizations, while the UK has 3; Germany, the Netherlands, and Switzerland each have 2, and Japan, Australia and China each having 1 (see Table 27.2).

When considering private organizations that publish in N&N, the landscape is dominated by the US, followed closely by Japan (Table 27.3).

Further information on global N&N publication analysis can be found in the observatoryNANO report *Benchmarking Global Nanotechnology Scientific Research: 1998–2007* [2].

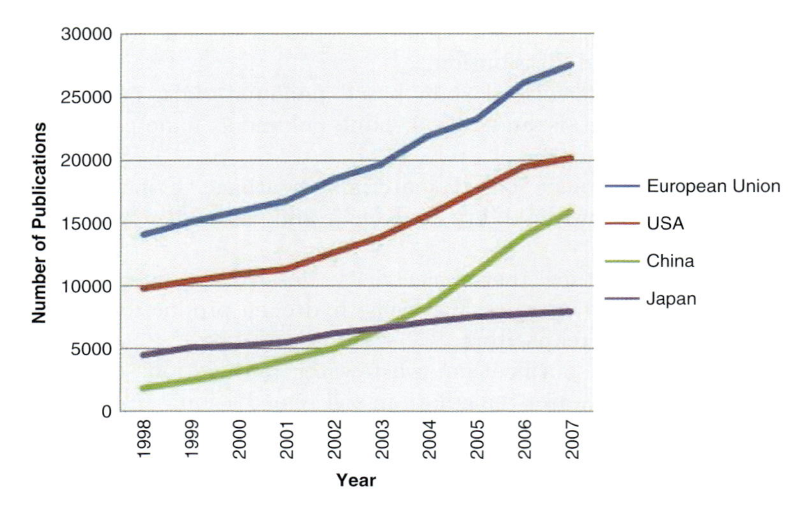

Fig. 27.4. Publication trends in N&N per country/region between 1998 and 2007, based on Web of Science

27.4.2 Patent Analysis

When considering patents, the largest portion of the 132 000 nanotechnology patents in the PATSTAT database is allocated to the Chemistry and Materials sector, followed closely by ICT (see Fig. 27.6).

Again, the US dominates, with the EU share at 20% overall, which is a poor reflection of its publication status. The EU leads in patenting in only two sectors: aerospace, automotive, and transport; and construction (Fig. 27.7). Further information on the patenting analysis can be found on the observatoryNANO website [3].

27.5 Economic Analysis

Assessment of the economic impact of nanotechnology RTD, both within the EU and globally, is closely linked with the ST analysis described above. However, specific technology developments do not always marry well with market applications, and so a major part of this work is focused on establishing appropriate indicators and methodologies to reflect the role that ST developments have in current and future socio-economic impacts. In doing so it will allow government to make strategic judgements on funding priorities, e.g., the framework programmes; allow industry to prioritize RTD programmes, and provide investors with an early indication of opportunities in different technology sectors.

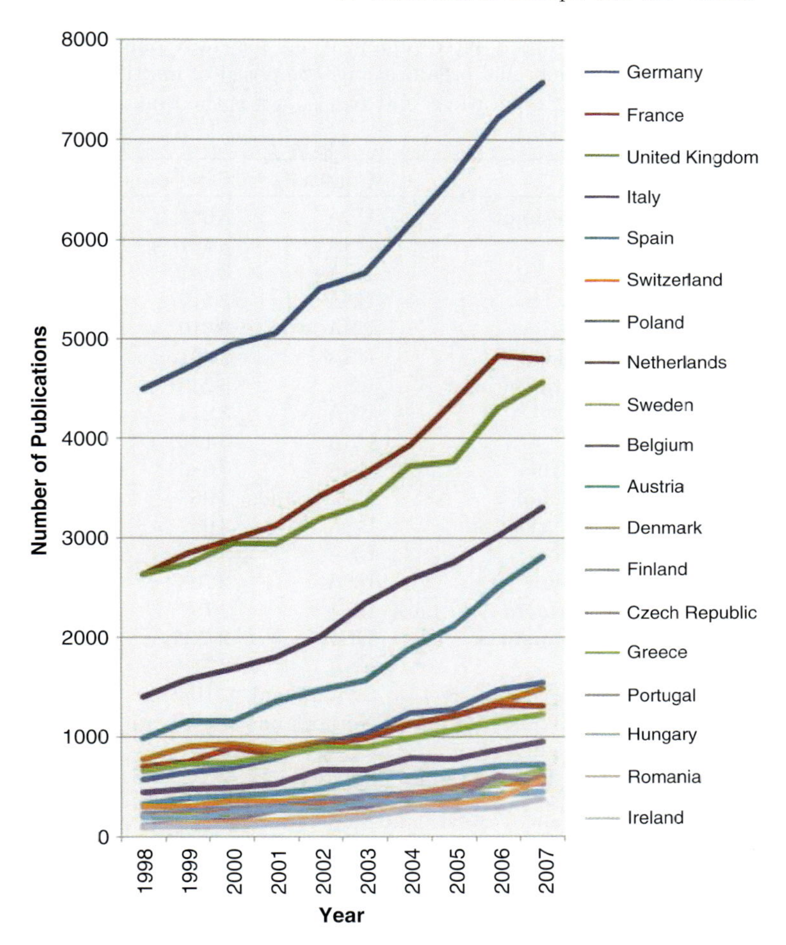

Fig. 27.5. Publication trends in N&N per European country between 1998 and 2007, based on Web of Science

This analysis builds upon existing raw data and results from published or 'grey' literature in combination with analysis of the implications of technology developments in terms of

- Global markets, for example, IPOs, current global trends, and nature and location of projected markets.
- Potential markets for EU RTD and the effect of competing technologies from other global regions.
- Potential economic risks, for example, re-insurance, disruption to existing industry.
- Socio-economic impacts, such as development of employment markets and opportunities for nanotechnology.

Table 27.2. Forty most highly cited organizations for N&N publications in 2006. Those in *bold type* are 'umbrella' organizations composed of multiple sites. R_{citation} is the rank by the number of citations and $R_{\text{publication}}$ is the rank by the number of publications in 2006

R_{citation}	Institution	Country	Citation score	$R_{\text{publication}}$
1	Georgia Inst Technol	USA	3 663	37
2	CalTech	USA	3 491	37
3	Harvard Univ	USA	3 441	26
4	Columbia Univ	USA	3 417	92
5	Rice Univ	USA	3 410	107
6	Univ Calif Berkeley	USA	3 391	20
7	Univ Calif Santa Barbara	USA	3 245	43
8	Stanford Univ	USA	3 214	40
9	MIT	USA	3 126	19
10	Northwestern Univ	USA	3 086	45
11	Delft Univ Technol	Netherlands	3 067	126
12	Univ Calif Los Angeles	USA	3 062	59
13	Univ Washington	USA	3 044	70
14	Penn State Univ	USA	2 768	41
15	Lawrence Livermore Natl Lab	USA	2 754	121
16	**Univ Massachusetts**	**USA**	**2 722**	**98**
17	Univ Penn	USA	2 718	79
18	ETH	Switzerland	2 705	50
19	Eindhoven Univ Technol	Netherlands	2 668	120
20	Arizona State Univ	USA	2 641	86
21	Ecole Polytech Fed Lausanne	Switzerland	2 584	64
22	Princeton Univ	USA	2 548	94
23	Cornell Univ	USA	2 546	55
24	Univ Erlangen Nurnberg	Germany	2 544	101
25	**Jpn Sci & Technol Agency**	**Japan**	**2 456**	**9**
26	Univ Manchester	UK	2 442	95
27	Univ Michigan	USA	2 420	33
28	Pacific NW Natl Lab	USA	2 409	147
29	Univ Munich	Germany	2 401	113
30	Univ Sydney	Australia	2 394	117
31	Univ N Carolina	USA	2 393	90
32	**Univ Minnesota**	**USA**	**2 391**	**44**
33	HK Univ Sci & Technol	Hong Kong	2 391	109
34	Univ Calif Davis	USA	2 387	65
35	Univ Tennessee	USA	2 383	111
36	Univ Cambridge	UK	2 349	22
37	Johns Hopkins Univ	USA	2 338	88
38	**Univ Texas**	**USA**	**2 322**	**15**
39	Duke Univ	USA	2 318	144
40	Imperial Coll London	UK	2 314	73

Table 27.3. The world's most prolific private organizations publishing in N&N between 1998 and 2007

Country	Number of publications
USA	35
Japan	26
Germany	8
Netherlands	3
South Korea	3
Switzerland	2
UK	2
New Zealand	1
Singapore	1
Taiwan, China	1

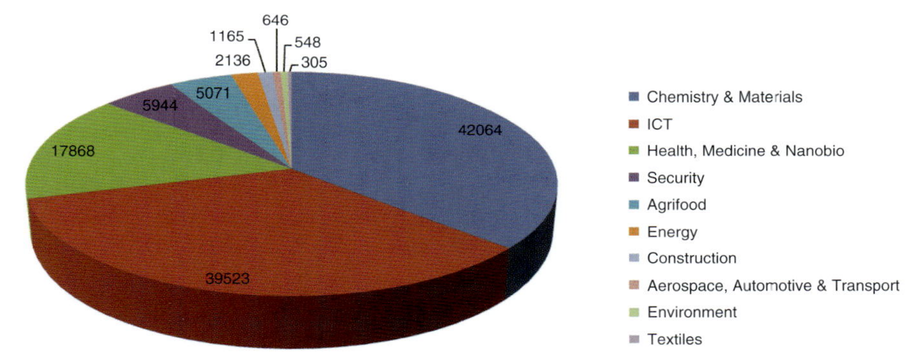

Fig. 27.6. Number of patents in the PATSTAT database per each of the 10 technology sectors

- Relationships between input indicators (funding, number of researchers, RTD infrastructures, number of nanotech research groups) and output indicators (nanotechnology markets, patents, publications, competitiveness, employment, health, energy efficiency, etc.).
- Public and private funding available per country, per region in the world.
- Patent use, licensing, exploitation activities.

This is summarized schematically in Fig. 27.8.

A major part of the economic assessment is to build working relationships with a number of different actors, such as high level industry management, venture capitalists, economists, programme managers in government funding agencies. The data (and opinion) provided by these individuals is used to elaborate the quantitative data provided through desk research such as patent analysis and review of published market information (such as stocks and shares, investments, company reports).

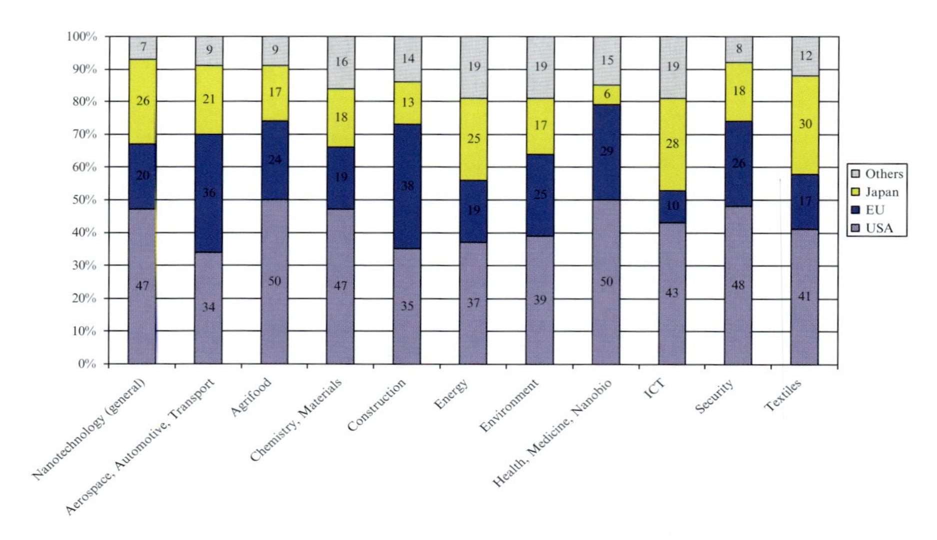

Fig. 27.7. Percentage country and regional shares of nanotechnology patents

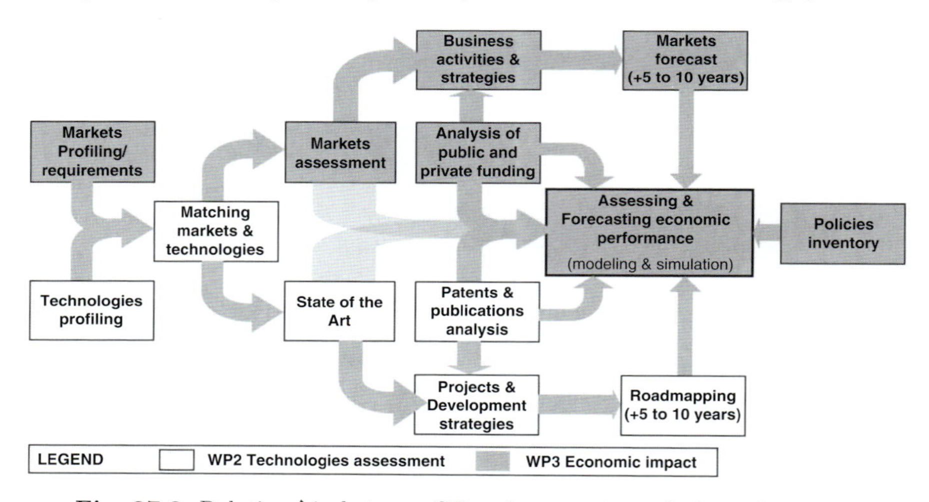

Fig. 27.8. Relationship between ST and economic analysis, and outputs

The strategy for the economic analysis is to take a more focused approach and examine market segments (chosen from knowledge of the landscape and through input from the expert community) where nanotechnology enabled developments are having an impact. In the first year of the project 35 different topics have been reported on [4]:

- *Aerospace, Automotive & Transport.* Structural parts/airframe; external panels/surfaces; powertrain; engine (ICE)/turbines.

- *Agrifood.* Nanocomposite packaging; coatings for packaging; edible coatings; biodegradable nanocomposites for packaging; delivery systems for nutraceuticals.
- *Chemistry & Materials.* Nanomagnetic materials; carbon nanotubes; nanodiamond; intrinsic conducting polymers.
- *Construction.* Cement based materials; construction ceramics; paints; Windows; insulation systems/materials.
- *Energy.* Photovoltaic; fossil fuel.
- *Environment.* Water treatment; soil remediation.
- *Health, Medicine & Nanobio.* Bone replacement materials; dental nanomaterials; in vivo imaging; drug delivery.
- *Information & Communication.* Memory; displays; materials.
- *Security.* Detection.
- *Textiles.* Water repellent/self-cleaning; anti-static; anti-bacteria; moisture absorption/wicking; filtration and UV protection.

Each report provides a general market description (impact of nanotechnology, drivers and barriers, sector segmentation and applications, possible future products and time range) and application profiles (short application description, functional requirements, boundary conditions, product examples, economic information and analysis, selected company profiles).

Funding Analysis

Globally, public funding has yet to peak [5]. While the rate of growth is slowing for those nations which have been funding N&N R&D for a number of years, there are several other new players (notably Russia) which have invested significantly in public funding. No matter which country, the major concern for policy makers is encouraging industrial investment and translating public funding into commercial products. The strategy varies- from supporting whole value chains (e.g., through networks and infrastructure) to direct investment in individual companies. Table 27.4 provides an overview of funding levels in selected countries.

Private funding is a different matter [6]. As a result of the economic downturn there has been a decrease in new venture capital fund raising in Europe and the US (some 61% decrease between the first quarter of 2007 and third quarter of 2008), and also restricted exit strategies. Investment in European nanotechnology companies has been historically low compared with the US (some 20–40 MEuro p.a. compared with in excess of 900 MEuro globally, the majority of which is in the US, Fig. 27.9).

Such private investment is vital for commercialization of new products and economic growth in countries. While the overall trend may be some cause for concern, it may be that the decrease is only a slight exacerbation of the cyclical nature of VC investment. Already in 2009 there has been substantial investment in one EU company (Oxford Nanopore),

Table 27.4. Public investment in nanotechnologies by selected countries. *Asterisk* indicates public–private funding

Country	Programme	Duration	Value
EU	FP7 NMP	2007–13	3.5 GEuro
Germany	Nano Initiative Action Plan 2010	2006–10	330 MEuro (+) p.a.
France	Nano2012 Programme	2008–12	2 GEuro*
Netherlands	NanoNed	To 2010	235 MEuro
UK	Micro and Nanotechnology Network (MNT)	2003–7	~ 130 MEuro (£90m)
Finland	FinNano	2005–10	70 MEuro
Norway	NANOMAT	2002–16	74.7 MEuro (to date)
Austria	Austrian Nano Initiative	2004–	35 MEuro
USA	National Nanotechnology Initiative (NNI)	2009–	~ 1.1 GEuro ($1.5bn)
Russia	Russian Corporation of Nanotechnologies (RUSNANO)	2007–	~ 3 GEuro (130bn roubles)*
Japan	Basic S&T Plan (Nanotechnology and Nanomaterials)	2006–10	> 500 MEuro p.a. (78bn yen)

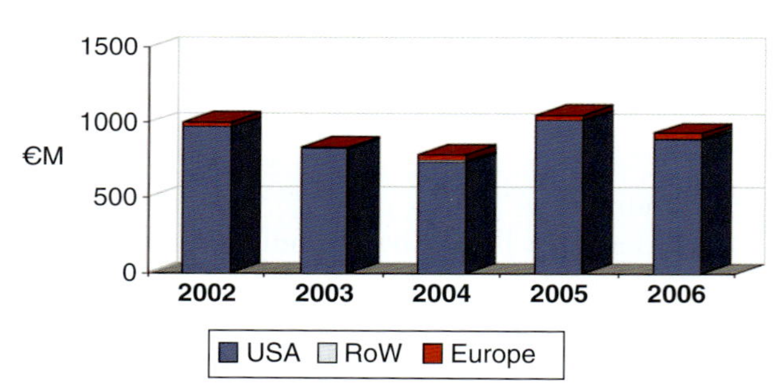

Fig. 27.9. Global Venture capital investment divided by target country. Source: Thomson VentureXpert database, analysed by Markku Maula (Helsinki University of Technology)

while another successfully negotiated an initial public offering (Nanoco Technologies).

27.6 Integrating an Analysis of the Wider Aspects of Nanotechnology Development

Although the major analysis performed by the observatoryNANO project is on ST and economic developments, this is placed in the context of wider aspects of nanotechnology, namely ethical and societal aspects;

environment, health and safety issues; and developments in standards and regulations.

27.6.1 Ethical and Societal Aspects

The consortium is monitoring both the ethical and societal impact of N&N and the impact that societal developments and ethical reflection can have on N&N developments. This work looks at N&N development in four areas:

- individual and collective responsibility,
- nanobiomedical ethics,
- ICT,
- communication between scientists, technology, and society.

It reviews both the findings from the observatoryNANO, and published work by other projects and organizations (including work on codes of conduct).

In the first year of the project the report *Individual and collective responsibility for nanotechnology* was written. This identified a number of ethical and societal issues including:

- governance issues (such as codes of conduct),
- innovation and intellectual property issues,
- precaution, risk and dual use issues,
- global justice issues (divide between developed and developing countries).

Each of these issues affects choices in the research, development, application, and market entry of nanotechnology and its products. Different stakeholders have important roles to play in these debates, with government, industry and the scientific community involved in each. Issues of public engagement (on benefits versus risks, and deciding strategic research priorities) are still being debated by various stakeholders, several studies have shown that most members of the public show little interest in nanotechnology, but those who do expect more benefits than risks.

To supplement this work a number of opinion leaders have been interviewed to understand their views on different aspects of N&N development. In the first year of the project these were: (Arie Rip, University of Twente; Tony Musu, European Trade Union Institute; Lena Perenius, CEFIC; Richard Jones, University of Sheffield, and strategic adviser to the UK Engineering and Physical Sciences Research Council). Full information can be found in the appropriate section of the website [7].

27.6.2 Environment, Health and Safety (EHS) Issues

Environment, health and safety (EHS) issues have, in recent times, become one of the most reported and discussed aspects of nanotechnology development. There are still many unknowns, in particular there remains a lack of understanding regarding the influence that different physical-chemical properties of nanomaterials (size, shape, composition, reactivity, surface area and/or chemistry) on their own or in combination have on biological responses. To ensure the continued safe and responsible development of nanotechnology, most national governments and international organizations that support N&N R&D are also funding specific EHS research.

While this project is not undertaking major new analysis of environment, health and safety issues related to N&N development; it is reporting on new developments in EHS research from other organizations and projects (indeed the partners involved in this activity are leading organizations in their own right and participating in a number of these international initiatives), and analyzing the output from the ST and economic analysis to determine whether these raise any new issues. In the first year of the project a report describing seminal research in EHS related to N&N has been published, and a number of different organizations identified and information regarding their activities provided on the website [8], such as

- *SAFENANO.* UK resource on nanotechnology hazard and risk, led by IOM.
- *OMNT.* French observatory led by CEA with a theme on 'nanoparticles, nanomaterials, impacts on health and environment'.
- *KIR-nano.* 'Risks of Nanotechnology Knowledge and Information centre', a Dutch initiative led by RIVM.
- *ICON.* A major international database of published research on nanoparticle risk issues.
- *The Project on Emerging Nanotechnologies (PEN).* Based within the Woodrow Wilson Center for Scholars, which hosts a series of inventories, exploring various aspects of nanotechnology.
- *OECD.* Nanotechnology database currently in development that will provide information on different nanomaterials.

27.6.3 Developments in Regulation and Standards

A globally accepted definition awaits the outcome of work being carried out by the International Standards Organization Technical Committee 229 (ISO TC 229), however in the Royal Society and Royal Academy of Engineering report of 2004, nanoscience was defined as "the study of phenomena and manipulation of materials at atomic, molecular, and macromolecular scales, where properties differ significantly from those at a larger scale" and nanotechnologies as "the design, characterization, production and application of structures, devices, and systems by controlling shape and size at the nanometre scale". ISO TC 229 is structured into 4 working groups (terminology and nomenclature; measurements and characterization; health, safety, and environment; materials specification). To date ISO TC 229 has produced two documents:

- ISO/TS27687 (Technical Specification): Terminology and definitions for nano-objects- nanoparticle, nanofibre and nanoplates.
- ISO/TR 12885 (Technical Report): Health and safety practices in occupational settings relevant to nanotechnologies.

The European organization for standards (CEN) works closely with ISO TC 229 (indeed both are chaired by the same person, Dr Peter Hatto), and national standards authorities also participate. The British Standards Institute has one of the leading national committees (NTI/1) which has published ten documents to assist organizations with specifying and handling different nanomaterials [9].

As discussed above a large amount of research has been undertaken in investigating EHS issues of nanomaterials. However, it remains difficult to evaluate, model and predict the ecological and toxicological behaviour of nanomaterials and consequently develop appropriate risk management and regulatory frameworks. This is further compounded by the fact that nanotechnology impacts so many different sectors, each with its own existing regulatory system. At present there is no one system for regulating nanotechnology, instead existing schemes and agencies are involved such as REACH (Registration, Evaluation, Authorization and Restriction of Chemicals), EMEA (European Medicines Agency), EFSA (European Food Safety Authority), and the Cosmetics Directive.

Soft regulation in the form of voluntary reporting schemes (UK and US) and codes of conduct (e.g., from the EC) have been implemented to bridge any gaps and to place the onus on the developer and manufacturer. The EC Code of Conduct, for example, was first published in 2008 and is directed towards development (rather than manufacturing). It promotes sustainability, precaution, and inclusiveness. Full information can be found in the appropriate section of the website [10].

27.7 Supporting Research and Business

Both research and business can benefit from tools to assist in the wider reflection of the implications of their activities. In the project two such tools are being developed. The first, a toolkit for ethical reflection and communication has been designed around a series of scenarios complete with questions to allow scientists to think about the larger societal and ethical implications of their research, and how this impacts different sections of the community. The second is focused on corporate social responsibility. It takes the NanoMeter tool (developed in the FP6 funded Nanologue project) and adapts it to different technology sectors. Used as an online and internal tool, it allows business to evaluate social aspects; EHS issues; resource requirement, and product stewardship; and evaluate gaps in knowledge; risks, and perceptions. Both tools will be developed over the course of the project in collaboration with the communities that will make use of them.

27.8 Establishing a Permanent European Observatory on Nanotechnologies

The long-term goal of the project consortium is to establish a permanent European Observatory on Nanotechnologies. The project has initiated two activities to support it in this goal: a review of other ‘observatories’ and the establishment of a Governing Board of external stakeholders.

There are many different ‘observatories’ each differs in terms of how it is funded (e.g., for a fixed period, or per ‘activity’), who it reports to (e.g., government agencies, the public), how it is governed (e.g., by internal or external stakeholders), who are the observers (e.g., employees, external experts), what types of observations it makes (e.g., horizon scanning, gaps) and what type of data it produces (e.g., raw data, or analysis). In the first year of the project eight initiatives were studied.

The Governing Board of the observatoryNANO consists of a diverse group of stakeholders: industry leaders, academics, economists, regulators, and legislators; government decision-makers from research, enterprise, economics or industry departments; representatives from leading civil society organizations. The role of the Governing Board is to critically review the scope of activities and methodologies employed by the observatoryNANO and advise the consortium of relevant new developments and opportunities.

The conclusions drawn from the first year study, from the review of the structure and mechanisms employed by the observatoryNANO project itself and input from the Governing Board will be used to elaborate a business plan for the continuation of the observatory beyond the initial four year funding period.

27.9 Conclusion and Future Work

The observatoryNANO project is an ambitious project bringing together a diverse set of organizations each with expertise in different aspects of N&N development. It has produced a rich online resource of information in its first year, integrating the analyses into a series of linked reports, allowing policy and decision makers, and other interested parties to access information on all aspects of a particular topic. In its second and subsequent years it intends to integrate these analyses further by focusing efforts on selected topics that are showing higher levels of activity, based on publications, patents, strategies, and debates, and through its interaction with a wide expert community provide informed analysis for policy and decision makers to base recommendations upon.

27.10 About the Project Consortium

The ObservatoryNANO project is led by the Institute of Nanotechnology (IoN) (UK), and includes:

- VDI Technologiezentrum (DE),
- Commissariat à l'énergie atomique (CEA) (FR),
- Institute of Occupational Medicine (IOM) (UK),
- Malsch TechnoValuation (MTV) (NL),
- Triple Innova (DE),
- Spinverse (FI),
- Bax and Willems Consulting Venturing (B&W) (ES),
- Dutch National Institute for Public Health and the Environment (RIVM) (NL),
- Technical University of Darmstadt (TUD) (DE),
- Associazione Italiana per la Ricerca Industriale (AIRI) (IT),
- Nano and Micro Technology Consulting (NMTC) (DE),
- Swiss Federal Laboratories for Materials Testing and Research (EMPA) (CH),
- University of Aarhus (DK),
- MERIT–Universiteit Maastricht (NL),
- Technology Centre AS CR (CR).

References

1. www.observatorynano.eu/project/catalogue/2/ for all ST reports
2. www.observatorynano.eu/project/catalogue/2PU/
3. www.observatorynano.eu/project/catalogue/2PA/
4. www.observatorynano.eu/project/catalogue/3MR/ for all market reports
5. www.observatorynano.eu/project/catalogue/3PF/
6. www.observatorynano.eu/project/catalogue/3VC/
7. www.observatorynano.eu/project/catalogue/4/
8. www.observatorynano.eu/project/catalogue/6/
9. See the NTI/1 homepage to freely download these reports: www.bsi-global.com/en/Standards-and-Publications/Industry-Sectors/Nanotechnologies/Committee Activities/
10. www.observatorynano.eu/project/catalogue/5/

28

Situation in Europe and the World: Nanotechnology and Scientific Policy. Action of UN Agencies in Developing Countries

Shamila Nair-Bedouelle

The modern long-term economy is based on scientific progress and the subsequent technological achievements. Without this, the world would be the same as it was centuries ago, with populations living on the edge of survival, spending most of their time in search of food. Technology provides a way for societies to fight disease, to improve crop yields, to create new energy sources, to spread information, to favour the transport of goods and people, and much more!

However, it does not come for free and is barely accessible to those who most need it. Indeed, it is the result of a decisive social and financial investment in terms of education, scientific research, and above all technological development.

Nanotechnology has arisen as a very promising field of almost limitless potential applications, across a huge range of economic areas relating to the improvement of the quality of life and sustainable development. However, this utopic vision of a world transformed by nanotechnologies, free of problems of food supply, disease, poverty, and ignorance, is seriously challenged by ecologists, politicians, and other observers of this scientific revolution.

Scientific knowledge can only be successfully applied when industrial stakeholders and open access to technological knowhow are given equal importance. The aim in the following will be to define the various aspects of a scientific policy in relation to emerging technologies, such as nanotechnologies, their contribution to socio-economic development, and the requirements of the most needy, not forgetting the role of governments.

28.1 Scientific Policy for Sustainable Development

The growing pressure to respect the environment and the increasing threat to natural resources, biodiversity, and ecosystems have made it even more crucial to specify scientific objectives and technology policies. Harnessing the potential of science and technology for the purposes of sustainable development

P. Houdy et al. (eds.), *Nanoethics and Nanotoxicology*,

DOI 10.1007/978-3-642-20177-6_28,

will involve a determined effort to exchange knowledge and build an effective network for that exchange, but also a careful examination of plausibility and risks. The potential for global applications of nanotechnologies will require a detailed dissection of current scientific practices and the implementation of policies specific to those applications.

In 1972, the Stockholm Earth Summit judged it urgent to introduce measures to deal with the deterioration of the environment. In 1992, the Rio Conference took a significant step by setting up an agenda for sustainable development. At this United Nations conference on the theme of *Environment and Development*, the assembly agreed that environmental protection and economic and social progress were related in the quest for sustainable development. Finally, between the Rio Conference and the world summit on sustainable development in Johannesburg in 2002, the nations present laid down a strategic vision for the future of humanity, under the patronage of the United Nations. Among the conclusions of this summit was the decision to provide assistance to both developed and developing countries, ensuring that the latter gain access to advanced environmental technologies, and to increase scientific and technological capabilities with regard to sustainable development.

Nanoscience and nanotechnology are recent approaches to research and development which aim to interpret the structure and predict the behaviour of matter on the atomic and molecular scales. This opens the way to understanding new phenomena and designing new applications, e.g., in the field of energy on micro- and macroscopic scales.

Applications of nanotechnology will have an impact on the life of every citizen. However, it must be asked whether nations will be able to guarantee that the development of the so-called high-potential technologies will be favourable to sustainable development. Another question is whether these technologies will be reserved for the benefit of rich countries, depriving poor countries of the modern processes they will generate.

28.2 Missions of Specialised UN Agencies

With its mandate for intellectual collaboration and exchange of knowledge, the United Nations is uniquely placed to promote international cooperation. To support this cooperation, UNESCO organises its programmes and their offshoots on a global, regional, and national level, taking into consideration the specific features and geographical conditions of the different countries to which they apply. Although not all countries have the means to obtain the most advanced technologies, they must all be allowed to identify those that would be most profitable to their own population, while adapting them to local constraints and requirements. In fact, all governments have a growing need to increase their research and development capacities so that they can improve their own activities.

A suitable symbiosis between scientific capabilities, technology, and innovation for sustainable development and the combination of scientific discoveries to reduce poverty and to improve social peace can only be achieved in a favourable context, i.e., one that ensures a form of science transfer that respects societal problems. For this to work, all governments must have access to scientific knowledge and the possibility of using it freely.

UNESCO is concerned about the lack of recognition – in certain countries – of the importance of research and technology. One mission of the organisation is to consolidate the use of scientific and technological knowledge to improve social well-being. This will require an effective, common sense policy in which science and society are intimately bound together.

28.3 The Millennium Development Goals (MDGs)

28.3.1 Objectives

At the Millennium 2000 Summit, the 189 member countries of UNESCO adopted 8 objectives with the theme of inspiring, supporting, and guiding development. These objectives aim specifically to fight poverty, hunger, disease, illiteracy, and inequality, while favouring environmental protection. They also concern human rights and access to health care, education, housing, and safety for all. (In fact, these goals give substance to the aspirations of the 1990 summit. They are intended to directly affect the life and future ambitions of billions of people around the world.)

28.3.2 Current Status of MDGs

But alas! We are a long way from achieving these goals, although there has been some progress. Some of them have been hampered by climate change, environmental concerns, and food crises.

The UN Commission on Science and Technology for Development (CSTD) concludes that this slowness in achieving the goals has come about because several developing countries have not sufficiently coordinated their efforts to place science and technology at the center of their preoccupations. The commission insists on the fact that it is only by coupling current and emerging technologies that we will be able to reduce costs and eventually achieve these goals.

In 2002, the Secretary General of the United Nations set up an independent organisation called the Millennium Project to specify the most appropriate strategies for achieving the MDGs, noting that the sub-Saharan regions would require the highest investments, identifying them as genuine poverty traps that could only be helped by understanding recurrent problems such as the high cost of transport, reservoirs of disease, very small markets for the most common products, very low agricultural productivity, and a combination of limited access and slow dissemination of scientific and technological progress. In short, this report stresses that we are a long way from achieving the Millennium Development Goals in a great many underdeveloped countries. Indeed, billions of people live in a state of extreme poverty with less than 1 dollar per day, while there are around 20 000 deaths per day in these regions.

There is no problem of society to which science cannot bring some kind of solution. This implies that, in order to achieve the MDGs, we need to focus on the main sources of economic growth, and in particular those associated with scientific and technological innovations. Although science alone cannot fully resolve all these problems, without it, they will surely not be resolved. The main stumbling block for achieving the MDGs is a total lack of integration of science and technology into the development schedules of certain nations.

However, the vast majority of socio-economic advantages of nanotechnology should also benefit the poorest countries. In this respect, international cooperation between economically and industrially advanced countries (sharing knowledge and profits) and countries that are less well placed (access to knowledge, avoiding knowledge appartheid) is an absolute necessity. Innovative strategies must be identified to propagate the advantages of conventional knowledge, to protect against the potential risks of new processes, and to take into account the green revolution, given the growing potential of emerging technologies such as modern means of communication and nanotechnologies, among others.

28.4 Nanotechnology and Politics

In the complex interaction between science and society, there is a problem of consistency between on the one hand the use of available technological capabilities to promote equitable economic growth, and on the other hand the identification of sensible policies for measuring their impacts. By taking into consideration the standard logical references, namely scientific, technological, and innovative political references, these policies must have the following targets: the goals to be achieved, the results expected, and the decisive indicators for measuring and assessing the impacts of the chosen policy. In the case of the nanotechnologies, a notion of delay must also be included, associated with the uncertainties in the research and development of the systems needed to solve them. A general framework specific to a nanotechnology policy must implicitly involve a dynamic approach with optimal simulations of the results.

The basic condition for setting up a healthy nanotechnology policy to implement this emerging area is the careful conjunction of rational policy and decision-making. This must be taken into consideration at all levels, and in particular through a close international collaboration between scientists and sociologists, and between scientists and policy-makers.

Technological roadmapping and foresighting specify the methods to be defined and controlled, the advances, and also the processes that could usefully be introduced during industrialisation. The specification of objectives is in itself a useful stage which brings together all parties concerned with development to consider the obstacles to be overcome, the impacts, and future needs. However, a global and accurate long term prediction does not seem realistic for nanotechnologies in general, because the area is too great and the potential applications more or less unlimited. Instead, it seems more appropriate to attempt predictions in each market sector that has reached a certain level of maturity. More accurate technology foresighting must be consolidated by simultaneously anticipating future developments and scheduling appropriately.

Development policy – not only concerning nanotechnologies – must implement the five following features: research and development, infrastructure, education, practice, and societal aspects. The infrastructures needed for innovation in nanotechnology require a critical mass in science and technology and significant material investments, and this beyond the usual scales, even for the most developed countries. Political authorities must take on board the fact that infrastructures and their maintenance are a crucial issue in the development of a nation, and that the onerous nature of nanotechnology will involve investments leading to the creation of international sites (excellence clusters).

The key characteristic of nanoscience and nanotechnology is its cross-disciplinary nature, in a way that goes well beyond conventional concepts. Developed and developing countries must increase access to theoretical and practical higher education outside the usual methods, and promote cross-disciplinarity and a business spirit in research staff.

Among the main points to be considered when setting up a socio-economic policy framework specific to nanotechnology, and differing from conventional policies, we may mention: risks and opportunities, technical constraints such as monitoring, education, infrastructures, public information, legal approaches and associated risks, ethical aspects, risks to health and environment, intellectual property and patenting framework, and finally security.

Nanotechnology raises the problem of the hazards it may generate, and of course their management. This must be taken into account very early on in their development process, from the design stage up to their commercialisation, in order to ensure that they are made available in a thoroughly safe context. In parallel with technological development, toxicity and ecotoxicity studies must be conducted, so that risks and processes can be specified and adjusted to respect maximal allowed doses, and so that data corresponding to the responses to tolerable exposures can be provided.

A nomenclature and common testing methods must be established to allow regular global comparisons in a context of international harmonisation. In this respect, the OECD working group on nanomaterials has set up a forum on the coordination of international activities in this area, launching six specific projects to take into account differences in knowledge with regard to health and environmental impacts, general procedures to follow, and risk-taking between partners from different countries.

National parliaments have major responsibilities when faced with the increasing complexity of science and technology. In modern societies, these institutions have the duty to inform the public about ever more sophisticated, fast-changing, and often contentious developments. The recent debates and controversies in various countries about genetically modified foodstuffs, human cloning, genetic therapy and tests, new information technologies, and global warming exemplify the challenges faced by today's governments. In addition, in the case of nanotechnology, it is very difficult for policy-makers to make decisions when there is a lack of concrete information about nanoproducts, and about their potential societal impacts.

Most countries still make a distinction between policies that promote production rather than research capabilities. Very few countries actually have their own resources for nanotechnology innovation that could generate developments leading to applications. National strategies for funding innovation often require the involvement of different ministries, such as the environment, health, or agriculture ministries, making research more attractive and more stimulating because of the uncertainties regarding the inherent costs and benefits.

28.5 Other Political Considerations Regarding Nanotechnology: Patents

While an all-inclusive political approach to innovation is essential, three other factors must also be added: industrial property, regulations, and monitoring. In these days of market globalisation, long-term economic success depends on the generation, organisation, and exploitation of knowledge. Given the importance of knowledge when dealing with nanotechnologies, the problem is to determine what should or should not be patented – should we go right down to the molecular level? – and thus to redefine industrial property law. This scenario is very different from the problem of patenting genetic products, which often undergo only minor modifications with respect to existing products. Furthermore, nanotechnologies will result in researchers filing patents on everything from the atomic scale to molecular processes. Different approaches are being undertaken and examined in order to see whether they offer adequate protection or whether new rules must be laid down. In most developed countries, industrial protection still suffers from serious failings.

Nanotechnology must be developed in as safe and responsible a manner as possible. Ethical principles must be applied with regard to health and environmental risks to lay the ground for regulatory recommendations. In this respect, social dialogue is essential to ensure that real preoccupations are discussed rather than simply science fiction scenarios. In parallel, harmonious legislation must play its role to minimise distortion of the markets and ensure the protection of health and the environment. Current regulations make frequent reference to parameters bearing no relation to nanotechnological applications, such as free nanoparticles. For example, certain substances are not subject to any production limit below a certain threshold. Appropriate conjunctural regulations are essential, not only to protect consumers and the environment, but also to guarantee the trust of workers and investors.

28.6 Needs of Developing Countries

The ultimate aim for developing countries is to achieve a comparable level of development to the industrialised countries, with the guiding principle to make profitable use of knowledge and technology. However, for this to work, the transfer of knowledge from developed countries must be facilitated to reduce the existing gulf that separates them from less developed nations, while simultaneously avoiding environmental degradation.

The UN Millennium Project and the World Summit on Sustainable Development (WSSD) concluded that the capacity for scientific and technological growth in developing countries and the related international cooperation for transfer of knowledge were rather limited. They stressed the importance of sharing the benefits of economic growth and scientific progress in an equitable way. The 8 Millennium Development Goals were adopted to serve as indicators, in the hope of inspiring, guiding, and monitoring development efforts. Applications of science and technology are decisive factors for the success of the MDGs, especially in countries suffering from poverty, health problems, lack of education, and environmental problems.

Progress is possible in many areas with the help of research and innovations in nanoscience and nanotechnology. Three impact criteria are often cited when speaking of nanoscience: horizontal, key, or potential, since it may be relevant across all technological sectors. It often combines different convergent approaches relevant to several current societal problems. For example, nanotechnology is relevant to the environment through the use of more efficient catalysts, better batteries, and more effective light sources. Progress has also been noted in water purification and for several environmental issues, and the list of hopeful prospects for other improvements with the help of nanoscience is long. Recent publications establish the positive contributions of nanoscience to the sixth MDG, in the fight against HIV and aids, malaria, and other diseases that oppress developing countries. These contributions are the result of better diagnostics, the development of new medicines, and the use of purer water and healthier foods.

It is recognised that many developed countries will benefit indirectly from their mastery of nanotechnology in order to remain competitive worldwide in several economic sectors. While some of these are close to achieving the Millennium Goals and will benefit from their investments, it is clear that all countries could benefit from targeted investment to consolidate a return on investment in the accomplishment of the MDGs.

Increasing capabilities in nanoscience and nanotechnology will also provide better ways to assess the new risks they expose us to, when nanoparticles are released into the human environment. It is important to remember that even the most advanced countries still have very limited means for predicting the risks associated with genetically modified organisms (GMO). And these same countries still have great difficulty foretelling the risks relating to the toxicity of pesticides and residues from phytosanitary products which adulterate everyday products across the board.

The African Union has suggested setting up excellence clusters throughout Africa to contribute to the scientific development of the continent. These centres and associated high level institutions for nanotechnology are fundamental for building up national and regional technological capabilities suited to local situations and contexts. The investments needed are an order of magnitude greater than those devoted to biotechnologies. The role of international cooperation would be to improve basic understanding through research, to set up scientific committees, and to share information.

28.7 The Future: Nanotechnology and Development

These days, and more than ever before, science is a vital source of education and cultural enrichment, to obtain a good understanding of societies and their economies. Nanotechnology can make a considerable contribution to progress by providing weapons against disease, improving food safety, facilitating exchange, allowing better control of the environment to avoid conflict and stem natural disasters, and finally, by identifying new ways of using water, energy, and other natural resources. It is only by organising a network of science and innovative technology that knowledge can be used for sustainable economic development.

The challenges raised by the African continent, and certain other countries, highlight the need for a scientific approach to solving their problems, and this at all levels. Through the discovery and transformation of natural resources, science allows humans to improve their quality of life and obtain a pleasant existence. Science and technology are key factors in achieving peace and human progress.

References

1. UN Report of the Secretary General. Follow-up to the outcome of the Millennium Summit, ninth session, A/59/2005. In larger freedom: Towards development, security and human rights for all. Agenda items 45 and 55
2. United Nations, Report of the World Summit on Sustainable Development, Johannesburg, South Africa, 26 August to 4 September 2002, United Nations. New York, 2002 A/CONF.199/20

3. Report of the United Nations Conference on the Human Environment, Stockholm, 5–16 June 1972 (United Nations publication, Sales No. E.73.II.A.14 and corrigendum), Chap. I
4. Report of the United Nations Conference on Environment and Development, Rio de Janeiro, 3–14 June 1992 (United Nations publication, Sales No. E.93.I.8 and corrigenda), Vols. I–III
5. S. Holtz: Canadian Institute for Environmental Law and Policy, Discussion Paper on a Policy Framework for Nanotechnology, March 2007, ISBN 978-1-896588-59-9
6. Millennium project, Report to the UN Secretary General. Investing in development: A practical plan to achieve the Millennium development goals. ISBN 1-84407-217-7 (2005)
7. European Commission: NMP-linking with national policies, October 2008, ISSN 1725-8472
8. European Commission: Nanoscience and nanotechnologies. An Action Plan for Europe 2005–2009. ISBN 92-894-9597-9
9. EU Policy for Nanosciences and Nanotechnologies, 2004
10. J. Pelley, M. Saner: International approaches to regulatory governance of nanotechnology. ISBN 978-0-7709-0530-9 (2009)
11. I. Malsch: Which research in converging technologies should taxpayers fund? Exploring societal aspects. Technology Analysis and Strategic Management **20**, 137–148 (2008)
12. F. Salamanca-Buentello, D.L. Persad, E.B. Court, D.K. Martin, A.S. Daar, P.A. Singer: Nanotechnology and the developing world. Plos Medicine **4**, 2 (2005)

29

Nanotechnology and the Law

Sonia Desmoulin-Canselier and Stéphanie Lacour

Law and nanotechnology form a vast subject. The aim here will be to examine them from the societal standpoint of nanoethics, if necessary without due reference to the work that has been undertaken. For while law differs from ethics, as we shall attempt to explain throughout this reflection, it must also be studied in its relationship with social realities. Could we not say that the law simply reflects the evolving reality, an evolution it has no other option than to adapt itself to? But might this view not ignore some other functions of the law[1] [1], such as the expression of common values, a history, and a culture?

The development of nanotechnology raises questions today about what the law of nanotechnology might be, or should be. It is not easy to answer. Indeed, the present epoch is characterised by a highly ambivalent relationship to legal regulation, faltering between denouncement of the legislative logorrhea[2] [2, p. 218] and a fear of the legal vacuum in the absence of specific dispositions. Since the time has now come to make these legislative choices, it will be important to examine with care the potential disagreements over the construction of a suitable regulatory framework.

The first difficulty in this task is to set out what features of nanotechnology development should be the subject of specific laws. In actual fact, this investigation reflects two closely related problems: one concerns what exactly should be covered by the law, while the other concerns appropriate definitions.

[1] To quote C. Atias and D. Linotte in [1]: "The myth which has law adapt to fit the facts has the effect of concealing, even conjuring away an essential step in the legislative process. Having gathered and assessed the facts, but before the technical construction of the regulation, one must chose a legal policy. This period of choice, selecting the facts and the goals, is a necessary one."

[2] To quote A. Supiot [2]: "In order to accommodate every nook and cranny of social life, one must say everything, prescribe everything. A tireless desire to encompass the full complexity of social existence in a set of rules leads to a normative logorrhea which gradually makes these rules incomprehensible, rendering arbitrary the power to enforce them."

P. Houdy et al. (eds.), *Nanoethics and Nanotoxicology*,
DOI 10.1007/978-3-642-20177-6_29, © Springer-Verlag Berlin Heidelberg 2011

The choices here are further complicated by the need to take into account the broad range of disciplines, activities, and objects that enter into the issue. The term 'nanotechnology' reflects an attempt to specify the boundaries of an interdisciplinary research field by reference to a length scale. While this may be perfectly relevant for the purposes of science, such a criterion is not necessarily so for the law.

Furthermore, even if one were to limit the sphere of influence of the law to sanitary and environmental risks, it would still be necessary to take stock of legal resources and highlight certain maladjustments. For example, in the present state of positive law, understood as currently applicable law, certain notions and terminology would appear to be inadequate. For example, the regulatory notion of chemical substance as it is currently defined (in particular, in Article 3, Point 1, of the EC's REACH regulations [3]) seems to be somewhat off-target with regard to the relevant features of nanoparticles and nanomaterials.

Given the projects currently under adoption, it is also a delicate matter to construct ex abrupto an appropriate terminology. The shifts in meaning observed since the work of the Grenelle Environment Forum and up to the latest version of the bill presented to the French Parliament [4] perfectly illustrate the many stumbling blocks to be avoided here. Indeed, between false evidence and clumsy formulations, the legislator struggles to identify convincing turns of phrase to sculpt an effective text. The reports drawn up for the parliamentary debate on the bill for the planning law to implement the conclusions of the Grenelle Environment Forum [5, 6], and the discussions themselves (especially the session of 5 February 2009 at the Senate and the resulting modifications made to Article 37 of the bill), thus showed that the expression 'substances in the nanoparticulate state' could be interpreted as meaning either single nanoparticles or nanomaterials, or it could be taken to embrace all kinds of products resulting from nanotechnology or exhibiting some nanometric dimension.

At the present stage in the debate, it is impossible to describe the law applicable to nanotechnology in a completely definitive and exhaustive manner. However, it is well to point out the challenges confronting jurists when they try to respond to societal concerns over nano-objects and nanotechnology (see Sect. 29.1). Engaged as they are in a logic of joint regulation and disregard for territorial boundaries, nanotechnology development policies also put the law to the test (see Sect. 29.2).

29.1 The Law in the Face of Societal Concerns over Nano-Objects and Nanotechnology

The public greet the development of nanotechnology with a certain enthusiasm, thanks to the technical innovations that may be made in areas as varied as the fight against pollution, traceability of products and their sanitary properties, detection of health problems, and the development of new treatments. This positive opinion was expressed, for example, in the conclusions of the Citizens' Conference organised in the Ile-de-France region of France, where citizens declared themselves generally favourable to nanotechnology, and this for a range of different reasons. In their view [7–9]:

> Nanotechnology represents undeniable progress and even hope for the world of today and tomorrow, whether it be in the area of health, everyday life, the natural environment, or living conditions. Nanotechnology also gives hope of improving help to developing countries. Moreover, nanotechnology is unavoidable from an economic standpoint. Its development is expected to lead to the creation of wealth and employment.

This positive tone is nevertheless qualified by some doubts [10–13]. These concern in particular the fate of nano-objects and nanomaterials in the human body and in the environment, along with their possible harmful effects. These were brought out very clearly at the round table on environmental health organised in the context of the national consultation in France called the *Grenelle de l'environnement* (Grenelle Environment Forum) [15]. However, there were also questions about potential misuse of technical innovations relating to nanotechnologies across a range of different sectors: micro- and nanoelectronics and the consequences of ubiquitous computing,[3] brain implants and possible effects on people's sense of autonomy and identity, labs-on-chips and facilitated access to detection of sometimes incurable diseases, etc. In addition, a desire was expressed for more information about and monitoring of products circulating on the market, so that citizens and consumers can make better choices, but also so that public authorities can react quickly in the event of problems.

Today these questions and expectations concern mainly the public authorities and through them, the law. As an instrument for regulating human activities and protecting individuals, and as a repository for values considered to be fundamental by a given society at a given moment of time, the law does indeed stand on the front line. Of course, public decision-makers cannot turn a deaf ear to these appeals, but they are faced with major difficulties if they are to put forward relevant legal solutions. Taking into account uncertain, even fuzzy risks, and monitoring the life cycles of nanoproducts are the challenges that our legal system must face up to in order to meet the needs of the day.

[3] The term 'ubiquitous computing' was coined by Mark Weiser of Xerox Park in Palo Alto, California. It refers to an omnipresent system of invisible computers, distributed throughout our environment. The reader may consult [16] for more details.

29.1.1 How to Account for Uncertain, Even Fuzzy Risks

Existing law would not appear to be completely unequipped to deal with uncertain health and environmental risks. Indeed there are a number of resources, ranging from the precautionary principle to specific regulations organising certain activities, controlling certain market sectors, or protecting exposed workers. The situation seems less favourable when one considers risks relating to the convergence of nanotechnologies and other technologies deriving from biology, information science, and cognitive science (NBIC convergence), such as the risk of so-called transhumanist excesses,[4] aiming to 'improve' the human species. However, these initial observations need to be taken further if we are to ascertain the true state of the means available to positive law.

Since the issue here is to take into consideration uncertain health and environmental risks, we may be sure that the precautionary principle will be a decisive factor. This principle has been recognised to have legal scope in international law [17], EC law [18], and national law [19], since the beginning of the 1990s, initially in the area of environmental damage. Today, it has legal and constitutional value in French law [20] and features among the general principles of EC law [21]. Its field of application has now been extended to health risks [22].

While the definitions are still not perfectly uniform, the key feature of the precautionary principle is that it applies when scientifically presented concerns involve a risk of significant harm for health or the environment [23]. Concretely, it says that measures can be – and sometimes *must be*, depending on the wording – adopted despite uncertainties that may remain regarding the characteristics and the occurrence of a risk that could cause serious and irreversible damage to the environment or health. These measures will include increasing understanding of the risk, and anticipating and if possible avoiding realisation of the potential damage [24]. Regarding the industrial use of nanoparticles and the commercialisation of nanomaterials, application of the precautionary principle has been judged necessary by expert and advisory panels [25–27], ethical committees [28, 29], and jurists [30–32]. The European Commission and the French government have also expressed agreement on this point [33]. And it is for this reason that they have appealed to the appropriate expert bodies [34] and that public funds have been allocated for research on the health and environmental impacts of increased use of nano-objects.

However, this consensus over the relevance of the precautionary principle is not easy to translate into concrete and effective measures for preventing

[4] This movement advocates the use of science and technology to develop the physical and mental capacities of human beings, ultimately so that they may escape the limitations imposed by nature, and even escape from ageing and death. See, for example, the World Transhumanist Association, *The Transhumanist Declaration*, 2002.

the threatened harmful consequences. It is precisely because understanding is still limited and there remain significant unknowns that it is hard to work out what decisions should be taken. It even turns out that it would be difficult to produce an inventory of products, producers, importers, or exposed persons, because there does not yet exist any specific procedure for gathering the necessary information [35]. This is why, following up on demands expressed by associations present at the Grenelle Environment Forum, the French government presented a bill to Parliament in which one of the articles would require a declaration to the administrative authorities of any production, importation, or commercialisation of 'substances in the nanoparticulate state' (a declaration of quantities and uses) [4]. Another bill bearing on the national commitment to the environment aims to spell out and organise this declaration procedure [36]. While this kind of inventory is essential in order to implement the precautionary principle, existing law does not provide an adequate solution here.

Future measures may also reveal their limitations if the legislator is unable to spell out what is meant by 'substances in the nanoparticulate state' in such a way that all the most worrying hypotheses get covered, and not just some small fraction of them. Indeed, it would be unfortunate if nanomaterials or aggregates of nanoparticles were not covered by the new regulations, if it should turn out that their effects on health or the environment remain problematic. By an amendment adopted on 5 February 2009, the Senate extended this to substances in the nanoparticulate state or organisms containing nanoparticles or resulting from nanotechnology. While this modification was made through a desire for clarification, it does not directly cover the problem of nanoparticle aggregates or nanomaterials that do not constitute innovations made by specialised nanotechnology research bodies. Furthermore, it raises new questions about what should be understood by 'substances resulting from nanotechnology'.

Other questions about the ability of current legal solutions to implement the precautionary principle in an effective way have been formulated in the REACH regulation (Registration, Evaluation, Authorization and restriction of CHemicals) that governs commercialisation of chemical substances in the European Community [37]. In this respect, despite attempts at interpretation put forward by the European Commission [38], many uncertainties remain. It may be that this new regulation, adopted on the basis of the precautionary principle and which seeks to improve protection of health and the environment by compelling producers and importers (and consequently also users) to transmit information about these products, their uses, and their effects, will not in fact succeed in obtaining new and relevant information on nanoparticles and nanomaterials from the above-mentioned producers and importers. Indeed, and this even beyond REACH, French and European regulations in general were not designed to take into account the consequences of quantum physics. Effects due to change of scale, and more specifically the appearance of novel characteristics on the nanoscale, have up to now been ignored by positive law. Without becoming too pessimistic, we must therefore recognise the challenge that nano-objects and nanotechnology raise for the law. They undoubtedly reveal certain limitations and highlight the difficulties involved in applying the principle of precaution. The present lack of scientific knowledge clearly makes public decision-making extremely complex, while the economic and social stakes must also be taken into account.

The development of nanotechnology still raises many other questions. For example, various studies and reports on the implications of nanomedicine or nanobiotechnology, in particular with regard to the possible perturbation of individual identity, or humanist values concerning eugenics (implants – especially brain implants, integrated diagnostic devices, facilitated detection devices allowing faster testing such as labs-on-chips, etc.). Hence, in the opinion of the *Commission de l'éthique de la science et de la technologie du Québec*, we find the following analysis [39]:

> The convergence of nanotechnology with other disciplines like biology, information and communication technologies, and cognitive science brings with it many ethical and social challenges, particularly regarding human identity and the relationship between human beings and nature. [...] Nanotechnology might help to optimise certain physiological characteristics of human beings. The developments heralded by the convergence of knowledge and technology are almost without limit and may include cognitive capabilities. Certain developments will raise a good many fundamental questions regarding personal and social representations of human identity: what we understand and consider to be human, what we judge to be normal (or acceptable) and what is not, the subjective boundary between therapy and optimisation of human capacities, double talk on how to give disabled people a proper place in society, worship of performance, equity in the choice of services offered by the public health service, and a vision of individual responsibility and autonomy within the community.

The French *Comité consultatif national d'éthique* and the European Group on Ethics have also discussed the way these issues follow from the development of certain applications of NBIC convergence [40–42].

Can the law provide answers and safety nets? There are articles in the Civil Code which assert the need to respect human dignity and the obligation to protect the integrity of the human species. However, jurists hold somewhat contradictory opinions on the practical scope of these texts. For some authors, an objective interpretation of dignity must be upheld. In this perspective, this dignity and the accompanying protection are attributed to all human beings, requiring them to respect this obligation of dignity even with regard to their own persons [43, 44]. For other authors, a subjective concept of dignity is more appropriate. Its protection would then be ensured by respecting the autonomy and freedom of choice of the individual [45–47]. Article 16 of the Civil Code, which states that "the law ensures the primacy of the person, forbids any affront to the dignity of the latter, and guarantees respect for the human being from the very inception of life", is not precise enough to decide this controversy over the legal scope of the concept [48]. It is then difficult to determine what it is that most affronts human dignity, between actions aiming to free individuals from subservience to their biological characteristics and the decision compelling them to respect the values of the society in which they live.

Regarding Article 16-4 of the Civil Code, which asserts at the outset that "none shall impair the integrity of the human species", it remains very hard to interpret. Indeed, how can one reconcile a concept whose very meaning implies evolution – that of the species – with an obligation to maintain integrity? Indeed, the French Constitutional Council has refused to recognise any principle of constitutional value that would consecrate the protection of human genetic heritage [49].

Looking beyond principles, there are regulations controlling experimental practices on humans, and these should ensure certain fundamental values. By virtue of these texts, it should not be possible in France to carry out any research attempting to improve human performance without any regard for the advancement of knowledge or therapeutic progress. However, there can be no doubt that it would be very hard to draw the dividing line between detection, prevention, and treatment of pathologies on the one hand and the aim of improving human performance on the other. The extreme case of the mad scientist is rarely encountered, and reality is of course much more complex. The opportunities offered by nanotechnology make the work of ethicists and jurists much more delicate.

29.1.2 Taking into Account the Life Cycle of Nanoproducts

The concerns formulated about the consequences of widespread use of nanoparticles and nanomaterials have been accompanied by a request to monitor the nano-objects released onto the market, whatever their intended use (e.g., as medicines, consumer goods, building materials, etc.) [50]. The intention here is to go beyond a simple inventory of production and producers and to organise a full-scale traceability of the products, including the stage where they get dispersed in the environment. The hope is to set up a virtuous and transparent chain of production and use, extending from the creation of the product to its final destruction, with each step being thought out in relation with the previous and following steps, and each link in the chain being aware of its role and its responsibilities beyond its immediate contracting parties.[5] This ambitious idea aims without doubt to increase protection

[5] See in particular the AFSSET recommendation of July 2006 [26] to develop tools for defining industrial responsibility, to organise an independent reflection on the possibility of a procedure for ensuring the traceability of engineered nanomaterials, and to study the consequences of industrial secrecy for the assessment of

of health and the environment, by encouraging the invention or creation of products designed at the outset to be recycled.[6]

Public policy-makers have partly integrated this concern. For example, the EC communication entitled *Nanosciences and Nanotechnologies: An Action Plan for Europe 2005–2009* expresses the need to develop, hand in hand with EC Member States, international organisations, European agencies, industry, and other stakeholders, the terminology, guidelines, models, and standards required for assessing risks right through the life cycle of nanoscience and nanotechnology products [52]. However, the desire to set up this kind of integrated monitoring and traceability from invention or production right through to recycling or destruction, turns out to be extremely difficult to achieve in the current legal framework. Indeed, the present legal system is organised essentially around spheres of activity or types of product, and responsibilities are limited to direct causality (poor execution of the contract, damage of some kind), even though the causal relationship is sometimes interpreted broadly when it comes to civil responsibility. The legal requirement of traceability is rather recent and has only been fully developed in certain limited areas, such as foods and medication. Even in these sectors, the fate of the object in the environment remains disconnected from the rest of the chain. The only exceptions are special rules relating to genetically modified organisms, regarding in particular coexistence and traceability [53].

However, the case of nanoparticles and nanomaterials illustrates the relevance of such an approach. It is quite possible that the most harmful effects, both for ecosystems and for public health, may arise when nanoparticles and nanomaterials accumulate in the environment. Now the uses, hence the occurrences of dispersion, of nano-objects are much more varied than those of GMOs. Quite generally, it seems that this concept of production chain would meet the requirements for environmental health, i.e., that it would satisfy our rising awareness that the environment plays a key role in the good state of health of the population. However, it remains to invent this way of

health and environmental risks of engineered nanomaterials. This kind of concern is not unique to France or the European Union. See, for example, the opinion of the *Commission de l'éthique de la science et de la technologie* on ethics and nanotechnology [39, p. XXI, p. 40]: "The Commission recommends that the government of Quebec, guided by the precautionary principle in a perspective of sustainable development, should take into consideration all stages in the life cycle of a product resulting from nanotechnology or containing nanometric elements, and that to this end it should integrate the notion of 'life cycle' into all its policies regarding such an approach, in such a way as to avoid any harmful consequence of a technological innovation on human health or on the environment."

[6] The problem of designing a product while taking into consideration other requirements than those relating to the good operation of the product goes well beyond the health and environmental issues. As an example, electronic or computer systems can be built taking into account the requirements for respect of individual rights (on this point, see [51]).

designing regulation in its generality. The fragmentation of areas of competence among different administrative organisations, expert bodies, and other decision-making authorities shows that it is already difficult to coordinate actions and harmonise decisions. These difficulties are not unique to French and European law. For example, the Council of Canadian Academies advised the Canadian government to adopt a regulatory approach based on the life cycle of nanomaterials, explaining that past experience with chemical substances shows that the simple examination of manufactured nanoproducts and their immediate uses is not sufficient to predict their long term effects on health and the environment [54]. The general release of nano-objects through international trade will thus force public policy-makers around the world to rethink their legislative and regulatory planning.

29.2 The Law in the Face of Nanotechnology Development Policies

Nanotechnologies are being developed on a worldwide scale. This is clearly revealed by the fact that the OECD has set up two working groups to examine this issue since 2007.[7] In every case, these products are presented as the spearhead of competition and economic competitiveness[8] in the industrialised countries, fitting perfectly with the philosophy promoted by the EU's Lisbon strategy [57] and the various reports dealing with them[9] [58]: the research we do today, we are told, will bring us wealth tomorrow.

While the new technologies lie at the heart of this competitive vision of globalisation, they are nevertheless subject to regulation which traditionally remains, from the legal angle, rooted at the national or EC level. In the mind of the public, and sometimes also the legislator, it can be hard to situate legal standards in relation to other means of ethical, economic, or technical regulation.

The law is thus put to the test by the development of nanotechnologies, firstly by its territorial associations (Sect. 29.2.1) and secondly by its specificity as a form of regulation (Sect. 29.2.2).

[7] The working group on manufactured nanomaterials was set up in 2006 and the one on nanotechnology in 2007. For more information, the reader may consult the OECD Internet site [55].

[8] There is thus an international understanding to encourage the development of business plans and start-up companies in this area [56].

[9] In this sense, consider the following remark from 2004 [58]: "Public policy must meet a major challenge over the next thirty years: the challenge raised by nanotechnology. An ambitious reorganisation of scientific and technical programmes is on the agenda in France and the European Union to stimulate employment and competitiveness. In this context, nanotechnology is likely to play a significant role."

29.2.1 The Territoriality of the Law in a Context of International Competition

Ubi societas, ibi jus; ibi societas, ubi jus.[10] Law is a phenomenon that cannot even be conceived without a society and its members. As has been stressed by certain authors [59], individual states are still the legal entities best placed to give structure to societies as a whole, even if the existence of European law and frequent calls for international or global legal entities, compounded with the requirements of human rights and world trade, would seem to threaten this model.

In the area of science and technology, this specificity of legal regulation is confronted with contrasting logics and realities. Indeed, as pointed out by Pasteur, while the scholar has a homeland, science does not, and it has been argued by R.K. Merton that this reference to universality became an integral part of the scientific ethos as early as 1942 [60]. Economic globalisation, in which the policies for funding and supporting nanotechnology are systematically rooted, further strengthens this tendency to go beyond territorial boundaries.

There are precedents for such a confrontation between the imperatives of a market economy, in a highly internationalised area, and those arising through the territoriality of law. They help us to understand the extent to which the reconciliation of sometimes contradictory interests at the international level is a fine art, and often a fool's game.

This is the case, for example, in the field of industrial property rights. While deeply rooted in national traditions at the outset, these rights, with patent rights in the lead, became the subject of multilateral conventions from the end of the nineteenth century. The aim was to reconcile their national aspects with the interests of exploiting inventions, which for their part would extend well beyond the frontiers of member states. The Paris Union Convention, adopted on 20 March 1883 [61] and subsequently modified on many occasions, set up a system for the coexistence of national rights based on the principle of assimilation of union members to nationals, asserting the equality of union members with regard to access to and use of patents. It also established the Paris Convention priority right, which allows someone who files a patent in one member country to benefit from a period of one year to do the same in the other member countries without the first claim being used to oppose later ones on the grounds of anteriority. But above all, it was built upon the principle of independence of patents,[11] which shows the extent to which national laws retain their force.

[10] No law without society, no society without law.

[11] This principle states that the patent law of each country is applicable to all who would file or exploit their patent on its territory, whence no other regulations could be imposed on a given member state, in particular with regard to the duration of protection.

However, in the 1980s, in the face of growing economic globalisation, criticisms were voiced about this system for protecting inventions, and negotations were undertaken to harmonise intellectual property rights on the international scale in the broader framework of a mechanism for managing world trade. On 15 April 1994, these negotiations resulted in the adoption of the Marrakesh Agreement, setting up the World Trade Organisation (WTO) [62]. A whole annex is devoted to intellectual property rights. This annex, called Trade-Related Aspects of Intellectual Property Rights (TRIPS) [63], results from a complex recipe of compromises and mutual concessions, not only between the different notions of intellectual property upheld by the signatory countries, but also with the other interests and values borne by the WTO [64]. The problem of reconciling the general good with the rights of patentees was the subject of bitter negotiations between southern hemisphere and industrialised countries, even though the points of economic and ideological disagreement subsequently shifted during the talks.

The TRIPS agreement achieves a certain form of universalisation of invention patent rights, on its own ground to begin with, since the protection of industrial property is made compulsory, sometimes in a somewhat contrived way, for all WTO member countries.[12]

But not all the consequences of this form of universalisation of invention patent rights are entirely positive. Article 27 of TRIPS [65] illustrates the extent to which the patent, viewed from the specific angle of international trade, has become an instrument of hegemony, now covering all technical fields and even extending, as has often been criticised, to allow upstream patenting. This unfortunate trend is flagrant in the field of biotechnologies [66]. Patents seem to be moving away from their original calling, which excluded discoveries in their field of relevance, reserving them for industrial applications of novel technical inventions whose disclosure and publication for the benefit of society as a whole justified granting a monopoly.

This trend illustrates the perils involved in internationalising law. Among its long-term repercussions, after swallowing up the field of biotechnology, the extension of the field of patentability can be expected today to set upon the area of nanotechnology. The TRIPS agreement, ensuring the North American promotion of an omnipresent patent, has already allowed the filing of very wide-reaching patents concerning what are called the building blocks of matter.[13]

This situation has given rise to a lot of criticism. While its image may suffer from the negative consequences of the phenomenon, nanotechnology could provide an opportunity to challenge this internationalisation. If the ins and

[12] Among other reasons, this will delay the entry of certain countries, notably China, in the WTO.

[13] Some groups, critical of the development of nanotechnologies, have already pounced upon this trend to stigmatise the patenting of 'atomically modified organisms' (AMO). For example, see the work by the ETC Group [67].

outs of internationalisation are not suitably controlled, there is a risk of its destroying the tools the law offers to guarantee the development of nanotechnology. The case of the TRIPS agreements provides a good illustration, it seems to us, of the detrimental effects of international competition. The pressure imposed by economic regulation of world trade has in this case played a driving role in bringing together legislations with sometimes contradictory values and concerns. The result of this alleged reconciliation has been a harmful alteration of patentability criteria, although it was supposed to found this new international law.

Under the guise of a uniformisation and universalisation of the legal regulations applicable to invention patents, what has actually happened with TRIPS is rather the international grounding of a vision of the patent that is strongly linked to the economy of innovation in industrialised countries, and in particular, the United States [66]. As an instrument in the service of the values of a given society and its members, the law is indissociable from that society. It is this unshakeable relationship between the law and the society to which it is dedicated that gives its specificity to the rule of law, whether it be national or international. If in the search for a common rule primacy is given to values (or interests?) that are not shared by those upon whom the rule will be imposed, there is a risk of its losing all substance and becoming ineffective.

In other words, the need to internationalise the legal regulation, while it may be linked to world economic trends and, in the field that interests us, to the intrinsically universal nature of science, should not be deduced from these external elements, but rather from the observation of an international community of values and principles.

Careful observation of the development of nanotechnology in France and in the world does not reveal any such values in the present state of affairs. Quite the contrary, the axiological presuppositions used a basis for relevant science and technology policies show a high level of non-uniformity in the values that feed them.[14] This observation, combined with the terminological difficulties encountered in the field of nanotechnology [68,69] – which make it difficult to specify a clear dividing line between those elements whose development one wishes to promote and those that ought to be more tightly controlled – should not be neglected by public policy-makers thinking about the construction of a legal framework for these technologies.

In any case it would seem today that, in the case of the nanotechnologies, public decision-makers are quite ready to appeal to more flexible, sometimes negotiated forms of regulation, such as those used so far by the European Commission. Hence the need to reconsider the specificity of the rule of law in the face of alternative means of regulation.

---

[14] This is certainly true of NBIC convergence, the spearhead of public policy encouraging the development of nanotechnology in the first reports of the National Nanotechnology Initiative in the United States, which are far removed from the values upheld by the European Union or France with regard to this issue.

29.2.2 The Specificity of the Law and Alternative Means of Regulation

There are several signs that can be interpreted as calls for flexible regulation in the area of emerging technologies in general, and nanotechnologies in particular: European advocacy of a code of good conduct for the responsible development of nanoscience and nanotechnology research [70], requests to organise a public debate by law [4], repeated calls for technical standardisation, and the development of a strong ethical commitment in this area. This immediately raises the question of how to distinguish legal regulation from the kind of regulation offered by ethics, economics, or technical standards.

Interpreted in the strictest sense, the law is a set of socially decreed and sanctioned rules of conduct imposed on members of society [71]. In the field of nanotechnology, there is little indication of any desire to submit their development specifically to rules of this kind. But this absence does not in our opinion reduce the relevance of the rule of law within a controlling framework. Indeed, the latter is present at all stages in the life cycle of a given nanotechnology, from research right through to waste management. It is just that the law applicable to nanotechnology is not yet specific to them.

So how could one reliably determine which features of nanotechnology should be subject to the law? The most common analysis by specialists in this area is to assert that the legal framework has involved, up to now, only the products resulting from nanoscience and nanotechnology, without yet concerning itself with these technologies in themselves. This argument is certainly supported by the fact that the law as we defined it above only exceptionally finds its place in the regulation of scientific research activities [72]. It is no less criticisable for that, if we observe that, just to consider the most salient features of regulation in this field, the principle of precaution, now constitutionalised, applies to every stage of nanotechnological development, including the stage of scientific research.

It is worth stopping for a moment to consider the method chosen by the European Commission to try to disentangle exactly which aspects of the field that interests us are covered by legal regulation and which escape for the moment from its dominion. Indeed, in the EC communication entitled *Regulatory aspects of nanomaterials* [38], the Commission chose to opt for an inventory of legal standards finding application in the field of nanomaterials, before asserting that, globally, it was safe to conclude that current legislation covers to a large extent the risks relating to nanomaterials, and that these risks could be managed with the help of the existing legislative framework. But they added that modifications might have to be made depending on further information that might become available, e.g., regarding the thresholds used in certain legislations.

So the European Commission favours adaptation of positive law. While it may seem reasonable to study the existing legal framework and assess its relevance in the face of scientific evolution and the appearance of new products and technologies, its application to this area by the European Commission nevertheless deserves some comments.

To begin with, the field of investigation covered by this EC communication seems somewhat restrictive. Since it is integrated into an action plan relating to nanoscience and nanotechnology [50, 73], which advocates responsible, safe, and integrated development, it is hard to see why the Commission limited the study of the applicable regulatory field to nanomaterials alone, especially as the clearly stated objective is to ensure that society can benefit from innovative applications of nanotechnologies, while maintaining a high level of protection for health, safety, and the environment [37]. Does this mean that the business of the law is to regulate only nanomaterials, excluding all other products resulting from nanotechnologies, and the scientific research undertaken in this field? Or more prosaically, does it imply that nanomaterials are the only things produced by these technologies that present any risk with regard to protection of health, safety, and the environment, without considering the necessary protection of consumers, workers, or researchers?

If it were really so, which is of course doubtful, this assertion would at least require some justification, which is not the case in the EC text. On the contrary, it seems that, if the aim is to encourage responsible development of nanotechnologies, it would have been better to envisage a broader legal framework. But that would have been a much bigger project.

Another point is that, having reviewed existing regulations in the field of nanomaterials, the Commission concludes that the protection of health, safety, and the environment must be strengthened mainly by an improved implementation of existing legislation. They add that, as a consequence, the Commission and the agencies of the European Union will begin by examining the documents that currently support this implementation, such as the arrangements for enforcing legislation (the implementing provisions), but also the standards and technical orientations, in order to assess their relevance and suitability for application to nanomaterials.

This kind of approach is perfectly in keeping with the desire to decompartmentalize the field of application of the law, and seems altogether positive. It is of course difficult to determine, in the area of nanotechnology, what belongs to the rule of law and what belongs to other forms of regulation. But the idea of linking up the different normative registers and the various forms of regulation to construct together a joint regulation appropriate to nanotechnology would appear to be a good way to reach favourable solutions.

On the other hand, we should not ignore the fact that, even when envisaged from the point of view of a softening and an adaptation of the rules of law, the involvement of alternative means of regulation would not satisfy, or at least would only rather roughly satisfy, democratic criteria. Rules of law are in principle conditioned by these democratic criteria when drawn up, decreed, or modified. Conversely, ethical and technical standards, or what the Commission calls implementing provisions, are not subject to the same democratic adoption processes as the law understood in its broadest sense. And neither are they characterised by social and state sanctions when violated, which is one of the characterising features of the law. Such observations should not make us disregard these forms of regulation, which remain necessary and useful. However, they should lead us to examine in more detail their interaction and their interrelationships with the rules of law.

Consequently, guiding the normative control of nanotechnology toward a jointly regulated system, appealing in a complementary way to the resources of the rule of law but also to other forms of regulation, seems to us to ensure strong guarantees in democratic terms.

In our view, this is the meaning that should be attributed to the dispositions promoted in the code of good conduct for responsible research in nanoscience and nanotechnology, which calls for the close association of all stakeholders in this development (and the definition of these stakeholders chosen by the Commission is particularly broad [74]), through the principles of inclusiveness and transparency. It could be that the development of nanotechnology is a further opportunity, and a vital one given its present and forecasted societal impacts, to try to build together a flexible normative framework suited to new forms of technology and respecting democratic values. If we could only do that, it would already be an achievement.

References

1. C. Atias, D. Linotte: *Le mythe de l'adaptation du droit au fait.* Recueil Dalloz Sirey, I, p. 251 ff. (1977)
2. A. Supiot: *Critique du droit du travail.* PUF, Paris (1994)
3. Article 3, Point 1 of Regulation (EC) no. 1907/2006 of 18 December 2006 concerning the registration, assessment, and authorisation of chemical substances, together with the restrictions applying to these substances (REACH): For the purposes of the present regulations, the term 'substance' refers to a chemical element and its compounds in the natural state or obtained by man-made synthesis, including all additives required to maintain stability and all impurities resulting from the production process, but excluding any solvent that can be separated without affecting the stability of the substance or modifying its composition.
4. Bill for the planning law implementing the conclusions of the *Grenelle de l'environnement* (bill known as *Grenelle 1*), Article 37
5. Report by Christian Jacob on behalf of the *Commission des affaires économiques, de l'environnement et du territoire* of the National Assembly, presented on 1 October 2009
6. Report by Bruno Sido on behalf of the *Commission des affaires économiques* of the Senate, presented on 14 January 2009
7. Citizens' conference on nanotechnology, Ile-de-France region: Explorons les enjeux de l'infiniment petit. espaceprojets.iledefrance.fr/jahia/Jahia/NanoCitoyens
8. D. Bourg, D. Boy: *Conférence de citoyens, mode d'emploi.* Editions Charles Léopold Mayer (2005). On citizens' conferences and conditions for implementing them

9. D. Boy, D. Donnet-Kamel, P. Roqueplo: Un exemple de démocratie participative: 'la Conférence de citoyens' sur les organismes génétiquement modifiés. Revue Française de Science Politique **50** (2000)
10. Opinion of previously cited citizens' conference, pp. 5–6. This observation is valid for other consultations of civil society members: Démocratie locale et maîtrise sociale des nanotechnologies: les publics grenoblois peuvent-ils participer aux choix scientifiques et techniques? Report for the *Mission de la Métro*, Grenoble, 22 September 2005
11. Citizens' consultation EPE-APPA on environmental and sanitary issues relating to the development of nanotechnology, October 2006
12. Positions established within the FNE (federation of environmental protection groups), by the board of its health and environment network, February 2007, in the contributor's report entitled *Nanotechnologies: le point sur les débats, des orientations pour demain*, Cité des sciences et de l'industrie, 19 and 20 March 2007. www.cite-sciences.fr/francais/ala_cite/college/v2/html/2006_2007/cycles/cycle_252_ressources.htm
13. World Forum on Science and Democracy: *Principes de surveillance des nanotechnologies et nanomatériaux* (April 2008). fm-sciences.org/spip.php?article184
14. European Trade Union Confederation (ETUC): ETUC resolution on nanotechnology and nanomaterials, ETUC Executive Committee, Brussels, 24 and 25 June 2008. www.etuc.org/a/5162
15. Grenelle Environment Forum: Document summarizing the round tables held at the Hôtel de Roquelaure on 24, 25, and 26 October 2007 (Round table 3), Paris, November 2007
16. A. Greenfield: *Everyware: The Dawning Age of Ubiquitous Computing*, New Riders Publishing (2006)
17. Declaration on the environment and development, adopted in Rio, 1992, Principle 15
18. Article 130R of the Maastricht Treaty in 1992, then Article 174 in the consolidated version of the treaty founding the European Community
19. Article L. 110-1, II-1, of the *Code de l'environnement* resulting from the Barnier Law of 1995
20. Article L. 110-1 of the *Code de l'environnement*; Article 5 of the *Charte de l'environnement* (resulting from the French constitutional law of 1 March 2005)
21. See note from the Court of First Instance of the European Community (CFI), 26 November 2002, Artedogan affair: T-74/00, point 184; CFI 11 September 2002, Alapharma affair: T-70/99, point 171
22. C. Noiville: Science, décision, action: trois remarques à propos du principe de précaution. Petites Affiches **218–219**, 10 (2004)
23. *Le Principe de précaution*, special issue of the journal *Recueil Dalloz*, ed. by C. Noiville, p. 1515 ff. (2007)
24. EC communication on application of the precautionary principle, Brussels, 2.2.2000, COM (2000) 1 final; EC jurisprudence cited in [21]. Article 5 of the *Charte de l'environnement* cited in [20] for French law
25. Comité de la prévention et de la précaution (CPP): *Nanotechnologies, Nanoparticules. Quels dangers, quels risques?* Paris, May 2006
26. AFSSET: *Les nanomatériaux. Effets sur la santé de l'homme et sur l'environnement.* Paris, July 2006
27. AFSSET: *Nanomatériaux et sécurité au travail.* May 2008

28. COMETS: *Enjeux éthiques des nanosciences et nanotechnologies.* 12 October 2006
29. CCNE: Avis no. 96, *Questions éthiques posées par les nanosciences, les nanotechnologies et la santé.* January 2007
30. M.A. Hermitte: Relire l'ordre juridique à la lumière du principe de précaution. Recueil Dalloz **22**, 1518 (2007)
31. N. Herve-Fournereau: La sécurité sanitaire et écologique vis-à-vis des nanomatériaux. Cahiers Droit, Sciences et Technologies, CNRS Editions **1**, 57 (2008)
32. C. Lepage: *Mission sur la gouvernance écologique.* Final report of the first phase (proposition no. 38), p. 36, Paris (2008)
33. EC communication: Nanoscience and Nanotechnology: Action Plan for Europe 2005–2009. First implementation report 2005–2007, COM (2007) 505 final; Recommendation 2008/345/CE of the European Commission concerning a code of conduct for responsible research in nanoscience and nanotechnology, 7 February 2008
34. *Agences françaises de sécurité sanitaire de l'environnement et du travail* (AFSSET), *des aliments* (AFSSA), *des produits de santé* (AFSSAPS), and the *Haut Conseil de la Santé publique* for the French government; Scientific Committee on Emerging and Newly Identified Health Risks (SCENIHR) and Scientific Committee on Consumer Products (SCCP) for the European Commission
35. *Agence française de sécurité sanitaire de l'environnement et du travail* (AFSSET), *Nanomatériaux et sécurité au travail*, May 2008
36. French bill: *Portant engagement national pour l'environnement*, Article 73
37. Regulation (EC) no. 1907/2006 of the European Parliament and Council of 18 December 2006 concerning the registration, assessment, and authorization of chemical substances, together with the restrictions applying to these substances (REACH), setting up a European Chemicals Agency, modifying the directive 1999/45/CE, and abrogating the regulation (EEC) no. 793/93 of the council and the regulation (EC) no. 1488/94 of the Commission, together with the directive 76/769/CEE of the Council and the directives 91/155/CEE, 93/67/CEE, 93/105/CE, and 2000/21/CE of the Commission
38. EC communication: Regulatory aspects of nanomaterials, COM (2008) 366
39. Commission de l'éthique de la science et de la technologie: Opinion on *Ethique et nanotechnologies: se donner les moyens d'agir*, Québec, 14 June 2006, p. XXIV
40. European Group on Ethics: Opinion on the ethical aspects of nanomedicine, Opinion no. 21, 17 January 2007. ec.europa.eu/european_group_ethics/activities/docs/opinion_21_nano_en.pdf
41. European Group on Ethics: Ethical aspects of ICT implants in the human body, Opinion no. 20, 16 March 2005. ec.europa.eu/european_group_ethics/docs/avis20_en.pdf
42. CCNE: Opinion no. 96, *Questions éthiques posées par les nanosciences, les nanotechnologies et la santé*, 7 March 2007. www.ccne-ethique.fr/docs/fr/avis096.pdf
43. B. Edelman: La dignité de la personne humaine, un concept nouveau. Recueil Dalloz, Chronique, p. 185 (1997)
44. B. Mathieu: La dignité de la personne humaine: quel droit? quel titulaire? Recueil Dalloz, Chronique, p. 282 (1996)
45. O. Cayla: Le coup d'Etat de droit? Le Débat, May–August, p. 123 (1998)

46. N. Deffains: Les autorités locales responsables du respect de la dignité de la personne humaine. Sur une jurisprudence contestable du Conseil d'Etat. Revue Trimestrielle des Droits de l'Homme, p. 673 (1996)
47. Y. Thomas: Le sujet de droit, la personne et la nature. Sur la critique contemporaine du sujet de droit. Le Débat **100**, 85 (1998)
48. B. Maurer: La dignité humaine est-elle liberticide? A propos de la querelle sur 'l'affaire du lancer de nain'. Cahiers des Ecoles Doctorales, Faculty of Law, Montpellier, no. 1, Les controverses doctrinales, p. 187 (2000). For a synopsis of the doctrinal controversies arising around the concept of human dignity
49. Conseil constitutionnel, 27 July 1994, Recueil Dalloz, 1995, jurisprudence p. 237, note by B. Mathieu
50. See, e.g., Citizens' consultation EPE-APPA on environmental and sanitary issues relating to the development of nanotechnology, October 2006, Citizens' recommandation 1.4
51. S. Lacour: 'Ubiquitous computing' et droit: l'exemple de la radio-identification. In: S. Lacour (ed.), *La Sécurité de l'individu numérisé*, L'Harmattan (2008), p. 29, esp. p. 43 ff
52. European Commission: Nanosciences and Nanotechnologies: An Action Plan for Europe 2005–2009. COM (2005) 243 final, of 7 June 2005
53. M.A. Hermitte: Qu'est-ce qu'un droit des sciences et des techniques? A propos de la traçabilité des OGM. Tracés, Techno no. 16 (2009)
54. Council of Canadian Academies: Small is different: A science perspective on the regulatory challenges of the nanoscale. Report by the Expert Panel on Nanotechnology, July 2008, p. 15
55. www.oecd.org/document/36/0,3343,en_2649_34269_38829732_1_1_1_1,00
56. www.nanochallenge.com/files/index.cfm?id_rst=396
57. Lisbon Treaty of 13 December 2007, Official Journal of the European Union C 306, 17.12.2007
58. F. Roure, J.P. Dupuy: Les nanotechnologies, éthique et prospective industrielle. Report by the *Conseil général des Mines et du Conseil général des technologies de l'information* (2004)
59. R. Encinas de Munagorri, G. Lhuilier: *Introduction au droit.* Series *Champs Université*, Flammarion, Paris, p. 11 (2002)
60. R.K. Merton (1942): The normative structure of science. In: *The Sociology of Science: Theoretical and Empirical Investigations*, R.K. Merton (ed.), University of Chicago Press, 1973
61. Paris Union Convention: www.wipo.int/portal/index.html.fr. The Paris Union currently has 173 member states
62. www.wto.org/indexfr.htm
63. www.wto.org/english/docs_f/legal_f/27-trips.pdf
64. T.L. Wasescha: L'accord sur les ADPIC, un nouveau regard sur la propriété intellectuelle. In: *Droit et économie de la propriété intellectuelle*, M.A. Frison-Roche and A. Abello (eds.), LGDJ (2005)
65. World Trade Organization: Agreement on aspects of intellectual property law affecting trade, Article 27: Patentable subject matter: Subject to the provisions of paragraphs 2 and 3, patents shall be available for any inventions, whether products or processes, in all fields of technology, provided that they are new, involve an inventive step and are capable of industrial application

66. M. Vivant: Le système des brevets en question. In: *Brevets, innovation et intérêt général. Le brevet: pourquoi et pour faire quoi?* B. Remiche (ed.). Editions Larcier, p. 19 ff. (2007)
67. www.etcgroup.org
68. S. Desmoulin-Canselier: Les difficultés terminologiques de l'encadrement juridique des nanosciences et nanotechnologies et des nano-objets. In: *Actes de l'atelier résidentiel 'Quelle régulation pour les nanosciences et les nanotechnologies?'* S. Lacour (ed.), Le Tremblay, 27–30 January 2009
69. S. Lacour: Problèmes de définition juridique dans le champ des nanotechnologies. Quels mots pour le faire et pour le dire? In: *Actes du colloque 'Comment appréhender les risques des nanoparticules d'aujourd'hui et de demain?'* D. Vinck (ed.), Grenoble, 9 October 2008, Editions Lavoisier, Cachan (2008)
70. Recommendation 2008/345/CE of the European Commission concerning a code of conduct for responsible research in nanoscience and nanotechnology of 7 February 2008
71. G. Cornu (ed.): *Vocabulaire juridique*, Association Henri-Capitant, PUF, 2000
72. R. Encinas de Munagorri: La communauté scientifique est-elle un ordre juridique? Revue Trimestrielle de Droit Civil, 247 (1998); M.A. Hermitte: *La liberté de la recherche et ses limites. Approches juridiques.* Editions Romillat (2001)
73. European Commission: COM (2004) 338 final du 12.5.2004
74. EC recommendation 2008/345/CE [46], Article 2b of the code of good conduct for responsible research in nanoscience and nanotechnology. Stakeholders in N&N research: Member states, employers, research funders, researchers, and more generally all individuals and civil society organisations involved in or interested in nanoscience and nanotechnology research

Part VI

Nanoethics and Social Issues

30

How the Risks of Nanotechnology Are Perceived

Daniel Boy and Solange Martin

In the ongoing debate about new technologies, from bioethics to GMOs and nanotechnologies, risk perception – by individuals – is understood by opposition to objective assessment of risk – by science. The absence of objective risks and the presence of perceived risks are often stressed by one side or the other, the first by those who support the development of such technologies, the second by those who insist upon regulatory control. However, objective risks and subjective risks are not only different, they are also asymmetric in their ability to conclude a debate. Whereas objective risk is collectively established, perceptions are always those of specific individuals who experience them with a suspicion of ignorance or ideology. In order to glimpse some possibilities for dialogue between the two types of risk, i.e., between the public and scientists, we shall give a brief overview of the main features structuring risk perception in the case of nanotechnology. We shall then sketch an inventory of the perceptions of the general public in France, and also among young people in Europe. We shall complement these results by a study of perception among a population of European experts. Finally, we shall outline what can be done collectively with these individual perceptions once they are known.

30.1 Criteria Giving Structure to Perception

Nanotechnology lends itself particularly well to disaster scenarios, to the point where some of the most famous of these have been written by two eminent scientists who were themselves involved in their early development. This situation is unusual enough to deserve mention. The first, Eric Drexler [1], recounts the tale of a 'grey goo', which comes into existence when scientists lose control of a theoretical molecule capable of reproducing itself. This molecule ends up absorbing all the matter in the world. The second case was Bill Joy who, in an article written in 2000 [2], revisits the Frankenstein story by inventing creatures, in fact intelligent robots, that escape from their creators, human beings of course, and end up superseding them not only intellectually but

P. Houdy et al. (eds.), *Nanoethics and Nanotoxicology*,

DOI 10.1007/978-3-642-20177-6_30, © Springer-Verlag Berlin Heidelberg 2011

also socially, in a relatively near future. In addition to these two rather dark visions, one must remember the cyborg (cybernetic organism) or cyberman, and by extension, the story of Faust,[1] where a man sells his soul to become all-powerful, and reinvents life and man himself. Finally, one could also mention 'Big Brother', popularised by George Orwell in his novel *1984*, where the population is relentlessly and ruthlessly controlled by nanosensors and other intelligent forms of dust. Compared with other new forms of technology, nanotechnology does not necessarily exhibit any features that would affect the way they are perceived. What is specific about them is rather that they concentrate and potentialise all the known kinds of hopes and fears. This is no surprise given that nanotechnology stems from and revolutionises conventional forms of science and technology, and in doing so also manages to make them converge.

In a context where GMOs are regularly in the news, following on from mad cow disease, asbestos, and, in France, contaminated blood, it is easy to see why the issue of risk perception is of major importance in the case of nanotechnology. Consider, for example, the NanoEthicsBank, a database conceived by the Center for Nanotechnology and Society, set up in the US by the National Science Foundation and the National Nanotechnology Initiative [4]. This lists 685 studies and reports (in October 2008) dealing with the perception and acceptability of nanotechnology, the establishment of a regulatory framework, and good practices recommended for researchers and industry. A significant effort is thus being made to assess and respond to any social opposition to nanotechnology. The anticipation of a possible rejection by the population as a whole is thus taken rather seriously, especially since it could result in significant financial losses in the event of a consumer boycott or a moratorium on industrial applications.

Researchers and institutions in charge of technological development and risks have built up a considerable knowledge of the criteria influencing individual risk perception.[2] Age, sex, culture, and level of education, but also social position and value system, are all factors influencing the way an individual perceives risk. As a general rule, women turn out to be more sensitive to risk, and young people rather less. Studies [5] also regularly show that the privileged classes are generally less apprehensive of most risks. At the other

[1] This originated from a traditional sixteenth century German story [3], and inspired a great many works of fiction from Marlowe to Goethe, including Jarry, Valery, Giono, Pessoa Mann, Butor, and others. It is the story of a doctor who, from his earliest days, dreams of possessing universal knowledge. Of course, he is unable to do this, and on the brink of suicide, he accepts a pact with the devil, in the form of Mephistopheles: he will achieve all his Promethean desires provided that he relinquishes his soul. He accepts.

[2] The first compilations of criteria structuring individual risk perception were assembled by Starr in 1969 and Slovic in 1977. They have since been supplemented, in particular by Covello [9].

end of the social scale, we also know that poorer populations tend to have more difficulty projecting themselves into the future, and are less concerned about chronic or invisible health risks [6]. We also have comparative studies between the perceptions of the general public and those of institutional experts, notably in France, with the PERPLEX study [7]. Globally, compared with experts, the public tends to rate risks as being higher. The public also tends to be more wary of the authorities and more often considers that the truth has been withheld. On the other hand, the public and experts tend to organise risks globally into the same hierarchy. However, the level of training or an occupational practice relating to assessment of the risk in question can explain some differences, and in particular a lesser sensitivity to the risk on the part of experts.

30.2 Nanotechnology: A Checklist of Risk Perception

So what about risk perception with regard to nanotechnology? Does the public have any particular opinion about this issue, or is the subject still too little publicised and inadequately discussed to make it worthwhile trying to identify public perceptions and attitudes?

30.2.1 The Attitude Toward Nanotechnology in France

In France, national opinion polls concerning the social representation of science have been carried out at relatively regular intervals [8]. In the most recent of these investigations, a question was asked about the perception of nanotechnology[3] (Question 1). In order to evaluate more precisely the current state of understanding of the public in this area, the open question method was chosen. With this technique, the interviewer asks a relatively general question, e.g., what do you think of when you hear the word 'nanotechnology'?, without proposing pre-established forms of response. Interviewees are thus allowed to respond freely, in their own words. The texts of the answers noted down by the interviewer are then 'coded', i.e., distributed among significant categories which can be counted up at the end. Table 30.1 shows the results obtained.

The first striking fact from this study is the high percentage of interviewees who gave no answer (53%). The second observation concerns the most frequent spontaneous association: for about one third of those questioned (35%), it is size that specifies the notion of nanotechnology (the infinitely small, microscopic technology, etc.). Two further types of answer reach the threshold of 5%: the medical theme (medicine, surgery, etc.) and the connection with

[3] The survey was carried out on behalf of the EDF and the *Palais de la Découverte* of 18–30 October 2007 on a sample of 1 000 people representative of the French population, aged 18 and above, questioned face to face at their home by TNS-Sofres interviewers. Quota method (sex, age, socio-professional category of the head of the household) and stratification by region and urban category.

Table 30.1. What do you think of when you hear the word 'nanotechnology'? The figure in the right-hand column is a percentage. The total is greater than 100% because those interviewed sometimes gave several answers

Kind of technology, interpretation of term	**39**
Including:	
Infinitely small/microscopic technology	35
It is a new technology/new	3
Technology/high-precision technology	2
Fields of application	**12**
Including:	
Medicine	5
Including:	
Used in medicine/research	3
Used in surgery/development of prostheses, implants	2
Computing/electronics/robotics/miniaturisation of electronic chip	5
Nuclear/atoms/study of matter	2
Atmospheric measurements/wind turbines	1
Consequences	**2**
Including:	
Represents progress/a major step forward for science/the future	2
Dangerous for human beings/harmful/destructive/hope this technology will be abandoned	1
Other (online work, food, techno music, hydrogen storage, etc.)	**2**
No answer	53

electronics (computing, robotics, electronic chips, etc.). The other themes mentioned are extremely wide-ranging. Finally, at the time of this survey (October 2007), positive or negative value judgements were very rare (progress 2%, danger 1%).

Answers here differ for the main part due to the level of education and the age of the people interviewed. For example, 81% of interviewees who have not gone beyond primary school studies and 71% of the older generations (65 and over) supply no answer to this question, compared with only 18% of those with a higher scientific education and 38% in the 18–24 age group. As often happens regarding scientific information, a difference is also observed between men and women: 62% of women do not answer this same question compared with only 43% of men.

To complete this first assessment, a second series of questions can be used (Question 2). It provides an indirect measure of the values attributed to various technological issues through the question of their opposition (see Table 30.2): Does the public consider protest movements directed against

Table 30.2. Here is a series of situations where people have taken action to oppose technical innovations. For each such action, would you say that it is altogether acceptable (AA), fairly acceptable (FA), totally acceptable (TA), rather unacceptable (RU), altogether unacceptable (AU), totally unacceptable (TA), or no opinion (NO)?

	AA	FA	TA	RU	AU	TU	NO
Boycott of food products containing GMOs	32	37	**69**	17	8	**25**	6
Fight against nuclear waste storage site	30	39	**69**	18	8	**26**	5
Fight against construction of mobile phone relay	13	42	**55**	30	8	**38**	7
Destruction of open field GMOs	22	27	**49**	28	17	**45**	6
Fight against development of nanotechnology	**4**	**18**	**22**	**22**	**16**	**38**	**40**

issues such as GMOs, nuclear waste, mobile phone relay stations, and nanotechnology to be 'acceptable'?

The results show that public judgement varies significantly depending on the issue, even though on the whole active opposition seems widely recognised as being legitimate. However, nanotechnology seems to be an exception to this rule since 22% of those questioned considered the fight against their development as 'acceptable, compared with 38% who considered this action as 'unacceptable'. There, too, the high level without opinion (40%) shows that nanotechnology does not particularly motivate the public one way or the other, due to a lack of basic information.

However, by analysing the responses in terms of the level of education of those questioned, we find that the refusal to contest is higher in educated categories: hence 61% of those with a higher scientific education consider the fight against the development of nanotechnology as 'unacceptable', compared with 38% on average. It is worth stressing this result, because it is by no means the general rule: if we consider anti-GMO actions, the educated are in fact inclined to support opposition slightly more often than the average, particularly when they can be situated on the left of the political spectrum. But this is not the case when we consider nanotechnology.

30.2.2 Attitudes Toward Nanotechnology in the European Union

A survey of the attitudes of young people to science was carried out in the European Union in September 2008.[4] The questionnaire enquired about the subject's interest in science and technology and opinions about these

[4] Flash Eurobarometer 239, *Young People and Science*, carried out from 9 to 13 September 2008 on a sample of 25 000 young people aged between 15 and 25, from 27 countries of the European Union. The interviews were mostly by telephone and included around 1 000 people per country.

Table 30.3. There is debate over whether the following scientific areas and technological innovations involve more risks than advantages for society or vice versa. For the following items, please indicate, in your opinion, whether they have more advantages than risks for society, more risks than advantages, the same, or no opinion

Issue	More advantages	More risks	Same level	No opinion
Brain research	74	9	13	5
Mobile phones	55	16	27	1
Computer surveillance techniques and CCTV	54	17	25	4
Human embryo research	50	22	21	7
Nanotechnologies	**43**	**11**	**19**	**27**
Nuclear energy	25	46	24	4
Genetically modified foodstuffs	17	49	29	5

issues, particularly regarding the risks induced by certain kinds of technology. Regarding risk perception, the following question can be used to order a certain number of technologies or research areas with different levels of risk (see Table 30.3).

The answers obtained here show that, on the whole in the European Union, nanotechnology is generally considered to cover an area that involves more advantages than risks (43% as compared with 11%). However, once again, this observation must be moderated by the fact that 27% of those questioned do not give any opinion on this question, while in the other areas mentioned, the percentages without opinion are much lower (from 1 to 7%).

As in the French survey examined above, it is observed that men have a more favourable attitude than women (55% as compared with 32%), while women tend more often not to respond to this question (35% as compared with 19% for men). We also note a more favourable attitude to nanotechnology among those young people who have gone furthest with their education.

It is difficult to give a direct interpretation of the classification of the different countries represented in the survey according to their level of appreciation of nanotechnology. No doubt the most favourable are those with the reputation for having a high general level of education (Denmark), but also some countries that are generally less well placed according to this criterion (Lithuania, France, Spain, Italy) (see Table 30.4).

In view of these two surveys, it is difficult to make an accurate assessment of the state of public opinion owing to the very high levels of ignorance about what nanotechnology really involves. On the other hand, when something is actually known about nanotechnology, it appears to be perceived

Table 30.4. Attitudes toward nanotechnology in different countries, classified in order of decreasing percentage of positive opinions

Country	More advantages	More risks	Same level	No opinion
Denmark	**58.0**	10.0	15.2	16.8
Lithuania	**57.1**	7.1	16.0	19.8
France	**50.8**	11.3	18.5	19.4
Spain	**48.9**	9.6	16.2	25.3
Italy	**48.8**	7.1	10.0	34.1
Finland	**48.7**	6.5	16.0	28.9
Slovenia	**46.0**	10.0	21.0	23.0
Portugal	**45.8**	4.8	10.2	39.2
Austria	**45.2**	12.1	20.7	22.0
Greece	**43.9**	10.0	15.9	30.1
Czech Republic	**43.7**	18.8	19.4	18.1
Estonia	**42.5**	8.8	21.3	27.5
Belgium	**42.2**	15.7	17.3	24.9
Latvia	**42.1**	7.1	16.4	34.3
United Kingdom	**42.0**	9.9	23.4	24.7
Germany	**41.5**	10.3	24.8	23.4
Poland	**41.2**	12.4	21.6	24.8
Slovakia	**40.5**	9.7	15.4	34.4
Eire	**39.9**	14.1	19.4	26.6
Luxembourg	**39.1**	8.7	21.7	30.4
Malta	**36.4**	4.5	18.2	40.9
Sweden	**35.8**	8.6	17.4	38.2
Romania	**33.9**	18.5	15.9	31.7
Netherlands	**33.2**	10.7	20.2	35.9
Cyprus	**31.9**	12.8	21.3	34.0
Hungary	**30.8**	13.4	20.6	35.2
Bulgaria	**29.9**	9.0	20.9	40.1
All	**43.5**	10.8	19.1	26.6

all the more positively as the level of education increases. This result can be interpreted as expressing an a priori positive and confident attitude toward science and these developments, in particular for those parts of the population that come closest to it through their level of education. In the majority of studies of the perception of science and technology, it remains true that the educated classes are most favourable toward science and its applications. But there are exceptions to this rule for technologies where the balance of risk and benefit is uncertain, and where it seems that the precautionary principle must apply, e.g., GMOs and nuclear energy.

30.2.3 European Experts and the Different Applications of Nanotechnology

What about those who are closest to the technologies themselves and thus have a better understanding, i.e., the experts who are actually in charge of developing nanotechnology?

As part of the preparation for a European call for research proposals in the European Research Area Network Scientific Knowledge for Environmental Protection (ERA-net SKEP), a European network of environmental research funders [10], an opinion poll [11] was carried out among 157 experts[5] from the areas involved in NBIC convergence, viz., nanotechnology, biotechnology, information and communication technologies, and cognitive sciences. The small size of the sample makes it difficult to extrapolate the results, or to compare several criteria such as sex, membership of the scientific community, academic background, nationality, etc. However, the main trends, and also the main differences (20%) remain significant. In addition, two months after the end of the questionnaire, a two-day conference brought together about 50 European experts, most of whom had participated. Reference will be made to these discussions [12] when qualitative elements are needed to interpret the quantitative results of the questionnaire.

A first series of questions deals with risks induced by different types of technological applications resulting from nanotechnology: free and incorporated nanoparticles, nanostructured materials, electronic nano-objects, and nanobiological objects. Different types of risk are proposed: health risks, environmental risks,[6] and, for electronic applications, risks relating to personal data protection. Other questions refer to the degree of irreversibility, i.e., the impossibility of counteracting or correcting any damage or risk that has been realised, and also the degree of social acceptability of the different technological applications. Each item is qualified from 'very low' to 'very high', with

[5] The European experts (748 professionals in one of the relevant areas: nanotechnology, biotechnology, ICT, and cognitive sciences) were identified through a review of the literature (SKEP, D6.2, 2008) and communication of the staff in charge of nanotechnology by the institutional partners of ERA-net SKEP. Those questioned were mainly scientists (66%, including 57.5% in the public sector and 8.5% in the private sector). Staff from the public administration of these technologies (ministries, state agencies, or European Commission) constitute the other large group making up the sample (23.5%). The civil society, including associations, unions, and elected representatives, together with the world of industry, are poorly represented. Note also that 82% of the sample had had an education in the hard sciences and 18% in the human sciences. Women represent 21% of the sample. A total of 92% of those questioned declared that they were expressing a personal opinion, and not the general opinion of their organisation or institution.

[6] This covers the risk of harmful elements resulting from nanotechnology being disseminated in the environment (particularly if they are persistent or prone to bioaccumulation) during production, use, or the final stage of the life cycle.

the possibility of not answering or answering 'unknown' if the interviewee considers that scientific knowledge is too limited to do so.

Quite surprisingly, risks relating to personal data protection are judged to be the highest, with 63%, i.e., almost two-thirds of the experts describing them as high or very high. So it is the uses and sociopolitical consequences of electronic nanotechnology that cause the most concern. One might predict, although the question has never been asked as such in another survey, that the result would be different for a representative sample of the general public, more sensitive to health and environmental issues. One may also expect the indirect risks of the new technologies to be more easily taken into account than the direct risks when it comes to experts who study, develop, and regulate these same technologies, as suggested by the PERPLEX study [7]. But it remains true that the concern over uses and awareness of sub-political effects [13] of the new technologies are to a large extent shared by experts.

Regarding the more 'conventional' risk of negative impacts on health, free nanoparticles are considered to be the most risky applications (43.5% responses of 'high' and 'very high'), well ahead of nanobiological objects (23%), nanostructured materials (22%), incorporated nanoparticles (21%), and electronic nano-objects (9%). In parallel, it should be stressed that the risks are also considered by many to be 'unknown' in the current state of scientific knowledge. This is the case at 34.5% for free nanoparticles, 23% for nanobiological objects, 21.5% for nanostructured objects, 21% for incorporated nanoparticles, and 18% for electronic nano-objects.

The precedent provided by asbestos, which is chemically neutral but toxic merely as a consequence of the filamentary shape of its components, certainly plays an important role in the concern that free nanoparticles may be toxic for humans. The fact that there have already been a certain number of studies [14] proving the toxicity of this type of particle could also explain this concern. However, that does not explain why it is the technological application for which the risks are also considered to be the least well known (34.5% of responses). This implies that the issues on which the experts are in possession of the most studies is not necessarily the one where certainty is best established.

In comparison with health risks, environmental risks are generally judged to be less high in the case of nanobiological objects (18 versus 23%) and especially free nanoparticles (32.5 versus 43.5%). In contrast, environmental risks are considered to be much higher for incorporated nanoparticles (31 versus 21%), nanomaterials (27.5 versus 21.5%), and electronic nanotechnologies (16.5 versus 9%). Astonishingly, there seems to be a partly inverse relation between the perception of a high health risk and the perception of an equally high environmental risk. Indeed, the score for environmental risks increases for applications which were judged the least risky from the health standpoint, to the extent of changing the order of the most risky technologies. During the SKEP conference, it became clearly apparent that this result could be explained by taking into account the end of the life cycle of the given applications. The presence and realisation of the risk then depend on regulatory, economic, and social systems set up to recover and process scrapped nano-objects. Once again, it is clearly the context in which nanotechnologies are actually deployed that is targeted by the experts in their estimation of risk. Note that the proportion of answers 'unknown' is particularly high regarding environmental risk (18.5–40%). The very small number of studies [14] devoted to ecotoxicology and life-cycle analyses of nanotechnology products could explain this result. Furthermore, these two areas of research are clear priorities for the experts answering the questionnaire and those present at the SKEP conference [12].

Another interesting result is that the irreversibility of a risk that has been realised is also judged to be very high (from 25.5 to 41%), systematically higher than the environmental risk itself, but with another high proportion answering 'unknown' (from 20 to 36%). Irreversibility is an important component of the dangerousness of potential damage. It is a criterion that structures individual risk perception [9]. Recall also that irreversibility is one of the conditions, along with radical uncertainty,[7] for the concept of 'option value' introduced by Claude Henry [15] to be relevant. Whenever there is a reasonable doubt about the presence of potentially irreversible risks, even in the absence of scientifically established certainties, Claude Henry shows that optimising gains and benefits means optimising utility over time and not at a particular time t. In other words, more is obtained if we leave open as many future ways of taking action as possible. It then becomes rational to introduce measures now to maintain future utility. As a consequence, the perception of irreversibility is a strong incentive to apply the precautionary principle.

Appreciation of the social acceptability of these same technologies is only partly correlated with the previously estimated level of risk. Hence, 44% of the experts considered it to be high to very high in the case of free nanoparticles and 40% in the case of electronic nano-objects, whereas the value is 60% for nanostructured materials and 61.5% for incorporated nanoparticles. Most of the experts were confident about the social acceptance of the technologies that they consider to be safest. There was much greater disagreement for free nanoparticles, which might reactivate recent recollections of asbestos in the public mind, but also for electronic nano-objects owing to the issue of personal data protection. However, the application considered to be the most likely subject of social opposition remains nanobiological objects: 54.5% of the experts felt that these objects would be difficult or very difficult to accept by society. Taking into account the 25.5% who did not decide one way or the other, that leaves only 25.5% who consider that nanobiological objects will not lead to social opposition.

This result is easily explained if we recall the precedent provided by GMOs. Indeed, the latter were hotly contested and resulted in a significant regulatory

[7] This is uncertainty that cannot be described probabilistically because the list of possible future states of the world is incomplete.

effort, which led to a moratorium at the European level. Nanobiological objects thus inherit from the perception of genetically modified organisms, just as free nanoparticles do from the case of asbestos. However, the SKEP conference [12] showed that relatively new, or at least more pressing ethical questions also come into play in the case of nanobiological objects. Life is no longer simply manipulated, it is also denatured by merging it with the non-living. The dividing line between the two becomes fuzzy, thereby throwing doubt on the boundaries between human and machine. This question arises in Bill Joy's disaster scenario [2], where intelligent and sensitive robots replace humans in a relatively near future, and also in Eric Drexler's [1], where the biological ability to self-replicate is associated with the manipulation of matter and results in the end of the world, thanks to the grey goo.

Relatively concerned about social acceptability, the experts consider that the main task of the public authorities, coming before research funding (22%) and regulatory effort (31%), is to organise actions involving the public (46.2%). Of these 46%, the vast majority (62%) advocate informing the public, while the others (38%) support public participation in research and development for these technologies. In the first case, information is provided by experts and delivered to the public. In the second, the public is also a source of information. The public also plays an active role in the collective decision-making process bearing on the development and regulation of nanotechnology.

30.3 What Should Be Done with Perceptions?

In the final reckoning, the general public has little knowledge of nanotechnologies and turns out to be fairly confident and favourable toward them. For their part, the experts consider nanotechnologies to have significantly unknown, risky, and potentially irreversible sanitary, environmental, or sociopolitical consequences, but whose occurrence will largely depend on the regulations and uses made of them. At the same time, they are also concerned about social acceptability and consider that the first priority of the public authorities is to organise actions involving the public, which leads us to our last question: how should we deal collectively with individual perceptions of technological risks, once these perceptions exist and are identified?

The main issue is whether we should communicate with the public or associate the public in some form of technical democracy. These two approaches relate to two paradigms: risk and radical uncertainty. In the risk paradigm, knowledge is sufficiently stabilised to be able to make probability models of the risks and introduce preventive measures. Of course, individual perceptions are not so much based on what is probable as on what might be particularly dangerous. However, it is often possible to bring perceived risks closer to scientifically established risks by making an effort of communication, and this all the more effectively if the individuals in question have confidence in the communicating institutions. In the radical uncertainty paradigm, it is impossible to know all future states of the world. It is the whole business of making probabilistic models that is called into question. In this case, precaution takes precedence over prevention.

In a situation of radical uncertainty and precaution, it is still important to inform the public. However, the issue is not so much how to reduce the risk, but rather to justify whether the risk should be taken or not. The evaluation of risk is then based on a prior assessment of the utility of the technologies. The requirements of democracy soon imply that the public must be involved in the decision-making process when it comes to assessing the benefits of major technological choices, and this not only through political representation, but also through a more direct engagement, e.g., citizens' conferences and public debates. This participative technical democracy is particularly recommended to increase trust between the population and its institutions.

In July 2004, the British Department of Trade and Industry (DTI) set up three committees, all involved in consultative actions with stakeholders (industry, research, civil society) in the form of forums:

- The Nanotechnology Issues Dialogue Group (NIDG) which coordinates activities resulting from the Royal Society and the Royal Academy of Engineering report of July 2004, entitled *Nanoscience and nanotechnologies: Opportunities and uncertainties*.
- The Nanotechnology Research Coordination Group (NRCG) which develops a research programme on risk.
- The Nanotechnology Engagement Group (NEG) which supports various activities relating to public participation, including *Small Talk*, *Nanodialogues*, and *Nanojury* (see below).

Small Talk. Twenty events between September 2004 and November 2006 combining popular science conferences and panel debates. Organised by Think-Lab, in partnership with the British Association for the Advancement of Science, Ecsite-UK, the Royal Institution, and the Cheltenham Science Festival, these events brought together some 1 200 participants. At the end, the participants did not produce formal recommendations, but declarations to scientists and the research minister.

Nanojury. A citizens' jury was organised in June and July of 2005 at the initiative of the Cambridge University Nanoscience Center, Greenpeace U.K., the Guardian, and the Politics, Ethics and Life Center of the University of Newcastle. Twenty British citizens chosen randomly met for a period of 5 weeks to discuss nanotechnology. In 10 working sessions, these citizens were informed about nanotechnologies by a group of experts. During the last sessions, they drew up recommendations which were subsequently presented publicly in London in September 2005 [16].

Nanodialogues. Organised by Demos and the University of Lancaster between May 2005 and May 2006, the idea here was to experiment with the public's ability to take active part in decision-making processes relating to the development of emerging technologies in four different contexts:

- *A People's Inquiry on Nanotechnology and the Environment* brought together three focus groups comprising stakeholders' representatives and thirteen citizens to consider the use of nanoparticles to decontaminate soils, in partnership with the British Environment Agency (EA).

- *Engaging Research Councils* explored public participation in research management through a 3-day workshop bringing together scientists, research administrative managers, and citizens, in partnership with the Engineering and Physical Science Research Council and the Biotechnology and Biological Sciences Research Council.
- *Nanotechnology and Development* was concerned with the contribution of nanotechnology to world development, particularly regarding universal access to drinking water by 2015. This initiative was run in partnership with Practical Action in Zimbabwe. It brought together politicians, administrators, and representatives of the two communities for 3 days.
- *Corporate Up-Stream Engagement* (in partnership with Unilever) set up four focus groups to consider the use of nanotechnology in the production of three types of consumer goods: hair products, dental health products, and food products. Various scenarios were discussed.

In France, there have only been two debates, the first initiated by the regional council of the Ile-de-France region of France during the winter of 2007, and the second in the context of conferences at the *Cité des Sciences de la Villette* at the request of the research ministry in the spring of 2007. In 2008, however, Group 3 of the Grenelle Environment Forum, entitled *Instaurer un environnement respectueux de la santé*, recommended that a public debate be organised on the subject of nanotechnologies. The submission of the *Commission nationale du débat public* was ratified by the outline law known as *Grenelle 1*.

References

1. E. Drexler: *Engines of Creation: The Coming Era of Nanotechnology.* (Anchor Books, New York 1986)
2. B. Joy: Why the future doesn't need us. Our most powerful 21st-century technologies – robotics, genetic engineering, and nanotech – are threatening to make humans an endangered species. Wired Magazine, San Francisco, California (April 2000)
3. Anonymous: *Historia von D. Johann Fausten.* (Johann Spies, Frankfurt 1587)
4. hum.iit.edu/~csep/NanoEthicsBank/intro/intro.html
5. D. Boy: *Pourquoi avons-nous peur de la technologie?* (Presses de Sciences Po, Paris 2007)
6. P. Peretti-Watel: *Sociologie du risque.* (Armand Colin, Paris 2000)
7. IRSN, PERPLEX (perception des risques par le public et les experts): Observatoire de l'opinion sur les risques et la sécurité (2007)
8. D. Boy: Les attitudes du public à l'égard de la science. Sofres, L'Etat de l'Opinion (2002), pp. 167–182
9. V.T. Covello, F.W. Allen: Social and behavioral research on risk: Uses in risk management decision-making. In: *NATO ASI Series G, Environmental Impact Assessment, Technology Assessment and Risk Analysis*, Vol. 4. (Springer, Berlin Heidelberg New York 1985), pp. 1–14

10. www.skep-era.net/
11. SKEP ERA-net D6.3: Summary of perceptions and science needs of policy makers, operational staff, scientists, experts and stakeholders. ADEME, MEDAD (2008)
12. SKEP ERA-net D6.4: Nanotechnology, biotechnology, information technology and cognitive sciences: Environmental opportunities and risks of converging technologies. ADEME, MEEDDAT (2008)
13. U. Beck: *La société du risque, sur la voie d'une autre modernité.* (Paris, Aubier 2001)
14. SKEP ERA-net D6.2: Converging technologies and environmental regulations. Literature review. ADEME, MEDD (2008)
15. C. Henry: An existence theorem for a class of differential equations with multivalued right-hand side. Journal of Mathematical Analysis and Applications **41**, 179–186 (1973)
16. www.nanojury.org.uk

31

Robotics, Ethics, and Nanotechnology

Jean-Gabriel Ganascia

31.1 Preliminaries

It may seem out of character to find a chapter on robotics in a book about nanotechnology, and even more so a chapter on the application of ethics to robots. Indeed, as we shall see, the questions look quite different in these two fields, i.e., in robotics and nanoscience. In short, in the case of robots, we are dealing with artificial beings endowed with higher cognitive faculties, such as language, reasoning, action, and perception, whereas in the case of nano-objects, we are talking about invisible macromolecules which act, move, and duplicate unseen to us. In one case, we find ourselves confronted by a possibly evil double of ourselves, and in the other, a creeping and intangible nebula assails us from all sides. In one case, we are faced with an alter ego which, although unknown, is clearly perceptible, while in the other, an unspeakable ooze, the notorious grey goo, whose properties are both mysterious and sinister, enters and immerses us. This leads to a shift in the ethical problem situation: the notion of responsibility can no longer be worded in the same terms because, despite its otherness, the robot can always be located somewhere, while in the case of nanotechnologies, myriad nanometric objects permeate everywhere, disseminating uncontrollably.

On the other hand, it is by no means a pointless exercise to discuss roboethics – that is, as we shall see later, the ethics of robots – in this book, because this will help, by contrast and analogy, to understand what nanoethics – i.e., the ethics of nanotechnology – actually is, or might be. But it should be stressed at the outset that ethics, whether of robots or of nanotechnology, cannot be reduced to a mere list of behavioural rules. Here, ethics differs from deontology or what some call morals, that is to say, it differs from the law. But that does not make ethics any the less a practical matter, for it bears upon our acts and our motives.

In the case which concerns us, viz., roboethics, this means that we shall be interested in what underpins the moral constraints we impose upon ourselves when designing and building robots, and the conceptual devices that were

P. Houdy et al. (eds.), *Nanoethics and Nanotoxicology*,
DOI 10.1007/978-3-642-20177-6_31,

deployed to lay those foundations. We shall thus survey the different aspects of robot ethics, and in conclusion, we shall examine the relevance of these aspects in the context of nanotechnology. The chapter is organised accordingly: after a brief prehistory, then history of robot ethics, we shall discuss current affairs in roboethics. Finally, we shall examine the lessons that nanoethics might draw from roboethics, and therewith end the chapter.

31.2 Prehistory and History of Robot Ethics

Robot ethics is ancient history. It even pre-exists robots themselves, and not only their material reality, but also their name. Recall that the word 'robot' comes from the Czech *robota*, which means 'hard work, chores'. It was invented by a Czech writer by the name of Karel Čapek, in a play entitled *RUR – Rossum's Universal Robots* [1]. It refers to artificial workers ready to do whatever, and however much, is asked of them. They suffer from our indifference to them. According to Karel Čapek, these beings that we have manufactured to serve us deserve our attention from the moment they become conscious. With the help of a sensitive and intelligent young woman, the inventor's own daughter, these humanoid robots revolt against a social order they consider unfair, and obtain human recognition.

This play, written in 1920, raised a great deal of interest, picking up on burning social issues of the day. When he came to power, Hitler was worried about it. It seems that Karel Čapek was even a favourite to win the Nobel Prize for Literature and that it was only through fear of upsetting the dictator that the Swedish Academy felt obliged to withhold this distinction from him. Very soon, there was general concern over human responsibility toward automatons.

Note that, in 1921, when Karel Čapek's play was published, robots lived an essentially phantasmagoric existence. Of course, many automatons were built in the eighteenth and nineteenth centuries, but these mechanical replicas of ourselves remained clumsy and awkward beings. On the other hand, humans had long been trying to build artificial workers. Hence, in Book XVIII of *The Iliad* [2], we find a strange passage in which Hephaestus, the god of fire, and in particular the blacksmith's fire, is served by robots:

> On this the mighty monster hobbled off from his anvil, his thin legs plying lustily under him. He set the bellows away from the fire, and gathered his tools into a silver chest. Then he took a sponge and washed his face and hands, his shaggy chest and brawny neck; he donned his shirt, grasped his strong staff, and limped towards the door. There were golden handmaids also who worked for him, and were like real young women, with sense and reason, voice also and strength, and all the learning of the immortals; these busied themselves as the king bade them.

Closer to our own time, the Jewish cabalistic tradition reports the existence, toward the end of the sixteenth century, of a clay statue called the Golem, which was made by Rabbi Loew, better known as the Maharal of Prague [3]. Like contemporary computers, this machine came to life when a message was passed behind its teeth. Usually, it busied itself with everyday household tasks, like an eager and diligent servant.

This extraordinary statue inspired many legends. According to one of these, one Saturday, day of prayer, Rabbi Loew had forgotten to remove the message from behind the Golem's teeth, whereupon it began to get agitated, shouting and terrifying all the neighbours, while the master was fulfilling his holy duties down at the synagogue. When he got back, Rabbi Loew destroyed his creation for fear that it might resume its troublesome initiatives. According to another story, the word EMETH appeared on the Golem's forehead. In Hebrew, this means 'truth'. Now it is said that, one day, the Golem picked up a knife in order to remove the first letter of this word. This would have left 'METH', which means death in Hebrew.

All these mythologies leave an aura of ambivalence about the Golem which foretells the ambivalence of contemporary technical achievement. On the one hand, Rabbi Loew, who had the knowhow to create such a perfect object, was widely praised, even worshipped, to the extent that the chair on which he used to sit is still on display in the old synagogue in Prague. On the other hand, such a thing as the Golem sometimes runs the risk of escaping its masters and creators, who must of course prevent such a thing from ever happening. Our general responsibility with regard to human technical creations, and in particular machines, is so clearly stated here that Norbert Wiener refers explicitly to it in *God and Golem* [4], a work entirely devoted to the ethical issues of cybernetics and the first computers.

To cut a long story short, the threat that robots raise for humanity has always been present. In 1938, tired of reading so many poorly conceived stories of invasive and aggressive robots, the Russian-born biologist Isaac Asimov put together a series of short stories and novels [5], organised around three immutable laws of robotics, to which he adjoined the necessary add-ons as required for the development of his undertaking. These laws underlying the creation of androids are intended to prevent them from ever harming human beings:

- A robot may not injure a human being or, through inaction, allow a human being to come to harm.
- A robot must obey any orders given to it by human beings, except where such orders would conflict with the First Law.
- A robot must protect its own existence as long as such protection does not conflict with the First or Second Law.

To sum up, this hasty foray into the prehistory and the history of robotics has shown us the long existence of the robot, or more exactly, its long existence in the human mind, which goes back at least as far as Ancient Greece. And jointly with the phantasmic presence of Pygmalion, Pinocchio, and all kinds of animated statue, ethical preoccupations were also born from the earliest times: how can we ensure that whatever controls its own animation and moves by its own means does not become autonomous? What limits should we impose to ensure that our creations do not swallow us up? Sometimes one must have the courage to destroy what one has made. This is the lesson we learn from the legend of the Golem.

In every case, the robot acquires an analogous status to our own. It looks like our double, but we are to the robot in the same situation as God – or Nature in an atheist perspective – is to us. Indeed, robots are our creatures just as we are the creatures of God or Nature. This makes the whole affair all the more daring and risky. It also explains why we think of imposing on these artificial beings the same restrictions as we impose upon ourselves before God. Note finally that, symmetrically with this requirement of the robot's deference before humans, we sometimes speak of human responsibility with respect to robots, analogous to that of God with respect to humans. On this point, the enigma of *RUR – Rossum's Universal Robots* [1], the play by Karel Čapek, is eloquent. And other works of science fiction also adopt this line of thinking. We might also wonder whether these perspectives are relevant to nanotechnology. However, that is not our subject. Here we are hardly concerned with imaginary representations of robots, but must focus rather on the contemporary reality of robotics and on the many questions it raises.

31.3 Roboethics

By roboethics we understand here anything that touches upon the ethics of robots or the ethics of humans with respect to robots. The term was invented in 2002 by Veruggio, and officialised by the first roboethics symposium, held in San Remo in January 2004 [6]. It was deliberately coined to resemble the word 'bioethics' and now seems to be used by scientists in official publications, by universities, by professional associations, and so on. The field of application of roboethics has grown considerably with the increase in the number of robots and with the ever greater role they play in contemporary economics. In this respect, it is worth noting that the active robot population in manufacturing industries is now something like a million 'individuals', and to this one must add housekeeping robots, companion robots, space robots, medical robots, drones, robot soldiers, etc., not to mention an uncountable number of virtual robots zipping back and forth on the Web. In short, we are living today in a new world where humans coinhabit more and more often with robots. This raises many questions. How should we assume this new human condition, living in symbiosis with robots? Is there no risk of becoming, if not the victims, at least the slaves of the machines which we originally designed to serve us?

31.3.1 A Roadmap for Roboethics

These are perfectly legitimate questions. They have been tackled in several different ways. We shall not give an exhaustive description of all the deliberations that have been brought to bear. Instead, we shall simply discuss a report drawn up by the European Robotics Research Network [6] to deal with ethical questions. This report resulted from a workshop held from 27 February to 3 March 2006. Its very existence attests to the need felt by scientists to tackle these ethical issues. The same can be said in the context of military robotics in the US, where analogous reflections have been conducted.

Having mentioned Asimov's laws and recalled what is meant by ethics, the report shows that, even if robots can be reduced to assemblages of deterministic mechanisms, that does not imply that they have no ethical dimensions. More precisely, insofar as they are able to act autonomously, they are agents. Of course, they do not have autonomy of the will, in the ethical sense, and do not possess consciousness. In this respect, they are merely machines. But they appear so complicated that we would not be able, during the time of action, to anticipate all their determinations. From this point of view, we must treat them as being endowed with autonomy. The understanding that we have of them no longer passes solely by the relevant physics, even though they are constructed using material constituents all of whose properties we are able to control perfectly. There is a sphere of intelligibility of robots which helps us to apprehend them by thinking of them as intentional systems, in other words, as agents with goals, desires, emotions, etc. It is by reference to this sphere that we can perceive of them as moral agents. More precisely, since robots are viewed as agents, we attribute their actions to them. But the consequences of their actions can greatly influence the lives of human beings in society. They sometimes improve our lives, but there is a risk that they may be detrimental to them. We thus make the distinction between those that act for the good, i.e., for human happiness, and those that cause harm. And here we may attribute morality to them. It is in this sense that robots are qualified as moral agents.

To establish the moral value of robots, the report reviews all contemporary applications of robotics. It constructs a taxonomy and for each category it indicates the risks, before making recommendations. This classification of contemporary robots arranges them in six families which are themselves divided up into kinds and species. Here is a synopsis of the classification:

- *Humanoids.* These are characterised by their resemblance to humans. They are subdivided into artifical minds and artificial bodies.
- *Production Systems.* These mainly concern industrial manufacturing and remotely controlled work in hostile environments, such as the reactors in nuclear power stations and flexible workshops. They include many kinds of robot, depending on the application and depending on whether the robots are autonomous or remotely controlled, but these technical distinctions are irrelevant from our present point of view, which concerns the ethics of robots.

- *Adaptive Domestic Robots and Intelligent Houses.* This class of robots covers indoor robots and ubiquitous robots, also called onboard or mobile robotics.
- *Outdoor Robotics.* This covers land robots, marine robots, airborne robots, and space robots.
- *Medical Robotics.* This includes surgical robotics, biorobotics, intelligent medical assistants, including aid and diagnostic systems and monitoring systems, and robotics for biocomputing, e.g., protein and genome sequencing robots.
- *Military Robotics and Intelligent Weapons.* This includes autonomous vehicles, on land, in the sea, or in the air, intelligent bombs, automatic surveillance, etc., robot soldiers and systems for improving the motor and/or perceptive performance of humans, especially exoskeletons which considerably increase human stamina capabilities.

For each of these categories of material robots, lists were drawn up to produce an inventory of the perils awaiting us and the elementary precautions required to save us from them. For example, it is easy to imagine a domestic robot taking unfortunate initiatives or doing silly things, such as burning a shirt it is ironing or swallowing the wire from an electric lamp when doing the vacuuming. We are all afraid of seeing a machine which, at the patient's bedside, decides without proper medical consultation to greatly increase the dose of a given medicine. Our knees shake at the thought of a robot soldier arriving to carry out its 'mission', while systematically eliminating all those that might get in its way.

But these are precisely the kind of quite ordinary situations we will soon have to face up to. Laboratory studies and military projects in Europe and the USA should soon convince us of this. In this respect, the above-mentioned report [6] is eloquent, being written by robotics researchers who are fully aware of the state of the art, the work that is currently underway, and the projects in the pipeline. Worse still, it happens that, in quite unexpected situations, robots make wholly disconcerting, even shocking decisions. When the risks are known, we should be able to protect ourselves from their consequences. But how can we guard against a hazard that we do not know? And if a robot were guilty of reprehensible acts, who should be treated as responsible? Should we incriminate the robot's designer, its maker, or its owner?

Some say that the situation is changing now that machines are becoming more and more autonomous. For these commentators, the law must change too, to define the status of complex material systems whose behaviour may now escape both their designer's and their owner's control. Others think that legal fictions have long existed, and that with suitable adaptation they will allow us to handle these contemporary realities. In the opinion of these commentators, we could for example attribute to intelligent robots an analogous status to that of slaves in Ancient Rome. Indeed, the slave could be punished, but his acts engaged his owner financially, the latter being held legally responsible.

Whatever is done, a new approach must be put together, involving clear principles, laws, and a jurisprudence. It would not be possible to envisage all the applications of robotics, and hence all the risks we run. So the intention is to establish new rules as soon as the new threats are identified. This is the subject of the roboethics roadmap put forward by the above-mentioned roboethics working group: to set up protocols for establishing these rules.

31.3.2 Ethics of Virtual Robots

Apart from material robotics whose inventory we have just outlined, there is also a field of virtual robotics. As the term implies, it is not deployed in the external world, but solely in universes which, like *Second Life* [7], are qualified as virtual because they consist merely of digital data flows.

As an illustration, US researchers have designed intelligent agents which they call elves, because they follow you around everywhere from dawn to dusk, like benevolent spirits. The elves read your emails, manage your timetable, and record your phone calls. They send you well-meaning SMSs to draw your attention to various issues. They assist their masters as best they can. For example, they make appointments like diligent secretaries, they book plane tickets, they do the shopping at the supermarket, they deal with administrative matters, they reserve seats at the theatre for family outings in the evening, and so on and so forth.

However, from time to time, these agents make mistakes despite themselves and create problems for their charges. There are tales of these wonderful elves carrying out quite reprehensible tricks on their owners by simple inadvertence [8]. For example, a university bod had a paper to finish by the end of the day, but the list of people who wanted to meet him kept growing longer and longer in his diary that particular day, out of all proportion, simply because he was unable to explain to his agent that, although he was in his office, he wished on no account to be disturbed. Another was woken at 3 a.m. by his elf, who wished to inform him that the plane he would take that same day at 11 a.m. would be delayed by one hour. Well, these are minor discomforts and there is nothing there to offend our ethics. But that may not always be the case.

Let us return to the example of the elves. They offer their charge the possibility of setting up an order of priority among their appointments, in such a way that it is always possible to postpone some of them. For example, if you get a phone call from the director who needs to see you urgently, whereas you had planned to meet your secretary, the elf takes it upon itself to postpone the latter engagement. In this perspective, how would you explain to one of your students that she does not have absolute priority when she is just in the process of finishing her doctoral thesis? Perhaps one should not allow elves to reveal their orders of priority? But if this is the case, it means that we must make dissembling robots. Is that ethical? More generally, elves know a considerable amount about us. Should they communicate it when asked? If not, how do we justify their holding it back?

In addition to these general questions regarding the discretion of virtual robots [9–11], there are ethical problems with the Internet itself. Should we authorise robots to systematically exploit all the data they come across, disclose it to as many people as possible, and disseminate it unreservedly on the Web? This raises questions relating to the protection of privacy. But it also bears upon what is allowed or not allowed on the Web.

To deal with the first point, let us stress that more and more data about private individuals are stored on the Web. Today all sorts of data relating to our movements, our health, our purchases, and our taxes are going through the Web. Anybody succeeding in bringing them together would have a considerable power over us. In this respect, recall that during the election campaign for the United States presidency, Barak Obama built up a data base which collected information about most American voters. By examining this data base in detail, he targeted those who were likely to swing, and for each such voter, he chose the militants best placed to convince them. Should it be forbidden to create this kind of data base? Does the protection of privacy necessarily mean storing personal data in unbreakable safes? Today many people consider that we are the owners of all information about ourselves, in other words that we have an inalienable moral right over photos, images, or recordings that refer to us. As a consequence, we should be able to control the dissemination of information relating to our person and preside over what others have the right to know about us, giving or withholding our explicit consent to any request that concerns us specifically. However, such theoretical principles come up against the problems of everyday usage and obvious material stumbling blocks.

The second important point concerning virtual robotics relates to the proliferation of robots on the Web. We have all been the victims of computer viruses, Trojan horses, or other electronic threats. This evil-doing bestiary of injurious robots must of course be wiped out. It is no longer the doing of talented or facetious youngsters. Computer delinquency is now rife. Today, organised groups control this virtual zoo of worms, viruses, and other deleterious software. These groups blackmail the major industrial companies. If they do not pay the ransom, they suffer massive attacks that temporarily disable their computer system. This kind of racketeering or warlike behaviour must of course be condemned.

Another example are the search engines which continually aspire to assimilate absolutely the whole content of everything available on the Web using virtual robots, and then index these contents. Naturally, no one is at issue with search engines, which have become quite indispensable today. However, we need to avoid such robots deploying in such a massive way that they actually saturate the Web. Moreover, insofar as possible, we would like to protect certain private data on the Web. To this end, there is an ethical code for robots which has been perfectly formalised [12] and which has indeed been implemented throughout the Internet. A file called `robot.txt` is associated with each Website that contains confidential information. This is where you may explicitly declare that you wish to exclude your site, or part of your site, from the field of action of robots. Associated with these files is an exchange protocol which robots exploring the Web are supposed to respect. If they do not, they risk being qualified as harmful, then pursued, or even excluded from the system.

31.3.3 Responsibility Toward Robots

To conclude this section on roboethics, let us also mention an important dimension from a symbolic point of view, although it does not bear upon the ethics of robots, i.e., on the rules that humans must respect when they build or use robots, but rather upon the ethics of humans with respect to robots considered as autonomous beings. Should we be allowed to treat robots just as we like, on the pretext that we made them and that they are our creatures? The play *RUR – Rossum's Universal Robots* [1] written by Karel Čapek illustrates this point very clearly. And likewise the film *AI – Artificial Intelligence* made by Steven Spielberg in 2001 in homage to Stanley Kubrick. The question may seem absurd in view of the feeble performance of today's robots. But it is a matter of principle: if we manage to build automatons able to make decisions, possessing an artificial conscience, and able to suffer, would we then not have obligations toward them?

A second, more acute and more pressing question concerns virtual robotics and the respect we owe to our intelligent agents. The contemporary philosopher Luciano Floridi claims the existence of a new stratum of intelligibility, the infosphere [13, 14], which is defined by analogy with the biosphere, the environment of living beings, as the environment of all informational entities. These include search engines exploring the Web, automatons populating virtual worlds, and avatars through which we may interact in video games or with digital universes.

According to Luciano Floridi, since the theory of information governs the infosphere, the fundamental ethical criterion of the infosphere should be based on the concepts arising from this theory. The philosopher thus founded what he calls information ethics, basing it on the notion of information entropy introduced by the mathematician Claude Shannon (1916–2001) [15] at the end of the 1940s as the basic feature of his theory of information. Just as entropy in the physical sense measures the disorder of a system, in other words the absence of knowledge we have about it, the entropy of information measures the disorder of an information system. And in this context, the interest of a message is measured by the extent to which it tends to reduce the information entropy of the whole.

In physics, the entropy of a closed system always increases in time, which means that such a system always tends to become more disorganised. It is this unavoidable increase in entropy on our planet that leads some to say, incorrectly, that we lack energy, while the total energy remains the same. What we refer to as energy consumption should strictly speaking be called entropy increase. And it is this growing entropy that is detrimental. Likewise for the infosphere, that is, the environment of the informational entities within which we are now condemned to live. For this, too, undergoes an increase in entropy, which means that we have less and less control over the information that diffuses through it. Once again, it is not that there is a lack of information, far from it. The problem is the degradation of its quality.

An ethical attitude in the infosphere is therefore measured, according to Luciano Floridi, by the reduction it can bring to the information entropy. In the name of this ethics, we thus reprove the spreading of false rumours and the scrambling of news by tidal waves of meaningless messages. Likewise, any destruction of an informational entity that would lead to a permanent loss of information would be clearly condemned. In short, information ethics as formulated by Luciano Floridi would have us respect all informational entities as such, whether they be simple avatars of our fellow humans, representatives of their interests in the infosphere, or artificial agents carrying information.

31.4 Extrapolation to Nanoscience

In order to extend the discussion from the subject of this chapter, which is roboethics, to the field of nanoscience, we shall not be describing or alerting against the risks involved in nanoscience, and nor shall we try to put forward solutions there. We shall simply tackle the question reflexively, by examining what guided robotics specialists in their ethical investigations and seeing how this might lead to a similar investigation in the field of nanoscience. For this purpose, we shall tackle three crucial issues. One bears upon the reality of the risks, the second discusses the possibility of a nanoethics roadmap comparable to its counterpart for roboethics, and the third investigates the possible distinguishing features of nanoethics.

31.4.1 Reality and Virtuality

As we have seen, much of the discussion about roboethics, as stimulated for example by Rabbi Loew at the end of the Middle Ages [3], or more recently by Isaac Asimov [5] and Karel Čapek [1] at the beginning of the twentieth century, took place before robots even achieved any real existence. These imaginary constructions underpin our current feelings about robotics. They are the starting point for our reflections on the ethics of robots. For example, the protocols laid down today for virtual robots [12] are inspired by the laws of roboethics invented by Asimov [5]. The world of imagination thus plays an important role in our consideration of ethical questions relating to the development of science and technology.

As a consequence, we should pay careful attention to the projections we make on nanoscience. These announce risks due to anarchic proliferation of self-reproducing nanoscale objects. There is no doubt that such risks should be examined with due care. Indeed, the comparison serves as an intellectual stimulus and in this respect is therefore a good thing. However, such announcements also require a circumspect and critical attitude, because if we are not careful, there is a risk of paralysing any further developments for no valid reason. Recall, for example, the scaremongering predictions of Bill Joy, founder of Sun Microsystems, in 2000 [16], or the fiery declarations of so-called demiurges, like Hugo de Garis who got his name into the main Paris newspapers by announcing the inevitable creation of artificial beings far superior to humans [17], or the prophecies of Hans Moravec [18] who claimed soon to be able to couple the human brain to computers to transform us all into cyborgs, that is, cybernetic organisms.

Faced with this kind of proclamation, is there not a danger of concealing the reality behind unfounded fears, or being blinded by such? In other words, referring back to Floridi's principles of information ethics [13, 14] as discussed above, are we not confronted here by willful and quite undesirable amplification of information entropy? When we consider such ill-considered claims, it seems urgent to make a meticulous and complete inventory of the risks, like the one produced by European robotics specialists. In this area as in any other, an ethical attitude consists in discussing and elucidating the issues, rather than stirring up unfounded fears.

31.4.2 Do We Need a Roadmap for Nanoethics?

However, even if we manage to build a complete inventory of the risks involved in the development of nanoscience, that does not mean that we shall be able to predict the unpredictable. What actually happens sometimes escapes prediction. We need to prepare for that and attempt to react. The role of ethics is not to end all discussion by imposing some incontrovertible rule. Quite the opposite. Ethics should open our minds to what may come upon us.

The idea of a roadmap is precisely to satisfy this requirement. It is not a catalogue, nor a digest, nor a compendium, nor a treaty. It does not purport to assemble all knowledge, all laws, or all rules. It sets out waymarkers, it provides indications, it suggests principles which, at the opportune moment, will help us to cope with a situation. It should be useful to industry and public authority alike. It will allow us to send out warning signals and make decisions.

To illustrate its role, here is an anecdote. A few years ago, I was invited to assess projects for the European Commission. At the time, I was certainly less aware than I am today about the ethical questions relating to the development of new technologies. But I was nevertheless extremely disturbed by a project for an 'intelligent house' where, for the safety of its occupants, every movement of every individual was continuously recorded and analysed by computers. We were asked to tick a box if we felt that any ethical problems might be raised by the development of such a project. But when I exposed my concern, explaining that I wished to tick the box and ask the project's architects for further explanations, I found myself sharply scolded by the EC representative. In his opinion, much worse things already existed in the United Kingdom, where pedestrians were being filmed whenever they stepped into the street. So there was no question of letting trifles stand in the way of strategic European industrial developments. At the time, I gave in, but today I regret that decision. I now believe that the existence of a roadmap, agreed by representatives of the scientific community who had carefully considered the ethical consequences of the applications of their work, would have been of great assistance.

31.4.3 Collision and Contamination Between Spheres of Intelligibility

The roboethics roadmap is certainly useful, and the same may well be true for nanoethics. On the other hand, the roadmap may not be sufficient for roboethics. Indeed, it has proved necessary to introduce new concepts in order to tackle the issues raised by virtual robotics. This is what justified the introduction of the notions of infosphere and information ethics by Luciano Floridi. Will the same be true for nanoethics? It is with this open question that we would like to continue the parallel between roboethics and nanoethics.

To get a good grasp of Floridi's ideas, I believe it important to read him from a Spinozan perspective, which happens to be his own. From this standpoint, information corresponds to what Spinoza calls, in *The Ethics* [19], a mode, that is, a particular way of being of substance, or in more contemporary terms, a sphere of intelligibility of reality. So just as for a human beings the extension mode is the body and the mode of thought is the mind, so the mode of information is the informational entity. In short, a given thing can be simultaneously viewed as being in different modes. As we have just seen, virtual robotics can be viewed both in the extension mode, as belonging to the sphere of intelligibility of physical phenomena, and in the informational mode, as belonging to its own sphere of intelligibility, which can be apprehended via concepts from the theory of information.

Let us now reconsider nanoethics, and try to transpose the conclusions we have just drawn. Then we may ask ourselves what sphere of intelligibility nanoscience and its progeny belong to. Clearly, they take on a meaning when viewed as physical (or chemical) matter, when viewed from the standpoint of the environment of living beings, i.e., with reference to the biosphere, and when viewed from the standpoint of information, i.e., with rerference to the infosphere. Among these different points of view, is there one more fertile than the others for laying down the principles of nanoethics, or is there a sphere of intelligibility intrinsic to nanoethics? Or should we appeal to several spheres at once? And if so, is there not a risk of these spheres of intelligibility colliding and contaminating one another, and thereby generating confusion? In concrete terms, that would mean that, in the field of nanoscience, any attempt to consider a phenomenon as belonging to a single order of intelligibility, for example, that of physics, biology, or information science, would be perfectly ineffectual, because the objects, by their essence, would systematically elude it. Does the claimed convergence between nanoscience, biology, computing, and cognitive science not betray the disquiet that is felt, rightly or wrongly, before the imminence of this contamination between spheres of intelligibility? If the risks of contagion should be realised, this would mean that nanoethics is in fact unique and in this respect could not benefit from experience acquired in roboethics. If not, for each of the hazards due to development of nanoscience, it suffices to identify the sphere it belongs to. In any case, before talking about nanoethics, the first thing is to clarify which sphere or spheres of intelligibility the nanosciences belong to and then establish the boundaries of each such sphere.

References

1. K. Čapek (1921): *RUR – Rossum's Universal Robots*. Play translated from Czech and presented by Jan Rubeš. (Editions de L'Aube, Regards croisés 1997)
2. Homer: *The Iliad*. Translation by Samuel Butler. (Dover Thrift Editions, Dover, New York 1999)
3. M. Idel, M. Atlan, C. Aslanof: *Le Golem*. (Editions le Cerf, Paris 1992)
4. N. Wiener: *God & Golem Inc.: A Comment on Certain Points where Cybernetics Impinges on Religion*. (MIT Press 1966)
5. I. Asimov: *I, Robot*. (Spectra, New York 2004)
6. G. Veruggio: EURON Roboethics Roadmaps, EURON Roboethics Workshop, 27 February–3 March 2006
7. A 3D virtual world where users can socialise: secondlife.com
8. C. Knoblock, J. Ambite, M. Carman, M. Michelson, P. Szekely, R. Tuchinda: Beyond the elves: Making intelligent agents intelligent. AI Magazine **29**, 33–39 (2008)
9. S. Bringsjord, K. Arkoudas, P. Bello: Toward a general logicist methodology for engineering ethically correct robots. IEEE Intelligent Systems **21**, 38–44 (2006)
10. J.G. Ganascia: Using non-monotonic logics to model machine ethics. In: Computer Ethics and Philosophical Enquiry (CEPE) 2007 Conference Proceedings, July 2007, San Diego, California, USA (2007)
11. T. Powers: Deontological machine ethics. American Association of Artificial Intelligence, in Fall Symposium 2005 Proceedings, Washington, D.C. (2005)
12. Encyclopedia of robots: www.annuaire-info.com/robots.html
13. L. Floridi, J. Sanders: On the morality of artificial agents. Minds and Machines **14**, 349–379 (2004)
14. L. Floridi: Information ethics: From case-based analyses to theoretical foundations. In: ETHICOMP98 Proceedings, republished in ETHICOMP Journal (1998), www.ccsr.cse.dmu.ac.uk/journal
15. C. Shannon: A mathematical theory of communication. Bell System Technical Journal **27**, 379–423, 623–656 (1948)
16. B. Joy: Why the future doesn't need us. Wired Magazine, April 2000, www.wired.com/wired/archive/8.04/joy.html
17. H. de Garis: 2000 Débats pour le siècle à venir, Hugo de Garis, chercheur en intelligence artificielle. *Le Monde*, *Horizon-débat* pages (9 November 1999)

18. H. Moravec: When will hardware match the human brain? Journal of Evolution and Technology Vol. 1 (1998), www.transhumanist.com/volume1/moravec.htm
19. B. Spinoza: *L'Ethique*. Introduction, translation, notes, and comments by Robert Misrahi. PUF, Paris, Philosophie d'aujourd'hui, 1990. Editions de l'Eclat (2005)

32

Ethics and Industrial Production

Daniel Bernard

The development of nanotechnology seems inevitable, for it alone would be able to solve or circumvent the huge difficulties to be faced by industrial and post-industrial societies, in both their private and their public aspects, and including the ageing population and its expectations with regard to health, the evolution of the climate, pollution, the management of food resources and raw materials, access to drinking water, control of energy production and consumption, equitable and sustainable development, etc.

But the very viability of nanotechnology is open to a whole range of conceptual, technical, industrial, economic, and social doubts and uncertainties. While more and more seminars and conferences are devoted to the hazards and potential risks due to nanotechnology in general, and nanomaterials in particular, industry is already trying to bring answers to the issues raised by these new techniques for manipulating the infinitely small, in the minds of consumers, (part of) the scientific community, environmental groups, and public authorities.

In a context where many bodies are advocating a 'responsible development' of nanotechnology and nanomaterials, industry would like to find a fair balance between development, production, and use of nanotechnology and nanomaterials in items and systems, and its ethical, legal, and financial responsibilities.

Industry has a duty to itself to present the expected benefits of nanotechnology and nanomaterials, but without concealing the possible risks, and in particular the impact on humans and the environment, and hence to emphasise the issues relating to the choice of nanotechnology and nanomaterials as development priorities, if possible without referring only to the economic and industrial stakes.

Nanotechnology and nanomaterials do indeed bring many prospects for development. But manipulating life or matter at the scale of atoms or molecules is not wholly without risks.

While nanotechnologies can be qualified as transverse technologies, since they involve skills from different scientific and technological disciplines, they

P. Houdy et al. (eds.), *Nanoethics and Nanotoxicology*,

DOI 10.1007/978-3-642-20177-6_32,

are mainly relevant to innovative areas which transcend traditional industrial markets and will contribute to building a new technological world. As declared by Eric Drexler, founder of the Foresight Institute, nanotechnology is not going to improve the industrial world as it is today. Instead it will simply replace it.

Nanotechnology already has a direct impact on the major areas of industry:

- *Information and Communication Technologies.* Beginning by increasing data storage densities, nanotechnology has now turned to the development of molecular and biomolecular nanoelectronics, spintronics, and quantum computing, suggesting a major technological breakthrough in this area.
- *Energy.* Nanotechnology is already present in the production of renewable energies (photovoltaic or thermal solar energy, wind energy) and non-renewable energies, as well as energy storage (by electrochemical processes such as lithium ion batteries and supercapacitors, or by hydrogen or methane adsorption in nanoporous materials that can store these fuels in a reversible way), and generation by fixed or mobile energy sources (fuel cells for cars, cell phones, portable computers, etc.). It is also contributing to energy saving (thermal insulation, improved efficiency of lighting systems, reduced energy consumption by household and industrial equipment, etc.).
- *Health.* Nanotechnology will help us to meet the increasing needs of the population, due notably to its ageing, by developing pharmaceutical, medical, and surgical technologies, such as DNA chips for early diagnosis of disease, targeted treatment of tumour cells, tissue engineering, biomimetic materials, bioactive and biocompatible implants, neuroprostheses, etc.
- *Quality of Life.* Apart from food quality, comfort in the home, textiles and clothing, cosmetics, individual mobility (dependants, etc.), collective mobility (transport, etc.), and leisure, nanotechnology will bring benefits to environmental technologies, and will contribute in particular to decontamination of soils and water, allowing the detection and neutralisation of micro-organisms and pollutants. They will help to reduce the production of waste and participate in managing the final stage in the life cycle of manufactured products through nanomarking.
- *Safety and Security.* Nanotechnology will allow the development of selective molecular-scale sensors that can be implanted in the environment or in hostile conditions to detect the presence of chemical or biological agents, but it will also provide ways of ensuring the traceability of products, including food products, security on private, professional, or public property, etc.

Nanotechnology will contribute significantly to the development of clean technologies (clean tech), with low energy and matter consumption, and low or zero levels of waste products cast off into the environment.

All these areas of activity depend upon the basic chemical and metallurgical industries which produce nanomaterials like nanoparticles, nanofibres, and nanosheets, thermoplastic, thermohardening, and elastomeric nanostructured polymer materials, metal alloys, metal oxides, ceramics, and glasses.

Nanomaterials will thus witness a particular development due to the fact that they constitute the basic building blocks for manufactured nano-articles and nanosystems. All economic sectors are thus beginning to benefit from the remarkable properties of nanomaterials, and will benefit more and more from the development and use of functional nanomaterials in the design and manufacture of new articles and systems.

On the other hand, the synthesis of nanomaterials, including nanoparticles, can sometimes lead to substances whose novel physical and chemical properties give them also biological properties whose impact on human health, but also on fauna and flora, remains unknown.

32.1 Some Observations

The first observation we can make regarding nanotechnologies is the obvious lack of information about what they actually are. But how can we make enlightened decisions as industrialists, legislators, researchers, workers, or consumers if there is no common understanding about what comprises nanotechnology and nanomaterials?

Indeed, there has only existed an internationally recognised definition of nanotechnology for a few months at the time of writing. The first elements of a terminology for naming the various aspects of nanotechnology and nanomaterials are currently being established by the International Standards Organization (ISO).

Ensuring that the products made by nanotechnology are innocuous is a major preoccupation. But at the present time, it is impossible to know whether the products already commercialised may be harmful in some way. The data available for the time being regarding their effects on animals, the environment, or humans are not beyond dispute.

In particular, toxicological and ecotoxicological studies on nanomaterials, and in particular on perfectly characterised nanoparticles, are still few and far between, and their conclusions are not accepted without significant reservation by the scientific and technical community. Indeed, many studies are conducted on nanomaterials and nanoparticles that are at best only poorly characterised.

But which quantities can be used to provide an unequivocal physical, chemical, and structural description of nanomaterials and nanoparticles? Another issue is the validity of conventional protocols for assessing the toxicity of chemical substances when it comes to studying nanomaterials. Can the traditional OECD protocols for toxicological and ecotoxicological studies of chemical products be transposed to nanomaterials, and if so, are they necessarily meaningful? Work is underway to try to answer this question. However, at the present time, there is no internationally accepted standard protocol to establish the specific toxicology of nanomaterials and nanoparticles.

While there are still very few indisputable publications on the acute toxicity of nanoparticles, there are even fewer data on the chronic toxicity and ecotoxicity of chemical substances in the nanoparticulate form. The results of the available studies are globally divergent and cannot be used to draw conclusions regarding a specific contribution of the nanoparticulate aspect to the hazards exhibited by these substances. And this despite the fact that certain results have been given a huge amount of media space, particularly when they can be exploited to incite alarmist comments.

There is still no epidemiological data relating to exposure to nanoparticles, except some data referring globally to carbon blacks and certain noncrystalline silicas made up of aggregates of elementary nanoparticles whose sizes are distributed throughout the micro- and nanoscales. The current lack of techniques for generating aerosols that would be representative of exposure to engineered nanoparticles, aerosolisation techniques that would make it possible to carry out nanoparticle inhalation studies, is one reason for the absence of data regarding their chronic toxicity. Likewise, there is no standardised protocol to assess the environmental impacts of nanomaterials and nanoparticles.

There are thus no indisputable experimental results regarding the specific toxicity due to the 'nano' aspect of materials and particles, whether exposure is by inhalation, contact with the skin or mucous membrane, or by ingestion. It should also be stressed that there is no simple and quick method for specifically detecting a given kind of nanoparticle in the atmosphere. This raises problems regarding the detection and monitoring of occupational exposure to engineered nanoparticles.

Finally, while there do exist technologies for characterising and quantifying nanomaterials and nanoparticles, the experimental protocols for implementing them are not yet uniformised, and major differences can thus arise between the results obtained by the laboratories carrying out the analyses, e.g., when determining size distributions. It is thus important to develop and standardise measurement techniques and instruments, calibration procedures, and certified reference materials, in order to be able to validate measurement and characterisation methods for nanomaterials on an international level. Multifunctional nanotechnological items and systems will also require the elaboration of specific standards.

However, while there are currently no standards or regulations specific to nanotechnology and nanomaterials, e.g., in the European REACH regulations (Registration, Evaluation, Authorization and restriction of CHemicals), or the US EPA regulations (Environmental Protection Agency), the existing regulations for chemical products, ultrafine particles, and occupational safety do of course apply, and a certain number of international standards (ISO) covering the areas of ultrafine (submicron) particles can be applied to nanomaterials, within the limitations imposed by their field of validity.

The authorities are even now wondering whether they should set up specific regulations in each country, or across broader economic zones, starting by attributing a new Chemical Abstract Safety (CAS) registration number to the nanoparticulate forms of chemical substances already known in the bulk state.

Many opinions and synopses have been drawn up by various French, European, and international institutes, agencies, committees, and organisations (e.g., CPP, AFSSET, NIOSH, EPA, IRSST-Robert-Sauvé). These opinions and reports are now more numerous than the original indisputable studies on which policy-makers could base their decisions.

The emergent nature of nanotechnologies and the fact that many different disciplines contribute to this emergence means that certain actions will be necessary premises for their systematic and responsible development. No scientific study has yet detected a hazard specific to the nanometric scale. On the other hand, it is not yet possible to show that nanomaterials are or will be strictly without risk. Unfortunately, we can still only speculate and extrapolate when specific hazards due to the nanoparticulate dimension are at issue. Fortunately though, the risk of exposure to nanomaterials remains a perceived risk rather than a real one. As we are reminded by M. Proust, facts do not enter the world our beliefs inhabit.

The lack of data about the hazards associated with nanomaterials and nanoparticles, data which would be used to assess the potential risks from continuous or accidental exposure to them, thus creates a decision-making problem for industry, when confronted with such uncertainty. This can only lead them to apply the precautionary principle, starting from the stage where these new materials are actually designed, in order to prevent exposure of researchers, workers, and consumers, precluding their dissemination into the work atmosphere and the environment, and continuing right through their life cycles.

In companies, risk managers are also in the spotlight, and insurers are wondering how best to insure nanotechnologies where the risks are not yet identified and cannot therefore be evaluated. For insurers and reinsurers, the problem can be phrased in these terms: to reduce the uncertainties associated with nanotechnology, the analysis and management of risk and the options for acceptable transfer of risk must be investigated from a common vantage point shared by industrialists, scientists, public policy-makers, and insurers. Subjective responsibility could then be substituted for objective responsibility, making risk rather than fault the basis of responsibility.

Finally, nanotechnologies will replace existing technologies, which will become obsolete, and hence produce an evolution in the needs of the employment market and in worker skills. These changes may be a major challenge for worker training, and more generally for public education. This constitutes a key ethical issue for industry and for society since, more often than not, the most vulnerable workers are victims of any kind of transformation of the employment market brought about by the development of emergent technologies.

It thus seems crucial to pursue investigations through studies and through an extended reflection on the future of nanotechnology and nanomaterials. Likewise, it is essential not to draw subjective conclusions about their potential benefits and the possible risks.

32.2 Strategy

Three things stand out as indispensable premises for responsible management of the development of nanotechnology and nanomaterials: deciding upon the terminology, together with a common scientific and technical nomenclature, setting up procedures and standards, and pursuing research and dissemination of results.

Nanomaterials can be defined as materials containing or composed of nano-objects which confer upon these materials improved properties or properties specific to the nanometric dimension (1–100 nm). Nano-objects occur in the form of free, agglomerated, or aggregated nanoparticles, of fibres, tubes, wires, nanosheets, nanocrystals, or materials with porosity structured on the nanoscale.

In the first place, there is reason to be concerned about the very small amount of research so far undertaken regarding the possible consequences of nanomaterials for occupational health and safety, given that the first available data suggests that exposure levels may be rather high when these things are manipulated. This may in part be explained by the poor suitability for nanoparticles of the tools normally used to assess the exposure of workers in the work environment. Secondly, specialists disagree over the appropriacy of existing regulations. It will be difficult to deal with this issue until more precise data on the potential effects of nanotechnologies become available.

Faced with these uncertainties over the hazards, steps must be taken to prevent worker and consumer exposure to nanomaterials and nanoparticles, in order to control the risks during production, handling, and transformation, and during the use of the articles containing them.

Until research has made progress and more complete regulations, better suited to the specific features of nanotechnology, have been devised, the precautionary principle must guide industrial strategy regarding the steps to be taken in the context of a responsible approach to risk, to protect the health and safety of workers, and to preserve the environment.

To begin with, this strategy will be based on existing regulations, which bear upon:

- chemical products,
- ultrafine particles,
- occupational hygiene,
- environmental protection,
- consumer protection.

It will also take into account the documents and recommendations produced by industry itself, professional unions, and occupational health and safety organisations:

- safety data sheets for chemical products,
- guidelines for good conduct in the production and handling of nanomaterials,
- data, information, and recommendations.

And finally, it will be based on three principles:

- *Safety.* Deploying the material means for protection and specifically designed organisational procedures, in order to prevent worker and public exposure.
- *Acquisition of Knowledge.* In order to develop the knowledge needed for the exploitation of nanomaterials and nanoparticles in full possession of the facts and in perfect safety by scientific and technical work, at the initiative of industrialists, either in-company or outside, by public authorities in the context of national or international collaboration, and also by exchange of experiences between the different stakeholders.
- *Transparency.* By improved communications designed to inform, listen to, discuss, educate, and train all those involved.

32.3 Safety

Emissions of manufactured nanoparticles into the atmosphere during production may come from the process:

- In the case of faulty operation of the unit, during a change in the process parameters outside nominal operating conditions, or when incidents occur on a unit, e.g., leaks.
- During maintenance and cleaning operations on the installations.

Emissions can also occur downstream of production, when the nanoparticles are recovered, handled, packaged, transported, transformed (compounded, extruded, tooled, etc.), and also during any operation that might lead to aerosolisation of the nanoparticles and their dissemination in the atmosphere (e.g., when non-secured packaging is opened after transport).

The safety systems recommended and deployed by industry involve the application of a certain number of principles in order to prevent prolonged or accidental exposure of staff working with nanoparticles:

- Design production and transformation units where nanoparticles remain confined within the installations.
- Use production processes operating at atmospheric pressure or negative pressure.
- Limit the amounts of nanoparticles available in powder form in a given building or storage area.
- Develop and commercialise liquid or solid formulations to avoid handling nanoparticles in powder form on premises where the risk of dissemination could not be controlled.
- To introduce safety systems on premises where nanoparticles will be handled in powder form by the following means:

1. Collective safety systems, such as well ventilated atmospheres with extraction at source and treatment of gas effluents by incineration or filtration, collection and treatment of aqueous effluents, specific waste management, special airlock systems, separate changing rooms equipped with specific cleaning systems, such as showers, etc.
2. Individual safety systems, such as disposable suits, anti-dust masks, fresh-air masks, gloves, goggles, airtight suits with fresh air supply, etc.

All risk of dissemination into the environment is avoided by the following measures:

- Collection of effluents, waste, and rejected products in laboratories and workshops.
- Packaging of nanoparticles in specially designed secure containers.
- Secured transport of containers.
- Informing and training staff.
- Involving staff representatives, including the site's committee for health and safety at work, the work doctor, and the safety engineer.
- Setting up specific medical follow-up with an exposure register and medical files conserved for the whole period needed to ensure perfect traceability for both permanent and temporary employees (monitoring to be carried out in conjunction with the public health surveillance authorities).
- Restricting access to premises dedicated to the production and transformation of nanoparticles so that only those duly authorised can enter.
- Defining job descriptions, work procedures, and procedures for taking action when operations are not running to standard, as well as during maintenance of installations and cleaning of the premises.
- Monitoring the work premises and the environment using nanoparticle detectors (although no specific detector is yet available).
- Analysing work stations to optimise their configuration.
- Regularly checking that all safety measures and procedures are being applied correctly by means of audits.

The strategy for developing and commercialising nanomaterials and nanoparticles must also take into account the uncertainties over the associated hazards by proposing a responsible approach to prospects and customers.

We should favour the supply of nanoparticles predispersed in polymer matrices, or in liquid formulations. Their supply in powder form should be reserved for customers (who would transform the product) with a long experience of safety enforcement and in particular safety methods designed for handling nanoparticles.

We must also draw up guidelines for good practice in handling and provide advice for customers on how to set up the means and procedures for safe use of production units, so that processing companies and users can apply suitable safety measures.

All these safety measures must be continually kept up to date, with the possibility of modifying them as our knowledge of the intrinsic hazards of a given nanoparticle is improved, applying the principle of managing and controlling risks according to zones of hazard (control banding), developed at the initiative of the *Agence française de sécurité sanitaire de l'environnement et du travail* (AFSSET), in the framework of the *Association française de normalisation* (AFNOR) and ISO.

32.4 Acquisition of Knowledge

Given the lack of available data, industrialists have engaged upon an active policy of knowledge aqcuisition, particularly regarding the monitoring and characterisation of nanomaterials and nanoparticles representative of industrial products, but also regarding the possible impacts of these products on health and the environment, and on the safety aspects of producing and implementing nanomaterials and nanoparticles.

Just as industrialists are studying the means to protect their employees in the operations of producing and transforming nanomaterials and nanoparticles, they are also anticipating the management of the whole life cycle of items containing their nanomaterials and their nanoparticles, while financing and contributing to studies of their impact on health, safety, and the environment. In particular, they carry out toxicity and ecotoxicity studies on nanomaterials and nanoparticles using protocols that comply with the guidelines laid down by the OECD for chemical substances.

However, given the uncertainties over the applicability of these guidelines to nanomaterials and nanoparticles, the OECD has set up the Working Party on Manufactured Nanomaterials, which comprises eight working groups on the following themes:

- OECD database on safety research,
- research strategies on manufactured nanomaterials,
- safety testing of a representative set of manufactured nanomaterials,
- manufactured nanomaterials and test guidelines,
- voluntary schemes and regulatory programs,
- risk assessment,
- alternative methods of toxicity testing,
- exposure measures and mitigation.

Their missions are essentially to make an inventory of available data on nanomaterials and nanoparticles, to test the validity of existing guidelines, to propose if necessary alterations or new protocols, and to specifically study the toxicology of fourteen families of nanoparticles.

Beside the studies conducted according to OECD guidelines, many more fundamental studies are carried out by industrialists in collaboration with university laboratories, and more specifically in France with the laboratories of the *Institut national de la santé et de la recherche médicale* (INSERM), the *Centre national de la recherche scientifique* (CNRS), the *Commissariat à l'énergie atomique* (CEA), and the *Institut national de l'environnement industriel et des risques* (INERIS), etc. These studies are financed by industry, jointly financed by Europe as part of the EC's Sixth and Seventh Framework Programmes for Research and Technological Development (Nanosafe 1 and 2, Saphir, ENPRA, etc.), or under the Joint Action for Safety of Nanomaterials under the Public Health Program (DG SANCO), and in France through the *Agence de l'environnement et de la maîtrise de l'energie* (ADEME), the *Agence nationale de la recherche* (ANR), *Oséo* (the French innovation agency), the Genesis project, and competitive clusters. On the international level, all industries producing or using nanotechnology and nanomaterials are currently running projects to assess the hazards.

Studies are being carried out in France on industrial hygiene and safety, in collaboration with INERIS and the *Institut national de recherche et de sécurité* (INRS). An effort is made on behalf of small and medium sized companies through the structure *Action collective transrégionale nanomatériaux* (ACT Nano) set up by the DGCIS (industry ministry), and by the establishment of platforms in the NANO Innov framework, or directly supported by regional councils, like the CANOE project (*Consortium aquitain d'innovation nanomatériaux et électronique organique*) in the Aquitaine region of France.

Industry is also actively supporting the establishment of standards for nanotechnology and nanomaterials by taking part in the work of their national standards organisation. AFNOR's X457 Nanotechnologies Committee thus brings together all French stakeholders and represents France on international bodies, such as the TC 352 committee of the European Committee for Standardization (CEN) and ISO's TC 229 committee, which devise international standards in the field of nanotechnology and nanomaterials.

The standards will help to show that nanotechnologies are developed and commercialised in a safe, open, and responsible way by supporting the following features:

- health and safety of workers, consumers, and the environment,
- development and commercialisation,
- industrial property,
- communication concerning associated benefits, opportunities, and problems,
- REACH and TSCA (EPA) registration
- legislation and regulation,

and by providing recognised and validated protocols for

- naming, describing, and specifying,
- measuring and testing,
- determining the impact on health and the environment,
- assessing and managing risks,
- specifying products and their application performance.

Finally, industry supports the evolution of national and international regulations governing production, implementation, and use of nanomaterials and nanoparticles.

32.5 Transparency

Informing stakeholders is a key priority. Indeed, an attitude of minimising, even denying the possible health and environmental risks of nanotechnology and nanomaterials would look irresponsible and reduce credibility in the eyes of the public, and the latter is already broadly aware of the potential problems in this area, whence it could only lead to a general rejection. The situation will certainly evolve very quickly and industries involved in nanotechnology and nanomaterials must be ready to respond to public cross-examination.

This implies that they must organise opportunities for debate and discussion, using methods and rules which for the main part remain to be invented, because the acceptance of nanotechnologies can only come through broad and transparent communication among all stakeholders, communication which must be based on informing, listening, discussing, and educating.

Industrialists must therefore engage in communication in such a way as to

- Avoid a situation where the feeling that a small number of *cognoscente* had appropriated all knowledge of nanotechnology and nanomaterials might lead to their rejection by those who would thereby perceive this as a loss of control over their environment.
- Avoid the fear of domination that science and technology exert on those who, in their majority, have no access to scientific and technical culture.
- Encourage engineers to communicate, so that they can convince others of the acceptability of nanotechnology and nanomaterials.

This communication should involve the following:

- Explain the benefits of nanotechnologies, and the possible risks, the concrete problems encountered, the practical solutions that have been implemented to resolve them, along with the questions that remain unanswered.
- Inform and train all players, including company employees (whether or not they are involved in nanotechnology and nanomaterials), their unions, staff representative bodies (in France, the CHSCT, etc.), customers, consumers and consumer societies, public authorities and elected representatives, public health and hygiene authorities (INRS, CRAM, InVS, etc.), students, teachers, and scientific researchers.
- Engage in public debate through information meetings, conferences, citizens' forums, and round tables, to reply to cross-examination, and listen to the opinions and advice of all stakeholders.

32.6 Conclusion

Despite these questions and uncertainties, nanotechnology represents a great opportunity for society. It will help to solve the major technological challenges of the day, and at the same time bring new jobs to meet the needs of the ever evolving employment market.

Nanotechnology and nanomaterials open up a whole new field that industry must standardise if it hopes to give it meaning and purpose. It will need an added dose of will-power and conscience to determine, not what it is able to do, but what it must do.

The relationship between ethics and social acceptability can be given a range of answers. Realists, and cynics, will insist that the logic of profit will win out in the end. But experience shows that, when the divergence between technology and ethics becomes too great, the resulting social discontent will have damaging effects that are difficult to undo.

So it is down to all players to do their bit in ensuring a responsible development of nanotechnology and nanomaterials and to thereby encourage their acceptance, guarantee of successful social and industrial progress.

References

1. Guide de bonnes pratiques nanomatériaux et HSE, Union des industries chimiques (UIC), March 2009
2. Responsible production and use of nanomaterials, Verband der Chemischen Industrie (VCI), March 2008
3. Les nanomatériaux, sécurité au travail, AFSSET, July 2008
4. Nano Risk Framework Nanopartnership Environmental Defense, DuPont, June 2007, www.nanoriskframework.com
5. Nanotechnologies Part 2: Guide to safe handling and disposal of manufactured nanomaterials, PD6699-2:2007, BSI 2007, ISBN 978-0580-60832-2I
6. Guide de bonnes pratiques favorisant la gestion des risques reliés aux nanoparticules de synthèse, Guide IRSST, 2008, www.irsst.qc.ca
7. Nanosafe 2: www.nanosafe.org
8. T. Schneider et al.: Evaluation and control of occupational health risks from nanoparticles. Norden, TemaNord, 581 (2007)
9. Document technique UIC DT 80: Evaluation et prévention des risques professionnels liés aux agents chimiques (November 2008)
10. SCENIHR: Opinion on the scientific aspects of the existing and proposed definitions relating to products of nanoscience and nanotechnologies (29 November 2007), ec.europa.eu/health/ph_risk/committees/04_scenihr/docs/scenihr
11. Dossiers INRS

Index

P. Houdy et al. (eds.), *Nanoethics and Nanotoxicology*,
DOI 10.1007/978-3-642-20177-6,